GIANT MICELLES
PROPERTIES AND APPLICATIONS

SURFACTANT SCIENCE SERIES

GIANT
MICELLES
PROPERTIES AND APPLICATIONS

Edited by

Raoul Zana
Institut Charles Sadron
Strasbourg, France

Eric W. Kaler
University of Delaware
Newark, Delaware

CRC Press
Taylor & Francis Group
Boca Raton London New York

CRC Press is an imprint of the
Taylor & Francis Group, an **informa** business

CRC Press
Taylor & Francis Group
6000 Broken Sound Parkway NW, Suite 300
Boca Raton, FL 33487-2742

First issued in paperback 2019

© 2007 by Taylor & Francis Group, LLC
CRC Press is an imprint of Taylor & Francis Group, an Informa business

No claim to original U.S. Government works

ISBN-13: 978-0-8493-7308-4 (hbk)
ISBN-13: 978-0-367-40355-3 (pbk)

Library of Congress Cataloging-in-Publication Data

Zana, Raoul, 1937-
   Giant micelles : properties and applications / Raoul Zana and Eric W. Kaler.
      p. cm. -- (Surfactant science series ; 140)
   Includes bibliographical references and index.
   ISBN-13: 978-0-8493-7308-4 (alk. paper)
   ISBN-10: 0-8493-7308-5 (alk. paper)
   1. Micelles--Research. I. Kaler, Eric W. II. Title.

QD549.Z36 2007
541'.345--dc22   2007002452

Visit the Taylor & Francis Web site at
http://www.taylorandfrancis.com

and the CRC Press Web site at
http://www.crcpress.com

Table of Contents

Introduction

Amphiphilic molecules such as surfactants have the remarkable ability to self-assemble in aqueous solutions to form a variety of microstructures including liquid crystals, vesicles, small micelles, and of particular interest here, long worm- or threadlike "giant" micelles. The history of the physical chemistry of micellar solutions dates back to a paper by J.W. McBain[1] in 1913, in which he proposed the then-controversial idea that self-assembled aggregates could form in aqueous solution. New ideas in science are often greeted skeptically, and this was no different. When, at a meeting of the Royal Society of Chemistry, McBain pointed out that changes in the colligative properties of surfactant solutions around the critical micelle concentration could be explained by the formation of micelles, the distinguished meeting chair responded,[2] "Nonsense, McBain." Scientists persevered, nonetheless, and micellization is now a widely studied phenomenon.

The giant micelles discussed in this volume refer to the long micelles that occur in many types of surfactant solutions, most frequently upon increasing the concentration of the surfactant or of an added salt, or in the presence of specific counterions. These micelles may be microns long and have either a circular or elliptical cross section of a few nanometers. These micelles can be discrete "worms" or "threads," or may crosslink into a network. In all cases the micelles can break and reform spontaneously or under stress, and so are often called "living polymers." To the best of our knowledge the expression "giant micelles" was first coined by G. Porte[3] in 1983.

The transition from spheroidal micelles to giant micelles has been directly visualized by transmission electron microscopy at cryogenic temperature, and the images show coexistence of spherical and elongated micelles. Giant micelles also form in organic liquids and on solid surfaces. The rheological behavior of solutions containing giant micelles has been studied in detail, and these rheological properties are the key to the many successful applications of these solutions. Giant micelles are *equilibrium* or *living polymers* as the micelle molecular weight fluctuates with time, and so are distinct from conventional polymers where the repeat units are covalently bonded and the molecular weight is constant. In some cases the giant micelles can close upon themselves (forming *ringlike* micelles) or *branch* rather than continue to grow in one dimension when a parameter (e.g., temperature or salt concentration) is modified. These branched micelles can ultimately form a network of micellar junctions. All these possible behaviors give rise to a variety of phenomena that have been and still are the subject of extensive investigations. Recent reports show that giant micelles can be formed by amphiphilic block copolymers, or by surfactants synthesized to respond to external stimuli by reversibly forming giant micelles from spheroidal micelles.

If the expression "giant micelle" is relatively recent, the concept of very long and linear micelles is surely not. The first experimental study, by light scattering, of elongated or "rodlike" micelles was reported by P. Debye and E.W. Anacker[4] in 1951. They concluded that the variation of the light measured as a function of scattering angle (the dissymmetry) agreed only with the presence of long rodlike micelles. They said:

> The micelles of n-hexadecyltrimethylammonium bromide in … potassium bromide are large enough to produce measurable dissymmetry in the scattered light. Dissymmetry measurements showed conclusively that these micelles are not spherical or disk-like in shape; analysis of the data indicates that the micelles are rod-like. The cross section of such a rod will be circular, with the polar heads of the detergent lying on the periphery and the hydrocarbon tails filling the interior. The ends of such a rod would most certainly have to be rounded off with polar heads.

Thus, both the concept of one-dimensional growth and the importance of micellar "endcaps" were established.

The impact of that micellar microstructure on flow or rheological properties of surfactant solutions was also noted early on, and was discussed in 1954 during a report of the study of the viscoelasticity of surfactant solutions. The following excerpt from a paper by N. Pilpel[5] in 1956 compels some modesty about recent achievements in the study of giant micelles. He said:

> At low concentration of electrolyte the soap (sodium oleate) is believed to be in the form of small spherical micelles but at higher electrolyte concentration these pack together into cylinders which then interlink to form a network structure.

This single short sentence describes both the sphere-to-rod transition of micelle shape and the formation of networks in solutions of giant micelles. Both topics have been investigated at great length and described in probably thousands of papers since. Two names emerge from the large number of workers involved in this effort: Candau[6] was the first to point out the analogy in the rheological behaviors of giant micelles and polymers; Hoffmann[7] was the first to emphasize that this rheological behavior can be completely dominated by the rate of reversible scission of giant micelles.

One of the important reasons that this field of study has been able to move forward quickly is that a powerful theoretical framework has evolved in parallel with a range of sophisticated experimental tools. The theoretical approaches also reach back to P. Debye,[8] who suggested that the thermodynamics of the micelle would reflect a balance of repulsions between the (ionic) surfactant head groups and van der Waals's attractive forces between the hydrocarbon tails. Although couched in terms of an incorrect lamellar micellar geometry, this "opposing force model" provided a seed from which much more detailed thermodynamic models have grown over the past 50 years. Of particular importance is the work of C. Tanford and coworkers[9] in the 1970s, who introduced a robust thermodynamic framework and used it to make a strong connection between micelle formation and the more general hydrophobic effect.

Theory has also contributed significantly to our understanding of the rheological properties within the framework of a mean field model developed by M. Cates[10] and co-workers. They opened the way to interpreting measurements of rheological properties by showing how the dynamic processes of micellar diffusion and the kinetics of micellar breakage and recombination can interact to set the time scales for stress relaxation in entangled wormlike micellar solutions. In many cases there is only one characteristic relaxation time, so many of the rheological properties can be interpreted in terms of the remarkably simple Maxwell model of a viscoelastic fluid.

The variety of behaviors encountered in solutions of giant micelles, the present and past interest in these systems, and their applications in important industrial processes led us to the idea of summing up in a volume the "state of the art" in this area of science. The present volume comprises three parts. The first four chapters focus on theoretical aspects of the formation of giant micelles from different points of view (molecular thermodynamics, packing, and computer simulations) and of their rheological behavior. The following 10 chapters deal with experimental aspects: transmission electron microscopy at cryogenic temperature, scattering methods, phase diagrams, linear and nonlinear rheology, and relaxation. They also review the properties of giant micelles on solid surfaces and systems of smart micelles that respond to external stimuli by a change of shape. Another chapter describes giant micelles formed from amphiphilic block copolymers. The last chapter in this part reviews noncovalent polymers stabilized by hydrogen bonds that show rheological behavior very similar to that of giant surfactant micelles. The third part comprises four chapters that review the applications of giant micelles in oil and gas production (a growing application), in drag reduction (an energy saving process), in shampoos, and in other products used by consumers daily, such as hard surface cleaners, personal care products, sunscreens, and drug delivery formulations.

We believe that having this range of information together in one volume will be of considerable help to the community studying and using these fascinating materials.

The Editors

REFERENCES

1. McBain, J.W. *Trans. Faraday Soc.* 1913, *9*, 99.
2. McBain, J.W. *Colloid Chemistry,* vol. 5, Alexander, J., Ed., Reinhold, New York, 1944, p. 102.
3. Porte, G. *J. Phys. Chem.* 1983, *87*, 3541.
4. Debye, P., Anacker, E.W. *J. Phys. Colloid Chem.* 1951, *55*, 644.
5. Pilpel, N. *J. Phys. Chem.* 1956, *60*, 779.
6. Candau, S., Hirsch, E., Zana, R. *J. Colloid Interface Sci.* 1985, *105*, 521.
7. Löbl, M., Thurn, H., Hoffmann, H. *Ber. Bunsenges. Phys. Chem.* 1984, *88*, 1102.
8. Debye, P. *J. Phys. Colloid Chem.* 1949, *53*, 1.
9. Tanford C., *The Hydrophobic Effect*, Wiley, New York, 1973.
10. Cates, M.E. *Macromolecules* 1987, *20*, 2289.

Editors

Dr. Raoul Zana is directeur de recherches emeritus at the Institut C. Sadron of the CNRS. He earned his degree in chemistry from the University of Strasbourg and that of *ingenieur chimiste* from the Ecole Nationale Superieure de Chimie de Strasbourg, both in 1958. He obtained his Doctorate es Sciences in 1964 at the Centre de Recherches sur les Macromolecules (CRM) of the CNRS, working with Professors R. Cerf and H. Benoit on ultrasonic relaxation in polymer solutions. From early 1965 to mid-1967 he was a postdoctorate investigator at Case Western Reserve University (Cleveland, Ohio), working with Professor E. Yeager. There he brought to use a method of measurement of ultrasonic vibration potentials, an effect predicted by P. Debye. This work was in part at the origin of the now widely used method of measurement for zeta potentials of colloidal systems. He returned to Strasbourg at the CRM (then Institut Charles Sadron) as a *director de recherches* and started his work on polyelectrolytes and micellar systems. He created the research group "physicochemistry of colloids" and directed it for many years.

Dr. Zana has directed for 4 years the so-called Coordinated Research Group "Microemulsions" that included 80 scientists. He has served on the editorial boards of the journals *Langmuir*, *Journal of Colloid and Interface Science*, *Current Opinions in Colloid Interface Science*, and *Southern Brazilian Journal of Chemistry*. He has held many appointments as an invited professor: University of Sherbrooke (Canada), Drexel University (Philadelphia, Pennsylvania), Science University (Tokyo), University of Palermo (Italy), Technion (Haifa, Israel), University of Fukuoka (Japan), University of Auckland (New Zealand), University of Sydney (Australia), and California State University (Northridge, California). He has been consulting with several companies and has served in various positions in the CNRS and the University L. Pasteur of Strasbourg. He has published more than 375 papers, reviews, or chapters in volumes, and edited three volumes of the Surfactant Science Series: *Surfactant Solutions: New Methods of Investigation*, *Gemini Surfactants*, and *Dynamics of Surfactant Self-Assemblies*. He contributed much to each of these volumes. The last two volumes describe a good part of his studies.

Eric W. Kaler earned a BS degree in chemical engineering (with honors) from the California Institute of Technology in 1978 and a PhD in chemical engineering from the University of Minnesota in 1982 working with L. E. Scriven and H. T. Davis. Dr. Kaler joined the faculty of the Department of Chemical Engineering at the University of Washington in 1982 and was promoted to associate professor in 1987. He moved to the Department of Chemical Engineering at the University of Delaware in 1989, became a professor there in 1991, department chairman in

1996, and was appointed the Elizabeth Inez Kelley Professor of Chemical Engineering in 1998. He became dean of the College of Engineering in 2000. He was also a visiting professor at the Universität of Graz in Austria in 1995. His research interests are in the areas of surfactant and colloid science, statistical mechanics, and thermodynamics.

Dr. Kaler received one of the first Presidential Young Investigator Awards from the National Science Foundation in 1984, the Curtis W. McGraw Research Award from the American Society of Engineering Education in 1995, the 1998 American Chemical Society Award in Colloid or Surface Chemistry, and the 1998 ACS Delaware Section Award. He was elected a fellow of the American Association for the Advancement of Science in 2001, and received the Chilton Award from the Wilmington AIChE Section in 2002. In 2005, Dr. Kaler was awarded the E. Arthur Trabant Institutional Award for Women's Equity by the University of Delaware and received the Lectureship Award from the Division of Colloid and Surface Chemistry of the Chemical Society of Japan. He received the Kash Mittal Award from the Surfactants in Solution Symposium in 2006. He has chaired three Gordon Research Conferences and serves or has served on the editorial boards of the journals *Langmuir*, *Colloids and Surfaces*, *Colloid and Interface Science*, and *AIChE*, and was an associate editor of the *European Physical Journal*. He is also the founding coeditor-in-chief of the international journal *Current Opinions in Colloid and Interface Science*. He has authored or coauthored more than 200 peer-reviewed papers and holds 10 U.S. patents. He has been a consultant to numerous companies, and has served in a variety of positions in several professional societies.

Contributors

Nicholas L. Abbott
Department of Chemical and
  Biological Engineering
University of Wisconsin-Madison
Madison, Wisconsin, United States

Valerie Anderson
Schlumberger Cambridge
  Research Ltd.
Cambridge, England

Abraham Aserin
Casali Institute of Applied Chemistry
The Institute of Chemistry
Hebrew University of Jerusalem
Jerusalem, Israel

Rob Atkin
School of Chemistry
University of Sydney
Sydney, Australia

Frank S. Bates
Department of Chemical Engineering
  and Materials Science
University of Minnesota
Minneapolis, Minnesota,
  United States

Fernando Bautista
Departamento de Ingeniería Química
Universidad de Guadalajara
Guadalajara, Mexico

Avinoam Ben-Shaul
Department of Physical Chemistry and
  the Fritz Haber Research Center
Hebrew University of Jerusalem
Jerusalem, Israel

Annabelle Blom
School of Chemistry
University of Sydney
Sydney, Australia

Laurent Bouteiller
Universite Pierre et Marie
  Curie–CNRS
Paris, France

Wim J. Briels
Faculty of Science and Technology
University of Twente
Enschede, The Netherlands

Luigi Cannavacciuolo
Institute of Solid State Research
Jülich, Germany

Michael Cates
School of Physics
University of Edinburgh
Edinburgh, Scotland

Shmaryahu Ezrahi
Materials and Chemistry Department
Israel Defense Force
Tel-Hashomer, Israel

Suzanne Fielding
School of Mathematics
University of Manchester
Manchester, England

Nissim Garti
Casali Institute of Applied Chemistry
The Institute of Chemistry
Hebrew University of Jerusalem
Jerusalem, Israel

Wu Ge
Department of Chemical and
  Biomolecular Engineering
Ohio State University
Columbus, Ohio, United States

Frank Pierce Hubbard, Jr.
Department of Chemical and
  Biological Engineering
University of Wisconsin-Madison
Madison, Wisconsin, United States

Trevor Hughes
Schlumberger Cambridge
  Research Ltd.
Cambridge, England

Martin In
Laboratoire des Colloïdes, Verres
  et Nanomatériaux
CNRS-University of Montpellier 2
Montpellier, France

Eric W. Kaler
Department of Chemical Engineering
University of Delaware
Newark, Delaware, United States

Octavio Manero
Instituto de Investigaciones
  en Materiales
Universidad Nacional Autonoma de
  Mexico
Mexico City, Mexico

Sylvio May
Department of Physics
North Dakota State University
Fargo, North Dakota, United States

Ramanathan Nagarajan
Department of Chemical Engineering
Pennsylvania State University
University Park, Pennsylvania,
  United States

Erik B. Nelson (retired)
Schlumberger Ltd.
Sugar Land, Texas, United States

Florian Nettesheim
Department of Chemical Engineering
University of Delaware
Newark, Delaware, United States

Luc Nicolas-Morgantini
L'Oréal Recherche
Clichy, France

Wouter K. den Otter
Faculty of Science
  and Technology
University of Twente
Enschede, The Netherlands

Johan T. Padding
Faculty of Science and Technology
University of Twente
Enschede, The Netherlands

Jan Skov Pedersen
Department of Chemistry and iNANO
  Interdisciplinary NanoScience
  Center
University of Aarhus
Aarhus, Denmark

Jorge E. Puig
Departamento de Ingeniería Química
Universidad de Guadalajara
Guadalajara, Mexico

Peter Schurtenberger
Department of Physics
University of Fribourg
Fribourg, Switzerland

J. Felix Armando Soltero
Departamento de Ingeniería Química
Universidad de Guadalajara
Guadalajara, Mexico

Phil Sullivan
Schlumberger Ltd.
Sugar Land, Texas, United States

Yeshayahu (Ishi) Talmon
Department of Chemical Engineering
Technion-Israel Institute of
  Technology
Haifa, Israel

Eran Tuval
Materials and Chemistry Department
Israel Defense Force
Tel-Hashomer, Israel

Gregory G. Warr
School of Chemistry
University of Sydney
Sydney, Australia

Gilles Waton
Institut Charles Sadron (CNRS)
Strasbourg, France

You-Yeon Won
School of Chemical Engineering
Purdue University
West Lafayette, Indiana, United States

Jacques L. Zakin
Department of Chemical and
  Biomolecular Engineering
Ohio State University
Columbus, Ohio, United States

Raoul Zana
Institut Charles Sadron (CNRS)
Strasbourg, France

Ying Zhang
Department of Chemical and
  Biomolecular Engineering
Ohio State University
Columbus, Ohio, United States

1 Molecular Thermodynamics of Giant Micelles

Ramanathan Nagarajan

CONTENTS

1.1   INTRODUCTION

The first molecular theory of micelle formation was proposed by Debye.[1,2] For the lamellar micelle suggested by McBain,[3,4] Debye formulated two competing energies controlling micelle formation: the repulsive interactions between ionic head groups on the micelle surface and the attractive van der Waals interactions between the tails. By minimizing the energy of a micelle, its equilibrium size and the critical micelle concentration (cmc) were obtained. The theory was extended by Hobbs[5] to account for electrolyte effects. Halsey[6] amended the energy expression to consider cylinders and concluded that one-dimensional aggregate growth will be as rods and not as disks. Ooshika[7] and Reich[8] added a surface energy contribution to micelle formation. They also argued that the energy of the entire system, and not the energy of a micelle, should be minimized. Later, partition functions based on statistical thermodynamics were developed by Hoeve and Benson[9] and Poland and Scheraga.[10,11] Subsequently, Tanford[12–14] proposed his influential "principle of opposing forces" with three free energy terms: (i) a negative contribution due to removal of tail-water contact, which is responsible for micelle formation, (ii) a positive contribution due to residual tail-water contact which promotes micelle growth, and (iii) a positive contribution due to head group repulsions which limits micelles to finite sizes. Taking advantage of these earlier theoretical studies, our work has focused on developing *a priori* predictive models of aggregation behavior of surfactants starting from knowledge of their molecular structures and the environmental conditions in which they are present. Our first attempt[15] involved statistical thermodynamic formulations influenced by the work of Hoeve and Benson[9] and Poland and Scheraga[10,11] and led to the development of partition functions to represent various contributions to the free energy of micellization. Subsequently, we recast[16–20] the partition functions in the form of free energy contributions similar to those employed in Tanford's formalism. In our approach, the physicochemical factors controlling self-assembly are first identified by examining all the changes experienced by a singly dispersed surfactant molecule when it becomes part of an aggregate. To calculate the contribution to the free energy of aggregation associated with each of these factors, analytical equations explicitly connected to the molecular features of the surfactant are formulated. The chemical structure of the surfactant and the solution conditions are sufficient for estimating all the molecular constants appearing in these equations. The main advantage of our theoretical model lies in the analytical form of free energy expressions, the need to estimate only one, two, or three model parameters depending upon the type of surfactant (see Section 1.4.2 below), and the ability to make *a priori* predictions requiring simple numerical calculations. Further, a wide range of self-assembly phenomena such as solubilization,[17,20] mixed micelles,[18,19,21] and microemulsions[22] are treated by simple extensions of the basic micellization theory allowing for a unified approach to be developed. In this chapter, we describe our quantitative approach to predicting the formation and properties of rodlike micelles and the shape transitions they undergo. In Section 1.2, we summarize the geometrical relations for spherical, globular, and rodlike micelles, consistent with molecular packing considerations, and develop general thermodynamic equations that govern the formation of rodlike micelles. In Section 1.3, the molecular theory of rodlike micelle formation is described, with explicit equations to calculate the free energy of formation of aggregates. In Section 1.4, we show how to estimate the molecular constants appearing in the free energy expressions, describe the computational scheme for making predictive calculations, and demonstrate the predictive power of the molecular theory via illustrative calculations. In Section 1.5, we consider how rodlike micelles can be manipulated in various ways such as by the addition of a second surfactant, by the addition of an alcohol, and by the addition of a nonionic polymer. For this purpose, we present extensions of the molecular theory described in Section 1.3 to the problems of mixed micelles, and surfactant–polymer association. In each case, illustrative calculations are presented to show how rodlike micelles are manipulated. The last section presents some conclusions.

1.2 THERMODYNAMIC PRINCIPLES OF GIANT MICELLES

1.2.1 GEOMETRY OF GIANT MICELLES

The shapes of surfactant aggregates formed in dilute solutions are schematically shown in Figure 1.1. The small micelles are spherical. When large rodlike micelles form, they are visualized as having a cylindrical middle part and spherical end-caps, with the two parts allowed to have different diameters. For conditions when spherical micelles are not possible (namely for those aggregation numbers corresponding to the sphere that will have a radius larger than the extended tail length ℓ_s) and at the same time, the formation of rodlike micelles is also not favored, small nonspherical globular aggregates form. Israelachvili et al.[23] evaluated various shapes based on the constraint that molecular packing requirements should be satisfied locally everywhere on the aggregate. On this basis, they have suggested shapes generated by ellipses of revolution for these globular aggregates. The average area per surfactant molecule at the surface of the ellipse of revolution is practically the same as that of prolate ellipsoids, for aggregation numbers up to three times that of the largest spherical micelle. Therefore, we will calculate the average geometrical properties of globular aggregates in the sphere-to-rod transition region as for prolate ellipsoids (without implying the shape to be an

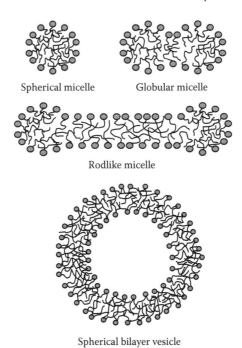

Spherical micelle   Globular micelle

Rodlike micelle

Spherical bilayer vesicle

FIGURE 1.1 Schematic representation of surfactant aggregates in dilute aqueous solutions. Structures formed include spherical micelles, globular micelles, rodlike micelles with spherical endcaps, and spherical bilayer vesicles. One characteristic dimension in each of these aggregates is limited by the length of the surfactant tail.

ellipsoid). Some surfactants pack into a spherical bilayer structure called a vesicle, which encloses an aqueous cavity. We will not consider vesicles in this chapter even though a transition between vesicles and rodlike micelles is possible and has been predicted by our theory.

For spherical, globular, and cylindrical aggregates containing g surfactant molecules, the volume of the hydrophobic core of the aggregate, V_g, the surface area of contact between the aggregate core and water, A_g, and the surface area of the aggregate at a distance δ from the aggregate core–water interface, $A_{g\delta}$, are listed in Table 1.1. Also given in the table is a packing parameter p, defined in terms of the geometrical variables characterizing the aggregate. The packing parameter p is used in the computation of the free energy of tail deformation (Section 1.3.3) and the area $A_{g\delta}$ is employed in the computation of the free energy of electrostatic interactions between surfactant head groups (Sections 1.3.6 and 1.3.7). Given a surfactant molecule, the geometrical properties of spherical or globular micelles depend only on the aggregation number g. For rodlike micelles, there are two independent variables, the radius of the cylindrical part and the radius of the spherical endcaps.

TABLE 1.1
Geometrical Properties of Surfactant Micelles

Spherical micelles: (Radius $R_S \leq \ell_S$)

$$V_g = \frac{4 \pi R_S^3}{3} = g \, v_S, \quad A_g = 4 \pi R_S^2 = g \, a$$

$$A_{g\delta} = 4 \pi (R_S + \delta)^2 = g \, a_\delta$$

$$p = \frac{V_g}{A_g R_S} = \frac{v_S}{a R_S} = \frac{1}{3}$$

Globular micelles: (Semiminor axis $R_S = \ell_S$, Semimajor axis $b \leq 3\ell_S$, Eccentricity E)

$$V_g = \frac{4 \pi R_S^2 b}{3} = g \, v_S, \quad A_g = 2 \pi R_S^2 \left[1 + \frac{\sin^{-1} E}{E (1-E^2)^{1/2}} \right] = g \, a, \quad E = \left[1 - \left(\frac{R_S}{b} \right)^2 \right]^{1/2}$$

$$A_{g\delta} = 2 \pi (R_S + \delta)^2 \left[1 + \frac{\sin^{-1} E_\delta}{E_\delta (1 - E_\delta^2)^{1/2}} \right] = g \, a_\delta, \quad E_\delta = \left[1 - \left(\frac{R_S + \delta}{b + \delta} \right)^2 \right]^{1/2}$$

$$p = \frac{V_g}{A_g R_S} = \frac{v_S}{a R_S}, \quad \frac{1}{3} \leq p \leq 0.406, \quad R_{eq} = \left(\frac{3 V_g}{4 \pi} \right)^{1/3}$$

Cylindrical part of rodlike micelles: (Radius $R_C \leq \ell_S$, Length L_C)

$$V_g = \pi R_C^2 L_C = g \, v_S, \quad A_g = 2 \pi R_C L_C = g \, a$$

$$A_{g\delta} = 2 \pi (R_C + \delta) L_C = g \, a_\delta$$

$$p = \frac{V_g}{A_g R_C} = \frac{v_S}{a R_C} = \frac{1}{2}$$

Endcaps of rodlike micelles: (Endcap radius $R_S \leq \ell_S$, Cylinder radius $R_C \leq \ell_S$)

$$V_g = \left[\frac{8 \pi R_S^3}{3} - \frac{2 \pi}{3} H^2 (3 R_S - H) \right] = g \, v_S, \quad A_g = \left[8 \pi R_S^2 - 4 \pi R_S H \right] = g \, a$$

$$A_{g\delta} = \left[8 \pi (R_S + \delta)^2 - 4 \pi (R_S + \delta)(H + \delta) \right] = g \, a_\delta, \quad H = R_S \left[1 - \left\{ 1 - (R_C / R_S)^2 \right\}^{1/2} \right]$$

$$p = \frac{V_g}{A_g R_S} = \frac{v_S}{a R_S}$$

1.2.2 SOLUTION MODEL OF MICELLE FORMATION

The surfactant solution is a binary mixture of the surfactant and water. However, we can view it as a multicomponent system by treating the aggregate of each size and shape as a distinct chemical component described by a characteristic chemical potential.[12,24,25] We consider a solution consisting of N_W water molecules, N_1 singly dispersed surfactant molecules, and N_g aggregates of all possible shapes and aggregation number g, with g allowed to take all values from 2 to ∞. The subscript W refers to water, 1 to the singly dispersed surfactant, and g to the aggregate containing g surfactant molecules. The Gibbs free energy of the solution G is the number-weighted sum of the chemical potentials μ_i of the various species i.

$$G = N_w \mu_w + N_1 \mu_1 + \sum_{g=2}^{g=\infty} N_g \mu_g \qquad (1.1)$$

The equilibrium condition of a minimum in the free energy leads to

$$\frac{\mu_g}{g} = \mu_1 \qquad (1.2)$$

This equation stipulates that the chemical potential of the singly dispersed surfactant molecule is equal to the chemical potential per molecule of an aggregate of any size and shape. For a dilute surfactant solution, one can write (for all values of g including $g = 1$)

$$\mu_g = \mu_g^o + k_B T \ln X_g \qquad (1.3)$$

where μ_g^o is the standard state chemical potential of the specie g, X_g is its mole fraction in solution, k_B is the Boltzmann constant, and T is the absolute temperature (expressed everywhere in this chapter in °K). The standard state of the solvent is defined as the pure solvent whereas the standard states of all the other species are taken to be infinitely dilute solution conditions. Combining the above two equations, we get the aggregate size distribution

$$X_g = X_1^g \exp - \left(\frac{\mu_g^o - g\,\mu_1^o}{k_B T} \right) = X_1^g \exp - \left(\frac{g\,\Delta\mu_g^o}{k_B T} \right) \qquad (1.4)$$

where $\Delta\mu_g^o$ is the difference in the standard chemical potentials between a surfactant molecule present in an aggregate of size g and a singly dispersed surfactant in water. To calculate the aggregate size distribution, we need an explicit equation for $\Delta\mu_g^o$.

The critical micelle concentration (cmc) can be calculated from the aggregate size distribution[26,27] by constructing a plot of any one of the functions X_1, ΣX_g, $\Sigma g X_g$, or $\Sigma g^2 X_g$ (which are proportional to different experimentally measured properties of the surfactant solution such as surface tension, electrical conductivity, dye solubilization, light scattering intensity, etc.) against the total surfactant concentration X_{tot} ($= X_1 + \Sigma g X_g$) and equating the cmc to the total surfactant concentration at which a sharp change in the plotted function occurs. The cmc can also be estimated[27] as the value of X_1 for which the amount of surfactant in the micellized form is equal to that in the singly dispersed form. Yet another approach[27] to estimate the cmc is to equate it to the total surfactant concentration for which about 5% of the total surfactant is in the form of micelles.

From the micelle size distribution we can compute average aggregation numbers using the definitions

$$g_n = \frac{\Sigma g \, X_g}{\Sigma X_g}, \quad g_w = \frac{\Sigma g^2 \, X_g}{\Sigma g \, X_g}, \quad g_z = \frac{\Sigma g^3 \, X_g}{\Sigma g^2 \, X_g} \tag{1.5}$$

where g_n, g_w, and g_z denote the number-average, the weight-average, and the z-average aggregation numbers, respectively, and the summations extend from 2 to ∞. The ratios g_w/g_n and g_z/g_w are unity for monodispersed systems and are equal to 2 and 3/2 for systems exhibiting very high polydispersity.[25] Thus either of these ratios can be used as an index of polydispersity. For a surfactant with any kind of head group, we can show[28] from the size distribution that the average aggregation numbers g_n and g_w depend on the concentration of the micellized surfactant as follows:

$$\partial \ln g_n = \left(1 - \frac{g_n}{g_w}\right)\partial \ln \sum g \, X_g, \quad \partial \ln g_w = \left(\frac{g_z}{g_w} - 1\right)\partial \ln \sum g \, X_g$$

$$g_n \propto \left[\sum g X_g\right]^{(1-g_n/g_w)}, \quad g_w \propto \left[\sum g X_g\right]^{(g_z/g_w-1)} \tag{1.6}$$

This equation states that the average aggregation numbers g_n and g_w must increase appreciably with increasing concentration of the micellized surfactant if the micelles are polydispersed; the average aggregation numbers must be virtually independent of the total surfactant concentration if the micelles are narrowly dispersed. Further, Equation 1.6 shows that the exponent relating the average micelle size to the total surfactant concentration is a direct measure of the aggregate polydispersity. These are purely thermodynamic results independent of any models for micellization.

1.2.3 FORMATION OF RODLIKE MICELLES

Rodlike micelles are visualized as having a cylindrical middle part with two spherical endcaps as shown in Figure 1.1. Mukerjee[29–33] was the first to treat the thermodynamics of rodlike aggregates by recognizing that two characteristic equilibrium constants are necessary. Mukerjee used a dimerization constant and, for aggregates larger than dimers, a constant stepwise association equilibrium constant. However, we can now identify the two relevant equilibrium constants with the molecules in the cylindrical part and those in the endcaps, respectively. The thermodynamic formulation for rodlike micelles has been presented by many authors[23,34–36] using somewhat different terminologies, but the equivalence between all of these treatments and their connection to Mukerjee's pioneering work has been discussed in our previous work.[37] The standard chemical potential of a rodlike micelle of size g with g_{cap} molecules in the two spherical endcaps and $(g - g_{cap})$ molecules in the cylindrical middle can be written as

$$\mu_g^o = (g - g_{cap})\,\mu_{cyl}^o + g_{cap}\,\mu_{cap}^o \tag{1.7}$$

where μ_{cyl}^o and μ_{cap}^o are the standard chemical potentials of the molecules in the two regions of the rodlike aggregate, respectively. Introducing the above relation in the aggregate size distribution we get

$$X_g = \left[X_1 \exp - \left(\frac{\Delta\mu_{cyl}^o}{k_B T} \right) \right]^g \exp\left[-g_{cap}\left(\frac{\Delta\mu_{cap}^o - \Delta\mu_{cyl}^o}{k_B T} \right) \right] \tag{1.8}$$

where $\Delta\mu_{cyl}^o$ and $\Delta\mu_{cap}^o$ are the differences in the standard chemical potentials between a surfactant molecule in the cylindrical middle or the endcaps of the rodlike micelle and a singly dispersed surfactant molecule. Equation 1.8 can be rewritten as

$$X_g = \frac{1}{K} Y^g, \quad Y = \left[X_1 \exp - \left(\frac{\Delta\mu_{cyl}^o}{k_B T} \right) \right], \quad K = \exp\left[g_{cap}\left(\frac{\Delta\mu_{cap}^o - \Delta\mu_{cyl}^o}{k_B T} \right) \right] \tag{1.9}$$

where K is a measure of the free energy penalty for the molecules present in the spherical endcap compared to those in the cylindrical portion. The average aggregation numbers can be computed from Equation 1.9 by analytically summing the series functions.[20,23]

$$g_n = g_{cap} + \left(\frac{Y}{1-Y} \right), \quad g_w = g_{cap} + \frac{Y}{1-Y}\left(1 + \frac{1}{Y + g_{cap}\,(1-Y)} \right) \tag{1.10}$$

Equation 1.10 shows that for values of Y close to unity, very large aggregates are formed. The total concentration of surfactant present in the aggregated state is given by the expression[20,23]

$$\sum g\, X_g = \frac{1}{K}\frac{g_{cap}Y^{g_{cap}}}{1-Y}\left(1+\frac{Y}{g_{cap}\,(1-Y)}\right)=X_{tot}-X_1 \qquad (1.11)$$

Here, X_{tot} and X_1 refer to the total amount of surfactant and the amount of surfactant present as singly dispersed molecules (practically, the cmc), so that the difference between them is the amount of surfactant in the micellar form. In the limit of Y close to unity and $g_{cap}(1-Y) \ll 1$, Equation 1.11 reduces to

$$\sum g\, X_g = \frac{1}{K}\left(\frac{1}{1-Y}\right)^2 = X_{tot}-X_1 \qquad (1.12)$$

Introducing Equation 1.12 in Equation 1.10 yields the dependence of the average aggregation numbers on the surfactant concentration.

$$g_n = g_{cap}+\frac{1}{1-Y}=g_{cap}+[K\,(X_{tot}-X_1)]^{1/2},$$

$$g_w = g_{cap}+\frac{2}{1-Y}=g_{cap}+2[K\,(X_{tot}-X_1)]^{1/2} \qquad (1.13)$$

Since a realistic value for X_{tot} is less than 10^{-2} (that is, a surfactant concentration of 0.55 M or below), it is evident that K must be in the range of 10^8 to 10^{12}, if large rodlike micelles ($g \sim 10^3$ to 10^5) are to form at physically realistic surfactant concentrations. The polydispersity index (g_w/g_n) goes from unity to 2 as the micelles grow from the smallest rods at low surfactant concentration to giant rodlike micelles at high surfactant concentrations. The cmc is approximately estimated from Equation 1.9 based on the condition that Y is close to unity.

$$X_{cmc} = \exp\left(\frac{\Delta\mu^o_{cyl}}{k_B T}\right) \qquad (1.14)$$

The thermodynamic results obtained so far are independent of any specific expression for the standard free energy change $\Delta\mu^o_g$ associated with aggregation and constitute general theoretical principles governing the aggregation behavior of surfactants. However, to perform predictive calculations of the aggregation behavior, specific expressions for $\Delta\mu^o_g$ are needed and these are developed in the following section.

1.3 FREE ENERGY CHANGE ON GIANT MICELLE FORMATION

1.3.1 CONTRIBUTIONS TO FREE ENERGY CHANGE ON AGGREGATION

The standard free energy difference $\Delta\mu_g^o$ between a surfactant molecule in an aggregate of size g and one in the singly dispersed state can be decomposed into a number of contributions on the basis of molecular considerations.[20] First, the hydrophobic tail of the surfactant is removed from contact with water and transferred to the aggregate core which is like a hydrocarbon liquid. Second, the surfactant tail inside the aggregate core is subjected to packing constraints because of the requirements that the polar head group should remain at the aggregate–water interface and the micelle core should have a hydrocarbon liquidlike density. Third, the formation of the aggregate is associated with the creation of an interface between its hydrophobic domain and water. Fourth, the surfactant head groups are brought to the aggregate surface, giving rise to steric repulsions between them. Last, if the head groups are ionic or zwitterionic, then electrostatic repulsions between the head groups at the aggregate surface also arise. Explicit analytical expressions for each of these free energy contributions have been developed in our earlier studies[20] in terms of the molecular characteristics of the surfactant, and they are briefly discussed in this section.

1.3.2 TRANSFER OF THE SURFACTANT TAIL

The contribution to the free energy from the transfer of surfactant tail from water to the micelle core is estimated by considering the core to be like a liquid hydrocarbon. This allows us to estimate the transfer free energy using compiled experimental data on the solubility of hydrocarbons in water.[38,39] *Note that no information derived from experiments involving surfactants is used anywhere in our model.* We have estimated[20] the group contributions to the transfer free energy for a methylene and a methyl group in an aliphatic tail to be

$$\frac{\left(\Delta\mu_g^o\right)_{tr}}{k_B T} = 5.85 \ln T + \frac{896}{T} - 36.15 - 0.0056\,T \quad \text{(for } CH_2\text{)}$$

$$\frac{\left(\Delta\mu_g^o\right)_{tr}}{k_B T} = 3.38 \ln T + \frac{4064}{T} - 44.13 + 0.02595\,T \quad \text{(for } CH_3\text{)}$$

$$(1.15)$$

Since the micelle core differs from a liquid hydrocarbon, a contribution to the free energy from this different state is considered next.

1.3.3 DEFORMATION OF THE SURFACTANT TAIL

In the surfactant tail, the methylene group attached to the polar head group is constrained to remain at the aggregate–water interface. The other end (the terminal methyl group) is free to occupy any position inside the aggregate as long as uniform liquidlike density is maintained in the aggregate core. It is necessary for the tail to deform nonuniformly along its length in order to satisfy both the packing and the uniform density constraints. The positive free energy contribution resulting from this conformational constraint on the tail is referred to as the tail deformation free energy. By adapting the method suggested for block copolymers by Semenov,[40] we have derived a simple analytical expression for the tail deformation energy.[20] For spherical micelles

$$
\frac{\left(\Delta\mu_g^o\right)_{def}}{k_B T} = \left(\frac{9\, p\pi^2}{80}\right)\left(\frac{R_S^2}{N\,L^2}\right) \tag{1.16}
$$

where p is the packing factor defined in Table 1.1, R_S is the core radius, L is the segment length and N is the number of segments in the tail ($N = \ell_S/L$, where ℓ_S is the extended length of the tail). As suggested by Dill and Flory,[41,42] a segment is assumed to consist of 3.6 methylene groups (hence $L = 0.46$ nm). L also represents the spacing between alkane molecules in the liquid state, namely, L^2 ($= 0.21$ nm^2) is the cross-sectional area of the polymethylene chain. Equation 1.16 is used also for globular micelles and the spherical endcaps of rodlike micelles. For the cylindrical middle of rodlike micelles, the coefficient 9 is replaced by 10, the radius R_S is replaced by the radius R_C of the cylinder, and $p = 1/2$.

1.3.4 FORMATION OF AGGREGATE CORE–WATER INTERFACE

Micelle formation generates an interface between the hydrophobic core and the surrounding water medium. The free energy of formation of this interface is calculated as the product of the surface area in contact with water and the macroscopic interfacial tension σ_{agg} characteristic of the interface[20]

$$
\frac{\left(\Delta\mu_g^o\right)_{int}}{k_B T} = \left(\frac{\sigma_{agg}}{k_B T}\right)(a - a_o) \tag{1.17}
$$

Here, a is the surface area of the micelle core per surfactant molecule, and a_o is the surface area per molecule shielded from contact with water by the polar head group of the surfactant. Expressions for the area per molecule, a, corresponding to different aggregate shapes are provided in Table 1.1. The area a_o depends on the extent to which the polar head group shields the cross-sectional area L^2 of the surfactant tail. If the head group cross-sectional area a_p is larger than L^2, the tail cross section is shielded completely from contact with water and $a_o = L^2$. If a_p is smaller than L^2, then the head group shields only a part of the cross-sectional area of the tail from contact with water, and in this case $a_o = a_p$. Thus, a_o is equal to the smaller of a_p and L^2.

The aggregate core–water interfacial tension σ_{agg} is taken equal to the interfacial tension σ_{sw} between water (W) and the aliphatic hydrocarbon of the same molecular weight as the surfactant tail (S). The interfacial tension σ_{sw} can be calculated in terms of the surface tensions σ_S of the aliphatic surfactant tail and σ_W of water[43]

$$\sigma_{sw} = \sigma_s + \sigma_w - 2.0\psi(\sigma_s\sigma_w)^{1/2}$$

$$\sigma_s = 35.0 - 325\,M^{-2/3} - 0.098\,(T - 298) \qquad (1.18)$$

$$\sigma_w = 72.0 - 0.16\,(T - 298)$$

In Equation 1.18, ψ is a constant[43] equal to 0.55, M is the molecular weight of the surfactant tail, and the expressions for σ_S and σ_W (in mN/m) are based on literature data.[20,43,44]

1.3.5 HEAD GROUP INTERACTIONS—STERIC

Upon micelle formation, the surfactant head groups are brought to the micelle surface where they are crowded when compared to the infinitely dilute state of the singly dispersed surfactant molecules. This generates steric repulsions among the head groups. For compact head groups, by analogy with the repulsion term in the van der Waals equation of state, we have proposed[15,16,20] the expression

$$\frac{\left(\Delta\mu_g^0\right)_{steric}}{k_B T} = -\ln\left(1 - \frac{a_p}{a}\right) \qquad (1.19)$$

where a_p is the cross-sectional area of the polar head group. This equation is used for spherical and globular micelles as well as for the cylindrical middle and the spherical endcaps of rodlike micelles. This approach to calculating the steric interactions is inadequate when the polar head groups are not compact, as in the case of nonionic surfactants with oligoethylene oxide head groups. An alternate treatment[20] for head group interactions in such systems, considering the head groups to be polymerlike, has been developed by us, and it showed mixed success in predicting the micellar properties.

1.3.6 HEAD GROUP INTERACTIONS—DIPOLAR

If the surfactant has a zwitterionic head group with a permanent dipole moment, then dipole–dipole interactions arise at the micelle surface. The dipoles are oriented normal to the interface and stacked such that the poles of the dipoles are located on parallel surfaces. The interaction free energy is estimated by considering that the poles of the dipoles generate an electrical capacitor and the distance between the planes of the capacitor is equal to the distance of charge separation d (or the dipole length) in the zwitterionic head group. Consequently, one gets for spherical micelles[20]

$$\frac{\left(\Delta\mu_g^o\right)_{dipole}}{k_BT} = \frac{2\pi e^2 R_S}{\varepsilon a_\delta k_BT}\left[\frac{d}{R_S+\delta+d}\right] \tag{1.20}$$

where e is the electronic charge, ε the dielectric constant of the solvent, R_S the radius of the spherical core, and δ the distance from the core surface to the place where the dipole is located. This equation is employed also for globular micelles and the endcaps of rodlike micelles. For the cylindrical part of the rodlike micelles, the capacitor model yields

$$\frac{\left(\Delta\mu_g^o\right)_{dipole}}{k_BT} = \frac{2\pi e^2 R_C}{\varepsilon a_\delta k_BT}\ln\left[1+\frac{d}{R_C+\delta}\right] \tag{1.21}$$

where R_C is the radius of the cylindrical core of the micelle. The dielectric constant is taken to be that of pure water[44] and is calculated using the expression[20]

$$\varepsilon = 87.74\exp[-0.0046(T-273)] \tag{1.22}$$

1.3.7 HEAD GROUP INTERACTIONS—IONIC

If the surfactant has an anionic or cationic head group, then ionic interactions arise at the micellar surface. An approximate analytical solution to the Poisson–Boltzmann equation derived by Evans and Ninham[45] for spherical and cylindrical micelles is used in our calculations.

$$\frac{\left(\Delta\mu_g^o\right)_{ionic}}{k_BT} = 2\left[\ln\left(\frac{S}{2}+U\right)-\frac{2}{S}(U-1)-\frac{2C}{\kappa S}\ln\left(\frac{1}{2}+\frac{1}{2}U\right)\right] \tag{1.23}$$

$$U = \left[1+\left\{\frac{S}{2}\right\}^2\right]^{1/2}, \quad S = \frac{4\pi e^2}{\varepsilon\kappa a_\delta k_BT}$$

In Equation 1.23, the area per molecule a_δ is evaluated at a distance δ from the hydrophobic core surface (see Table 1.1) where the center of the counterion is located. The first two terms on the right-hand side of Equation 1.23 constitute the exact solution to the Poisson–Boltzmann equation for a planar geometry and the last term provides the curvature correction. The curvature-dependent factor C is given by[20]

$$C = \frac{2}{R_S + \delta}, \quad \frac{2}{R_{eq} + \delta}, \quad \frac{1}{R_C + \delta} \tag{1.24}$$

for spheres/spherical endcaps of spherocylinders, globular aggregates (with an equivalent radius R_{eq} defined in Table 1.1) and cylindrical middle part of spherocylinders, respectively. κ is the reciprocal Debye length and is related to the ionic strength of the solution via

$$\kappa = \left(\frac{8 \pi n_o e^2}{\varepsilon k_B T} \right)^{1/2}, \quad n_o = \frac{(C_1 + C_{add})N_{Av}}{1000} \tag{1.25}$$

In equation 1.25, n_o is the number of counterions in solution per cm^3, C_1 is the molar concentration of the singly dispersed surfactant molecules, C_{add} is the molar concentration of any salt added to the surfactant solution, and N_{Av} is Avogadro's number.

1.4 MODEL PREDICTIONS OF GIANT MICELLES

1.4.1 COMPUTATIONAL APPROACH

The equation for the size distribution of aggregates, in conjunction with the geometrical relations in Table 1.1 and the expressions for the various contributions to the free energy of micellization summarized in Section 1.3, allow us to make predictive calculations of micellization behavior. The calculations for spherical and globular micelles are carried out based on the maximum-term method[20,46] instead of computing the detailed distribution of sizes. This approach is built on the recognition that for spherical or globular micelles the size dispersion is usually narrow and g_n and g_w are practically the same. g_n can be taken as the value of g for which the number concentration X_g of the aggregates is a maximum, while g_w can be taken as the value of g for which the weight concentration gX_g of the aggregates is a maximum. Only the aggregation number g is the independent variable in the case of spherical or globular micelles, and the maximization of either X_g or gX_g can be carried out with respect to g. The cmc is estimated as the value of X_1 for which the amount of surfactant in the micellized form is equal to that in the singly dispersed form.

In the case of rodlike micelles, by minimizing $\Delta\mu_{cyl}^o$ for the cylindrical part, the equilibrium radius R_C of the cylindrical part of the micelle is determined.

Given the radius of the cylindrical part, the number of molecules g_{cap} in the spherical endcaps is found to be that value which minimizes $\Delta\mu^\circ_{cap}$. Given g_{cap}, $\Delta\mu^\circ_{cyl}$, and $\Delta\mu^\circ_{cap}$, the sphere-to-rod transition parameter K is calculated from Equation 1.9, the average aggregation numbers at any total surfactant concentration from Equation 1.13, and the cmc from Equation 1.14.

The search for the parameter values that maximize X_g or minimize $\Delta\mu^\circ_{cyl}$ and $\Delta\mu^\circ_{cap}$ was carried out using the IMSL (International Mathematical and Statistical Library) subroutine ZXMWD. This subroutine is designed to search for the global extremum of a function of many independent variables subject to any specified constraints on the variables. This subroutine has been used for all the calculations described in this chapter.

1.4.2 ESTIMATION OF MOLECULAR CONSTANTS

Illustrative calculations in this chapter have been performed for a few surfactant types. The molecular constants associated with the surfactant tail are the volume v_S and the extended length R_S of the tail. For the head group, one needs the cross-sectional area a_p for all types of head groups, the distance δ from the core surface where the counterion is located in the case of ionic head groups, the dipole length d, and the distance δ from the core surface at which the dipole is located, in the case of a zwitterionic head group. All these molecular constants can be estimated from the chemical structure of the surfactant molecule.

1.4.2.1 Estimation of Tail Volume v_S

The molecular volume v_S of the tail containing n_C carbon atoms is calculated from the group contributions of $(n_C - 1)$ methylene groups and the terminal methyl group. The group molecular volumes are estimated from the density vs temperature data for aliphatic hydrocarbons.[47]

$$v_S = v_{CH_3} + (n_C - 1)\, v_{CH_2}$$

$$v_{CH_3} = 0.0546 + 1.24 \times 10^{-4}\,(T - 298)\; nm^3 \qquad (1.26)$$

$$v_{CH_2} = 0.0269 + 1.46 \times 10^{-5}\,(T - 298)\; nm^3$$

1.4.2.2 Estimation of Extended Tail Length ℓ_S

The extended length of the surfactant tail ℓ_S at 298°K is calculated using a group contribution of 0.1265 nm for the methylene group and 0.2765 nm for the methyl group.[12,20] Given the small volumetric expansion of the surfactant tail over the range of temperatures of interest, the extended tail length ℓ_S is considered as temperature independent and the small volumetric expansion is accounted for by increases in the cross-sectional area of the tail.

TABLE 1.2
Molecular Constants for Surfactant Head Groups

Surfactant Head Group	a_p (nm²)	a_o (nm²)	δ (nm)	d (nm)
β-Glucoside	0.40	0.21	—	—
Dimethylphospheneoxide	0.48	0.21	—	—
Sodium sulfate	0.17	0.17	0.545	—
Trimethylammonium bromide	0.54	0.21	0.345	—
n-alcohol	0.08	0.08	—	—

1.4.2.3 Estimation of Head Group Area a_p

The head group area a_p is calculated as the cross-sectional area of the head group near the hydrophobic core–water interface. The glucoside head group in β-glucosides has a compact ring structure[16,29] with an approximate diameter of 0.7 nm, and hence, the effective cross-sectional area of the polar head group a_p is estimated as 0.40 nm². For sodium alkyl sulfates, alkyl trimethyl ammonium bromides, alkyl dimethyl phosphene oxides and n-alcohols, the cross-sectional area of the polar group a_p has been estimated taking the bond lengths and bond angles into account as equal to 0.17, 0.54, 0.48, and 0.08 nm², respectively (see Table 1.2).

1.4.2.4 Estimation of δ and d

For ionic surfactants, the molecular constant δ depends on the size of the ionic head group, the size of the hydrated counterion, and also the proximity of the counterion to the charge on the surfactant ion. Visualizing that the sodium counterion is placed on top of the sulfate anion, we estimate $\delta = 0.545$ nm for sodium alkylsulfates. For alkyltrimethyl ammonium bromide, the surfactant cation is nearer the hydrophobic core, and we estimate $\delta = 0.345$ nm.

All the molecular constants characterizing the head groups of surfactants considered in this chapter are listed in Table 1.2. *It is important to note that while the molecular theory obviously needs many molecular constants, there are no free parameters involved in this theory. No information from experimentally determined properties of surfactant solutions has been used to deduce any molecular constant. Hence, the calculations are completely predictive in nature. Most of the required molecular constants such as dielectric constant, surface tension, molecular volumes, and so forth, are already well known in the literature. For all practical purposes, we need to estimate only one model parameter for nonionic surfactants (a_p), two for ionic surfactants (a_p and δ) and three for zwitterionic surfactants (a_p, δ, and d). Clearly any perception that a molecular theoretical approach requires many parameters is entirely incorrect.*

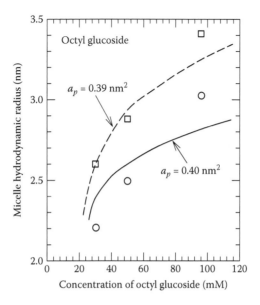

FIGURE 1.2 Dependence of the micellar size (expressed as the hydrodynamic radius) on the concentration of the surfactant for nonionic octyl glucoside. Both lines correspond to predicted values but for marginally different values of the molecular constant a_p. Squares denote reported experimental data.[48] Circles correspond to modified experimental data if the reported hydrodynamic radius had included one layer of water. See text for discussion.

1.4.3 ILLUSTRATIVE PREDICTED RESULTS

In Figure 1.2, the predicted hydrodynamic radius of nonionic octyl β-glucoside micelle is plotted against the surfactant concentration. The hydrodynamic radius of the micelle is calculated from the predicted weight-average aggregation number using the expression

$$R_H = \left(\frac{3 g_w v_S}{4 \pi} \right)^{1/3} + \ell_p \tag{1.27}$$

where ℓ_p is the length of the polar head group which for the ring structure of the β-glucoside has been estimated[16,29] to be 0.7 nm. One may note from Equation 1.9 that the sphere-to-rod transition parameter K, which determines g_w (see Equation 1.13), can be dramatically altered by small changes in the free energy difference $(\Delta\mu^\circ_{cap} - \Delta\mu^\circ_{cyl})$, since g_{cap} that appears in the definition of K is much larger than unity. For example, assuming a typical value of 90 for g_{cap}, a small change of 0.05 $k_B T$ in the free energy difference $(\Delta\mu^\circ_{cap} - \Delta\mu^\circ_{cyl})$ will cause a change in K of $e^{4.5} = 90$, which in turn can change the predicted value for g_w by a factor of about 10. Therefore, the predicted average aggregation numbers are very sensitive even to small changes in the free energy estimates when rodlike micelles form. This is illustrated by the calculations carried out for two marginally different values of the parameter a_p, which affects the magnitude of the head group steric interaction energy. The predictions are compared with the data provided by dynamic light scattering measurements.[48] It is not clear whether the experimental hydrodynamic radii correspond to dry or hydrated aggregates. Therefore, both the reported hydrodynamic radii and the radii obtained by sub-tracting the diameter of a water molecule are plotted in Figure 1.2. Given the sensitivity of K to the free energy estimates, the agreement between the measured and predicted aggregate sizes is satisfactory. We note that although rodlike micelles are formed, they are still not very large for octylglucoside. When the alkyl chain length increases, g_{cap} will increase and, consequently, K will increase significantly, leading to the formation of giant rodlike micelles.

The predicted values for the sphere-to-rod transition parameter K for anionic sodium dodecyl sulfate as a function of the added NaCl concentration are plotted in Figure 1.3 and the predictions are compared against results from light scattering measurements.[49] The largest deviation between predicted and experimental value of K is less than a factor of 10 at the highest ionic strength. As already noted,

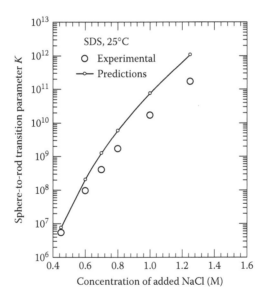

FIGURE 1.3 Dependence of the sphere-to-rod transition parameter K for sodium dodecyl sulfate on the concentration of added electrolyte NaCl. Points are from light scattering measurements[49] at 25°C and predictions are shown with the same symbols (in smaller size) and the connecting line.

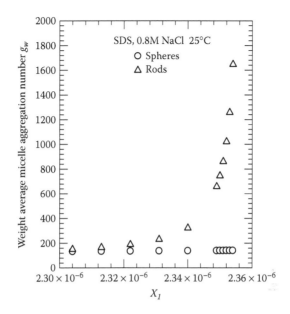

FIGURE 1.4 Predicted weight-average aggregation numbers of coexisting populations of globular micelles and cylindrical micelles as functions of the singly dispersed SDS concentration. Calculations are for 0.8 M added NaCl and 25°C. For concentrations near the cmc, the micelles are relatively small and globular, while at higher concentrations, the rodlike micelles show significant growth in size.

small variations in the free energy estimates of the order of 0.05 $k_B T$ are adequate to cause a variation in K by a factor of 100. Considering this, the deviations between experimental data and predictions can be considered satisfactory. The predicted radius of the cylindrical part of the aggregate is smaller than the fully extended length of the surfactant tail. It increases from 1.45 nm to 1.49 nm as the electrolyte concentration is increased from 0.45 M to 1.25 M at 25°C. For this range of ionic strengths, the radius of the endcap remains unaltered and is equal to the extended length of the surfactant tail. The predicted cmc, as defined by Equation 1.14, decreases from 0.38 mM to 0.204 mM over this range of added salt concentration.

When rodlike micelles form, smaller spherical/globular micelles can also coexist. The relative concentration of these two populations of aggregates determines the average solution properties of the surfactant system. Figure 1.4 presents the weight-average aggregation numbers corresponding to both smaller globular micelles and larger cylindrical micelles formed in solutions of sodium dodecyl sulfate at 0.8 M added NaCl. The micelle average aggregation numbers are plotted against the concentration of the singly dispersed

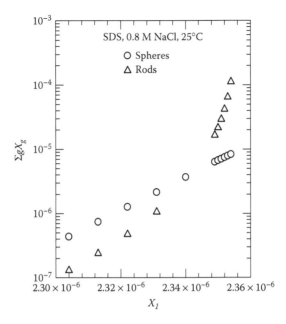

FIGURE 1.5 Predicted amount of surfactant incorporated into coexisting populations of globular micelles and cylindrical micelles as a function of the singly dispersed SDS concentration. Calculations are for 0.8 M added NaCl and 25°C. For concentrations near the cmc, the globular micelles dominate, while at high concentrations the rodlike micelles dominate.

surfactant (which controls the chemical potential of the surfactant in solution). In Figure 1.5, the amount of surfactant incorporated into the globular micelles and that in the rodlike micelles are plotted, also against the singly dispersed surfactant concentration. Clearly, in the region near the cmc and below, the globular micelles dominate. Just beyond the cmc, the large rodlike micelles dominate.

The temperature dependence of the sphere-to-rod transition parameter K has been calculated for sodium dodecyl sulfate for two concentrations of added NaCl. The predicted values of K are plotted in Figure 1.6 along with the experimental values determined from dynamic light scattering measurements.[49] The largest deviation in K is about a factor of 100, and the theory predicts a somewhat stronger dependence of K on the temperature than that observed experimentally. Figure 1.7 presents the predicted values of K as a function of added electrolyte concentration for the homologous family of sodium alkyl sulfates with 10 to 13 carbon atoms in their hydrophobic tails. The figure shows that the largest deviation in K between the predicted and the experimental values is smaller than a factor of 5 for C_{11}, C_{12}, and C_{13} sodium sulfates, and is about a factor of 100 for the C_{10} sodium sulfate.

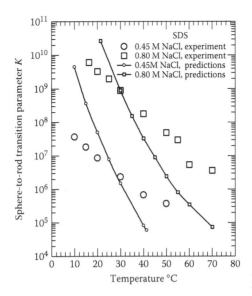

FIGURE 1.6 Influence of temperature on the sphere-to-rod transition parameter K of sodium dodecyl sulfate in solutions containing 0.45 M and 0.80 M added NaCl electrolyte. Points are experimental data[49] and the predictions are shown with the same symbols (in smaller size) and the connecting line.

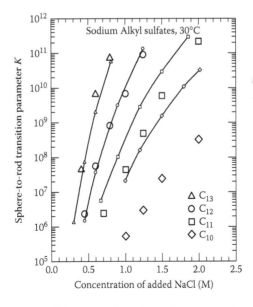

FIGURE 1.7 Dependence of the sphere-to-rod transition parameter K on the surfactant tail length and on the concentration of added electrolyte NaCl for sodium alkyl sulfates. Points denote the experimental data[49] at 30°C, and predictions are shown with the same symbols as the corresponding experimental points (in smaller size) and the connecting line.

1.5 TUNING STRUCTURAL TRANSITIONS IN GIANT MICELLES

1.5.1 Surfactant Mixtures

Transformation of other aggregate structures to rodlike micelles and vice versa is possible by the addition of a second surfactant. Such addition can cause all solution properties of the surfactant such as the cmc and micelle size, as well as the aggregate shapes to change. We have formulated a molecular theory for surfactant mixtures to predict the wide range of variations caused by different combinations of binary surfactant mixtures.[18,21] The treatment is a direct extension of the molecular theory developed for a single surfactant. For binary surfactant mixtures, the size distribution of aggregates has the form

$$X_g = X_{1A}^{g_A} X_{1B}^{g_B} \exp - \left(\frac{\mu_g^o - g_A \, \mu_{1A}^o - g_B \, \mu_{1B}^o}{k_B T} \right) = X_{1A}^{g_A} X_{1B}^{g_B} \exp - \left(\frac{g \, \Delta \mu_g^o}{k_B T} \right) \quad (1.28)$$

The mole fractions of A and B in the singly dispersed surfactant mixture are denoted as α_{1A} and α_{1B} while the mole fractions of A and B in the mixed aggregates are denoted by α_{gA} and α_{gB}. Expressions for the standard free energy change on aggregation are obtained by a simple extension of the equations developed in Section 1.3 for single component surfactant systems. Only the modifications necessary for the treatment of surfactant mixtures are described below. More details can be found in our earlier papers.[18–21]

1.5.1.1 Free Energy of Formation of Mixed Micelles

1.5.1.1.1 Transfer of the Surfactant Tail
For a mixed micelle of composition (α_{gA}, α_{gB}), the transfer free energy per surfactant molecule is the weighted average of the contributions from the two different tails.

$$\frac{\left(\Delta \mu_g^o \right)_{tr}}{k_B T} = \alpha_{gA} \frac{\left(\Delta \mu_g^o \right)_{tr,A}}{k_B T} + \alpha_{gB} \frac{\left(\Delta \mu_g^o \right)_{tr,B}}{k_B T} \quad (1.29)$$

This transfer free energy does not include contributions from the mixing of A and B tails inside the micellar core, which are accounted for separately.

1.5.1.1.2 Deformation of the Surfactant Tail
If surfactants A and B have different tail lengths, portions of both molecules may not be simultaneously present everywhere in the micellar core. Let us assume

that surfactant A has a longer tail than surfactant B, $\ell_{SA} > \ell_{SB}$. If the micelle radius R is less than both ℓ_{SA} and ℓ_{SB}, then even the shorter tail can reach everywhere within the core of the micelle. If $\ell_{SA} > R_S > \ell_{SB}$, then the inner region of the micellar core, of dimension $(R_S - \ell_{SB})$, can be reached only by the A tails. Taking into account the different extent to which the A and the B tails are stretched for the two situations described above, we have obtained the expression

$$\frac{\left(\Delta\mu_g^o\right)_{def}}{k_BT} = B_g\left[\alpha_{gA}\frac{R_S^2}{N_AL^2} + \alpha_{gB}\frac{Q_g^2}{N_BL^2}\right], \quad B_g = \left(\frac{9\,p\pi^2}{80}\right) \tag{1.30}$$

$$Q_g = R_s \text{ if } R_S \langle\ell_{SA}, \ell_{SB} \qquad Q_g = \ell_{SB} \text{ if } \ell_{SA}\rangle R_S\rangle \ell_{SB}$$

N_A and N_B stand for the number of segments in the tails of surfactants A and B, respectively, and p is the packing factor defined in Table 1.1. This equation is used for spherical and globular micelles and for the spherical endcaps of rodlike micelles. For the cylindrical part of the rodlike micelles, the coefficient 9 in B_g is replaced by 10, the radius R_S is replaced by the radius R_C of the cylindrical core, and the packing parameter $p = 1/2$.

1.5.1.1.3 Formation of Aggregate Core–Water Interface

The free energy associated with the formation of an interface is given by the expression

$$\frac{\left(\Delta\mu_g^o\right)_{int}}{k_BT} = \frac{\sigma_{agg}}{k_BT}(a - \alpha_{gA}\,a_{oA} - \alpha_{gB}a_{oB}) \tag{1.31}$$

Here, a_{oA} and a_{oB} are the areas per molecule of the core surface shielded from contact with water by the head groups of surfactants A and B. Since the interfacial tension against water of various hydrocarbon tails of surfactants are close to one another, σ_{agg} is approximated by the micelle composition averaged value.

1.5.1.1.4 Head Group Steric Interactions

Extending the expression used for single surfactant systems to binary surfactant mixtures, one can write

$$\frac{\left(\Delta\mu_g^o\right)_{steric}}{k_BT} = -\ln\left(1 - \frac{\alpha_{gA}\,a_{pA} + \alpha_{gB}\,a_{pB}}{a}\right) \tag{1.32}$$

1.5.1.1.5 Head Group Dipole Interactions

The dipole–dipole interactions are computed for spherical and globular micelles and the spherical endcaps of rodlike micelles from

$$\frac{\left(\Delta\mu_g^o\right)_{\text{dipole}}}{k_BT} = \frac{2\pi e^2 R_S}{\varepsilon\, a_{\text{dipole}}\, k_BT} \left[\frac{d}{R_S+\delta+d} \right] \alpha_{g,\text{dipole}} \tag{1.33}$$

For the cylindrical part of the rodlike micelles,

$$\frac{\left(\Delta\mu_g^o\right)_{\text{dipole}}}{k_BT} = \frac{2\pi e^2 R_C}{\varepsilon a_{\text{dipole}}\, k_BT} \ln\left[1+\frac{d}{R_C+\delta} \right] \alpha_{g,\text{dipole}} \tag{1.34}$$

In the preceding relations, $\alpha_{g,\text{dipole}}$ is the fraction of surfactant molecules in the aggregate having a dipolar (zwitterionic) head group. If both A and B are zwitterionic, then

$$\alpha_{g,\text{dipole}} = \alpha_{gA} + \alpha_{gB} = 1, \quad a_{\text{dipole}} = a_\delta \tag{1.35}$$

If surfactant A is zwitterionic and surfactant B is nonionic or ionic, then

$$\alpha_{g,\text{dipole}} = \alpha_{gA}, \quad a_{\text{dipole}} = \frac{a_\delta}{\alpha_{g,\text{dipole}}} \tag{1.36}$$

The dipole–dipole interactions may arise even if neither of the surfactants is zwitterionic. Such a situation occurs when the surfactant mixture consists of an anionic and a cationic surfactant. The two oppositely charged surfactants may be visualized as forming ion pairs. Depending upon the location of the charges on the two surfactant head groups, these ion pairs may be assigned a dipole moment. The distance of charge separation d in the zwitterionic head group, now refers to the distance between the locations of the anionic and the cationic charges, measured normal to the micelle core surface. Thus, for such systems

$$\alpha_{g,\text{dipole}} = \text{the smaller of } \frac{(\alpha_{gA}, \alpha_{gB})}{2}, \quad a_{\text{dipole}} = \frac{a_\delta}{\alpha_{g,\text{dipole}}}, \quad d = \left| \delta_A - \delta_B \right| \tag{1.37}$$

The factor 2 in the expression for $\alpha_{g,\text{dipole}}$ accounts for the fact that a dipole is associated with two surfactant molecules, treated as a pair. δ_A and δ_B represent the distance normal to the hydrophobic core surface at which the charges are located on the A and B surfactants.

1.5.1.1.6   Head Group Ionic Interactions

The ionic interactions at the aggregate surface are calculated using Equation 1.23, in conjunction with the curvature factor C given by Equation 1.24, with the modification that S is now given by

$$S = \frac{4\pi e^2}{\varepsilon \kappa a_{\delta\,ion} k_B T}, \quad a_{\delta ion} = \frac{a_\delta}{\alpha_{g\,ion}}, \quad \delta = \alpha_{gA}\,\delta_A + \alpha_{gB}\,\delta_B \qquad (1.38)$$

$\alpha_{g\,ion} = 1$, if A and B are ionic of the same kind
$\alpha_{g\,ion} = \alpha_{gA}$, $\delta_B = 0$, if only A is ionic and B is nonionic or zwitterionic
$\alpha_{g\,ion} = |\alpha_{gA} - \alpha_{gB}|$, if A and B are ionic and oppositely charged

1.5.1.1.7   Free Energy of Mixing of Surfactant Tails

This is the only contribution that is not present in the free energy model for single component surfactant solutions. This contribution accounts for the entropy and the enthalpy of mixing of the surfactant tails of A and B in the hydrophobic core of the micelle, with respect to the reference states of pure A or pure B micelle cores. This contribution is calculated using the Flory–Huggins expression.[50]

$$\frac{\left(\Delta\mu_g^o\right)_{mix}}{k_B T} = [\alpha_{gA}\,\ln\eta_A + \alpha_{gB}\,\ln\eta_B]$$

$$+ \left(\alpha_{gA}\,v_{SA}\left(\delta_A^H - \delta_{mix}^H\right)^2 + \alpha_{gB}\,v_{SB}\left(\delta_B^H - \delta_{mix}^H\right)^2 \right)\Big/ k_B T \qquad (1.39)$$

$$\delta_{mix}^H = \eta_A\,\delta_A^H + \eta_B\,\delta_B^H$$

where η_A and η_B are the volume fractions of A and B in micelle core, δ_A^H and δ_B^H are the Hildebrand solubility parameters[51] for the tails of surfactants A and B, and δ_{mix}^H is the volume fraction–averaged solubility parameter of the two components within the micelle core.

1.5.1.2   Estimation of Molecular Constants and Computational Approach

The only molecular constant that is new for calculations involving surfactant mixtures is the Hildebrand solubility parameter for the hydrocarbon tails of the surfactants. This can be estimated using a group contribution approach based on the properties of pure components.[51,52] For aliphatic hydrocarbon tails, the solubility parameters can be estimated from

$$\delta^H = \frac{0.7 + 0.471\,(n_C - 1)}{v_S}\ \text{MPa}^{1/2}, \quad v_S\ \text{in nm}^3, \quad 1\,\text{MPa} = 1\frac{\text{J}}{\text{cm}^3} \qquad (1.40)$$

The computational approach follows the description in Section 1.4.1. For the binary mixtures, the concentration of the singly dispersed surfactant mixture X_1 (= $X_{1A} + X_{1B}$) and the composition of this mixture α_{1A} are used as inputs.

1.5.1.3 Predicted Results

Calculations have been carried out for binary mixtures of anionic sodium dodecylsulfate (SDS) and nonionic decyldimethylphospheneoxide ($C_{10}PO$) at 25°C with 0.8 M added NaCl. Results for pure SDS presented in Section 1.4.3 show that SDS forms large rodlike micelles with $K \sim 7 \times 10^9$ at this condition. In contrast, pure $C_{10}PO$ at this condition is predicted[21] to form small, globular micelles with aggregation number of about 50. The two surfactants have slightly different lengths of hydrophobic tails, the sizes of their head groups are different, and one head group is ionic. Hence the properties of mixed aggregates will depend strongly on the composition. The predicted cmc of the mixture is plotted in Figure 1.8 as a function of both the micelle composition and the composition of the singly dispersed surfactant. From this figure we note that the fraction of SDS in micelles is always larger than the fraction of SDS in the singly dispersed surfactant. The predicted sphere-to-rod transition constant K is plotted against the composition of the micelle and the composition of the singly dispersed surfactant in Figure 1.9. We find that on the addition of the nonionic $C_{10}PO$ to the anionic SDS rodlike micelles, the rods grow (K increases) significantly

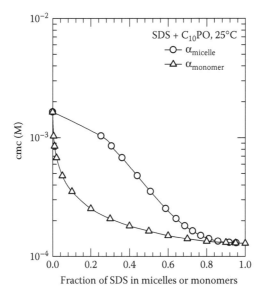

FIGURE 1.8 cmc of SDS + $C_{10}PO$ binary mixture as a function of the composition of micelles (circles) and that of singly dispersed surfactants (triangle) at 25°C. Fractions refer to mole fraction in the binary surfactant mixture. Both points and lines are predicted values.

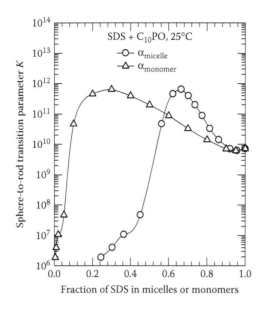

FIGURE 1.9 Sphere-to-rod transition parameter K for SDS + $C_{10}PO$ binary mixture as a function of the composition of micelles (circles) and that of singly dispersed surfactants (triangle) at 25°C. Both points and lines are predicted values.

initially, reach a maximum in size, and then decrease in size, eventually transforming into small globular micelles. This behavior can be interpreted in terms of the various interactions. The head group of SDS (0.17 nm²) is smaller in size compared to the head group of $C_{10}PO$ (0.48 nm²). When small amounts of nonionic $C_{10}PO$ are incorporated in the SDS micelles, the ionic head group repulsions at the micelle surface decrease without at the same time increasing the steric repulsions significantly. In the free energy model, this implies the lowering of both $\Delta\mu^o_{cyl}$ and $\Delta\mu^o_{cap}$, but the lowering of $\Delta\mu^o_{cyl}$ is larger in magnitude compared to the lowering of $\Delta\mu^o_{cap}$, contributing to an increase in K. When larger amounts of $C_{10}PO$ are incorporated, the steric interactions become more significant. Under these conditions, the lowering of $\Delta\mu^o_{cyl}$ is smaller in magnitude compared to the lowering of $\Delta\mu^o_{cap}$, contributing to a decrease in K. When the fraction of SDS in the micelle becomes smaller than about 0.50, the rods cease to exist and micelles become globular. Thus we see both a growth of rods and a rod-to-sphere transition on the addition of a nonionic surfactant.

Figure 1.10 shows the predicted aggregation numbers of SDS-$C_{10}PO$ micelles as a function of micelle composition when the total surfactant concentration is in the proximity of the cmc and when the micelles are small and globular. The dependence of the size of these small micelles on micelle composition shows a maximum similar to the dependence of K on micelle composition. When the nonionic molecules are added to the ionic micelle, the resulting reduction in ionic

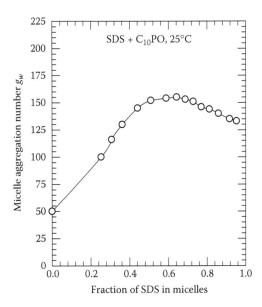

FIGURE 1.10 Average aggregation number of $C_{10}PO + C_{10}SO$ mixed micelles at cmc and at 25°C, as a function of the composition of micelles. Both points and lines are predicted values.

head group interactions is larger than the increase in the steric interactions between the head groups. This causes the micelles to become larger. Similarly, when a small amount of ionic SDS surfactants are added to the nonionic micelles, the resulting decrease in the steric interactions is larger than the increase in the ionic head group interactions. This causes the micelles to grow. Hence, we observe the maximum.

Another type of binary mixture investigated is a mixture of anionic and cationic surfactants. When present together, the anionic and cationic surfactants are expected to form ion pairs with no net charge; this decreases their aqueous solubility and results in precipitation.[53,54] These surfactant mixtures can also generate mixed micelles or mixed spherical bilayer vesicles in certain concentration and composition ranges.[55–58] As noted earlier, depending upon the location of the charges on the anionic and the cationic surfactants, one can associate a permanent dipole moment with each ion pair. Consequently, these surfactant mixtures behave partly as ionic single chain molecules and partly as zwitterionic paired chain molecules. We have calculated the aggregation characteristics of binary mixtures of decyltrimethylammonium bromide (DeTAB) and sodium decylsulfate (SDeS). The calculated cmc values are presented in Figure 1.11 as a function of both the composition of the micelle and the composition of the singly dispersed surfactant. Note that over a range of SDeS mole fraction in the micelle around 0.5, the micelles are rodlike,

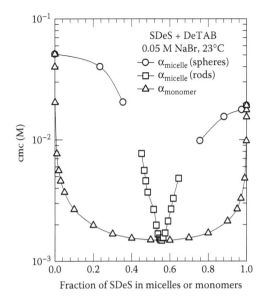

FIGURE 1.11 cmc of SDeS + DeTAB mixtures as a function of the composition of singly dispersed surfactants (triangles) and the composition of micelles (circles denote that micelles formed are spherical/globular, while squares denote that the micelles formed are rodlike). Solution includes 0.05 M added NaBr and temperature is 23°C. Points and lines are predicted values.

while micelles that are rich in either the cationic or the anionic surfactant are smaller and globular. Mixed micelles exist at all compositions although we have shown a region of discontinuity in the figure corresponding to the composition range where the sphere-to-rod transition occurs. The average aggregation numbers of globular micelles in the regions rich in either of the two surfactants are shown in Figure 1.12. One can observe that both the pure anionic micelle and the pure cationic micelle grow on incorporating some molecules of the other ionic kind. The sphere-to-rod transition parameter K is plotted against the composition of the micelle and that of the singly dispersed surfactant in Figure 1.13. The increase in K is quite significant implying that the mixed aggregates are indeed giant rodlike micelles. Note that the mixed rodlike micelles possess roughly equal number of the anionic and cationic surfactants. The small deviation from the micelle composition value of 0.5 arises because of the different sizes of the polar head groups.

Some binary mixtures of anionic and cationic surfactants have been observed to give rise to spherical bilayer vesicles in aqueous solutions.[55–58] In our earlier paper,[59] we have presented the predicted results for binary mixtures of cationic cetyltrimethylammonium bromide (CTAB) and anionic sodium dodecylsulfate (SDS), in the presence of 1 mM NaBr as electrolyte. We have identified regions

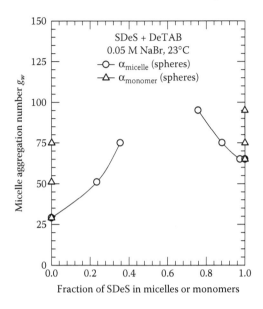

FIGURE 1.12 Average aggregation number of spherical/globular micelles formed from SDeS + DeTAB mixtures as a function of the composition of singly dispersed surfactants (triangles) and the composition of micelles (circles). Solution includes 0.05 M added NaBr and temperature is 23°C.

of spherical micelles, rodlike micelles, bilayer vesicles, as well as precipitation in this system.

1.5.2 Addition of an Alcohol

Transformation of rodlike micelles can be effected also by the addition of alcohols. We treat the alcohol as a nonionic surfactant, and predict the behavior of rodlike micelles formation. Here, we consider the addition of n-hexanol to anionic SDS and investigate the change in micellar properties on the addition of the alcohol. All predictive calculations have been carried out using the equations outlined in Section 1.5.1. The only molecular constant needed is the head group area a_p for the alcohol which is estimated to be 0.08 nm². Calculations have been performed in the presence of 0.45 M NaCl as added salt and at 25°C. Pure SDS under these conditions forms relatively small rodlike micelles, with $K \sim 6 \times 10^6$ (see Figure 1.3 or Figure 1.6).

Figure 1.14 presents the calculated cmc as a function of the fraction of hexanol in the cylindrical part of the micelle (for all practical purposes this is the composition of the micelle since the endcaps of rodlike micelles will make a negligible contribution when the micelles are large). The cmc data have been plotted both as the sum of SDS and n-hexanol concentrations and also as the concentration of SDS alone. The cmc expressed as a function of the SDS concentration decreases,

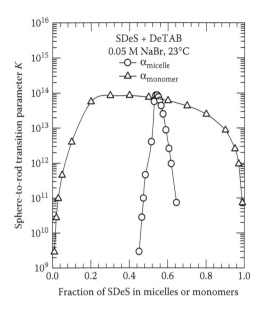

FIGURE 1.13 Sphere-to-rod transition parameter K for micelles formed from SDeS + DeTAB mixtures as a function of the composition of singly dispersed surfactants (triangles) and the composition of micelles (circles). Solution includes 0.05 M added NaBr and temperature is 23°C. Note that the composition of micelles is close to an equimolar presence of the anionic and the cationic surfactants over the entire range of singly dispersed surfactant composition. All points and lines are predicted values.

a well-known effect due to the addition of alcohol.[60] The calculated results show a limiting value for the fraction of hexanol in the micelle, at about 0.72, corresponding to the saturation of the system with n-hexanol.

The sphere-to-rod transition parameter K is plotted in Figure 1.15 as a function of the fraction of n-hexanol in the cylindrical part. We observe a dramatic growth of micelles, with K increasing from 10^7 to 10^{23}. Note that most experimental determinations of K have been limited to conditions where K is in the range 10^6 to 10^{12} as in Figure 1.7. Taking a C_{12} tail, and a total surfactant mole fraction X_{tot} of 10^{-4}, we get the length L of the cylindrical micelle to be about 7.7 nm when $K \sim 10^8$, 770 nm when $K \sim 10^{12}$, and 7700 μm, when $K \sim 10^{20}$. Thus the larger K values obtained in the presence of hexanol imply immensely long cylinders, much larger than the approximately 1 μm size wormlike micelles observed microscopically. It is quite possible that the conditions for the formation of these very long rodlike micelles in the 10^2 to 10^3 μm length scale are not realized because lamellar aggregates become preferred over rodlike aggregates. Although the present theory allows for the examination of the lamellar structures, the calculations have not been carried out for this system as yet.

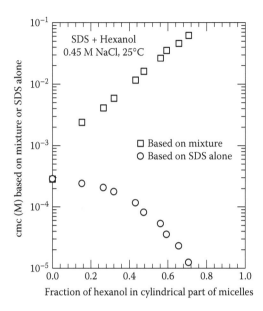

FIGURE 1.14 Critical micelle concentration of SDS + hexanol mixtures as a function of the fraction of hexanol in the cylindrical part of rodlike micelles. Solution is at 25°C and includes 0.45 M added NaCl. Squares denote the cmc represented as the sum of both SDS and hexanol concentrations, while the circles denote the cmc represented only in terms of SDS concentration. Circles show the decrease in cmc caused by the addition of hexanol. The fraction of hexanol in the cylindrical part has an upper bound of about 0.71, which corresponds to the condition of saturation of the solution with hexanol.

The predicted distribution of alcohol molecules between the endcaps of the rodlike micelles and the cylindrical middle portion is shown in Figure 1.16. There is preferential incorporation of alcohol in the cylindrical part compared to the endcaps. The area per head group in the cylindrical middle is smaller than that in the spherical endcaps. The incorporation of alcohol in both regions contributes to a decrease in the head group repulsions. However, the preferential partitioning in the cylindrical part leads to a larger decrease in the head group repulsions in the cylindrical part compared to that in the endcaps. The resulting decrease in $\Delta\mu^o_{cyl}$ is larger in magnitude compared to the resulting decrease in $\Delta\mu^o_{cap}$. Hence the sphere-to-rod transition parameter increases significantly on the addition of alcohol.

1.5.3 Addition of Nonionic Polymers

An interesting approach to tune rodlike micelles is by the addition of uncharged polymers to the surfactant solution. One may anticipate that, in surfactant solutions

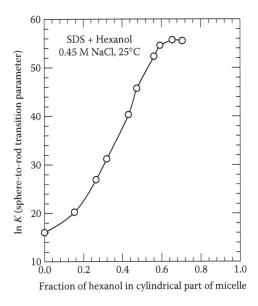

FIGURE 1.15 Sphere-to-rod transition parameter K of SDS + hexanol mixtures as a function of the fraction of hexanol in the cylindrical part of rodlike micelles. Solution is at 25°C and includes 0.45 M added NaCl.

containing nonionic polymers, aggregation of the surfactant molecules would occur, but these aggregates may be bound by physical interactions to the polymer. We visualize a polymer–micelle association structure wherein the polymer adsorbs at the micelle–water interface (Figure 1.17). The adsorbed polymer segments shield a part of the hydrophobic domain of the micelle from being in contact with water. The physicochemical properties of the polymer molecule determine the area of mutual contact between the polymer molecule and the hydrophobic surface of the micelle. A characteristic parameter a_{pol} is defined to represent this area of mutual contact per surfactant molecule in the micelle. Although it is not very realistic to anticipate that the polymer molecule can provide a uniform shielding of the micelle surface from water by the amount a_{pol} per surfactant molecule, this area parameter is defined in the mean-field spirit to serve as a quantitative measure of the nature of polymer–micelle interactions. Based on such a visualization of polymer decorated surfactant aggregates, we have developed a molecular theoretical model and have treated the problem of polymer interactions with spherical and rodlike micelles,[61,62] spherical vesicles,[63] and droplet and bicontinuous type microemulsions.[63,64] Therefore, only a brief account is presented here focusing on the rodlike micelles. A somewhat different approach to modeling has been presented by Ruckenstein et al.[65]

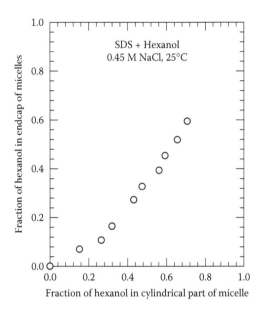

FIGURE 1.16 Distribution of hexanol between the cylindrical part and the endcap of a rodlike micelle for SDS + hexanol mixtures as a function of the fraction of hexanol in the cylindrical part of the rodlike micelles. Solution is at 25°C and includes 0.45 M added NaCl. Preferential incorporation of hexanol in the cylindrical part compared to the endcap is shown on the figure.

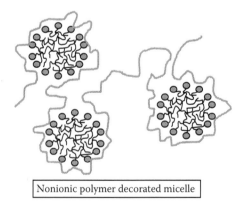

Nonionic polymer decorated micelle

FIGURE 1.17 Schematic representation of an association structure involving surfactant micelles and a nonionic polymer molecule. The structural model visualizes the polymer molecule to be adsorbed at the hydrophobic–hydrophilic interface of the surfactant aggregate.

1.5.3.1 Free Energy of Micellization in the Presence of Polymer

In polymer-free surfactant systems, the free energy of micellization for an ionic surfactant has the form (see Section 1.3)

$$\frac{\left(\Delta\mu_g^o\right)}{k_BT} = \frac{\left(\Delta\mu_g^o\right)_{tr}}{k_BT} + \frac{\left(\Delta\mu_g^o\right)_{def}}{k_BT} + \frac{\sigma_{agg}}{k_BT}(a-a_o) - \ln\left(1-\frac{a_p}{a}\right) + \frac{\left(\Delta\mu_g^o\right)_{ion}}{k_BT} \quad (1.41)$$

where the five terms account respectively for the transfer free energy of the surfactant tail, the deformation free energy of the tail, the free energy of formation of the micelle–water interface, the steric repulsions between the head groups, and the electrostatic repulsions between the ionic head groups. The last term is absent for nonionic surfactants and is replaced by the dipole interaction free energy term for zwitterionic surfactants.

Only small modifications are needed to represent the standard free energy of formation of aggregates decorated by a nonionic polymer molecule. As mentioned above, we assume that the polymer segments shield the hydrophobic domain from water by an area a_{pol} per surfactant molecule. This gives rise to three competing contributions to the free energy of aggregation. First, a decrease in the hydrophobic surface area of the aggregate exposed to water occurs. This decreases the positive free energy of formation of the aggregate–water interface and thus favors the formation of polymer-bound aggregates. Second, steric repulsions arise between the polymer segments and the surfactant head groups at the aggregate surface. This increases the positive free energy of head group repulsions and thus disfavors the formation of the polymer-bound aggregates. Finally, the contact area a_{pol} of the polymer molecule is removed from water and transferred to the surface of the aggregate, which is concentrated in the surfactant head groups. This alters the free energy of the polymer and can favor or disfavor the formation of the polymer-bound micelles depending upon the type of interactions between the polymer segments and the interfacial region rich in head groups. Taking these factors into account, one may write

$$\frac{\left(\Delta\mu_g^o\right)}{k_BT} = \frac{\left(\Delta\mu_g^o\right)_{tr}}{k_BT} + \frac{\left(\Delta\mu_g^o\right)_{def}}{k_BT} + \frac{\sigma_{agg}}{k_BT}(a-a_o-a_{pol})$$

$$- \ln\left(1-\frac{a_p+a_{pol}}{a}\right) + \frac{\left(\Delta\mu_g^o\right)_{ion}}{k_BT} - \frac{\Delta\sigma_{pol}a_{pol}}{k_BT} \quad (1.42)$$

The first two terms and the fifth term are identical to those appearing in Equation 1.41. The modified third term accounts for the enhanced shielding of the micellar core from water provided by the polymer. The modified fourth term accounts for the increase in the steric repulsions due to the presence of the polymer. The sixth term is new and represents the change in the interaction free energy of the polymer molecule. This interaction free energy is written as the product of an interfacial tension difference and the area of the polymer that is removed from water and brought into contact with the micellar surface. Here, $\Delta\sigma_{pol}$ is the difference between the macroscopic polymer–water interfacial tension and the interfacial tension between the polymer and the aggregate head group region. The factor $\Delta\sigma_{pol}$ is obviously affected by both the hydrophobic character of the polymer molecule and by its interactions with the surfactant head groups that are crowded at the interface. One may note that $\Delta\sigma_{pol}$ affects only the cmc and not the aggregation number, since the polymer interaction free energy contribution is not dependent on the aggregation number g. In contrast, a_{pol} will influence both the cmc and the aggregate properties. Based on the above free energy equation, the equilibrium aggregation number, the surface area per surfactant molecule of the bound micelle, and the cmc are calculated as described in Section 1.4.

Two new molecular constants, a_{pol} and $\Delta\sigma_{pol}$, that depend on the type of the polymer are necessary for predictive calculations. At present, there are no *a priori* methods available for estimating a_{pol} while $\Delta\sigma_{pol}$ can, in principle, be determined from interfacial tension measurements as suggested by Ruckenstein et al.[65] For the illustrative calculations discussed below, we take $a_{pol} = 0.20$ nm^2 and $\Delta\sigma_{pol} = 15, 0,$ or -15 mN/m. Positive and negative values for $\Delta\sigma_{pol}$ correspond to polymer having an affinity or dislike for the micellar head group region compared to water.

1.5.3.2 Polymer Induced Rod-to-Sphere Transition

Illustrative calculations are presented here for an ionic surfactant and a nonionic surfactant. The calculated results are summarized in Table 1.3 for a surfactant solution containing the anionic SDS at 25°C in the presence of 0.8 M NaCl. In the absence of the polymer, large polydispersed rodlike micelles are formed in solution as shown by the large value of $K = 6 \times 10^9$. When the polymer molecules

---

TABLE 1.3
Polymer Association Behavior with Surfactant Micelles at 25°C

Surfactant	cmc (mM)	a (nm^2)	g	K	Shape
$C_{12}PO$	0.243	0.582	—	5.6×10^9	Rod
$C_{12}PO + P*$	0.128	0.795	28	4.5×10^2	Sphere
$C_{12}SO_4Na + 0.8M$ NaCl	0.313	0.477	—	6.0×10^9	Rod
$C_{12}SO_4Na + 0.8M$ NaCl $+ P*$	0.057	0.561	88	6.1×10^4	Globular

* P refers to polymer-bound aggregates; the corresponding cmc values listed are for $\Delta\sigma_{pol} = 0$ mN/m. These cmc values should be multiplied by 0.49 if $\Delta\sigma_{pol} = 15$ or by 2.043 if $\Delta\sigma_{pol} = -15$ mN/m.

are present, the magnitude of K is dramatically reduced. Consequently, only small globular micelles are formed. The cmc corresponding to the formation of the polymer-bound globular micelles (0.057, 0.028, and 0.116 mM for $\Delta\sigma_{pol} = 0$, 15, and −15 mN/m, respectively) is lower than that for the formation of polymer-free rodlike micelles (0.313 mM). Therefore, the equilibrium favors the formation of the smaller globular micelles in the presence of the polymer. Thus a rod-to-globule transition is induced by the addition of the polymer.

In the case of a nonionic surfactant dodecyl dimethyl phosphene oxide (designated as $C_{12}PO$), large rodlike micelles are formed in polymer-free solutions as reflected in the large value of $K = 5.6 \times 10^9$, at a cmc of 0.243 mM. In the presence of the nonionic polymer, K is dramatically decreased and the allowed aggregate is small and globular. The calculations show that if $\Delta\sigma_{pol} = 0$ or 15 mN/m, the cmc is 0.128 or 0.063 mM in the presence of the polymer. This implies that equilibrium will favor the formation of small polymer-bound globular micelles. In contrast, if $\Delta\sigma_{pol} = -15$ mN/m, the cmc corresponding to polymer-bound micelles is larger (0.262 mM) than the cmc corresponding to the polymer-free micelles. In such a case, the binding with polymer does not occur and polymer-free rodlike micelles coexist with free polymer molecules in solution. In the free energy model represented by Equation 1.42, the nature of polymer-induced transitions is a consequence of an increase in the area per molecule of the aggregate. Therefore, a surfactant system which generates globular micelles will never produce rodlike micelles on the addition of the polymer. However, it would be possible to transform spherical bilayer vesicles to rodlike or globular micelles and rodlike micelles to globular micelles.

1.6 CONCLUSIONS

A description of the molecular theory to predict the formation and properties of giant rodlike micelles is presented here. A number of illustrative calculations are detailed. Most importantly, the model calculations show how sensitive is the calculation of the rod-to-sphere growth parameter K. The theory for surfactant solutions is extended to solutions of surfactant mixtures, surfactant–alcohol systems, and surfactant solutions containing nonionic polymer, to illustrate numerous ways by which one can create rodlike micelles from spherical micelles, allow rodlike micelles to grow larger, or transform rodlike aggregates into spherical shapes. Although bilayer vesicles are not described in this chapter, we have shown in other papers how they can also be described by the same molecular approach and how one can cause transitions between vesicles and rodlike aggregates (see Ref. 59).

It is obvious that we have focused in this chapter entirely on our own work since the emphasis was on presenting a unified molecular theory of giant micelles formed in a variety of self-assembling systems. The only other molecular theoretical approach in the literature that is designed to obtain quantitative *a priori* predictions is that of Blankschtein. The work of Blankschtein has appeared in numerous papers, with rodlike micelles being emphasized in references.[66–70]

A number of other modes of tuning rodlike micelles, not considered here, have been treated in our previous papers. One approach is based on solubilizing hydrocarbons in surfactant micelles. We have shown[20,59] that both growth of rods and also a rod-to-sphere transition are possible depending on the type and amount of hydrocarbon solubilized. Another approach is based on the use of polar organic solvents as additives to water.[71] This would result in an increase in the area per molecule and thereby allow vesicles to transform to rodlike micelles and make giant rodlike micelles into smaller rods or spheres. Yet another approach is by the design of surfactant molecules such as the gemini surfactants with appropriate spacer lengths to obtain giant micelles.[72] Giant micelles can also be generated from polymeric analogs of surfactants, namely, block copolymers. We have constructed[73–75] molecular theoretical models for the formation of rodlike block copolymer micelles and for structural transitions between spherical, rodlike, and bilayer aggregates induced by the solubilization of hydrocarbons. Another interesting approach is the formation of rodlike aggregates at solid surfaces in contact with surfactant solutions. We have treated the formation of rodlike aggregates and transitions between various aggregate shapes for hydrophobic[76] as well as hydrophilic solid surfaces.[77]

ACKNOWLEDGMENTS

Professor Eli Ruckenstein has been an active collaborator in much of the work described in this paper, as the references show. The author has benefited from numerous discussions with him over the last 30 years.

REFERENCES

1. Debye, P. *J. Phys. Colloid Chem.* 1948, *53*, 1.
2. Debye, P. *Ann. NY Acad. Sci.* 1949, *51*, 575.
3. McBain, J.W. *Trans. Faraday Soc.* 1913, *9*, 99.
4. McBain, J.W. *Colloid Science*, D.C. Heath and Company, San Francisco, 1950.
5. Hobbs, M.E. *J. Phys. Colloid Chem.* 1951, *55*, 675.
6. Halsey, Jr., J.D. *J. Phys. Chem.* 1953, *57*, 87.
7. Ooshika, Y. *J. Colloid Sci.* 1954, *9*, 254.
8. Reich, I. *J. Phys. Chem.* 1956, *60*, 257.
9. Hoeve, C.A.J., Benson, C.G. *J. Phys. Chem.* 1957, *61*, 1149.
10. Poland, D.C., Scheraga, H.A. *J. Phys. Chem.* 1965, *69*, 2431.
11. Poland, D.C., Scheraga, H.A. *J. Colloid Interface Sci.* 1966, *21*, 273.
12. Tanford, C. *The Hydrophobic Effect,* Wiley, New York, 1973.
13. Tanford, C. *J. Phys. Chem.* 1974, *78*, 2469.
14. Tanford, C. In *Micellization, Solubilization and Microemulsions*, Mittal, K.L., Ed., Plenum Press, New York, 1977, p. 119.
15. Nagarajan, R., Ruckenstein, E. *J. Colloid Interface Sci.* 1977, *60*, 221.
16. Nagarajan, R., Ruckenstein, E. *J. Colloid Interface Sci.* 1979, *71*, 580.
17. Nagarajan, R., Chaiko, M.A., Ruckenstein, E. *J. Phys. Chem.* 1984, *88*, 2916.
18. Nagarajan, R. *Langmuir* 1985, *1*, 331.

19. Nagarajan, R. *Adv. Colloid Interface Sci.* 1986, *26*, 205.
20. Nagarajan, R., Ruckenstein, E. *Langmuir* 1991, *7*, 2934.
21. Nagarajan, R. In *Mixed Surfactant Systems*, Holland, P.M., Rubingh, D.N., Eds., ACS Symposium Series 501, American Chemical Society, Washington, D.C., 1992, Chapter 4, p. 54.
22. Nagarajan, R., Ruckenstein, E. *Langmuir* 2000, *16*, 6400.
23. Israelachvili, J.N., Mitchell, J.D., Ninham, B.W. *J. Chem. Soc. Faraday Trans. II* 1976, *72*, 1525.
24. Ruckenstein, E., Nagarajan, R. *J. Phys. Chem.* 1975, *79*, 2622.
25. Nagarajan, R. *Colloids Surf. A Physicochem. Eng. Aspects* 1993, *71*, 39.
26. Ruckenstein, E., Nagarajan, R. *J. Colloid Interface Sci.* 1976, *57*, 388.
27. Nagarajan, R., Ruckenstein, E. *J. Colloid Interface Sci.* 1983, *91*, 500.
28. Nagarajan, R. *Langmuir* 1994, *10*, 2028.
29. Mukerjee, P. *Adv. Colloid Interface Sci.* 1967, *1*, 241.
30. Mukerjee, P. *J. Phys. Chem.* 1969, *73*, 2054.
31. Mukerjee, P. *J. Phys. Chem.* 1972, *76*, 565.
32. Mukerjee, P. *J. Pharm. Sci.* 1974, *68*, 972.
33. Mukerjee, P. In *Micellization, Solubilization and Microemulsions*, Mittal, K.L., Ed., Plenum Press, New York, 1977, p. 171.
34. Tausk, R.J.M., Oudshoorn, C., Overbeek, J.Th.G. *Biophys. Chem.* 1974, *2*, 53.
35. Tausk, R.J.M., Overbeek, J.Th.G. *Biophys. Chem.* 1974, *2*, 175.
36. Mazer, N.A., Benedek, G.B., Carey, M.C. *J. Phys. Chem.* 1976, *80*, 1075.
37. Nagarajan, R. *J. Colloid Interface Sci.* 1982, *90*, 477.
38. Abraham, M.H. *J. Chem. Soc. Faraday Trans.* 1984, *80*, 153.
39. Abraham, M.H., Matteoli, E. *J. Chem. Soc. Faraday Trans.* 1988, *84*, 1985.
40. Semenov, A.N. *Soviet Phys. JETP* 1985, *61*, 733.
41. Dill, K.A., Flory, P.J. *Proc. Natl. Acad. Sci. USA* 1980, *77*, 3115.
42. Dill, K.A., Flory, P.J. *Proc. Natl. Acad. Sci. USA* 1980, *78*, 676.
43. Girifalco, L.A., Good, R.J. *J. Phys. Chem.* 1957, *61*, 904.
44. *CRC Handbook of Chemistry and Physics*, CRC Press, Boca Raton, 1980, 60th Edition.
45. Evans, D.F., Ninham, B.W. *J. Phys. Chem.* 1983, *87*, 5025.
46. Rao, I.V., Ruckenstein, E. *J. Colloid Interface Sci.* 1987, *119*, 211.
47. Daubert, T.E., Danner, R.P. *Physical and Thermodynamic Properties of Pure Chemicals*, Design Institute for Physical Property Data, American Institute of Chemical Engineers, Hemisphere Publishing Corporation, New York, 1988.
48. Focher, B., Savelli, G., Torri, G., Vecchio,G., McKenzie, D.C., Nicoli, D.F., Bunton, C.A. *Chem. Phys. Lett.* 1989, *158*, 491.
49. Missel, P.J., Mazer, N.A., Benedek, G.B., Young, C.Y., Carey M.C. *J. Phys. Chem.* 1980, *84*, 1044.
50. Flory, P.J. *Principles of Polymer Chemistry*, Cornell University Press, Ithaca, NY, 1962.
51. Hildebrand, J.H., Prausnitz, J.M., Scott, R.L. *Regular and Related Solutions*, Van Nostrand Reinhold Company, New York, 1970.
52. Barton, A.F.M. *Handbook of Solubility Parameters and Other Cohesion Parameters*, CRC Press, Boca Raton, 1983.
53. Malliaris, A., Binana-Limbele, B., Zana, R. *J. Colloid Interface Sci.* 1986, *110*, 114.
54. Stellner, K.L., Amante, J.C., Scamehorn, J.F., Harwell, J.H. *J. Colloid Interface Sci.* 1988, *123*, 186.

55. Kaler, E.W., Murthy, A.K., Rodriguez, B.E., Zasadzinski, J.A.N. *Science* 1989, *245*, 1371.
56. Brasher, L.L., Herrington, K.L., Kaler, E.W. *Langmuir* 1995, *11*, 4267.
57. Bergström, M. *Langmuir* 1996, *12*, 2454.
58. Kondo, Y., Uchiyama, H., Yoshino, N., Nishiyama, K., Abe, M. *Langmuir* 1995, *11*, 2380.
59. Nagarajan, R., Ruckenstein, E. In *Equations of State for Fluids and Fluid Mixtures*, Sengers, J.V., Kayser, R.F., Peters, C.J., White Jr., H.J., Eds., Elsevier Science, Amsterdam, 2000, Chapter 15, p. 589.
60. Rao, I.V., Ruckenstein, E. *J. Colloid Interface Sci.*1986, *113*, 375.
61. Nagarajan, R., Kalpakci, B. *Polymer Preprints* 1982, *23*, 41.
62. Nagarajan, R. *Colloids Surf.* 1985, *13*, 1.
63. Nagarajan, R. *J. Chem. Phys.* 1989, *90*, 1980.
64. Nagarajan, R. *Langmuir* 1993, *9*, 369.
65. Ruckenstein, E., Huber, G., Hoffmann, H. *Langmuir* 1987, *3*, 382.
66. Carale, T.R., Blankschtein, D. *J. Phys. Chem.* 1992, *96*, 459.
67. Thomas, H.G., Lomakin, A., Blankschtein, D., Benedek, G.B. *Langmuir* 1997, *13*, 209.
68. Zoeller, N., Lue, L., Blankschtein, D. *Langmuir* 1997, *13*, 5258.
69. Srinivasan, V., Blankschtein, D. *Langmuir* 2003, *19*, 9946.
70. Srinivasan, V., Blankschtein, D. *Langmuir* 2005, *21*, 1647.
71. Nagarajan, R., Wang, C.C. *Langmuir* 2000, *16*, 5242.
72. Camesano, T.A., Nagarajan, R. *Colloids Surf. A. Physicochem. Eng. Aspects* 2000, *167*, 165.
73. Nagarajan, R. In *Solvents and Self-Organization of Polymers*, Webber, S.E., Munk, P., Tuzar, Z. Eds., NATO ASI Series E: Applied Sciences, Vol. 327, Kluwer Academic Publishers, 1996, p. 121.
74. Nagarajan, R. *Colloids Surf. B. Biointerfaces* 1999, *16*, 55.
75. Nagarajan, R. *Polym. Adv. Technol.* 2001, *12*, 23.
76. Johnson, R.A., Nagarajan, R. *Colloids Surf. A. Physicochem. Eng. Aspects* 2000, *167*, 31.
77. Johnson, R.A., Nagarajan, R. *Colloids Surf. A. Physicochem. Eng. Aspects* 2000, *167*, 21.

2 Molecular Packing in Cylindrical Micelles

Sylvio May and Avinoam Ben-Shaul

CONTENTS

2.1 INTRODUCTION

The elongated shape of linear, wormlike, micelles is among the most common aggregation geometries of amphiphilic molecules in aqueous solutions. The cylindrical body of these uni-dimensional micelles (1D) can be regarded as an intermediate packing geometry, of a higher *growth dimensionality* than that of a spherical micelle and lower than that of a planar bilayer. In all these three canonical structures, the hydrophobic tails of the constituent amphiphiles form a compact, liquid-like, hydrocarbon core, with their polar headgroups residing on its surface, thus largely shielding the hydrocarbon tails from direct contact with water. The planar bilayer is a two-dimensional (2D) object; it can grow laterally along $d = 2$ directions, but its third dimension is always microscopic. Namely, its thickness, or, more precisely, the distance ($2b$) between its two hydrocarbon-water interfaces cannot exceed $2b_{max}$, where b_{max} is the length of the fully extended hydrocarbon tail. Similarly, the growth dimensionality of a wormlike micelle is $d = 1$. It can elongate along the cylindrical axis, but its diameter ($2b$) — and hence the two perpendicular dimensions — cannot exceed $2b_{max}$. The growth dimensionality of a spherical micelle (keeping its spherical symmetry) is, of course, $d = 0$, since to maintain a compact hydrophobic core all its three dimensions must be smaller than $2b_{max}$. Note that this last statement is only relevant to systems where a spherical micelle is the intrinsically preferred (or, "spontaneous") packing environment of the amphiphiles. As briefly discussed later in this chapter, many other amphiphiles, in fact most of those that prefer packing into long cylindrical micelles, first assemble — for entropic reasons — into small spherical micelles. However, above the second *critical micelle concentration* (second cmc, to be distinguished from the first cmc above which micelles form), driven by the lower packing energy in the cylindrical geometry, the added amphiphiles incorporate into the middle of the aggregate, forming a gradually elongating cylindrical midsection, capped by two approximately hemispherical micellar caps. In these systems the spherical micelle is just the low concentration limit of a linear aggregate.

The reversible self-assembly of amphiphiles in aqueous environments is a fundamental process, omnipresent in living organisms and underlying numerous technological applications.[1] Thermodynamically, micellar solutions belong to the wider class of systems known as complex fluids.[2,3] One of the special characteristics of these solutions is that the solute particles, unlike in ordinary molecular solutions, can modify their size, shape, and even their aggregation geometry. Such changes occur generally in response to varying ambient conditions, like the total concentration, temperature, or ionic strength. For instance, upon increasing the amphiphile concentration in a solution of cylindrical (rodlike) micelles, the average micelle size increases monotonically, resulting in substantial intermicelle interactions. Eventually, these interactions become strong enough to drive a transition from an isotropic to a nematic phase of rodlike aggregates. However, unlike in ordinary liquid crystalline solutions, the micelle size in the nematic phase is much larger than in the coexisting isotropic phase, demonstrating a subtle coupling between the intermolecular (or, intraaggregate) and intermicelle interactions.[4] This coupling arises because the (noncovalent) forces holding together the amphiphiles in a micelle are relatively weak, typically on the order of $\approx 10k_BT$ per molecule, where k_B is Boltzmann's constant and T is the absolute temperature. At high total amphiphile concentrations, interaggregate interactions, often dominated by packing (excluded volume) interactions of entropic origin, may be of comparable magnitude, and thus induce structural and morphological changes of the aggregates. There are many examples for the coupling between intermolecular and interaggregate interactions besides the isotropic-nematic transition of rodlike micelles mentioned above. These include, for instance, the formation of branched micelles and interconnected phases resembling polymer gels, or the transition between lamellar and bicontinuous (sponge) phases in surfactant solutions.[3,4]

Above the first cmc but still in the dilute solution regime, adding amphiphiles to a micellar solution of (intrinsically preferred) spherical micelles will simply result in the formation of new micelles, all of the same size. On the other hand, amphiphiles whose intrinsically preferred aggregation geometry is the planar bilayer will spontaneously assemble into macroscopic bilayer sheets. This process, which typically occurs already at extremely small amphiphile concentrations is, in fact, a first-order transition from a very dilute phase of isolated amphiphiles into a condensed 2D bilayer phase. As compared to these two cases, the phase behavior associated with amphiphiles whose spontaneous aggregation geometry is cylindrical is considerably more interesting, even in dilute solution. Like other 1D systems, linear molecular aggregation is a continuous, gradual process, as opposed to a first-order phase transition. That is, the average micellar size increases with the total concentration, yet it is always finite. Although this topic has been intensively discussed in the literature, including in other chapters in this volume, we shall briefly review its basic thermodynamic aspects in Section 2.2, mainly in order to emphasize the intimate relationship between structural-molecular characteristics of amphiphile organization to the thermodynamics of micelle formation and growth.

The tail region of amphiphiles consists typically of one or more short hydrocarbon chains. In a spherical micelle the hydrocarbon tail occupies, on average, a conelike section of the hydrophobic core, whose base is at the hydrocarbon-water interface and whose tip is in the center of the micelle. Similarly, the average shape of the tail volume in a cylindrical micelle is wedgelike, and in a planar bilayer it is cylindrical. Other average shapes characterize more complex structures, such as the saddle geometry encountered in certain inverted phases and intermicellar junctions. In bilayers above the chain-melting transition temperature, and in all micellar aggregates, the hydrocarbon tails are thermally excited, fluctuating between numerous accessible chain conformations. This conformational flexibility is responsible for the ability of amphiphiles to form a surprisingly large variety of aggregate geometries, including micelles, membranes, and bicontinuous structures.[1,3,4] Owing to this conformational freedom, the tails enable the

Most chapters in this volume are concerned with the structural, rheological, and thermodynamic characteristics of giant, "wormlike," micelles. Once the size of these micelles exceeds the persistence length of a cylindrical micelle, their conformational behavior is similar to that of flexible polymers. We note, however, that owing to the reversible nature of amphiphile assembly, wormlike micelles are "living polymers." As noted above, they can form and break in response to intermicelle interactions, as well as external fields, for example, flow fields. The self-assembled nature of wormlike micelles thus affects many physical properties and contributes to a remarkable rheological behavior of their solutions.[5] The structure and stability of these micelles, as well as their growth characteristics, flexibility, and ability to form intermicellar junctions (as observed in certain systems) depend on the intermolecular interactions governing their packing into micellar aggregates. These, molecular, aspects of amphiphile assembly are the main focus of the present chapter.

formation of uniform, liquidlike, hydrocarbon cores for all these aggregation geometries. In the simplest phenomenological theory of amphiphile self-assembly, known as the opposing forces model (OFM),[6,7] this is known as *the hydrocarbon droplet assumption*. In fact, in this model it is further assumed that, owing to the liquid-like nature of the hydrophobic core, the packing free energy of the tails is independent of the aggregation geometry; namely, it is the same, for instance, in a cylindrical micelle and a planar bilayer. Following this assumption the preferred aggregation geometry in dilute solution, or, in other words, the spontaneous packing curvature of amphiphiles in a micellar aggregate, is dictated by the balance of two forces, both operative at the micelle's interfacial region. These opposing forces, are (i) the repulsion between amphiphile headgroups which tends to maximize the interfacial area per molecule, and (ii) the hydrocarbon-water surface energy, which acts in the opposite direction.

The OFM picture provides a convincing qualitative explanation to various self-assembly phenomena. For example, the preference of strongly charged surfactants to form curved micelles, or the tendency of double-tailed phospholipids to assemble into lipid membranes.[6,7] On the other hand, in order to account for certain, more subtle though no less important properties of micellar aggregates, one must take into account the different packing constraints imposed on the *hydrocarbon tails* in different packing geometries. The issue of modeling the molecular packing in cylindrical micelles is discussed in Section 2.3. It includes an approach for treating the conformational properties of amphiphile chains in the hydrophobic cores of different micellar geometries, namely a mean field, molecular-level, theory of chain packing which has been widely applied to various issues pertaining to amphiphile self-assembly.[8] In the subsequent sections we shall show that conformational chain packing statistics are also crucial for explaining the *bending elasticity* of wormlike micelles (see Section 2.4), the *second cmc* observed in various micellar solutions (see Section 2.5), and the energetics of *intermicellar junctions* (see Section 2.6).

Wormlike micelles are formed by numerous ionic and nonionic, single-tailed surfactants, but also by certain double-tailed amphiphiles,[9] as well as block copolymers.[10] Nonionic micelles are not subject to the many complications that arise due to the long-range nature of electrostatic interactions. Yet, in the case of ionic surfactants one can regulate the preferred aggregation geometry by varying the nature of the counterions or the salt content in solution. Frequently, the growth of ionic micelles can be enhanced by adding a large amount of salt; but certain wormlike micelles also exist under salt free conditions.[11,12] Because charged wormlike micelles bear similarities to polyelectrolytes,[13] concepts developed for charged polymers, such as the electrostatic contribution to the persistence length,[14,15] can often be applied to wormlike micelles. As noted above, however, differences arise from the self-assembled nature of the micelles, and hence their ability to adjust their length distribution, to form transient junctions or even to cross each other (ghost crossing). We finish the present chapter in Section 2.7 with a brief discussion of the effects that electrostatic interactions have on the properties of linear micelles.

2.2 MICELLAR AGGREGATION THERMODYNAMICS

In this section we briefly outline the thermodynamics of amphiphile self-assembly in dilute solution. The system of interest is an aqueous solution of volume V, containing M amphiphiles, in both monomeric and aggregated form. In the dilute solution regime M is much smaller than the number, M_w, of solvent (water) molecules. We shall use M_1 to denote the number of monomeric amphiphiles in solution, and M_N ($N \geq 2$) for the number of aggregates of aggregation number N. Thus, M_2 is the number of dimers, M_3 the number of trimers, etc. On the other hand, NM_N is the total number of molecules incorporated into aggregates comprising $N \geq 2$ amphiphiles. (In micellar solutions, as discussed in the next section, there is typically a minimal aggregation number, $\bar{N}$, so that aggregates of intermediate sizes, $2 \leq N \leq \bar{N}$ are highly unlikely to appear in solution.) It is common to express the aggregate size distribution on the *mole fraction* scale, $X_N = NM_N/M_t$, where in the dilute solution limit $M_t = M_w + M \approx V/v_w$, where v_w is the volume per molecule in pure water. The sum $\Sigma_{N=1}^{\infty} X_N = X$, is, of course, the overall mole fraction of amphiphiles in solution, $X = M/M_t$.

The Helmholtz free energy, $F = F[\{X_N\}]$, corresponding to any given distribution, $\{X_N\}$, of aggregate sizes is given by[1,3,16]

$$F = M_t \sum_{N=1}^{\infty} \left[X_N \tilde{\mu}_N^0 + k_B T \frac{X_N}{N} \left(\ln \frac{X_N}{N} - 1 \right) \right] \tag{2.1}$$

where $\tilde{\mu}_N^0$ is the standard chemical potential per amphiphile in an aggregate of size N. The second term in the sum, or more precisely $-k_B (M_N/M_t)[\ln(M_N/M_t) - 1]$, is the translational entropy of micelles of size N in solution. The sum of these terms is the contribution of (monomeric amphiphiles and micelles) to the mixing entropy of the solution. Note that Equation 2.1 does not include the contribution of the solvent to the free energy (namely, $M_w[\tilde{\mu}_w^0 + k_B T X_w \ln X_w]$), because in the dilute solution limit this is a constant, independent of $\{X_N\}$. It may also be noted that $\mu_N^0 - N\tilde{\mu}_1^0 = N[\tilde{\mu}_N^0 - \tilde{\mu}_1^0]$ is the standard free energy of formation of an N-micelle from N monomers. Physically, this is the packing free energy of the N-amphiphiles into a micelle, measured with respect to the state of a fully dissociated aggregate.

Equation 2.1 is valid for any multicomponent solution of (fixed) composition $\{X_N\}$. In a solution of self-assembling molecules, all association-dissociation "reactions" are possible, e.g., $A_N \rightleftharpoons A_m + A_{N-m}$ etc., where A_m denotes a micelle of m amphiphiles. The equilibrium (or, equivalently, the most probable) distribution of average micellar sizes, $\{X_N^{eq}\}$, is the one which minimizes $F = F[\{X_N\}]$, subject to the condition of conserving the overall number of amphiphiles, M_t,

$$\sum_{N=1}^{\infty} X_N = X = constant. \tag{2.2}$$

The resulting, equilibrium, distribution is

$$X_N = N \exp\left[-N\left(\tilde{\mu}_N^0 - \mu\right)/k_B T\right] \tag{2.3}$$

where μ is the Lagrangian multiplier conjugate to the conservation constraint Equation 2.2; and where for notational brevity we have omitted the superscript eq. As in any macroscopic system, fluctuations of the X_N around their equilibrium values are negligible.

Substituting the equilibrium distribution, Equation 2.3, back into Equation 2.1 we obtain

$$\frac{F}{M_t} = \mu X - k_B T \sum_{N=1}^{\infty} \frac{X_N}{N} \tag{2.4}$$

The osmotic pressure of the micellar solution is $\Pi = -\partial F/\partial V = k_B T \Sigma_{N=1}^{\infty} M_N/V$, where the second equality is valid in the dilute solution limit. Thus, Equation 2.4 is the familiar relation, $F = G - \Pi V$, between the Helmholtz free energy, F, and Gibbs free energy, $G = \mu M$, of the solution, (again, apart from the constant part of the solvent).

The chemical potential per amphiphile in an N-mer is $\tilde{\mu}_N = \partial(F/M_t)/\partial X_N$, which, using Equation 2.1, yields $\tilde{\mu}_N = \tilde{\mu}_N^0 + (k_B T/N)\ln(X_N/N)$. At equilibrium, the chemical potential of all amphiphiles in solution should be independent of their state of aggregation, just as in phase equilibrium. Indeed, from Equation 2.3 it follows that

$$\tilde{\mu}_N = \tilde{\mu}_N^0 + (k_B T/N)\ln(X_N/N) = \mu \tag{2.5}$$

is a constant, $\tilde{\mu}_N = \mu$, for all N. As expected, this result confirms that the Lagrangian multiplier μ conjugate to the conservation condition of the number of amphiphiles (Equation 2.2) is, indeed, the chemical potential of *all* amphiphiles present in solution. Replacing μ by $\tilde{\mu}_1 = \tilde{\mu}_1^0 + k_B T \ln X_1$ in Equation 2.3 we find

$$\frac{X_N/N}{X_1^N} = \exp\left[-\frac{N}{k_B T}\left(\tilde{\mu}_1^0 - \tilde{\mu}_N^0\right)\right] \tag{2.6}$$

This expression is the familiar mass action law applied to the chemical reaction $NA_1 \rightleftharpoons A_N$.[6] Of course, this does not imply that micelle formation requires the simultaneous association of N monomers. Analogous expressions apply to any other set of such reactions, e.g., $A_{N-1} + A_1 \rightleftharpoons A_N$.

In the two following sections we briefly discuss how the dependence of $\tilde{\mu}_N^0$ on N dictates the cooperative formation of micelles from monomers at the *critical micelle concentration*, and their subsequent growth into elongated cylindrical aggregates.

2.2.1 THE CMC

In the limit of very low concentrations, the amphiphiles in solution are monomeric. Above a certain threshold critical concentration, the cmc, globular micelles composed typically of 30–100 molecules appear in solution, while the concentration of monomers remains essentially constant. (In the case of bilayer forming molecules, e.g., phospholipids, the transition generally occurs at vanishingly small concentrations, and directly to macroscopic aggregates.) As noted already in the introduction, if spherical micelles present the optimal packing geometry, then increasing the concentration above the cmc will result in the formation of additional micelles of this shape. On the other hand, if cylindrical packing is preferred, the globular micelles will gradually elongate into longer ones as the overall concentration increases. The cooperative nature of micelle formation at the cmc is a direct reflection of the existence of some minimal aggregation number $\bar{N}$, such that aggregates containing less than (about) $\bar{N}$ molecules are unstable, and thus either dissociate into monomers or associate to form $N \geq \bar{N}$-aggregates. To a first approximation, $\bar{N}$ is dictated by the size of a spherical micelle enabling amphiphile packing at their optimal area per headgroup. More formally, the existence of a minimal aggregation number is due to the fact that $\tilde{\mu}_N^0 \gg \tilde{\mu}_{\bar{N}}^0$ for all $N < \bar{N}$.

To demonstrate the cooperative behavior at the cmc one may conveniently assume that, apart from the monomeric state, $N = 1$, there is only one possible micelle size, $N = \bar{N} \gg 1$. The micellar size distribution X_N then contains only two nonvanishing entries: $X_1 = \exp[(\mu - \tilde{\mu}_1^0)/k_B T]$ and $X_{\bar{N}} = \bar{N} \exp[\bar{N}(\mu - \tilde{\mu}_{\bar{N}}^0)/k_B T]$, from which the chemical potential μ can be eliminated using $X_1 + X_{\bar{N}} = X$. A numerical calculation of X_1 and $X_{\bar{N}}$ for different choices of $\bar{N}$ is shown in Figure 2.1. The increased sharpness of the transition for large $\bar{N}$ is apparent. As the slope of the curve describing the function $X_1(X)$ changes abruptly from X to zero, one may conveniently define the critical micelle concentration[1] X_c through $(dX_1/dX)_{X_c} = 1/2$. A simple calculation reveals

$$X_c = \frac{1}{\bar{N}^{\frac{1}{\bar{N}-1}}} \exp\left[-\left(\frac{\bar{N}}{\bar{N}-1}\right)\left(\frac{\tilde{\mu}_1^0 - \tilde{\mu}_{\bar{N}}^0}{k_B T}\right) \right] \approx \exp\left(-\frac{\tilde{\mu}_1^0 - \tilde{\mu}_{\bar{N}}^0}{k_B T} \right) \qquad (2.7)$$

where the second equality is valid for $\bar{N} \gg 1$. It should be mentioned that alternative definitions for the cmc exist, either in terms of thermodynamic variables[17] or structural parameters. For example, the cmc may be defined as the concentration X for which a significant fraction (5–10%[18] or 50%[19]) of the surfactant is incorporated in micellar aggregates.

Physically, the increased sharpness of the transition with increasing $\bar{N}$ reflects the enhanced *cooperativity* of the aggregation process, approaching — in the

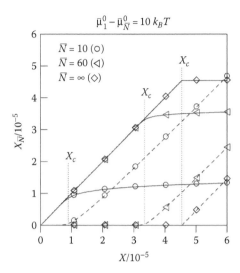

FIGURE 2.1 The mole fractions X_1 (solid curves) and $X_{\bar{N}}$ (dashed curves) as functions of the total mole fraction of amphiphiles X. The three different pairs of curves correspond to $\bar{N} = 10$, $\bar{N} = 60$, and $\bar{N} = \infty$, all for $\tilde{\mu}_1^0 - \tilde{\mu}_{\bar{N}}^0 = 10\, k_BT$. The corresponding cmc's according to Equation 2.7 are $X_c = 0.9 \times 10^{-5}$ for $\bar{N} = 10$, $X_c = 3.3 \times 10^{-5}$ for $\bar{N} = 60$, and $X_c = 4.5 \times 10^{-5}$ for $\bar{N} = \infty$; they are indicated by the three vertical dotted lines.

large $\bar{N}$ limit — the behavior of a first-order phase transition from monomers to a macroscopic condensed phase. (The constant monomer concentration above the cmc is then simply the solubility limit of the monomers in water.) Mathematically, the steep change in the slope of X_1 vs X at the cmc is evidenced by the second derivative $(d^2X_1/dX^2)_{X_c} = -\bar{N}^2$, calculated at the cmc $(X_1 = X_c)$ for $\bar{N} \gg 1$. In the limit $\bar{N} \to \infty$ the system exhibits a true first-order transition (see Figure 2.1). Of course, since one cannot pack an infinite number of molecules in a single spherical micelle, this is a hypothetical limit for this geometry.

2.2.2 One-Dimensional Growth

Linear micelles are generally modeled as aggregates comprising two packing environments: a cylindrical main body and two, roughly hemispherical, caps at the two ends of the cylinder, preventing direct contact of hydrocarbon tails with water. The micelle is often depicted as a *spherocylinder*, in which the radius of the caps is equal to the cylinder's radius, providing perfect matching between the two environments. This assumption can be relaxed, and in some cases must be modified, e.g., in systems exhibiting a second cmc, as discussed later in this chapter. However, such ramifications are mainly relevant for relatively short linear micelles. To account for the general growth characteristics of linear micelles it suffices to remember that the packing energy of amphiphiles in the end caps is different from that in the cylindrical body. The cylindrical geometry is obviously preferred over that of the caps, because otherwise there would be no thermodynamic incentive for micellar growth (since many spherical micelles involve higher entropy than fewer longer ones). The different packing energies thus imply a positive edge energy, δ, associated with each of the two end caps. Pictorially, 2δ is the energy required for breaking a linear micelle into two shorter micelles.

Regardless of the specific structure of the end caps, the standard chemical potential $\mu_N^0 = N\tilde{\mu}_N^0$ of an N-mer can be written as

$$\mu_N^0 = N\tilde{\mu}_{cyl}^0 + 2\delta \qquad (2.8)$$

where $\tilde{\mu}_{cyl}^0$ is the standard chemical potential per amphiphile in the cylindrical middle part, and δ is the excess end cap energy. (Of course, $\tilde{\mu}_{cyl}^0 \ll \tilde{\mu}_1^0$.) Large end cap energy, $\delta \gg k_B T$, provides the driving force for cylindrical micelles to grow into very long, wormlike, micelles. In this case, one can safely treat N as a continuous variable, i.e., $X_N \to X(N)$, and sums over aggregate sizes can be replaced by integrals: $\Sigma_{N=1}^{\infty} \to \int_0^{\infty} dN$. Setting $\mu_N^0 = N\tilde{\mu}_{cyl}^0 + 2\delta$ in Equation 2.3 for $X(N)$, integrating over all N and noting that $\int_0^{\infty} dN X(N) = X$, we find an explicit relationship between μ and X, i.e., $\mu = \tilde{\mu}_{cyl}^0 - k_B T/(e^{\delta/k_B T}\sqrt{X})$. Substituting this result back into Equation 2.3 then yields

$$X(N) = N\exp\left[-\frac{2\delta}{k_B T} - \frac{N}{e^{\delta/k_B T}\sqrt{X}} \right] \qquad (2.9)$$

We can now evaluate various properties of the micelle size distribution. For instance, the most probable aggregation number is $N^* = e^{\delta/k_B T}\sqrt{X}$. A more common characteristic of the size distribution is the weight average of micellar sizes, $\langle N \rangle = \int_0^{\infty} X(N) N \, dN / \int_0^{\infty} X(N) \, dN$, for which we find the familiar result

$$\langle N \rangle = 2e^{\delta/k_B T}\sqrt{X} = 2N^* \qquad (2.10)$$

For the standard deviation of the distribution we find $\sigma = \sqrt{\langle (N - \langle N \rangle)^2 \rangle} = \sqrt{2}\, N^*$, indicating a broad distribution of micellar lengths. Several examples of calculated size distributions are shown in Figure 2.2.

Substituting $X(N)$, as given by Equation 2.9 (with $\mu = \tilde{\mu}_{cyl}^0 - k_B T/(e^{\delta/k_B T}\sqrt{X})$) back into the Helmholtz free energy $F[\{X(N)\}]$ (see Equation 2.4) we find

$$F/M_t = X\tilde{\mu}_{cyl}^0 - 2k_B T\, e^{-\delta/kT}\sqrt{X} \qquad (2.11)$$

In the limit $\delta/k_B T \gg 1$ the amphiphiles strongly prefer the cylindrical packing environment, thus organizing in few very long aggregates. The only relevant contribution to the free energy in this limit is the packing energy of the molecules, namely, $F \to M_t X\tilde{\mu}_{cyl}^0$, consistent with the last equation.

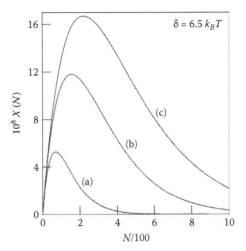

FIGURE 2.2 Size distributions, $X(N)$, of cylindrical micelles, calculated according to Equation 2.9 for: (a) $X = 0.1 \times 10^{-4}$, (b) $X = 0.5 \times 10^{-4}$, and (c) $X = 1.0 \times 10^{-4}$. The area under each curve, $\int_0^\infty X(N)\,dN = X$, is the corresponding total mole fraction X of amphiphiles in solution. In all examples, the excess energy of a single end cap is $\delta = 6.5\,k_B T$.

Beyond a certain length, known as the *persistence length*, l_p, which for cylindrical micelles is often on the order of several tens on nanometers (see Section 2.4), the micelles behave as flexible (living) polymers. This bending flexibility is associated with an additional free energy contribution to μ_N^0 which has not been included earlier. Note, however, that this contribution, which is only relevant for very long micelles, just adds to μ_N^0 a linear term in N, and hence a constant term to $\mu_N^0/N \approx \tilde{\mu}_{cyl}^0$. In the phenomenological description above it may be assumed that this term is already included in $\tilde{\mu}_{cyl}^0$. Another consequence of micellar flexibility is the ability of a micelle to close upon itself into a ring, thereby eliminating the excess packing free energy, 2δ, of the end caps. However, self-closing is unfavorable for short and long cylindrical micelles. In the former case, where the micellar rod length L is considerably smaller than the persistence length, l_p, the energy required to bend the rod is generally prohibitively high. (Forming a ring by a micelle of length $L = l_p$ involves a bending energy penalty of $2\pi^2 k_B T \sim 20 k_B T$.) In the latter case, where $L \gg l_p$, ring closure involves a substantial loss of configurational entropy. A simple way to estimate this loss is by treating a micelle of length L and persistence length l_p as a random walk of $L/l_p \equiv P$ steps on a 3D cubic lattice. The number of conformations available to the open micelle (ignoring self-crossing) is $\Omega_{open} = 2^{3P}$, whereas the corresponding number for the closed loop is $\Omega_{closed} = \{P!/[(P/2)!]^2\}^3$. The entropy loss upon closing the loop is thus, $\Delta S = k_B \ln[\Omega_{closed}/P\Omega_{open}]$, where the additional factor P

in front of Ω_{open} accounts for the fact that the closed micelle can be broken into an open one at any of its P segments. Using Stirling's approximation one finds $\Omega_{closed} = C\Omega_{open} P^{-3/2}$ where C is a constant. Hence, the free energy penalty of closing the chain, $\Delta F = -T\Delta S \approx (5/2)k_B T \ln P$, grows logarithmically with the micelle's length.

As discussed previously,[20] the entropically unfavorable reduction of the number of conformations eventually overcompensates the gain in end cap energy and thus leads to a suppression of ringlike micelles in the long micelle limit. Generally, theoretical work indicates that the conditions for ring formation are rather stringent.[21] Thus, to experimentally observe a dominating population of ringlike micelles requires — besides a large end cap energy — dilute but not too dilute conditions. It is therefore notable that In et al.[22] have recently synthesized a tetrameric surfactant which was found to form predominantly ringlike micelles under sufficiently dilute conditions. Ring-shaped morphologies were also observed for dimeric surfactants[23] and for amphiphilic block copolymer solutions.[24]

2.3 MOLECULAR PACKING FREE ENERGY

The molecular quantity governing the micellar size distribution, $\{X(N)\}$, is the standard chemical potential, $\mu^0(N) = N\tilde{\mu}^0(N)$, of the N-aggregate. This quantity, expressing the packing free energy of the constituent amphiphiles, also dictates the relative stability of different aggregation geometries. In the three canonical packing environments — planar bilayer membranes, cylindrical micelles, and spherical micelles — the standard chemical potential per amphiphile, $\tilde{\mu}^0(N) = \{\tilde{\mu}^0_{bil}, \tilde{\mu}^0_{cyl}, \tilde{\mu}^0_{sph}\}$, depends on the geometry but not on the number of molecules in the aggregate. In these uniform environments, all molecules are equivalent, and have the same packing properties. However, many amphiphilic aggregates comprise several, or gradually varying, packing geometries. Familiar examples of more complex geometries include nonspherical globular micelles, inverted aggregates such as the inverse hexagonal phase, and pores in lipid membranes.

In this chapter we shall specifically consider three nonuniform geometries related to cylindrical micelles. These are the curved geometry of a bent cylindrical micelle, the saddlelike structure of a trijoint intermicellar junction, and the end cap region of a rodlike micelle. In all these cases of locally varying packing geometries we may express $\tilde{\mu}^0(N)$ as an integral over all molecular contributions $\mu^0(N) = \int_{n=0}^{N} f(n)dn$, where $f(n)$ denotes the packing free energy of the nth amphiphile in the aggregate. As long as there are no long-range interactions present in the aggregate, $f(n)$ depends only on the nth amphiphile and its *local* environment. Rather than enumerating individual amphiphiles, it is more convenient to run the integration over the entire micellar (hydrocarbon-water) interface. This interface can be represented by a surface S (with surface element ds) that separates the hydrocarbon core of the micelle from the headgroup region. We can thus express $\mu^0(N)$ as

$$\mu^0(N) = \int_S ds\, \sigma(s) f(s) \qquad (2.12)$$

where $\sigma(s)$ is the local surface density of headgroups around point s on S. The local free energy $f(s)$ is a function of the position s on S. The aggregation number corresponding to the N-mer in Equation 2.12 is given by $N = \int_S ds\, \sigma(s)$.

2.3.1 FREE ENERGY CONTRIBUTIONS

There are three major contributions to the average packing free energy per molecule, f, of an amphiphilic molecule in a self-assembled aggregate: (i) f_h — the headgroup, or more precisely inter-headgroup interaction free energy, (ii) f_s — the interfacial energy associated with the (small) contact area (a) between the hydrocarbon tails and the surrounding aqueous solution, and (iii) f_c — the packing free energy of the hydrophobic tail. A common assumption in modeling the energetics of micellar aggregates is to ignore additional contributions to f and treat this function as a sum of the above three terms, $f = f_h + f_s + f_c$, for any point on S. All three contributions may depend on the local geometry, i.e., on the local cross-sectional area per molecule $a = a(s) = 1/\sigma(s)$ and on the local interfacial (principal) curvatures c_1 and c_2. The additivity of f is a most reasonable approximation, in view of the fact that the hydrophobic effect[1,6,18] creates a robust interfacial region which physically separates the headgroups from the tails. Hence, headgroup interactions, which include steric as well as electrostatic interactions between them, may sensitively depend on the local a, c_1, c_2, but are independent of the structure and conformational statistics of the tails.

The opposing forces model (OFM) mentioned in Section 2.1, has been suggested by Israelachvili et al. to account for the thermodynamic stability of micelles, membranes, and other self-assembled amphiphilic aggregates.[6,7,19] This phenomenological model proposes simple expressions for both the interfacial and headgroup energies, namely, $f_s = \gamma a$ and $f_h = B/a$. The first expression ($f_s = \gamma a$) accounts for the unfavorable contact energy between water molecules and hydrocarbon chain segments at the interface between the hydrophobic core and the surrounding solution. Thus γ is the effective surface tension corresponding to this interface, which is often approximated by the water-alkane surface tension $\gamma \approx 50\, erg/\text{Å}^2 = 0.12\, k_B T/\text{Å}^2$. The second term, $f_h = B/a$, is a phenomenological expression which lumps all headgroup interactions in a single parameter B. Because inter-headgroup forces are generally repulsive and thus tend to increase a, $f_h = B/a$ may be regarded as the first-order term in the expansion of f_h in powers of $1/a$.

In its simplest form the OFM assumes that both the interfacial and headgroup forces act within the same interaction surface, commonly regarded as located at the hydrocarbon-water interface. A simple extension of the model would assign to the headgroups their own interaction plane, at distance l_h away from this interface. The headgroup contribution to f is then given by $f_h = B/a_h$, where the cross-sectional area per headgroup $a_h = a[1 + (c_1 + c_2)l_h + c_1c_2l_h^2]$ is related to a through the two principal curvatures, c_1 and c_2, of the interface S. The two forces, interfacial tension and headgroup repulsion act in opposite directions, attempting, respectively, to minimize and maximize a, the interfacial area per molecule. They balance each other (i.e., minimize $f_s + f_h$) at the "optimal" headgroup area $a = a_0 = (B/\gamma)^{1/2}$. The free energy per amphiphile can thus be expressed in the form

$$f = \gamma a \left(1 - \frac{a_0}{a}\right)^2 + f_c \qquad (2.13)$$

Another major assumption made in most applications of the OFM is that f_c, the chain contribution to the molecular packing free energy, is a constant, independent of the aggregation geometry. This model thus suggests that amphiphiles always tend to pack into aggregates enabling $a = a_0$. More precisely, the model predicts that, for entropic reasons, amphiphiles in solution will preferentially assemble into the smallest aggregates allowing $a = a_0$. Underlying the assumption $f_c = constant$ is the notion that the hydrophobic core of amphiphilic aggregates is liquidlike, i.e., uniformly packed with hydrocarbon chain segments, at the density of a liquid phase of such chains. While this notion, known also as the *hydrocarbon droplet assumption*,[6,7] is supported by both experiment and theory, it does not mean that the conformational statistics of the tails, and hence f_c, are not affected by changes in the packing geometry of the chains. In fact, in the next section we shall employ the assumption that the hydrophobic region is a liquidlike core as a constraint on the probability distribution of chain conformation, thus deriving its dependence on the geometry of the packing environment. It should nevertheless be noted that the simple OFM (with $f_c = constant$) provides a convincing qualitative explanation for the preference of specific amphiphiles to aggregate into spheres, cylinders, or other structures. Still, as mentioned already in Section 2.1, in itself, it cannot reasonably account for properties like the curvature elasticity of lipid bilayers[16] or, as we shall see below, for the bending rigidity of cylindrical micelles. For these properties it is imperative to take into account both the interfacial ($f_s + f_h$) and chain (f_c) contributions to f.

In the next section we briefly outline a mean-field theory of chain packing in micellar aggregates. The theory has been widely applied for the calculation of thermodynamic functions (e.g., packing free energies, bending rigidities) and structural properties (e.g., bond orientational order parameters of amphiphile chains in bilayers and micelles), showing generally very good agreement with experimental and computer simulation data.[16,25]

2.3.2 Chain Conformational Statistics

A single hydrocarbon chain immersed in an aqueous solution perturbs the structure of water around it, resulting in unfavorable, "hydrophobic," interactions between chain segments and water molecules. The area of hydrocarbon-water contact is reduced when the chains condense to form a hydrocarbon droplet, such as the inner core of a micelle. This *hydrophobic effect*[18] constitutes the main driving force for amphiphile assembly into compact structures. Upon condensing into a liquid-like core each chain gains a certain cohesive (i.e., negative) energy, g, which depends on the segment density but not on the geometry of the aggregate. Assuming that this density is independent of the size and curvature of the hydrophobic core, one can safely assume that g provides a constant contribution to f_c, which for convenience may be set equal to zero. On the other hand, the conformational statistics, and hence the internal free energy, of the chains may be strongly affected by the packing geometry. The local packing geometry of a chain is fully specified by the area per molecule a, and the two principal curvatures c_1, c_2 of the hydrocarbon-water interface, implying $f_c = f_c(a, c_1, c_2)$. Below we outline a rather simple theoretical approach for calculating f_c.

The central quantity in this mean-field approach[8,16,26] is the probability distribution function (pdf), $P(\alpha)$, describing the probability of finding the chain in a given conformation, α. $P(\alpha, s)$ depends generally also on the local packing geometry at point s on S, as specified by the local a, c_1, c_2. In nonuniform, rapidly varying, packing geometries (e.g., of lipid molecules around an integral membrane protein), the calculation of $P(\alpha, s)$ is quite complicated, and depends also on the local environment s' at which *neighboring* chains originate.[27,28] Here, for brevity, we shall briefly outline the theory only for a uniform aggregate geometry.

The local conformational free energy per chain is given by

$$f_c = u_c - Ts_c = \sum_\alpha P(\alpha)\varepsilon(\alpha) + k_B T \sum_\alpha P(\alpha)\ln P(\alpha) \qquad (2.14)$$

where the summations run over all accessible chain conformations α. The energetic contribution u_c represents the conformational average over the internal energy $\varepsilon(\alpha)$ associated with each individual conformation α. For simple, saturated alkyl hydrocarbon tails, for instance, the $\varepsilon(\alpha)$'s are determined by the trans/gauche bond sequence along the chain. The second sum in Equation 2.14 is the conformational entropy of the chain.

The equilibrium pdf, $P(\alpha)$, is found by minimizing f_c in Equation 2.14 subject to the normalization condition $\sum_\alpha P(\alpha) = 1$ and subject to the relevant geometric packing constraints. These constraints express the average shape that the chain must adopt in order to fit into a specific aggregate geometry. The constraint can be formulated as

$$\sum_\alpha P(\alpha)\varphi(\alpha, z) = a(z) = a\left(\frac{z}{b}\right)^{2-d} ; \quad 0 \le z \le b \qquad (2.15)$$

where $\varphi(\alpha, z)dz$ is the volume occupied by a chain in conformation α within a shell $z, z + dz$, parallel to the interface S of the micelle; $a(z)$ denotes the area available per chain at position z. The second equality in Equation 2.15 specifies $a(z)$ for the three canonical packing geometries, with $d = 0, 1, 2$ denoting the growth dimensionalities of spherical, cylindrical, and planar bilayer aggregates, respectively. Here a is the cross-sectional area of the chain at S, and b is the thickness of the hydrocarbon core. Note that $ab/(3 - d) = v$ is the (constant) volume v of the hydrocarbon chain in the liquidlike core. The conditional minimization of $F[\{P(\alpha)\}]$ yields

$$P(\alpha) = \frac{1}{q} \exp \left\{ -\left[\varepsilon(\alpha) + \int_0^b dz\, \pi(z)\phi(\alpha, z) \right] / k_B T \right\} \qquad (2.16)$$

where the partition function q ensures the normalization of the equilibrium pdf, and where $\pi(z)$ is the lateral pressure profile that emerges as the set of Lagrangian multipliers conjugate to the packing constraints, Equation 2.15. Physically, $\pi(z)$ represents the lateral pressure imposed on an otherwise free (isolated) chain in order to fit into the volume available to it in the compact aggregate.

Substituting $P(\alpha)$ from Equation 2.16 back into the packing constraint, Equation 2.15, we obtain — as is often the case in mean-field theories — a self-consistency relation for the pdf; here a coupled set of equations for the $\{\pi(z)\}$. Any actual calculation of the lateral pressure profile must be based on the appropriate molecular chain model. An adequate representation of saturated alkyl (polymethylene) chains, for example, is provided by the *rotational isomeric state model*.[29] Its input parameters are the geometrical characteristics of the chain, i.e., the C-C bond length, methylene segment size, bond angles, and the trans/gauche isomerization energy. After evaluating the $\pi(z)$, and hence the equilibrium pdf, we can use Equation 2.14 in order to calculate f_c. The results of such numerical calculations for $-(CH_2)_{13}CH_3$ (C-14) chains, packed in the three basic geometries, are shown in Figure 2.3 as a function of the aggregate's half thickness b. Note that the maximal chain length for C-14 is $b_{max} \approx 18\,\text{Å}$, beyond which f_c would increase extremely steeply since $b > b_{max}$ implies the appearance of (a highly unfavorable) void space inside the hydrocarbon core. Interestingly, the minimum of f_c appears at very different thicknesses b for the different geometries: $b = 10\,\text{Å}, 16\,\text{Å}$, and $18\,\text{Å}$. These values correspond to the interfacial areas per molecules: $a = 40\text{Å}^2, 51\text{Å}^2$, and $68\,\text{Å}^2$, for the bilayer, cylinder, and sphere, respectively. It should be stressed that even though the changes in f_c for different b's are only fractions of $k_B T$ for individual molecules, they can amount to substantial free energy difference for a whole micelle.

2.3.3 MICELLAR STABILITY

In this section we consider the implications of Equation 2.13 with respect to the stability of the three canonical aggregation geometries. It is instructive to first

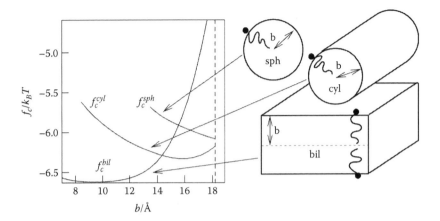

FIGURE 2.3 Chain conformational free energy in the three canonical geometries (sphere, cylinder, and planar bilayer; schematically displayed on the right hand side of the diagram) as a function of the hydrophobic half-thickness b. The calculations are for $-(CH_2)_{13}CH_3$ (C-14) tails, whose chain volume is $v = 405\text{Å}^3$. The dashed line denotes the length $b_{max} \approx 18\,\text{Å}$ of a maximally stretched (all-trans) chain, beyond which f_c increases drastically. The minimum of f_c^{sph} is achieved very close to that length. Adapted from May et al.[35] with permission.

investigate the predictions of the simple OFM model in which, as discussed above, f_c is treated as a constant,[6] (which for simplicity may be set equal to zero). Ignoring f_c, the OFM predicts a most stable aggregate if $a = a_0$, implying $f = 0$. In general there will be more than one packing geometry enabling $a = a_0$, in which case the preferred micellar geometry is the one which corresponds to the smallest possible (hence maximally curved) aggregate, because many small aggregates carry more translational entropy than fewer larger ones. To quantify this notion it is common to define the *packing parameter* $p = v/(a_0 b_{max})$, where, as before, v is the chain volume, and b_{max} is the length of the maximally extended tail. For minimal packing energy we require $a = a_0$ as well as $b \leq b_{max}$ in order to ensure a compact hydrophobic core. Noting that $v/a_0 b = 1/3, 1/2$, and 1, for spheres, cylinders, and planar bilayers, respectively, it follows that $p \leq 1/3$ allows optimal packing in all three geometries. As argued above, however, spherical micelles will be preferred on entropic grounds. Similarly, $1/3 < p \leq 1/2$ allows for both cylinders and bilayers, but cylinders are preferred because of their higher curvature. Bilayers prevail when $1/2 < p \leq 1$, and inverted structures for $p > 1$.

Let us now calculate f, including the chain conformational free energy f_c. We calculate the optimal aggregate half-thickness $b = (3-d)v/a$ and the corresponding free energy f as a function of the headgroup repulsion strength B for the three canonical aggregate geometries: bilayer (bil, $d = 2$), cylinder (cyl, $d = 1$), and sphere (sph, $d = 0$). Figure 2.4 shows the numerical results for C-14 chains, using the functions f_c shown in Figure 2.3. (To get a feeling for the order of magnitude of B values, recall that $B = \gamma a_0^2$, so that for, say, $\gamma = 0.12\,k_B T/\text{Å}^2$

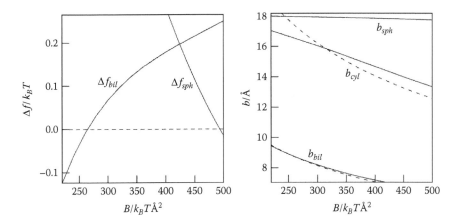

FIGURE 2.4 Left diagram: The difference in the free energy per molecule between a planar bilayer and a cylinder ($\Delta f_{bil} = f_{bil} - f_{cyl}$), and between a sphere and a cylinder ($\Delta f_{sph} = f_{sph} - f_{cyl}$), as a function of the headgroup repulsion strength B. Right diagram: The aggregate thickness b (solid lines) for which f is minimal. The dashed lines correspond to the optimal b, as implied by the requirement for optimal headgroup area ($a = a_0$), ignoring the tail contribution f_c. In this calculation, a small distance of $l_h = 1\,\text{Å}$ was assumed to separate the hydrocarbon-water and headgroup interaction surfaces. Adapted from May et al.[35] with permission.

and $a_0 = 50\,\text{Å}^2$ we obtain $B = 300\,k_BT\text{Å}^2$.) The quantities plotted in the left diagram are the free energy differences, $\Delta f_{bil} = f_{bil} - f_{cyl}$ and $\Delta f_{sph} = f_{sph} - f_{cyl}$ of, respectively, the bilayer and sphere compared to the cylindrical structure. The signs of Δf_{bil} and Δf_{sph} suggest that as B increases the stable aggregation structure undergoes — as expected — the sequence: planar → cylindrical → spherical. The corresponding equilibrium half-thicknesses, b, are shown in the right diagram (solid lines); the prediction from the OFM (with $f_c = 0$) are also displayed (dashed lines). In all cases, an increase in the headgroup repulsion strength, B, tends to increase a and thus lower b. The crossing points of the corresponding solid and dashed lines in the right diagram indicate the location of the minima of $f_c(b)$ (see Figure 2.3). The close correspondence of the two models argues that the aggregate thickness, b, and thus the cross-sectional area a, is indeed mainly determined by headgroup interactions.

2.4 BENDING ELASTICITY OF CYLINDRICAL MICELLES

Short cylindrical micelles, whose lengths are at most a few times larger than their diameter, are reasonably described as rigid rods. On longer length scales the micelles exhibit nonnegligible bending fluctuations along their contour, and their conformational statistics conform to the *wormlike chain*, or *persistent polymer,* model.[30,31] In this model, which is also adequate for describing thermal orientational fluctuations of relatively stiff biopolymers such as DNA or actin, the polymer is depicted as a homogeneous elastic filament undergoing moderate (harmonic) bending deformations. Thus, the energetic cost ΔF associated with the uniform bending of (a section of) a cylindrical micelle of length L is

$$\frac{\Delta F}{L} = \frac{1}{2}\kappa c^2 = \frac{1}{2}k_B T l_p c^2 \qquad (2.17)$$

where c is the curvature of the deformation, κ is the 1D bending modulus, and l_p is the *persistence length* of the micelle in question. The persistence length measures the length of angular correlations along the micelle's contour, namely, $\langle \cos\theta \rangle = \exp(-s/l_p)$, where s is the contour length between two points along the micelle's axis, and θ is the angle between the chain (tangential) directions at these points. For small curvature deformations this yields $\langle c^2 \rangle = 2s/l_p$. On the other hand, from the first equality in Equation 2.17 it follows (by Boltzmann averaging) that $\langle c^2 \rangle = 2sk_B T/\kappa$, explaining the relationship $l_p = \kappa/k_B T$ expressed by the second equality in this equation.[30] Consider now a segment of the micelle of length, say, $L = 2l_p$, and suppose it is uniformly bent to an arc of radius $c \sim 1/l_p$, implying a substantial change of directions, $\theta \approx 120°$, between the two ends of this segment. From Equation 2.17, the energetic cost of angular fluctuations of this order of magnitude is small, $\sim k_B T$. Such fluctuations are thus highly probable, consistent with the interpretation of l_p as the decay length of angular correlation along a worm-like chain. This also means that a long, wormlike, micelle of length L can be regarded as a freely jointed polymer composed of L/l_p (Kuhn) segments[30,31] of length l_p. The mean end-to-end distance of the micelle is thus $\sqrt{\langle R^2 \rangle} = l_p\sqrt{L/l_p} = \sqrt{l_p L}$.

Upon bending a micelle, the flexible tails comprising the hydrocarbon core can easily readjust their chain conformations, thus relieving much of the curvature stress. Similarly, possibly on a somewhat longer time scale, the polar headgroups can rearrange their positions on the micellar surface thus also lowering the deformation energy penalty. This ability to accommodate the curvature stress, while maintaining the micelle's integrity, reflects the relatively "soft" internal degrees of freedom of self-assembled amphiphilic aggregates, as compared to stiffer, semiflexible[31] filaments, which are governed by stronger cohesive forces, such as double stranded (ds) DNA, F-actin, or microtubules. The bending rigidity κ of a wormlike micelle is thus expected to be smaller than that of a comparably thick semiflexible filament. Indeed, bending constants of cylindrical micelles are typically in the range $\kappa = l_p k_B T = (100...500)k_B T\text{Å}$, smaller than that of dsDNA for example ($\kappa \approx 600 k_B T\text{Å}$ for B-DNA). This difference is more significant considering that the radius, $b = 10\,\text{Å}$, of the DNA rod is considerably smaller than typical micellar radii, e.g., $b = 16\,\text{Å}$ for C-14 micelles (see Figure 2.4). More explicitly, homogeneous rodlike materials can be characterized by $\kappa = (\pi/4)Yb^4$, where b is the radius and Y is Young's modulus with a typical value of $Y = 0.5\times10^9\ J/m^3$ for biomaterials.[32] This, indeed, roughly recovers the bending stiffness of B-DNA, $\kappa = (\pi/4)Yb^4 = 950k_BT\text{Å}$ for $b = 10\,\text{Å}$, but yields $\kappa = 6000\,k_BT\text{Å}$ for a $b = 16\,\text{Å}$ rod, much larger than the bending rigidity of a similarly thick micelle.

A better estimate of the bending rigidity of cylindrical micelles may be obtained based on the 2D elastic properties of other soft aggregates, especially 2D lipid bilayers or surfactant monolayers, whose curvature elasticity has been extensively studied for many systems using diverse techniques.[33] The elastic deformation energy, ΔF, associated with the (uniform) bending of such a layer of area A, to principal curvatures c_1 and c_2 is given by[34]

$$\frac{\Delta F}{A} = \frac{1}{2}K(c_1 + c_2 - c_0)^2 + \bar{K}c_1c_2 - F_{eq} \tag{2.18}$$

Here K and $\bar{K}$, are the splay and saddle-splay moduli, respectively, and c_0 is known as the spontaneous curvature. $F_{eq} = c_0^2\bar{K}K/(2\bar{K} + 4K)$ is the free energy per unit area in the equilibrium state, in which case $c_1 = c_2 = c_{eq} = c_0\,K/(2K + \bar{K})$. Even though Equation 2.18 is strictly valid only for small curvatures, we may use it to estimate the 1D bending stiffness of a cylindrical micelle.

The straight micelle of fixed length L and radius b has a mantle area $A = 2\pi bL$ and curvatures $c_1 = 1/b$ and $c_2 = 0$. Bending the micelle transforms it into a toroidal segment for which A and c_1 remain unaffected, and c_2 is positive (negative) on the outer (inner) part of the toroidal surface. Using Equation 2.18 for this model, it is easily shown that up to quadratic order in curvature $\Delta F = L\pi bKc^2/2$, and hence the simple relationship $\kappa = \pi bK$, independent of c_0 and $\bar{K}$, between the 1D and 2D bending constants emerges.[35] As the splay modulus of surfactant monolayers is typically $K \sim 10\,k_BT$, the prediction for the 1D bending stiffness of the cylindrical micelle is $\kappa = 320\,k_BT\text{Å}$. A very similar estimate would follow by considering the 1D bending (i.e., to $c_1 = 0, c_2 = c$, of a strip of a planar ($c_1 = c_2 = c_0 = 0$) bilayer of length L and width $\sim b$. Noting the additive contribution from the two bilayer leaflets one finds now $\kappa = 2Kb$. Simple dimensional analysis also suggests $\kappa \sim bK$. A comparison of this relation with measured values of κ and K for SDS micelles was recently made by Magid et al.[36]

A closer insight into the mechanism underlying the bending elasticity of cylindrical micelles may be gained using the simple molecular-level model, Equation 2.13. First we note that in a straight cylindrical micelle, all chains are directed, on average, towards the axis of symmetry, located exactly at the center of the rod. It is reasonable to assume that upon bending the tails will still point toward a common axis which, however, need no longer coincide with the midaxis of the micelle. It may be shifted by a distance εb towards the bending (concave) direction as depicted in Figure 2.5. The shift ε is dictated by the balance between two, generally competing, tendencies. One of these, as expressed by the first term in Equation 2.13, is the tendency to maintain the surface area per molecule, a, equal to the optimal value, a_0, as implied by the OFM. Since the overall area-to-volume ratio of a cylinder does not change upon bending, there is no change in the average surface area per molecule. Thus assuming (consistent with the simple OFM) that changes in f_c during bending are negligible, we would conclude $\kappa \equiv 0$.

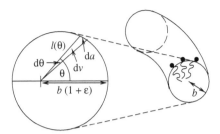

FIGURE 2.5 Schematic representation of a homogeneously bent cylindrical micelle, of radius b, forming a section of a torus. The zoom into its interior specifies the common axis towards which the tails of all amphiphiles are directed on average. This axis is located at distance εb away from the mid-axis. The local area and volume elements at angle θ are $da = \tilde{a}d\theta$ and $dv = \tilde{v}d\theta$, respectively. The average tail length corresponding to angle θ is denoted by $l(\theta)$.

For the molecular packing model depicted in Figure 2.5 it can be shown that $a = a_0$ is achieved for $\varepsilon = cb/3$.[35]

Unlike the interfacial forces which act towards $a = a_0$ (and hence $\varepsilon = cb/3$), the chain conformational free energy, f_c, tends to keep $\varepsilon = 0$. The balance of these tendencies will yield $0 < \varepsilon < cb/3$. To determine the dependence of ε on c it is convenient to use the linear form $\varepsilon = \eta c$ and determine the optimal *relaxation parameter* η by minimizing the bending deformation energy. Referring to Figure 2.5, let $l(\theta)$ denote the effective tail length of a molecule anchored at angular position θ. The corresponding cross-sectional area per molecule $a(\theta) = \tilde{v}\tilde{a}/\tilde{v}$ (recall v being the chain volume) can be calculated using appropriate parameterization of the local area and volume elements, $da = \tilde{a}d\theta$ and $dv = \tilde{v}d\theta$.[37]

For a given η, calculating the 1D bending stiffness based only on the OFM contribution to the bending energy (i.e., setting $f_c = 0$ and $f = \gamma a(1 - a_0/a)^2)$) one finds[35]

$$\kappa_h = (2/9)b\gamma\pi\,(b - 3\eta)^2 \qquad (2.19)$$

where $a = a_0 = 2v/b$ was used here for the straight cylinder. This is the OFM contribution to the 1D bending modulus $\kappa = \kappa_h + \kappa_c$ which, as noted above, yields $\kappa_h = 0$ for $\eta = \eta_h = cb/3$ (corresponding to $a(\theta) = a_0$ for all chains).

Clearly, however, if the straight cylinder offers optimal packing conditions for the hydrocarbon tails, they would prefer $l(\theta) \equiv b$ for all chains, implying $\eta = \eta_c = 0$. The optimal η should thus be determined by minimizing the sum of head and tail contributions. The chain conformational energy can be calculated numerically using the chain packing theory outlined in Section 2.3.2, or estimated using an approximate closed form expression for f_c. A reasonable approximate representation of $f_c(\theta)$ is provided by the quadratic form

$$f_c(\theta) = \tau\gamma a_0\,(1 - l(\theta)/b^*)^2 \qquad (2.20)$$

which appears consistent with numerical calculations of chain energies, such as those shown in Figure 2.3. Here b^* is the optimal tail length of a chain packed in a cylindrical geometry, and the dimensionless coefficient τ measures the resistance of the chain to deviations from b^*. For the C-14 chains with $B = 300\,k_B T\text{Å}^2$, Figures 2.3 and 2.4 suggest that b^* is actually equal to the radius $b = 2v/a_0$ of the straight cylindrical micelle. Based on the last equation it can be shown[35,37] that

$$\kappa_c = 2\pi b \gamma \tau\, \eta^2 \qquad (2.21)$$

where for simplicity we have set $b^* = b$. Minimizing $\kappa(\eta) = \kappa_h + \kappa_c$, using Equations 2.19 and 2.21, one finds that the optimal η and κ are given by

$$\eta = \frac{b}{3(1+\tau)}, \qquad \kappa = \frac{2}{9}\,\gamma\,\pi\,b^3\,\frac{\tau}{(1+\tau)} \qquad (2.22)$$

It may be noted that in addition to chain stretching, f_c may include a contribution due to tilt deformation. Tilt refers to the deviation of the average chain direction from the normal direction of the polar-apolar interface. The tilt deformation of amphiphilic layers has received some interest in recent years, particularly for lipid membranes[38–41] but its consequences with respect to the energetics of cylindrical micelles have not been studied so far.

Rather than using the phenomenological expression in Equation 2.20, the function $f_c(c)$ can be computed based on the formalism described in Section 2.3.2. Such a calculation has been performed for an approximate model of the bent cylindrical micelle,[35] according to which the cylinder is divided into an external (convex) and an internal (concave) region, as shown in the inset in Figure 2.6. Assuming that all chains in each of the two regions are equally distorted, the contribution κ_c was calculated based on a numerically performed curvature expansion. As a concrete example, the inset in Figure 2.6 shows the average conformational free energies of a chain in the external (E) and internal (I) parts of a bent cylindrical micelle with curvature $c = 1/90\,\text{Å}$. This figure also shows that the chain packing energy in the straight cylinder is always intermediate between those of the internal and external regions (dashed curve). The calculated dependence of κ_c on η for $B = 300\,k_B T\text{Å}^2$ (and thus $b = 16\,\text{Å}$) is shown in Figure 2.6. Also shown is κ_h, as derived from the OFM. The overall bending stiffness, $\kappa = \kappa_h + \kappa_c$, adopts its minimum $\kappa = 160\,k_B T\text{Å}$ at $\eta = 3.5$. These estimates compare well with the predictions of the phenomenological model according to Equation 2.22: $\eta = 2.7$ and $\kappa = 170\,k_B T\text{Å}$ for $\tau = 1$ and $b = 16\,\text{Å}$.

The scaling of the persistence length $l_p \sim b_{max}^\alpha$ with the length b_{max} of the surfactant tails has been investigated theoretically on the basis of a self-consistent field theory by Lauw et al.,[42] the predicted exponent was $\alpha = 2.4 - 2.9$. This is slightly smaller than $\alpha = 3$, predicted by Equation 2.22 (under the assumption that τ is independent of chain length).

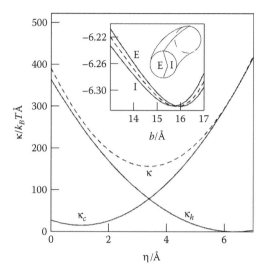

FIGURE 2.6 The bending modulus κ and its components κ_c and κ_h (all in units of $k_B T\,\text{Å}$) as a function of the relaxation parameter η for $B = 300\,k_B T\text{Å}^2$. Optimal relaxation is found for $\eta \approx 3\,\text{Å}$. The inset shows the average conformational free energy f_c in units of $k_B T$ in the external (convex, E) and internal parts (concave, I) of a bent cylindrical micelle with $c = 1/90\,\text{Å}$. The dashed line is f_c for a straight micelle with $c = 0$. Adapted from May et al.[35] with permission.

The persistence length l_p has been determined for a large number of cylindrical micelles. Rough estimates can be obtained using light scattering methods[43,44] or microscopic imaging such as cryo-TEM[22] or AFM.[45] Precise measurements[9,36,46,47] are based on small-angle neutron scattering which yields high-resolution scattering functions that can be compared with theoretical predictions. In fact, owing to the complexity in the behavior of wormlike micellar solutions, Monte Carlo simulations have proven to be the most appropriate tool to simulate these scattering functions.[46,48]

2.5 THE SPHERE-TO-ROD TRANSITION AND THE SECOND CMC

The minimal micelles formed at the cmc, even those that quickly elongate into wormlike micelles, are almost invariably spherical. If the optimal packing radius in the elongated cylindrical body were equal to that of the minimal spherical micelles, all micelles would grow as perfect spherocylinders. For this model the standard chemical potential (i.e., the amphiphile packing free energy) of the micelle is a simple sum of contributions from molecules packed in the spherical body and the two hemispherical caps: $\mu_N^0 = N\tilde{\mu}_{cyl}^0 + 2\delta$, with $\delta = (\bar{N}/2)(\tilde{\mu}_{sph}^0 - \tilde{\mu}_{cyl}^0)$ for $N \geq \bar{N}$; $\bar{N}$ denoting the number of molecules in the minimal micelle. According to this model micellar growth, or as it is often called the *sphere-to-rod transition*, is a continuous process, with the average micellar size increasing monotonically as a function of X (scaling with $\sqrt{X}$ at higher concentrations), as illustrated in Figure 2.2. Experimentally, however, it is often observed that the sphere-to-rod transition is discontinuous. That is, a second critical concentration, X_{2c}, must be surpassed in order to initiate the growth from globular into cylindrical micelles. X_{2c} marks the total amphiphile concentration at the onset of cylindrical growth. Between the two critical concentrations, X_c and X_{2c}, added amphiphiles merely increase the number of globular micelles without transforming them into elongated aggregates.

The first experimental evidence for this peculiar behavior was reported for cetylpyridinium bromide (CPBr) by Porte et al.[49] who also introduced the term "second cmc" for X_{2c}. Meanwhile, a second cmc has been observed for various other amphiphiles, including cationic dimeric (gemini)[23,50] and nonionic[51,52] amphiphiles, as well as various alkylpyridinium salts.[53] Cryo-TEM images indicate two distinct micellar populations above the second cmc:[23,51] globular and cylindrical. The cylindrical micelles appear to be semiflexible and with spherical end caps. Notably, both the globular micelles and the micellar end caps have larger diameters than the cylindrical main body (see Chapter 5, Figure 5.2 and Chapter 14, Figure 14.8). For diblock copolymers that self-assemble into worm-like micelles this feature is even more pronounced, as seen in cryo-TEM images.[54,55] Small cylindrical micelles, with midsections comparable to the size of the end caps, are generally not observed.

What then is the reason for the experimentally observed *discontinuous* transition from globular to elongated micelles? The presence of two distinct micellar populations clearly suggests that the packing properties of short cylindrical aggregates are energetically unfavorable. Their appearance is thus suppressed, resulting in a discontinous transition from spherical to long cylindrical aggregates. This basic notion underlies previous theoretical models of the second cmc,[56–58] and is also consistent with a recent simulation study.[59]

The suppression of short cylindrical micelles can be viewed as the result of an energetically unfavorable "repulsion" between the two (partly) spherical end caps. For nonionic amphiphiles this repulsion is likely a result of elastic interactions through the cylindrical main body, as illustrated in Figure 2.7. For charged micelles, long-range electrostatic interactions may induce similar effects as will be discussed in Section 2.7.

Amphiphiles residing within the spherical end caps are subject to spherical packing geometry, whereas those in the cylindrical main body experience cylindrical packing conditions. If the cross-sectional area per headgroup in both packing geometries were exactly the same, then the radius b_{sph} of the end cap would be larger than the radius b_{cyl} of the cylindrical main body by a factor of 3/2. On the other hand, assuming size-matching ($b_{sph} = b_{cyl}$) between end cap and cylindrical main body (as for a perfect sphero-cylinder) the area per headgroup within the end caps must increase to $a_{sph} = 3a_{cyl}/2 = 3v/b_{cyl}$, where v is the amphiphile's tail volume and a_{cyl} is the cross-sectional headgroup area within the cylindrical main body. Based on the OFM (see Equation 2.13) we then obtain an

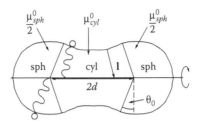

FIGURE 2.7 Illustration of a cross section along the axis of symmetry of a short cylindrical micelle with two end caps (denoted by "sph" and separated by distance $2d$). Each end cap induces an elastic (chain packing) perturbation within the cylindrical main body (denoted by "cyl"). Overlap of the perturbations results in effective repulsion between the two ends. The perturbation can be characterized by a position-dependent vector $\mathbf{l(r)}$ that specifies the length ($l = |\mathbf{l}|$) and orientation of the amphiphiles. The overall packing free energy $\mu^0 = \mu^0_{sph} + \mu^0_{cyl}$ is the sum of contributions from the two end caps (μ^0_{sph}) and from the cylindrical main body (μ^0_{cyl}). Adapted from May and Ben-Shaul[56] with permission.

excess end cap energy of $\delta = 2\pi\gamma b^2_{cyl}/9$, which amounts to more than $20\,k_B T$ for $b_{cyl} = 16\,\text{Å}$. The energetic incentive to minimize the excess end cap energy by increasing b_{sph} beyond b_{cyl} is thus substantial.

Geometrically, a structure composed of a cylindrical body of radius b_{cyl}, capped by two truncated spherical end caps (as in Figure 2.7) of a larger radius is not impossible. However, chain packing considerations imply that a micelle of this geometry is quite unlikely to form. First, the chains cannot be stretched beyond their maximum length. In Figure 2.3 we see that, e.g., for C-14 chains, the minimal chain packing energy in the cylindrical geometry is achieved for $b_{cyl} = 16\,\text{Å}$. Assuming that this is also the micellar radius preferred by the headgroups, then to ensure the same area per headgroup in the end caps their spherical radius should be $b_{sph} = (3/2)b_{cyl} = 24\,\text{Å}$, much larger than the length of a fully extended chain, which is $18\,\text{Å}$. Thus, $b_{sph} = b_{max} = 18\,\text{Å}$ (which also corresponds to the minimum chain packing energy in a sphere) is the largest end cap radius for the C-14 chains. According to the OFM, the corresponding energetic difference between a spherical end cap of $b_{sph} = 18\,\text{Å}$ compared to $b_{sph} = b_{cyl} = 16\,\text{Å}$ is still considerable; $\delta = 15\,k_B T$.

Another important chain packing consideration involves the transition zone between the end caps and the cylindrical main body. The chain stiffness (together with the OFM) defines a length scale that determines the spatial relaxation of aggregate thickness from the end caps toward the cylindrical main body. Typically, this length scale is comparable to the micellar thickness, thus being of the order of a few nm for most amphiphiles. The shape of the transition zone may be adjusted by the micelle so as to minimize its free energy. Recent modeling attempts have used various structural assumptions, such as conical[35] or catenoidal[57] transition zones, while others have calculated the shape of the transition region through functional minimization.[56] An example for the structure of a cylindrical

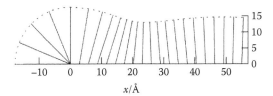

FIGURE 2.8 Calculated structure of a long cylindrical micelle terminated by a semi-spherical end cap. The dashed line marks the polar-apolar interface, the solid lines indicate the local directions of the surfactant tails. Size mismatch between end cap and cylindrical main body leads to an elastic perturbation that propagates into the cylinder as a damped oscillation with characteristic lengths ξ_1 and ξ_2 given in Equation 2.24. Adapted from May and Ben-Shaul[56] with permission.

micelle as calculated by functional minimization is shown in Figure 2.8. As can be seen, the presence of the hemispherical end cap induces a perturbation that propagates a few nm into the cylindrical main body. The shape shown in Figure 2.8 is not based on any structural assumptions other than the requirement that the cylindrical micelle should be end capped. As in the previous section, the free energy per molecule is modeled as a sum

$$f = \gamma a \left(1 - \frac{a}{a_0} \right)^2 + \tau \gamma a_0 \left(\frac{l - b^*}{b^*} \right)^2 \tag{2.23}$$

of an OFM term for the interfacial forces, and a quadratic term representing the chain stretching energy. In this equation a is the (locally varying) area per head-group within either the end cap or the perturbed cylindrical part; l is the corresponding *local* average chain length in the two regions, and b^* denotes the radius corresponding to the minimum of the chain conformational free energy $f_c(b)$ as given in Figure 2.3. This value happens to roughly coincide with the optimal tail lengths, b_{sph} and b_{cyl}, in the spherical and cylindrical parts, respectively. (Hence, $b^* = b_{sph}$ and $b^* = b_{cyl}$ in their respective regions.) The (dimensionless) chain stretching modulus τ is identical to that introduced in Equation 2.20. Note that due to the perturbation of the cylindrical part, the volume-to-surface ratio per amphiphile within the perturbed cylindrical part is not simply $l/2$, but rather depends on the rate at which the average (local) tail length $l(x)$ changes along the micellar axis x (details can be found in May and Ben-Shaul[56]). Minimization of the overall free energy shows that the shape of the micellar relaxation profile $l(x)$ is given by a damped oscillation $l(x)/b_{cyl} - 1 \sim \exp(-x/\xi_1)\cos(x/\xi_2)$. The two characteristic length scales are

$$\xi_1 = \frac{b_{cyl}\sqrt{2/3}}{\sqrt{\sqrt{1+\tau}-1}}, \quad \xi_2 = \frac{b_{cyl}\sqrt{2/3}}{\sqrt{\sqrt{1+\tau}+1}} \tag{2.24}$$

The first, ξ_1, characterizes the exponential decay length of the perturbation; with $b_{cyl} = 16\,\text{Å}$ and $\tau \approx 1$ (both extracted from Figure 2.3) we obtain $\xi_1 = 2\,\text{nm}$. The second, ξ_2, measures the wavelength of oscillations in micellar thickness. For finite τ these oscillations are only noticeable for $x < \xi_1$. In the limiting case that $\tau = 0$, we find that the presence of the spherical end caps induces a periodic oscillation ("pearling") along the micelle axis. This solution for $l(x)$ preserves a constant surface-to-volume ratio.

At this point we note the close similarity of the present approach with modeling interactions between inclusions in amphiphilic membranes, particularly in lipid membranes. Continuum elasticity theory has frequently been used to calculate the shape profile in the vicinity of one or several symmetric inclusions and the corresponding interaction free energy between inclusions.[60–62] Both the shape profile and the free energy between inclusions as function of their mutual distance are predicted to be described by damped oscillations. This situation is analogous for end capped cylindrical micelles where the end caps play the role of the inclusions and where their elastic interaction is mediated by the cylindrical main body. The damped oscillating free energy found for inclusion-containing membranes is also found for cylindrical micelles, as shown in Figure 2.9.

More specifically, Figure 2.9 (top) shows the excess free energy μ^0 (calculated according to Equation 2.12) of a cylindrical micelle comprising a total of N amphiphiles, relative to the packing free energy that these N amphiphiles would have if they were all packed in the cylindrical body of a very long micelle of optimal thickness b_{cyl}. Hence, μ^0 contains both the energy of forming end caps and their elastic interaction through the cylindrical part of the micelle. In the limit $N \to \infty$ we obtain the energy (2δ) of two individual end caps (corresponding to Figure 2.8), including the transition zone between the spherical and cylindrical regions. In the opposite limit, $N = \bar{N} = 60$, the two end caps have merged into a single spherical micelle. Generally, μ^0_{sph} denotes the contribution to the end cap energies stored within the two quasi-spherical cap regions, and $\mu^0_{cyl} = \mu^0 - \mu^0_{sph}$ is the contribution due to the perturbation of the cylindrical part. It is interesting to note that the two contributions are similar in magnitude, which underscores the need to include the cylindrical part into the calculation of end cap energies.

The distance $2d$ between the end caps is roughly proportional to the aggregation number N (for long micelles it is exactly proportional). The bottom half of Figure 2.9 shows the micellar shapes corresponding to three selected aggregation numbers, $N = 113$, $N = 121$, and $N = 146$. The geometrical shape of a micelle containing $N = 113$ molecules corresponds to a maximum in the aggregate's free energy μ^0. The energetic barrier around this number is about $10\,k_B T$, implying that micelles of aggregation numbers close to this value will be suppressed during micellar growth. This behavior is corroborated by size distributions calculated using Equation 2.3, and the function $\mu^0(N)$ in Figure 2.9.

In addition to being responsible for the lack of intermediate micellar sizes, the energetic barrier in $\mu^0(N)$ is also responsible for the appearance of a second threshold critical concentration, the second cmc, X_{2c}. To clarify this point it is

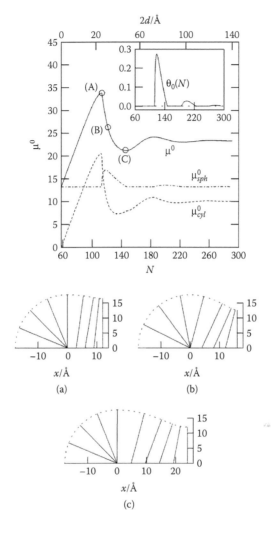

FIGURE 2.9 Top diagram: The excess packing free energy (in $k_B T$'s) of amphiphiles in a linear micelle, $\mu^0(N) = \mu_{sph}^0(N_{sph}) + \mu_{cyl}^0(N - N_{sph})$, relative to the energy of packing all N molecules in a cylindrical micelle of optimal thickness b_{cyl}. $\mu_{sph}^0(N_{sph})$ is the contribution due to the N_{sph} molecules constituting the end caps (relative to N_{sph} molecules in a cylinder of radius b_{cyl}). $\mu_{cyl}^0(N_{cyl})$, with $N_{cyl} = N - N_{sph}$, is defined analogously. Note that N_{sph} varies with N. The calculation is for micelles composed of C-14 amphiphiles, with $\tau = 1$ and preferred packing parameter of $p = v/(a_0 b_{max}) = 1/2.5$. The radius of the end cap is $b_{max} = 18$ Å. The inset shows the variation in the contact angle $\theta_0(N)$ (defined in Figure 2.7). The three circles on the $\mu^0(N)$ curve specify the aggregation numbers: $N = 113$ (a), $N = 121$ (b), and $N = 146$ (c). The corresponding micellar shapes are shown in the diagrams on the bottom, where the dashed lines mark the hydrocarbon-water interface. Chain orientations are shown at several arbitrary positions. Adapted from May and Ben-Shaul[56] with permission.

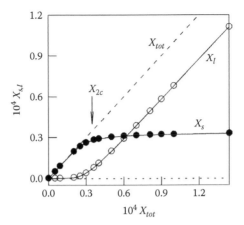

FIGURE 2.10 The change in amphiphile populations incorporated in small (X_s) and long (X_l) micelles with the total concentration, X. Small micelles correspond to aggregation numbers $\bar{N} \leq N \leq L = 250$. The dashed line shows $X = X_s + X_l$. The arrow marks the value of the second cmc, X_{2c}, as predicted by Equation 2.26. Adapted from May and Ben-Shaul[56] with permission.

instructive to divide the micellar size distribution into two populations, short (s) and long (l) micelles, according to

$$X_s = \int_{\bar{N}}^{L} X(N)\,dN, \quad X_l = \int_{L}^{\infty} X(N)\,dN \tag{2.25}$$

Here $\bar{N}$ denotes the size of the minimal (spherical) micelle ($\bar{N} = 60$ in Figure 2.9), and L is an aggregation number for which the two end caps no longer interact with each other (in Figure 2.9, $L = 250$ would be a convenient choice). Note that $X(N)$ is determined according to Equation 2.3 with $\mu^0(N) = N\tilde{\mu}^0(N)$, as given in Figure 2.9. The two micellar populations are plotted in Figure 2.10 as a function of the total amphiphile concentration X. The sharp transition observed for $X = X_{2c}$ is reminiscent of the first cmc as shown in Figure 2.1. Here, however, it separates small, spherelike micelles from long, cylindrical ones. The weight average of the latter grows, as expected, according to $\langle N \rangle = 2 \exp(\delta/k_B T)\sqrt{X}$ with an end cap energy $\delta = \mu_{sph}^0/2$ (measured relative to the energy of $\bar{N}$ molecules in the optimal cylindrical geometry). The second cmc, X_{2c}, is conveniently defined in analogy to the first, "monomer-to-micelle," cmc, i.e., as the value of X satisfying $(dX_s/dX)_{X_{2c}} = 1/2$. It can be shown that it exists if the excess energy upon approach of the two end caps $g(N) = 2\delta - \mu^0(N)$ has an energetic barrier so that $L^3 \ll \int_{\bar{N}}^{L} dN\, N^2 \exp[g(N)/k_B T]$. In this case the second cmc can be calculated[56] according to

$$X_{2c} = \int_{\bar{N}}^{L} dN\, N e^{-\mu^0(N)/k_B T} \tag{2.26}$$

The arrow in Figure 2.10 marks the position of the second cmc according to this definition. The steep onset of micellar growth at this point is evidenced by the leveling-off in the small micelle population and the linear growth of elongated micelles.

2.6 INTERMICELLAR JUNCTIONS

As briefly discussed in Section 2.2.2, flexible cylindrical micelles can eliminate their end cap energies by forming closed loops, a phenomenon which has been observed experimentally in dilute solutions of micelles involving large end cap energies.[22] At somewhat larger amphiphile concentrations, the formation of junctions (that is, branched micelles) provides another way to get rid of energetically unfavorable end caps. Moreover, for an entangled network of wormlike micelles, junctions provide a means to relax stress and thus to reduce viscosity.[63–65]

The formation of a (threefold) intermicellar junction may be viewed as a *fusion reaction* between the end cap of one cylindrical micelle with the main body of another, forming a Y-like joint, as illustrated in Figure 2.11. Unless they appear already at very low concentrations (thus also leading to network formation) junction formation is energetically unfavorable. Their appearance at higher concentrations may be favored on entropic grounds, due to intermicellar (primarily excluded volume) interactions, or enhanced by flow fields.[66] Clearly then, the relative magnitudes of the junction formation energy (ΔF) and the end cap excess energy (δ) are crucially important for the structural and dynamic behavior of a micellar solution.

Spherical end caps involve a large positive curvature, whereas Y-junctions contain a central part with bilayer-like packing geometry. In Figure 2.11 we show the results of molecular-level calculations of the junction formation energy, $\Delta F = F_Y - F_{cyl}$, i.e., the integrated packing energies of the molecules constituting the junction, relative to the energy of the same number of molecules when packed in a cylindrical micelle. Also shown in this figure is the free energy change $\Delta \tilde{F} = \Delta F - \delta$ in the fusion reaction of an end cap with a cylindrical micellar body. In this calculations, as in the two previous sections, the local molecular packing free energy is treated as a sum of a surface energy obeying the OFM and a quadratic chain stretching energy (see e.g., Equation 2.23). As shown in the inset, the junction is treated as composed of three cylinders, joining smoothly into a

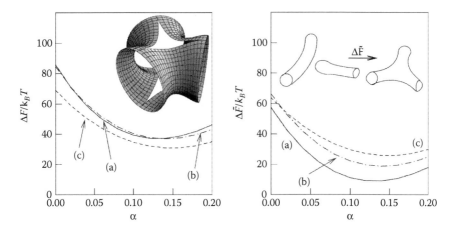

FIGURE 2.11 Left diagram: The energy ΔF of forming a Y-junction from an unperturbed cylinder of optimal thickness b. The three different curves correspond to different headgroup repulsion parameters: $B = 270 k_B T \text{Å}^2$ (a), $B = 330 k_B T \text{Å}^2$ (b), and $B = 450 k_B T \text{Å}^2$ (c). They are shown as function of the pinching factor α which specifies the decrease in thickness of the junction's middle part. The inset shows the structural model for the optimal Y-junction, corresponding to $B = 270 k_B T \text{Å}^2$ and $\alpha = 0.13$. Right diagram: The energy $\Delta \tilde{F}$ of forming a Y-junction via a fusion reaction between a cylindrical micelle and an end cap. The three curves correspond to the same B values as in the left diagram. The fusion reaction is schematically depicted in the right diagram. Adapted from May et al.[35] with permission.

saddle-like structure. The two structural variables of the junction are the radius r (in the junction plane) of the three semitoroidal sections connecting the cylinders, and the "pinching factor," α, measuring the thickness of the midpoint, $2b(1-2\alpha)$, relative to the diameter of the cylinder, $2b$. Both r and α are treated as variational parameters. Again we consider C-14 chains, but with varying optimal headgroup areas, as dictated by the interaction parameter B introduced in Section 2.3.

From Figure 2.4 we know that a rather wide range of headgroup repulsion parameters $B_{min} < B < B_{max}$ favor the formation of cylindrical aggregates. Values close to the transition to spherical micelles ($B \lesssim B_{max}$) imply small end cap energy and large Y-junction energy. On the other hand, proximity to the transition into a lamellar structure ($B_{min} \lesssim B$) suggests that the fusion reaction illustrated in the right diagram of Figure 2.11 could be energetically favorable. Consistent with that notion, several (mixed) bilayer systems that underwent a transition toward a micellar phase were recently found to disintegrate into a networklike structure of branched cylindrical micelles with large junction density and no apparent end caps.[67,68] Y-junctions have also been observed using cryo-TEM for several other surfactants, including alkylamine oxide and alkylethoxylate sulfate mixtures at high pH or salt conditions,[69] the nonionic surfactant $C_{12}EO_5$,[51] and a trimeric surfactant.[70] Moreover, imaging evidence exists for the formation of branched micelles and network-like structures for diblock copolymer blends (see Chapter 14, Section 14.5).[55,71]

Here, the individual micelles were found unable to exchange monomers but could relax into a state of local equilibrium subject to their quenched topological structure. Due to their larger size (as compared to surfactant micelles) cryo-TEM images of the copolymer micelles revealed a remarkable number of subtle structural details, including enlarged end caps that induce damped elastic undulations of the cylindrical main body, even more pronounced than those displayed in Figure 2.8. Moreover, short micellar branches exhibited a quantized length distribution, corresponding to discrete multiples of the undulation wavelength. This observation is consistent with our previous remark that when $\tau = 0$ micelles of integer multiples of ξ_2 (see Equation 2.24) are characterized by the same area-to-volume ratio.

Micellar phases that form randomly connected networks are able to exhibit a (first-order) transition into two coexisting phases of different surfactant concentration. Such a transition was already noted by Appell and Porte[72] for cetylpyridinium chlorate in high salt conditions and suggested to arise from network formation due to micellar branching.[20,73] Early theoretical and computer simulation studies supported this hypothesis.[74] Branching was also observed in cryo-TEM studies[51] for the nonionic micelles of $C_{12}EO_5$ (and for mixtures of $C_{12}EO_5$ with a nonionic phospholipid[75]). These studies reveal coexistence between a concentrated and a dilute network of interconnected cylindrical micelles. The generality of this transition for self-assembling networks that allow the formation of n-fold junctions ($n = 1,2,3,...$) was recently recognized by Tlusty and Safran.[76] Indeed, similar transitions (involving the coexistence of two networks of different density, of one network with a phase of disjointed chains, or of two disjointed chain phases with different degrees of branching[77]) are also found for microemulsions[51] and might be present in dipolar fluids.[78,79] The phase transition was suggested to be purely entropic;[74,77] its driving force is the large entropy of the junction-dominated phase. A complementary view, namely, the condensation of chains driven by a favorable energy of junction formation, was recently advanced and supported through Monte Carlo simulations by Kindt.[80]

The packing of the amphiphiles in a Y-junction is, on average, intermediate between that in a bilayer and a cylindrical micelle. This is because the junction may be, approximately, regarded as consisting of a small central bilayer-like region surrounded by three semi-toroidal sections. The latter can be viewed as three highly bent semi-cylindrical micelles, implying a cylinder-like packing geometry in these regions (the bending introduces saddle curvatures which further reduce the average curvature in these parts). Consistent with this notion, junctions predominantly appear for systems where amphiphile packing energies in cylinders and planar bilayers are not too different. In analogy to the micellar end cap (see Figure 2.8), there is a mismatch between the preferred thicknesses of the bilayer-like and the bent semi-cylindrical regions. As indicated in the right diagram of Figure 2.4, amphiphiles subject to cylindrical packing conditions tend to adopt a larger aggregate thickness $2b$ compared to the bilayer-like packing geometry. Hence, one expects the central bilayer part in a Y-junction to be pinched downward. Returning to Figure 2.11 we note that the optimal geometry of the Y-junction is indeed obtained for a positive pinching factor α, implying thinning of the middle part of the junction relative to the diameters of the joining cylinders.[35] The structures of the Y-junctions obtained by these variational calculations reveal a weak dependence of the junction energy on the headgroup repulsion parameter B. Furthermore, in all cases the minimum was adopted for a pinching parameter $\alpha \approx 0.13$ (see Figure 2.11). Notably, in all these calculations the fusion reaction appears to involve a small, yet positive, energetic penalty for the formation of a Y-junction, e.g., $\Delta \tilde{F} \approx 10 \, k_B T$ for $B = 270 \, k_B T \text{Å}^2$. Note, however, that even though junction formation may be energetically unfavorable, excluded volume interactions and network entropy considerations may drive their appearance at high concentrations or as transient, metastable, structures. It will be interesting to test other models (such as the continuum model[56] used to derive the end cap structure in Figure 2.8) with respect to predicting negative $\Delta \tilde{F}$.

2.7 ELECTROSTATIC PROPERTIES OF CYLINDRICAL MICELLES

Numerous surfactants carry ionic headgroups, such as sulphate and ammonium salts, carboxylic acids, and a multitude of others ionizable groups. Electrostatic repulsion between their charged headgroups tends to increase the optimal area per molecule, a_0, at the hydrocarbon-water interface of an amphiphilic aggregate. Consequently, single-tail ionic surfactants are typically characterized by a small packing parameter $p = v/(a_0 b_{max}) < 1/3$, explaining their pronounced preference to form small globular micelles. Nevertheless, many ionic surfactants are also involved in the formation of cylindrical, even wormlike, micelles. The growth of spherical into cylindrical micelles, indicating a packing parameter in the range $1/3 < p < 1/2$, may be due to a number of reasons, e.g., a large tail volume v, as in the case of gemini surfactants, or the use of strongly binding counterions.[13,81] Effective "charge dilution" by mixing anionic and cationic surfactants,[82] or ionic and zwitterionic (e.g., bile salt and lecithin) amphiphiles can also lead to the appearance of wormlike micelles. And, of course, growth of ionic micelles can also be enhanced by the classical method for screening electrostatic interactions, namely, by adding salt.

To account for the influence of electrostatic interactions on micellar growth we should include their contribution, f_{el}, in the overall free energy per amphiphile in a micelle, f. For simplicity, let us assume, as in the simple OFM, that f_c is constant, and suppose further that nonelectrostatic repulsions between headgroups (e.g., those due to excluded volume and hydration forces) are adequately described by the term B/a, so that $f = \gamma a + B/a + f_{el}$ (see Equation 2.13). Based on the linearized Poisson–Boltzmann theory, the electrostatic (charging) free energy per headgroup is given by $f_{el} = 2\pi k_B T l_B l_D / a$, and thus depends on the Bjerrum length, l_B, and the Debye screening length, l_D. (This result is valid for charges on a planar interface, such as that of a lipid bilayer. We ignore here the usually weak dependence of f_{el} on surface curvature.) The Bjerrum length,

$l_B = e^2/(4\pi\varepsilon_0\varepsilon_w k_B T)$, measures the distance over which the interaction between two elementary charges e equals the thermal energy $k_B T$ (note that ε_0 is the permittivity of free space). In an aqueous solution of dielectric constant $\varepsilon_w = 80$ its value is $l_B \approx 7$ Å. The Debye length $l_D = (8\pi l_B n_0)^{-1/2}$ is the distance beyond which electrostatic interactions (here in a symmetric 1:1 electrolyte present with bulk concentration n_0) are effectively screened. Using $f = \gamma a + B/a + 2\pi k_B T l_B l_D/a$, we now find

$$a_0 = [(B + 2\pi l_B l_D k_B T)/\gamma]^{(1/2)} \qquad (2.27)$$

As expected, the preferred area per headgroup is enhanced by electrostatic interactions, and depends strongly on the salt concentration in solution. For example, reducing n_0 from 1 M to 1 mM (corresponding to an increase of the Debye length from $l_D = 3$ Å to $l_D = 100$ Å) implies a corresponding change of the preferred headgroup area from $a_0 = 50$Å^2 to $a_0 = 200$Å^2. This crude estimate is certainly oversimplified, yet it clearly demonstrates the major role played by electrostatic interactions in determining the behavior of ionic surfactant solutions.

The long-range nature of the Coulomb potential plays a direct role in determining various structural characteristics (e.g., the persistence length) of wormlike micelles, as well as rheological and thermodynamic properties which are influenced by intermicellar interactions. Many of these properties have been studied in detail, both experimentally and theoretically. A number of authors have included electrostatic and nonelectrostatic interactions in self-consistent theoretical treatments of micelle formation and growth.[83–85] Below, we briefly outline some qualitative theoretical aspects concerning the influence of ionic interactions on the growth of cylindrical micelles.

The spatial distribution of counterions around the surface of a self-assembled macroion depends on its geometry. The counterions released from ionizable groups spread on planar surfaces, such as ionic lipid bilayers, are always largely immobilized. Even in the dilute limit they remain close to the surface, thus forming a diffuse layer of finite thickness. On the other hand, spherical macroions do not immobilize their counterions in this limit. Cylindrical macroions exhibit an intermediate behavior: depending on their line charge density, a certain fraction of the counterions may remain close to their surface. Quite generally then, *counterion condensation*[86,87] on charged surfaces is a geometry-dependent phenomenon.[88] Of greatest relevance to the behavior of ionic wormlike micelles is, of course, the distribution of counterions around charged cylindrical macroions.

For a (long) cylinder there is a critical line charge density λ (the number of charges per unit length along the cylinder) above which counterion condensation sets in. Specifically, if $\xi_M = \lambda l_B$ (known as the Manning parameter) exceeds unity ($\xi_M > 1$), a fraction of counterions $1 - 1/\xi_M$ stays in close proximity (i.e., closer than a distance R_M away from the rod) and cannot be diluted away. The effective average line charge density along the cylinder axis is thus $\hat{\lambda} = e/l_B$. The concept of counterion condensation remains valid for salt-containing solutions as long as R_M is smaller then the Debye length.[89] The implications of counterion condensation have been extensively studied for DNA, in which case $\xi_M = 4.2$ implies that a large fraction of the counterions are in fact condensed near the DNA rod. Cylindrical micelles are often more strongly charged and should thus exhibit substantial condensation. Consider for example a cylindrical micelle of radius $b = 16\,\text{Å}$ composed of amphiphiles whose area per head group is $a = 50\text{Å}^2$. If all these headgroups were ionized, the line charge density along the micelle axis would be $\lambda = 2\pi be/a$, implying $\xi_M = 2\pi bl_B/a = 14$, which is considerably larger than that of DNA.

The notion of counterion condensation appears also in the work of Safran et al.[90] and MacKintosh et al.,[91] who studied the influence of electrostatic interactions on the growth characteristics of wormlike micelles. In their model there are two populations of counterions: those that condense on the micelle surface thus reducing the "bare" line charge density along the micellar axis to $\hat{\lambda} = e/l_B$, the rest of the counterions are free and uniformly spread in solution and thus provide a constant contribution to the electrostatic free energy of the system, independent of the micellar size distribution. On the other hand, the size distribution is strongly affected by the Coulomb forces between the (effective) bare charges along the micelle backbone. The electrostatic interaction energy between two neighboring charges is small, just $1k_BT$, but owing to the long-range nature of the Coulomb interaction, the electrostatic free energy of a micelle is much larger due to significant contributions from distant charges along the micelle axis.

The length of a cylindrical micelle of radius b containing N molecules of area a is $L = (a/2\pi b)N$, and the number of effective bare charges is $L/l_B = \hat{\lambda}L/e$. Integrating over all pairwise Coulomb potentials in this micelle it can be shown that $\mu_{el}^0(N)$, the electrostatic contribution to the packing free energy of the micelle, is given by

$$\mu_{el}^0(N) = k_B Tc[N\ln(cN) - N] \qquad (2.28)$$

where $c = a/(2\pi bl_B)$ is a dimensionless constant. (The factor k_BT replaces here $e^2/4\pi\varepsilon_0\varepsilon_W l_B$ the electrostatic interaction energy between two neighboring charges.) The positive nonlinear electrostatic term $\sim N\ln N$ acts against $N\mu_{cyl}^0$, the (negative) linear contribution to $\mu^0(N)$ which arises from molecular packing in the cylindrical body and encourages micellar growth; as discussed in detail in Section 2.2.2. Adding the electrostatic contribution above, we now have $\mu^0(N) = N\mu_{cyl}^0 + k_B Tc[N\ln(cN) - N] + \delta$. Using this expression in the micellar growth formalism of Section 2.2, it can be shown that the two competing tendencies imply that in dilute solutions of ionic micelles most micelles are of roughly the same length L^*, in marked contrast to the polydispersed size distribution of neutral micelles. More explicitly, the size distribution is sharply peaked around a specific $N^* = (2\pi b/a)L^*$, which grows only weakly with X, according to the solution of the equation $N^* = [2\delta + \ln(X/N^*)]/ck_BT$.

This behavior prevails at low concentrations, as long as the micelles do not yet "overlap" in space. A measure for the overlap concentration is provided by the "mesh-size" of a solution of (long, randomly oriented) rodlike particles, which is $\tilde{L} \sim b/\sqrt{\phi}$ for rods of radius b, where $\phi = X(v/v_w)$ is the volume fraction of micelles in solution. Once L^* reaches $\tilde{L}$, the system enters the semi-dilute regime. Now, any given micelle charge sees, on average, an electrically neutral environment at distances $L > \tilde{L} = b/\sqrt{\phi}$, so that electrostatic interactions (along the micelle axis) beyond this distance are effectively screened. The functional form of Equation 2.28 for $\mu_{el}^0(N)$ is still valid, except that N should be replaced by $\tilde{N} = (2\pi b/a)\tilde{L} = (2\pi b^2/a)/\sqrt{\phi}$. Thus, now $\mu_{el}^0(N) = k_B T(l_B b/a)\sqrt{\phi}[\ln(l_B b/a)\sqrt{\phi})$ $-1]$ is independent of N and affects the size distribution only through the total amphiphile concentration $X = (v_w/v)\phi$. The electrostatic contribution now simply rescales the growth parameter δ to a lower value,

$$\delta_{eff} = \delta - k_B T \frac{l_B b \lambda^2}{\sqrt{X}} \tag{2.29}$$

resulting in lower average micellar size, $\langle N \rangle = 2\sqrt{X} \exp(\delta_{eff}/k_B T)$, as compared to that of neutral micelles.

MacKintosh et al.[91] have also analyzed the influence of added salt on the micellar growth characteristics. They predict that addition of salt (with intermediate or large concentration $n_0 \gg 1/(8\pi l_B L^2)$) is equivalent to increasing the total amphiphile mole fraction by $8\pi l_B b^2 n_0$, thus enhancing micellar growth. Qualitatively, this effect represents more efficient screening of the charges at the cylindrical body, as compared to the spherical end caps. Additional experiments are needed in order to test the various theoretical predictions outlined above. So far direct experimental information is rather limited. Under salt-free conditions it was found, for instance, that cationic gemini surfactants exhibit a multimodal population of aggregates.[12]

Another property that is affected by electrostatic interactions is the persistence length $l_p = l_p^0 + l_p^{el}$ of a wormlike micelle. The theoretical basis for assessing their influence has been provided by Odijk[92] and Skolnick and Fixman[93] (OSF). Based on the linearized, Debye Hückel, limit of Poisson–Boltzmann theory, these authors derived an explicit expression $l_p^{el} = l_B l_D^2/(4l^2)$ for the electrostatic contribution to l_p of an intrinsically rigid polymer (l_p^0 is the nonelectrostatic contribution to the persistence length). Nonlinear Poisson-Boltzmann theory was used by Le Bret[94] and Fixman[95] who considered charged cylinders with low inner dielectric constant and initially uniform surface charge density. The result for l_p^{el} turned out to depend on the relaxation parameter η (see also Equation 2.19) which describes how the charge density changes upon bending. Reasonable agreement with OSF was found for $\eta = b/3$. Generally, experimental determinations of the persistence length of charged wormlike micelles are also in agreement with the OSF predictions.[36,46,47] Yet, on the more theoretical level, there is no consensus regarding the correct scaling of l_p^{el} with the Debye length l_D

for flexible polyelectrolytes, as recently discussed by Dobrynin and Rubinstein.[15] At this point, computer simulations provide powerful tools to test the relation $l_p^{el}(l_D)$,[48,96] as well as to account for intermicellar interactions and the polydisperse character of solutions of reversibly assembling wormlike micelles.[46] What may also be important is the indirect influence of electrostatic interactions on l_p^0. This influence is expressed in a number of counterion-specific effects that have been observed experimentally.[13,81] To illustrate a possible mechanism recall that Equation 2.22 predicts the nonelectrostatic contribution to the persistence length l_p of a wormlike micelle to scale with the aggregate thickness $2b$ like $l_p^0 \sim b^3$. Changes due to electrostatic interactions of the cross-sectional area per amphiphile in a cylindrical micelle $a_0 \sim b^{-1}$ will thus be reflected in l_p^0.

ACKNOWLEDGMENTS

SM thanks the support from ND EPSCoR through NSF grant #EPS-0132289. ABS thanks the support of the Israel Science Foundation (ISF grant #227/02) and US–Israel Binational Science Foundation (BSF grant #2002-75), and the Archie and Marjorie Sherman Chair. The Fritz Haber Center is supported by the Minerva Foundation, Munich, Germany.

REFERENCES

1. Evans, D. F., Wennerström, H. *The Colloidal Domain, Where Physics, Chemistry, and Biology Meet*, 2nd ed.; VCH publishers, 1994.
2. Gelbart, W. M., Ben-Shaul, A. *J. Chem. Phys.* 1996, *100*, 13169–13189.
3. Safran, S. A. *Statistical Thermodynamics of Surfaces, Interfaces, and Membranes*; Addison-Wesley: New York, 1994.
4. Gelbart, W. M., Ben-Shaul, A., Roux, D., Eds.; *Micelles, Membranes, Microemulsions and Monolayers*; Springer: Berlin, 1994.
5. Cates, M. E., Candau, S. J. *J. Phys. Cond. Matter.* 1990, *2*, 6869–6882.
6. Israelachvili, J. N. *Intermolecular and Surface Forces*, 2nd ed.; Academic Press, 1992.
7. Israelachvili, J. N., Mitchell, J., Ninham, B. W. *J. Chem. Soc. Farad. 2* 1976, *72*, 1525–1568.
8. Ben-Shaul, A., Gelbart, W. M. In *Micelles, Membranes, Microemulsions, and Monolayers*, 1st ed.; Gelbart, W. M., Ben-Shaul, A., Roux, D., Eds.; Springer: New York, 1994, pp. 1–104.
9. Bombelli, F. B., Berti, D., Pini, F., Keiderling, U., Baglioni, P. *J. Phys. Chem. B* 2004, *108*, 16427–16434.
10. LaRue, I., Adam, M., Silva, M. D., Sheiko, S. S., Rubinstein, M. *Macromolecules* 2004, *37*, 5002–5005.
11. Oda, R., Lequeux, F., Mendes, E. *J. Phys. II France* 1996, *6*, 1429–1439.
12. Weber, V., Narayanan, T., Mendes, E., Schosseler, F. *Langmuir* 2003, *19*, 992–1000.
13. Magid, L. J. *J. Phys. Chem. B* 1998, *102*, 4064–4074.

14. Dobrynin, A. V. *Macromolecules* 2005, *38*, 9304–9314.
15. Dobrynin, A. V., Rubinstein, M. *Prog. Polym. Sci.* 2005, *30*, 1049–1118.
16. Ben-Shaul, A. In *Structure and Dynamics of Membranes*, Vol. 1; Lipowsky, R., Sackmann, E., Eds.; Elsevier: Amsterdam, 1995, pp. 359–402.
17. Paula, S., Süs, W., Tuchtenhagen, J., Blume, A. *J. Phys. Chem.* 1995, *99*, 11742–11751.
18. Tanford, C. *The Hydrophobic Effect*, 2nd ed.; Wiley-Interscience: New York, 1980.
19. Israelachvili, J. N., Mitchell, J., Ninham, B. W. *Biophys. Biochim. Acta.* 1977, *470*, 185–201.
20. Porte, G. In *Micelles, Membranes, Microemulsions, and Monolayers*, 1st ed.; Gelbart, W. M., Ben-Shaul, A., Roux, D., Eds.; Springer: New York, 1994, pp. 105–151.
21. Van Der Schoot, P., Wittmer, J. P. *Macrom. Theory Sim.* 1999, *8*, 428–432.
22. In, M., Aguerre-Chariol, O., Zana, R. *J. Phys. Chem. B* 1999, *103*, 7747–7750.
23. Bernheim-Groswasser, A., Zana, R., Talmon, Y. *J. Phys. Chem. B* 2000, *104*, 4005–4009.
24. Zhu, J., Liao, Y., Jiang, W. *Langmuir* 2004, *20*, 3809–3812.
25. Harries, D., Ben-Shaul, A. *J. Chem. Phys.* 1997, *106*, 1609–1619.
26. Benshaul, A., Szleifer, I. *J. Chem. Phys.* 1985, *83*, 3597–3611.
27. Fattal, D. R., Ben-Shaul, A. *Biophys. J.* 1993, *65*, 1795–1809.
28. May, S., Ben-Shaul, A. *Phys. Chem. Chem. Phys.* 2000, *2*, 4494–4502.
29. Flory, P. J. *Statistical Mechanics of Chain Molecules;* Wiley-Interscience: New York, 1969.
30. Grosberg, A., Khokhlov, A. *Statistical Physics of Macromolecules;* AIP Press: New York, 1994.
31. Rubinstein, M., Colby, R. *Polymer Physics;* Oxford University Press: New York, 2003.
32. Boal, D. *Mechanics of the Cell;* Cambridge University Press, 2001.
33. Lipowsky, R., Sackmann, E., Eds.; *Structure and Dynamics of Membranes*; Elsevier: Amsterdam, 1995.
34. Helfrich, W. *Z. Naturforsch.* 1973, *28*, 693–703.
35. May, S., Bohbot, Y., Ben-Shaul, A. *J. Phys. Chem. B* 1997, *101*, 8648–8657.
36. Magid, L. J., Li, Z., Butler, P. D. *Langmuir* 2000, *16*, 10028–10036.
37. May, S. Unpublished data.
38. Fournier, J. B. *Europhys. Lett.* 1998, *43*, 725–730.
39. Hamm, M., Kozlov, M. M. *Eur. Phys. J. E* 2000, *3*, 323–335.
40. Hamm, M., Kozlov, M. M. *Eur. Phys. J. B* 1998, *6*, 519–528.
41. May, S. *Eur. Biophys. J.* 2000, *29*, 17–28.
42. Lauw, Y., Leermakers, F. A. M., Stuart, M. A. C. *J. Phys. Chem. B* 2003, *107*, 10912–10918.
43. Imae, T. *Colloid Polym. Sci.* 1989, *267*, 707–713.
44. Von Berlepsch, H., Harnau, L., Reineker, P. *J. Phys. Chem. B* 1998, *102*, 7518–7522.
45. Geng, Y., Ahmed, F., Bhasin, N., Discher, D. E. *J. Phys. Chem. B* 2005, *109*, 3772–3779.
46. Jerke, G., Pedersen, J. S., Egelhaaf, S. U., Schurtenberger, P. *Langmuir* 1998, *14*, 6013–6024.
47. Schubert, B. A., Kaler, E. W., Wagner, N. J. *Langmuir* 2003, *19*, 4079–4089.

48. Cannavacciuolo, L., Pedersen, J. S., Schurtenberger, P. *Langmuir* 2002, *18*, 2922–2932.
49. Porte, G., Poggi, Y., Appell, J., Maret, G. *J. Phys. Chem.* 1984, *88*, 5713–5720.
50. Geng, Y., Romsted, L. S., Menger, F. *J. Am. Chem. Soc.* 2006, *128*, 492–501.
51. Bernheim-Groswasser, A., Wachtel, E., Talmon, Y. *Langmuir* 2000, *16*, 4131–4140.
52. Glatter, O., Fritz, G., Lindner, H., Brunner-Popela, J., Mittelbach, R., Strey, R., Egelhaaf, S. U. *Langmuir* 2000, *16*, 8692–8701.
53. González-Pérez, A., Varela, L. M., García, M., Rodríguez, J. R. *J. Colloid Int. Sci.* 2006, *293*, 213–221.
54. Zheng, Y., Won, Y., Bates, F. S., Davis, H. T., Scriven, L. E., Talmon, Y. *J. Phys. Chem. B* 1999, *103*, 10331–10334.
55. Jain, S., Bates, F. S. *Macromolecules* 2004, *37*, 1511–1523.
56. May, S., Ben-Shaul, A. *J. Phys. Chem. B* 2001, *105*, 630–640.
57. Bauer, A., Woelki, S., Kohler, H. H. *J. Phys. Chem. B* 2004, *108*, 2028–2037.
58. Kshevetskiy, M. S., Shchekin, A. K. *Colloid J.* 2005, *67*, 324–336.
59. Al-Anber, Z. A., Josep Bonet i Avalos, J., Floriano, M. A., Mackie, A. D. *J. Chem. Phys.* 2003, *118*, 3816–3826.
60. Dan, N., Pincus, P., Safran, S. A. *Langmuir* 1993, *9*, 2768–2771.
61. Nielsen, C., Goulian, M., Andersen, O. S. *Biophys. J.* 1998, *74*, 1966–1983.
62. May, S., Ben-Shaul, A. *Biophys. J.* 1999, *76*, 751–767.
63. Appell, J., Porte, G., Kathory, A., Kern, F., Candau, S. J. *J. Phys. II France* 1992, *2*, 1045–1052.
64. Khatory, A., Kern, F., Lequeux, F., Appell, J., Porte, G., Morie, N., Ott, A., Urbach, W. *Langmuir* 1993, *9*, 933–939.
65. Buhler, E., Munch, J. P., Candau, S. J. *Europhys. Lett.* 1996, *34*, 251–255.
66. Bruinsma, R., Gelbart, W. M., Ben-Shaul, A. *J. Chem. Phys.* 1992, *96*, 7710–7727.
67. Gustafsson, J., Oradd, G., Lindblom, G., Olsson, U., Almgren, M. *Langmuir* 1997, *13*, 852–860.
68. Gustafsson, J., Oradd, G., Nyden, M., Hansson, P., Almgren, M. *Langmuir* 1998, *14*, 4987–4996.
69. Lin, Z. *Langmuir* 1996, *12*, 1729–1737.
70. Danino, D., Talmon, Y., Levy, H., Beinert, G., Zana, R. *Science* 1995, *269*, 1420–1421.
71. Jain, S., Bates, F. S. *Science* 2003, *300*, 460–464.
72. Appell, J., Porte, G. *J. Phys. (France)* 1983, *44*, 689–695.
73. Gomati, R., Appell, J., Bassereau, P., Marignan, J., Porte, G. *J. Phys. Chem.* 1987, *91*, 6203–6210.
74. Bohbot, Y., Ben-Shaul, A., Granek, R., Gelbart, W. M. *J. Chem. Phys.* 1995, *103*, 8764–8782.
75. Kwon, S. Y., Kim, M. W. *Phys. Rev. Lett.* 2002, *89*, 258302/1–258302/4.
76. Tlusty, T., Safran, S. A. *Science* 2000, *290*, 1328–1331.
77. Zilman, A., Tlusty, T., Safran, S. A. *J. Phys. Cond. Mat.* 2003, *15*, S57–S64.
78. Camp, P. J., Shelley, J. C., Patey, G. N. *Phys. Rev. Lett.* 2000, *84*, 115–118.
79. Weis, J. J., Tavares, J. M., Gama, M. M. T. D. *J. Phys. Cond. Mat.* 2002, *14*, 9171–9186.
80. Kindt, J. T. *J. Phys. Chem. B* 2002, *106*, 8223–8232.
81. Magid, L. J., Han, Z., Li, Z., Butler, P. D. *J. Phys. Chem. B* 2000, *104*, 6717–6727.

82. Koehler, R. D., Raghavan, S. R., Kaler, E. W. *J. Phys. Chem. B* 2000, *104*, 11035–11044.
83. Evans, D. F., Mitchel, D. J., Ninham, B. W. *J. Phys. Chem.* 1984, *88*, 6344–6349.
84. Nagarajan, R., Ruckenstein, E. *Langmuir* 1991, *7*, 2934–2969.
85. Srinivasan, V., Blankschtein, D. *Langmuir* 2003, *19*, 9932–9945.
86. Oosawa, F. *Polyelectrolytes*, 2nd ed.; Marcel Mecker: New York, 1970.
87. Manning, G. S. *Q. Rev. Biophys.* 1978, *11*, 179–246.
88. Belloni, L. *Colloid Surf. A* 1998, *140*, 227–243.
89. Deserno, M., Holm, C., May, S. *Macromolecules* 2000, *33*, 199–205.
90. Safran, S. A., Pincus, P. A., Cates, M. E., MacKintosh, F. C. *J. Phys. France* 1990, *51*, 503–510.
91. MacKintosh, F. C., Safran, S. A., Pincus, P. A. *Europhys. Lett.* 1990, *12*, 697–702.
92. Odijk, T. *J. Polym. Sci. Part B: Polym. Phys.* 1977, *15*, 477–483.
93. Skolnick, J., Fixman, M. *Macromolecules* 1977, *10*, 944–948.
94. Le Bret, M. *J. Phys. Chem.* 1982, *76*, 6243.
95. Fixman, M. *J. Phys. Chem.* 1982, *76*, 6346–6353.
96. Everaers, R., Milchev, A., Yamakov, V. *Eur. Phys. J.* 2002, *8*, 3–13.

3 Computer Simulations of Wormlike Micelles

*Johan T. Padding, Wouter K. den Otter,
and Wim J. Briels*

CONTENTS

3.1 INTRODUCTION

Amphiphilic molecules may self-assemble in a great variety of morphological structures. A convenient parameter, which quantifies the importance of the molecular structure, is the packing parameter $p = v_s/(al)$, where v_s is the volume of the hydrophobic tail of the amphiphile, a is the area occupied by its hydrophilic head group, and l the length of its tail. In dilute solutions micellar structures tend to be spherical for values of the packing parameter around 1/3 and rodlike when p is close to 1/2.[1] With increasing surfactant concentration, rodlike micelles grow into long cylindrical structures or worms, whose lengths finally by far exceed their persistence lengths. The persistence length l_p is the length over which the worm may be considered to be a stiff rod. It is related to the bending rigidity coefficient κ by $l_p = \kappa/(k_B T)$, with k_B Boltzmann's constant and T the temperature. Already at relatively small concentration, wormlike micelles enter the semidilute regime and form an entangled network, very much like the entangled network occurring in a polymer solution. Such a network enhances the viscosity of the solution, sometimes by several orders of magnitude, and introduces elastic properties. Unlike polymers, however, the wormlike micelles can reversibly break, and the break-up rate is sensitive to temperature, flow or stress conditions, concentrations of various components, and other parameters (see Chapter 4, Sections 4.2.5 and 4.4; Chapter 8, Sections 8.2.2 and 8.4.2; and Chapter 10). As a result, wormlike micelles have a rich rheology, rendering them perfect functional fluids in many industrial applications.[2]

Rheological experiments produce a wealth of data, and theories can explain some of the experimental findings. Most macroscopic experiments, however, do not provide us with a detailed *fundamental* understanding of the underlying processes that lead to the emergent rheology. Often their interpretation is based on theories which contain uncontrolled approximations, and their range of applicability is limited to certain flow conditions. This is where simulations may contribute to our understanding of the peculiar rheology of wormlike micelles. With simulations we have the possibility to "zoom in" on the detailed processes and study their influence on the rheology. Simulations enable us to test the approximations made in theories, and contribute to a rational design of new viscoelastic materials based on wormlike micelles.

Now, simulations of wormlike micelles may be performed on many different length and time scales, from the atomistic to the mesoscopic. An overview is given in Figure 3.1. At the smallest scale we find the field of molecular dynamics. Molecular dynamics has witnessed a rapid development since the seminal hard sphere simulations of Alder and Wainwright in the 1960s.[3,4] Several factors have contributed to this rise, the most important of which is the immense performance improvement of integrated circuits, which has made computer power a cheap and abundant commodity. Versatile simulation packages and user-friendly graphical interfaces have been developed, in academia and by commercial enterprises.[5] And lastly, accurate force fields have been constructed to cater for anyone's research interests, provided they exclude chemical reactions and other phenomena of a quantum mechanical nature.[6] Nonbonded interactions between atoms are usually described by a Lennard–Jones potential, supplemented with Coulombic terms if the atoms are charged. Bonded interactions within molecules typically consist of a harmonic potential between 1-2 neighbors, a bending potential for 1-3 neighbors, and a dihedral potential for 1-4 neighbors, with an occasional improper dihedral to conserve chirality. The exact implementations of these terms, and their parameters, vary depending on the objectives adhered to in the construction of the force field. The computational demands of molecular dynamics, for which the interaction forces between every atom and its 25 to 50 nearest neighbors must be recalculated every time step, puts severe restrictions on the accessible length and time scales. A typical simulation box used for simulating surfactant systems has edges of the order of 10 nm and contains some 100,000 atoms, though occasionally even larger boxes are being used. The evaluation of a single time step then takes about a second of computer time, which is predominantly spent

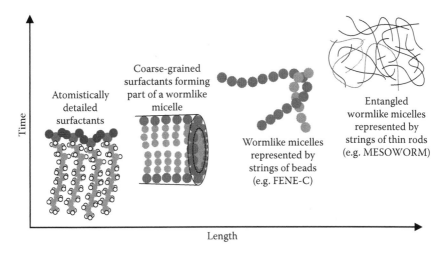

FIGURE 3.1 Simulations may be performed on many different length and time scales. With atomistic force fields (lower left) we can capture the influence of the specific surfactant chemistry on the properties of a piece of wormlike micelle. By coarse graining the surfactant molecules these calculations are greatly accelerated. More macroscopic properties, such as the chain length distribution and rheology, can be calculated with higher-level models, such as the FENE-C and MESOWORM models (upper right).

on the calculation of the nonbonded interactions between the atoms. With time steps in the order of a femtosecond, or up to 5 fs if constraints are used to eliminate the fastest intramolecular degrees of freedom, a week of computer time corresponds to about a nanosecond in real time. Of course, these numbers will vary considerably depending on the details of the system, the simulation setup, the force field (where in particular Ewald summations for long-range Coulombic interactions are demanding), the efficiency of the algorithm, and the available hardware.

The above discussion clearly illustrates that the capabilities of atomistic simulations of amphiphilic aggregates are limited, yet even within these boundaries interesting physical phenomena can be studied. In fact, by taking full advantage of the principles of statistical mechanics, it is possible to extract some macroscopic properties from these small simulation boxes. Other properties, especially those related to the rheology, require the simulation of larger time and length scales. These may be reached by treating groups of atoms, or sometimes even groups of amphiphiles, as one unit.

It is our aim to explain the concepts of atomistic and "coarse grained" simulations, including their successes and limitations, in the next sections. We will first explain the basic principles of computer simulations, and then concentrate on simulations of wormlike micelles. At this point, we would like to stress that we do not provide a complete review of all simulation activities in the field of wormlike micelles, but rather an overview of the authors' areas of interest.

3.2 PRINCIPLES OF COMPUTER SIMULATIONS

When performing numerical simulations, the aim is to calculate structural and thermodynamic properties, and possibly dynamical properties, of the system under investigation. Usually one simulates a box containing a number of particles. Potentials, and possibly frictions, are defined to describe the interactions between the particles. From the positions and velocities of the particles at a certain point in time, one can calculate the net forces on all of them. By integrating the equations of motion, this gives new positions and velocities, resulting in new forces, etc. In the following subsections we will describe the three most popular propagators, i.e., means of integrating the equations of motion: molecular dynamics, Langevin dynamics, and Brownian dynamics.

3.2.1 MOLECULAR DYNAMICS PROPAGATOR

In its simplest implementation, the molecular dynamics method generates a path on a constant energy surface, $H(\Gamma) = U$, by solving Newton's equations for N particles; Γ denotes a point in phase space $\Gamma = (\mathbf{r}^N, \mathbf{p}^N)$ with $\mathbf{r}^N$ denoting the positions $\mathbf{r}_1, ..., \mathbf{r}_N$ and $\mathbf{p}^N$ denoting the corresponding momenta $\mathbf{p}_1 = m\mathbf{v}_1, ..., \mathbf{p}_N = m\mathbf{v}_N$. According to Newton every particle feels a force exerted on it by all surrounding particles and possibly some agency external to the system under consideration, and is accelerated by this force. Neglecting externally applied forces we may write

$$m_i \frac{d^2 \mathbf{r}_i}{dt^2} = -\nabla_i \Phi, \tag{3.1}$$

$$\Phi = \sum_{i=1}^{N-1} \sum_{j=i+1}^{N} \varphi(r_{ij}), \tag{3.2}$$

where $\nabla_i = (\frac{\partial}{\partial x_i}, \frac{\partial}{\partial y_i}, \frac{\partial}{\partial z_i})$ and the potential energy Φ is assumed to be a sum of pair-terms, which only depend on the distance between the particles making up the pair (we will encounter more general expressions for the potential energy later on). Since Newton's equations are second-order equations, the paths of all particles are completely determined once the initial positions and initial velocities are given. Taking the scalar product of Equation (3.1) with $d\mathbf{r}_i/dt$ and summing over all particles yields

$$\frac{d}{dt}\left\{ \sum_i \frac{1}{2} m_i \frac{d\mathbf{r}_i}{dt} \cdot \frac{d\mathbf{r}_i}{dt} + \Phi \right\} = \frac{d}{dt}\{K + \Phi\} = \frac{dH}{dt} = 0, \tag{3.3}$$

where K is the kinetic energy and the Hamiltonian H is constant and equal to its initial value U. Note that K and Φ may change in the course of time, but that their sum is constant.

How should we solve Equation (3.1)? In order to concentrate on essential points, we simplify our notation for the time being and restrict ourselves to just one degree of freedom. Then we may write

$$\frac{dx}{dt} = v, \tag{3.4}$$

$$\frac{dv}{dt} = \frac{F}{m}, \tag{3.5}$$

where $F = -d\Phi/dx$ is the force acting on this particular coordinate. The simplest solution to this problem is

$$x(t + \Delta t) = x(t) + v(t)\Delta t, \tag{3.6}$$

$$v(t + \Delta t) = v(t) + \frac{F(t)}{m}\Delta t, \tag{3.7}$$

which must be repeated as often as needed to reach the time of interest. This algorithm is called the first order Euler algorithm. Applying it to the simple case of a harmonic oscillator reveals that this algorithm does not conserve energy; actually the energy grows exponentially with time.

An astonishingly simple solution to the above problem is to write:

$$x(t + \Delta t) = x(t) + \dot{x}(t)\Delta t + \frac{1}{2}\ddot{x}(t)(\Delta t)^2 + \frac{1}{6}\dddot{x}(t)(\Delta t)^3 + \cdots, \tag{3.8}$$

$$x(t - \Delta t) = x(t) - \dot{x}(t)\Delta t + \frac{1}{2}\ddot{x}(t)(\Delta t)^2 - \frac{1}{6}\dddot{x}(t)(\Delta t)^3 + \cdots. \tag{3.9}$$

After adding these two equations we obtain

$$x(t + \Delta t) = 2x(t) - x(t - \Delta t) + \ddot{x}(t)(\Delta t)^2 + \cdots. \tag{3.10}$$

This propagator is correct to order $(\Delta t)^4$. Instead of the position and velocity both at time t, we need the positions at time t and at time $t - \Delta t$ in order to advance the system by Δt. The price we pay for the better algorithm is that it doesn't look very nice, so let's make it look nicer. To this end we introduce new names

$$x(t) - x(t - \Delta t) = v\left(t - \frac{1}{2}\Delta t\right)\Delta t, \tag{3.11}$$

$$x(t + \Delta t) - x(t) = v\left(t + \frac{1}{2}\Delta t\right)\Delta t. \tag{3.12}$$

For the time being these equations only serve to define the new symbols $v(t - \frac{1}{2}\Delta t)$ and $v(t + \frac{1}{2}\Delta t)$. The propagator now reads

$$v\left(t + \frac{1}{2}\Delta t\right) = v\left(t - \frac{1}{2}\Delta t\right) + \frac{F(t)}{m}\Delta t. \tag{3.13}$$

$$x(t + \Delta t) = x(t) + v\left(t + \frac{1}{2}\Delta t\right)\Delta t, \tag{3.14}$$

The second line is nothing but repeating Equation (3.12), while the first line is Equation (3.10) rewritten in our new notation. Consequently the algorithm still is correct up to fourth order in Δt. It is called the Verlet leap-frog algorithm.

Obviously we now would like to know about velocities. From an analysis similar to the one just given we deduce that the velocity $\dot{x}(t)$ at time t is given by the obvious equation

$$\dot{x}(t) = \frac{1}{2}\left\{v\left(t + \frac{1}{2}\Delta t\right) + v\left(t - \frac{1}{2}\Delta t\right)\right\}, \tag{3.15}$$

correct to second order in Δt. It is important to stress again that this does not influence the propagator, which is correct to fourth order. For small enough time-steps Δt, the Verlet algorithm performs an excellent job at conserving total energy.

Extension of this method to a system with $3N$ Cartesian coordinates is trivial.

3.2.2 LANGEVIN DYNAMICS PROPAGATOR

Suppose we are studying a system described by many degrees of freedom, only part of which we are actually interested in. The prototypical example is a colloidal suspension containing solvent particles and colloidal particles. In the end we are only interested in the dynamics of the colloids. Let us simplify our system a bit by concentrating on a very dilute suspension, dilute enough that each colloid only "sees" solvent particles. The equation of motion for each colloidal particle then reads

$$m\frac{d\mathbf{v}}{dt} = -m\xi\mathbf{v} + \mathbf{f}^R. \tag{3.16}$$

In Equation (3.16), $-m\xi\mathbf{v}$ represents the systematic part of the forces exerted by the solvent molecules on the colloidal particle moving with velocity $\mathbf{v}$ (the friction force), and $\mathbf{f}^R$ the random part. For a colloid of radius a moving in a solvent of viscosity η, the friction frequency ξ is given by the well-known Stokes result $\xi = 6\pi\eta a/m$. Solving Equation (3.16) we find

$$\mathbf{v}(t) = \mathbf{v}(0)e^{-\xi t} + \frac{1}{m}\int_0^t d\tau e^{-\xi(t-\tau)}\mathbf{f}^R(\tau). \tag{3.17}$$

Let us next calculate the kinetic energy $\frac{1}{2}m\mathbf{v}(t)\cdot\mathbf{v}(t)$ at time t and average over a large ensemble of trajectories, all starting with initial velocity $\mathbf{v}(0)$. At large times t this should be equal to $\frac{3}{2}k_BT$. Since $\mathbf{f}^R(\tau)$ is random, its average is zero at all times. Then for large t, we have

$$\int_0^t d\tau\int_0^t d\tau' e^{-\xi(2t-\tau-\tau')}\langle\mathbf{f}^R(\tau)\cdot\mathbf{f}^R(\tau')\rangle = 3mk_BT. \tag{3.18}$$

Now, what can we say about the time correlation of the random forces? Since $\mathbf{f}^R$ is meant to represent fastly fluctuating forces, it is clear that $\langle\mathbf{f}^R(\tau)\cdot\mathbf{f}^R(\tau')\rangle$ is zero for $\tau - \tau'$ not very small. For convenience we will assume that the random force time correlations are proportional to $\delta(\tau-\tau')$. Since moreover the forces along different directions are uncorrelated we write $\langle f_\alpha^R(\tau)f_\beta^R(\tau')\rangle = C\delta_{\alpha\beta}\delta(\tau-\tau')$. Introducing this into Equation (3.18) and performing the necessary integrations we find, after taking the limit of t going to infinity

$$\left\langle f_\alpha^R(\tau)f_\beta^R(\tau')\right\rangle = 2mk_BT\xi\delta_{\alpha\beta}\delta(\tau-\tau'). \tag{3.19}$$

This is our first example of the fluctuation-dissipation theorem, which states that the random fluctuation characteristics are intimately linked to the friction characteristics.

Now what if the colloidal concentration becomes larger and the colloids start to push and pull on each other? Again, in order to concentrate on essential points we restrict ourselves to just one degree of freedom, in which case the equations of motion read

$$\frac{dx}{dt} = v, \tag{3.20}$$

$$m\frac{dv}{dt} = -m\xi v + f^R + F. \tag{3.21}$$

Here F is the (conservative) force due to interactions with all other colloids or possibly an external field. Now how will we numerically integrate this equation of motion? The singular nature of the delta-function occurring in the second moment of the random force introduces some difficulties. One way to circumvent this problem is to assume that the random force f^R is constant during one time step Δt, i.e., we are going to approximate the delta-function as

$$\delta(\tau - \tau') = \begin{cases} \dfrac{1}{\Delta t} & \tau' \in [\tau, \tau + \Delta t] \\ 0 & \text{otherwise.} \end{cases} \tag{3.22}$$

In order to generate a stationary Markovian Gaussian process, this algorithm is limited by the condition $\Delta t \ll \xi^{-1}$, i.e., the correlation time of the random force must be much smaller than the velocity relaxation time ξ^{-1}. If this is not a problem, we can use the Verlet algorithm, Equations (3.13)–(3.14), with f_α^R represented by a random variable with zero mean and variance $\sqrt{2mk_B T\xi / \Delta t}$, and approximating $-m\xi v(t)$ by $-m\xi v(t - \Delta t/2)$.

The purpose of coarse graining is to go to larger length and time scales. At some point the friction forces will start to dominate the conservative forces. Using the above algorithm, the time step Δt will then be severely limited by the friction ξ. The integration time step need not be restricted by ξ, however, since the influence of the friction and random force can be integrated over Δt. If we assume that the friction ξ is constant and that the systematic forces $F(t)$ remain approximately constant, this results in the following algorithm:[3]

$$x(t + \Delta t) = x(t) + \xi^{-1}(1 - e^{-\xi \Delta t})v(t)$$
$$+ \xi^{-2}(\xi \Delta t - 1 + e^{-\xi \Delta t})F(t)/m + \delta x^R, \tag{3.23}$$

$$v(t + \Delta t) = e^{-\xi \Delta t}v(t) + \xi^{-1}(1 - e^{-\xi \Delta t})F(t)/m + \delta v^R. \tag{3.24}$$

Note that Equation (3.24) directly follows from Equation (3.17). δx^R and δv^R are weighted stochastic integrals of the random force f^R. Each realisation of the stochastic process $f^R(\tau \in [t, t + \Delta t])$ will yield a different path, but all starting at the same initial coordinate and velocity (see Figure 3.2). A specific path is generated by producing a set of correlated numbers δx^R and δv^R. In a simulation they are sampled from a bivariate Gaussian distribution, with variances and correlation coefficient determined by the fluctuation-dissipation theorem, Equation (3.19). Details on this and similar algorithms can be found in textbooks.[3]

Up to this point we have assumed that the solvent causes friction and random forces on the colloids, but have neglected hydrodynamic interactions. If the

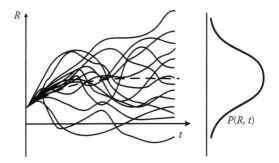

FIGURE 3.2 The ensemble of paths all starting at the same initial coordinate R and velocity $\dot{R}$. The probability distribution for R after a time t is $P(R,t)$. The dashed line is the average path of the ensemble.

hydrodynamic interactions are instantaneous, the Langevin equation in this more general case is given by[7–9]

$$m\frac{d\mathbf{v}_i}{dt} = -\sum_j \boldsymbol{\xi}_{ij}(\mathbf{r}^N)\cdot\mathbf{v}_j(t) + \mathbf{f}_i^R(t) + \mathbf{F}_i(t), \qquad (3.25)$$

where $\boldsymbol{\xi}_{ij}(\mathbf{r}^N)$ is a friction tensor which in principle depends on all coordinates $\mathbf{r}^N$. The fluctuation-dissipation relation now reads

$$\left\langle \mathbf{F}_i^R(\tau)\mathbf{F}_j^R(0) \right\rangle = 2mk_BT\boldsymbol{\xi}_{ij}\delta(\tau). \qquad (3.26)$$

As a simplification we may assume that $\boldsymbol{\xi}_{ij}$ only depends on the distance r_{ij} between particles i and j. Doing so, we arrive at the so-called dissipative particle dynamics (DPD) method.[10] Although costly, this method is rather popular since it conserves momentum, which is a prerequisite for hydrodynamic behavior, and it is easy to use in nonequilibrium simulations.

3.2.3 BROWNIAN DYNAMICS PROPAGATOR

Equation (3.23) is exact as long as the friction coefficients, and the systematic forces $F(t)$ can be assumed to be constant. Often the friction ξ is relatively high, in which case the position x hardly changes over a time interval ξ^{-1} in which the velocity v has lost its memory. The second moment of the random displacement δx^R is given by

$$\langle(\delta x^R)^2\rangle = \frac{k_BT}{m\xi}\Delta t\left[2 - \frac{1}{\xi\Delta t}\left(3 - 4e^{-\xi\Delta t} + e^{-2\xi\Delta t}\right)\right]. \qquad (3.27)$$

When $\xi\Delta t \gg 1$ this random term dominates the $\xi^{-1}v$ term in Equation (3.23), the latter being of order $\xi^{-1}(k_BT/m)^{1/2}$. In this limit we may therefore drop the velocity variable altogether and focus entirely on the coordinate x. Since moreover all terms $\exp(-\xi\Delta t)$ tend to zero, we find the well-known Brownian dynamics propagator:

$$x(t+\Delta t) = x(t) + \frac{D}{k_BT}F(t)\Delta t + \frac{\partial D}{\partial x}\Delta t + \delta x^R, \qquad (3.28)$$

$$\langle(\delta x^R)^2\rangle = 2D\Delta t, \qquad (3.29)$$

where $D = k_BT/(m\xi)$ is the diffusivity. In Equation (3.28) we have included the derivative $\partial D/\partial x$, which emerges in the more general case when the friction ξ is not a constant. Generalisation to $3N$ coordinates will lead to more complicated expressions when hydrodynamic interactions[7,8] or other forms of non-isotropic friction[11] are included. For more details the reader is referred to Refs. 3, 7, and 8.

3.2.4 BULK SYSTEMS

Bulk systems consist of about 10^{20} particles or more, while simulation boxes are limited to about 10^6 particles, but usually consist of tens of thousands of particles. In such small boxes most particles will be close to the system boundary and therefore not behave as a bulk particle at all. This problem is alleviated by the use of periodic boundary conditions.

First the simulation box is surrounded in all directions by copies of itself. Notice that we only have to keep track of the positions of the particles in the central box and that we then automatically know the positions of the particles in the other boxes. Next we do two things. (1) We remove the confining boundaries. Thus, each particle which leaves the central box automatically enters this box through the wall opposite the one through which it left the box. (2) We let each particle interact with all other particles within a certain cut-off radius, possibly including particles in one of the surrounding boxes. The cut-off radius is limited by the fact that we do not want a particle to interact with one of its own images. An even stronger restriction is that we do not want a particle to interact with a second particle and at the same time with some copy of this second particle. In this case the cut-off radius is at most $L/2$ with L the length of an edge of a cubic box.

3.2.5 SIMULATING DIFFERENT ENSEMBLES

As stressed in Section 3.2.1, the Verlet algorithm conserves energy and thereby samples the microcanonical (N,V,U) ensemble. The latter actually requires that the system is ergodic, meaning that a path, if long enough, visits all parts of the energy surface with equal probability. This we will assume to be the case. Then, what is the temperature? Strictly speaking $1/T = (\partial S/\partial U)_{NV}$. Although there is a way to proceed along these lines, it is rather complicated and we will not do so. We know that if the system is large enough, all ensembles are equivalent. Now, our systems usually are not very large, but the application of the periodic boundary conditions certainly makes them infinitely large in a special sense. Actually we perform a limited sampling, in which only checkerboard configurations are sampled, of an infinitely large system. So let us assume that our system, consisting of N particles in volume V and having energy U, may be considered to be a system consisting of N particles in volume V at temperature T, which we just don't know. By applying the canonical ensemble probability distribution, we may then calculate the average kinetic energy $\langle K \rangle$, obtaining

$$\langle K \rangle = (3N - 3)\frac{1}{2}k_B T. \tag{3.30}$$

Here $(3N - 3)$ is the number of degrees of freedom, 3 for each particle minus 3 because the center of mass of the box is fixed. In our (N,V,U) simulation the kinetic energy K is a fluctuating quantity. For large enough systems its average equals $(3N - 3)\frac{1}{2}k_B T$.

Now suppose we want to run a canonical (N,V,T) ensemble. We may use the kinetic temperature $\hat{T} \equiv 2K/[(3N - 3)k_B]$ as an estimate of T. We know that its average must be T and we even know its fluctuations. A rather ad hoc procedure would be to draw velocities from Maxwell's distribution for the initial velocities and subsequently to rescale all velocities at every time step by a factor

$$\chi = \sqrt{1 + \frac{\Delta t}{\tau}\left(\frac{T}{\hat{T}} - 1\right)}. \tag{3.31}$$

This will change the kinetic temperature $\hat{T}$ every time step according to

$$\hat{T} \longrightarrow \hat{T}\left[1 + \frac{\Delta t}{\tau}\left(\frac{T}{\hat{T}} - 1\right)\right]. \tag{3.32}$$

If the characteristic time τ is chosen to be equal to Δt this algorithm forces $\hat{T}$ to be equal to T at all times and we lose the natural fluctuations of the kinetic energy. Therefore τ is chosen by trial and error to be such that the fluctuations of the kinetic energy are about right. The procedure just described is usually referred to as applying the Berendsen thermostat. An alternative is the so-called Nosé–Hoover thermostat, which steers the temperature toward the required value in a much gentler way and correctly generates states in the canonical ensemble.[3,4,12]

Another alternative is the Langevin thermostat. Here it is envisioned that each particle, besides feeling forces due to the presence of the other particles, also is exposed to forces resulting from a heat bath to which it is coupled. The heat bath takes away momentum from the particle by means of a friction force, and adds new energy by randomly kicking it. The equations of motion are exactly as in Equation (3.21), but usually with a much smaller friction ξ in order not to perturb the dynamics too much. Given the friction ξ, the magnitude of the random forces should be chosen in accordance with the fluctuation-dissipation theorem to ensure that in the stationary state $\langle \hat{T} \rangle = T$.

Other ensembles may be simulated as well. For example, to control the isotropic pressure, the system box is allowed to change its volume (at fixed shape) during the simulation. For anisotropic fluids, in particular when a piece of worm-like micelle is simulated, it may even be desirable to control each component of the pressure tensor independently. This involves changing the shape of the simulation box. Details can be found in textbooks.[3,4]

3.3 WORMLIKE MICELLES IN ATOMISTIC DETAIL

As discussed in the introduction, the length scales attainable in atomistically detailed molecular dynamics simulations are limited, yet spherical micelles are so small that they fit in their entirety inside a periodic simulation box.[13,14] These simulations provide information on the overall structure of a micelle, the fluctuations of its shape, and on the conformations and dynamics of the constituent amphiphiles. Combining simulations with molecular-thermodynamic theory, it has become possible to predict thermodynamic properties like the free energy change associated with micellization, the critical micelle concentration (cmc), and the optimal size and composition of micelles formed, for both pure and mixed surfactant solutions.[15,16] Several authors have simulated the spontaneous self-assembly of micelles from a homogeneous solution, at concentrations far above the cmc, to extract rate laws for the growth process.[17,18]

But by far the most popular topic in simulations of surfactants are pre-assembled bilayers composed of lipids, which serve as models for biological membranes.[19,20] The important and versatile roles played by membranes in living nature has sprouted a series of research directions, including the mechanical properties of membranes, their permeability to water and several other small molecules, pore formation under tension, interactions with embedded proteins, and the influence of sterols.

Only a few atomistic simulations have reported on wormlike micelles. In an early massively parallel calculation, Maillet et al.[17] simulated four different pre-assembled cylindrical micelles of about 50 amphiphiles, with periodic boundary conditions set to make the worm effectively infinitely long, but none showed signs of a long-lived stable state. Marrink et al.[18] observed the self-assembly of a cylindrical micelle spanning across the entire width of a periodic simulation box. But as already emphasized by these authors, "the use of periodic boundary conditions means it is possible to form periodic aggregates as a way of avoiding the creation of an additional interface."[18] This situation is not likely to occur in large simulation boxes, but it is easily encountered for small boxes and high concentrations. In order for a wormlike micelle to be thermodynamically stable, it should retain its shape in the absence of external forces. For a finite length worm, terminated by caps at either end, this is easily tested by letting it freely relax to a tensionless state. An aggregate spanning the box is often under tension due to the length restraint imposed by the parallel dimension $L_\parallel$ of the periodic box. This tension is readily calculated from the difference between the pressures parallel ($p_\parallel$) and perpendicular ($p_\perp$) to the micelle, $\tau = -(p_\parallel - p_\perp)A_\perp$, with $A_\perp$ the area of tne box perpendicular to the worm. When using a barostat to equalize these pressures, by independently rescaling the parallel and perpendicular box dimensions, a simulation will automatically converge to the stable tension-less state of the model (provided such state exists). One may also calculate the tension as a function of the arc length L of the worm, which in good approximation is equal to $L_\parallel$ for short worms. For an elastic worm one then expects an elastic free energy $F_{elas} = (K_L/2L_0)(L_\parallel - L_0)^2$, hence the zero-tension intercept of the linear tension curve $\tau = (K_L/L_0)(L_\parallel - L_0)$ yields the equilibrium length L_0, and the elastic modulus K_L can be deduced from its slope.[21] In membrane simulations it is common practice to extract the elastic modulus from the area fluctuations in a barostatted simulation, but the equivalent expression for a worm, $K_L = k_B T L_0 / \langle (L_\parallel - L_0)^2 \rangle$, has so far not been used. Note that the small but systematic difference between the length $L_\parallel$ of the box and the actual arc length L of the worm is largely irrelevant at the currently accessible length scales in atomistic simulations, but will become important when the equilibrium length approaches or exceeds the persistence length.[21]

To the best of our knowledge, the only atomistic model that has been tested for the stability of its wormlike micelles is erucyl-bis(2-hydroxyethyl) methyl ammonium chloride (EHAC) in a 3% salt solution.[22] Snapshots of the system suggest that the simulations of 1–2 ns might not have been sufficiently long for the preassembled worm to fully relax to an equilibrated state, but extending some of the simulations to 5 ns or using differently produced starting configurations did not alter the results. The equilibrium length was determined to be nearly 0.5 Å per amphiphile, with an elastic modulus $K_L \approx 2 \times 10^{-9}$ Jm^{-1}. The latter number is an important input parameter for the mesoscopic simulations of Section 3.5, as there are no experimental values available for K_L.

A worm in a tensionless periodic simulation box is not straight, but gently undulates around this average shape in a perpetual Brownian motion. The bending free energy of a worm is, in good approximation, quadratic in the curvature, $F_{bend} = (\kappa/2)\int_0^L (d\hat{t}/ds)^2 ds$ with κ the bending rigidity, $\hat{t}$ a unit vector tangent to the worm, and s the path length along the worm. It is convenient now to express the undulations in one perpendicular direction, say x, as a Fourier series along the parallel direction, say z, in which case $x(z) = x_0 + \Sigma_q c_q \exp(-iqz)$ with wavenumber $q = 2\pi n/L_\parallel$ and n an integer. Using the equipartition theorem, one readily proves that the structure factors are given by $\langle |c_q|^2 \rangle = (k_B T/\kappa L_\parallel)q^{-4}$.[21] The exponent of -4 is a hallmark of tensionless worms and can be used as such to test for residual tension. It is a bit cumbersome to fit atomistic simulation results for EHAC[22] with this expression, because the length of the run is relatively short compared to the relaxation time of the slowest, i.e., longest wavelength, undulations. Furthermore, the dispersion relation will falter when the wavelength becomes comparable to the diameter of the worm, as the Fourier transformation then starts to sample the internal rather than the overall structure of the worm. The simulations yielded $\kappa \approx 2 \times 10^{-28}$ J/m, but this result is based on just a couple of wave numbers and additional simulations are needed to substantiate this value. A second number to express the rigidity of a worm is the persistence length, $l_P = \kappa/(k_B T)$. The simulations of EHAC yielded a value of about 55 nm, which corresponds to about 1200 amphiphiles and lies well within the range of 15–150 nm measured in a series of experiments on other wormlike micelles.[23]

3.4 COARSE-GRAINED WORMLIKE MICELLES

Although atomistic simulations are blessed with accurate and well-established force fields, the small length and time scales inherent to the atomistic level often make it a less attractive technique. These drawbacks are alleviated by coarse-grained (CG) simulation models, in which covalently bonded atoms are grouped together into CG particles. The head of an amphiphile is commonly reduced to 1–3 hydrophilic CG particles, while the tail(s) become short semi-flexible chains of 1–5 hydrophobic CG particles. Because these CG amphiphilic molecules are still independent entities, they have inherited the main properties of atomistic molecules, including the abilities to (dis)assemble and to diffuse within micelles. Coarse-grained solvent particles typically represent one to three water molecules. The obvious advantage of CG models over atomistic models is their improved efficiency, which allows for larger system sizes and/or longer run times, but CG models are also popular as quick test cases to put newly developed simulation techniques through their paces.

The early heuristic CG models are largely based on chemical intuition, like "hydrophobic particles repel solvent particles," yet these simple rules already sufficed to create stable micelles.[24] In principle, however, it should be possible to derive a force field for the particles based on an analysis of simulations with the underlying atomistic model. Although the bottom-up approach is popular in the simulation of melts of homopolymers,[25] the approach has received less attention in the context of surfactant systems. Because of the increased number of particle types and the nonhomogeneous conditions in the latter systems, where an amphiphile is surrounded by solvent in solution or by other amphiphiles in a micelle, it will presumably be more difficult to derive a consistent force field. Several force fields have exploited a top-down approach, tuning the interaction parameters between like particles to reproduce the experimental isothermal compressibilities and those between unlike particles to the mutual solvabilities of alkane/oil in water and vice versa.[26,27] The latter step is to a large extent inspired by the successes of the Flory-Huggins χ parameter in the description of polymer mixtures. Bonded and bending interactions between the amphiphilic particles are represented by simple harmonic springs. We note that coarse-graining the interactions should, in principle, lead to additional frictions and random forces (see Section 3.2.2). These are, however, neglected in most CG simulations of this level. The dynamic behavior will therefore be too fast, but thermodynamic and mechanical properties will not be affected.

The topics covered by CG simulations largely mirror those studied with atomistic models: a number of studies discussed micelles[28] and self-assembly,[29] but by far the most popular topic is again the simulation of biomimetic bilayers.[26,29] The increased length and time scales have also enabled studies of the formation of vesicles.[30] Wormlike micelles have again received relatively little attention. Box-spanning aggregates have been observed to self-assemble from homogeneous solutions,[31,32] but as remarked before, without additional stability tests it is unclear whether these wormlike structures are indeed a stable phase of the CG amphiphilic model under study.

The literature contains only two CG amphiphilic models that are known to possess a stable wormlike phase.[21,33] Both of these models were derived from existing models with a stable bilayer phase, by the simple expedient of enlarging the radius of the head particle. This constitutes a practical illustration of packing theory. The stabilities of both model worms were confirmed by the abilities of the micelles to retain their elongated shape without the aid of periodic boundary conditions. For our amphiphilic model,[21] which was based on the model by Goetz and Lipowsky,[31] we have also located the tensionless states under periodic boundary conditions, using a couple of worms with varying numbers N of amphiphiles. The advantage of this approach is that it provides direct access to the mechanical properties of the worm, as discussed previously for the atomistic simulations. From the structure factors then follows a bending rigidity of 5.8×10^{-29} Jm and a persistence length of 13 nm (about 250 amphiphiles), which lie at the lower ends of the experimental ranges. The elasticity calculations reveal that the tension is linear in the arc length of the worm, with elastic modulus $K_L = 4.6 \times 10^{-10}$ Jm^{-1} and an equilibrium length of $L_0/N = 0.51$Å per amphiphile. When plotted against the box length, we find that the tension is linear in $L_\parallel$ for stretched worms, but saturates for compressed worms. This plateau value is characteristic of a buckling transition, and its variation with N is well described by Euler's theory for a buckling rod. For longer worms than the $L_0 \leq 2l_p$ studied here, we expect a smooth transition to polymer-like elastic behavior.

When decreasing the size of the head groups, we find that the wormlike shape becomes unstable and the aggregate collapses into a flat disk-shaped bilayer. A similar transition may be induced by mixing a worm with bilayer-forming amphiphiles. Freely floating bilayers are often thermodynamically unstable, due to the unfavorable edge energy, and subsequently curl up to form a closed vesicle. Some experimental pathways to vesicle or liposome formation, in which a chemical or temperature trigger is used to destabilize the initial micelles,[34,35] follow the same chain of events.

But the main achievement of coarse-grained models is that they allow us to take the next step in studying micellar solutions: the simulation of two interacting worms. Theories on the rheology of entangled wormlike micelles require a description of the kinetic properties of micelles, like their abilities for scission and recombination, where the latter process may involve the end-points of two worms, one end-point and one midsection, or two midsections. Cates,[36] for instance, introduced living polymers capable of scission and end-to-end fusion into the DeGennes reptation theory,[37] to predict a characteristic time for the exponential decay of the shear relaxation modulus (see Chapter 4, Sections 4.2.5 and 4.4; Chapter 8, Sections 8.2.2 and 8.4.2; and Chapter 10). The simulation algorithms used in mesoscopic simulations, as discussed in the next section, are based on similar premises regarding feasible micellar processes. It is therefore important to understand which elementary processes are likely to happen under experimental conditions. Since direct experimental observations of the dynamics of an entanglement point are unfortunately not possible, we have used simulations at the coarse-grained level to study this process with as few preconceptions as possible.

The easiest way to create entanglements in a simulation box is to make a chain of interlocking circular wormlike micelles; the contact points between consecutive rings can then be put under tension by stretching the chain.[38] Since the rings of a chain along the z direction lie alternatingly in the xz and yz planes, the smallest repeated unit along z comprises two complete rings. By using twisted boundary conditions,[39] in which the face of the periodic box at $z = -L_z/2$ is rotated around the z-axis over 90° before being connected to the $z = L_z/2$ face, the simulation box is reduced to just one ring. This setup is illustrated in the first picture of Figure 3.3A, which shows two semicircles

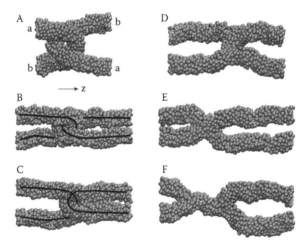

FIGURE 3.3 Entangled worms merge under tension, initially forming an "H-structure" (D), but eventually relaxing into a "2Y-structure" (F). The first snapshot shows a tension-less state, where the letters indicate the connections a-a and b-b made by the twisted periodic boundary conditions. The solid lines in B and C serve as guides to the eye. Head particles are dark gray, tail particles light gray, while solvent particles have been omitted for clarity. All pictures and particles are drawn to scale. The box is stretched between A and B, and again between D and E.

hooked behind one another in a stable tensionless configuration. Although the algorithmic changes are a bit involved, the ensuing halving of the simulation box, in our case to 840 amphiphiles and 45,800 solvent particles, makes the twisted boundary conditions worthwhile. Next, we bring the two worms into contact by elongating the box side L_z, allowing the system sufficient time to relax after every increment. The restraint imposed by the entanglement point then deforms the circular worm into an ellipse and stretches the worms (see Figure 3.3B), but they remain separate and the steps are reversible. Merging processes occurred in several independent simulation boxes[38] when the force on the worm reached about 60 pN, with the apex of one worm fusing with a point next to the apex of the other worm (see Figure 3.3C). Note that these activated events can only be detected if they happen within the timespan of the simulation, i.e., several nanoseconds, which necessitates a sufficiently strong tension on the worm in order to reduce the activation free energy barrier to several $k_B T$ at most. Under experimental conditions, the far longer contact times between worms will enable fusion at lower tensions, while slip at the entanglement point may also be of influence. After fusion, the formed network quickly relaxes because the branch points slide relatively easily along the worms. Within the conditions set by the simulation setup, the restructuring into the "H-structure" of Figure 3.3C reduces the tension by some three quarters, making this a long-lived state on the simulation time scale. But a similar reshuffle in a system of longer worms would hardly have reduced the tension, since the reduction of the total worm length is only moderate, and hence one would expect that a further relaxation step is likely to follow. To stimulate this process, its activation barrier had to be lowered again by stretching the box even further. We then find in several independent simulation boxes[38] that the cross-bar in the H-structure vanishes by a thermal fluctuation, creating the fourfold junction of Figure 3.3E, which immediately is elongated into two threefold junctions (see Figure 3.3F). The resulting "2Y-structure," consisting of one ring connected to its periodic image by a single worm, allows for a drastic reduction in the total worm length by simply exchanging amphiphiles between the two sections, and therefore easily accommodates to any imposed box length. Note that the second step occurred at a smaller pulling force, about 40 pN, than the first step, suggesting that in practice the two steps will follow one another in quick succession.

The dynamics at an entanglement point were later also simulated by Yamamoto and Hyodo.[33] In their simulations, two perpendicularly oriented linear worms are pulled toward one another by moving parts of the worms at fixed velocities. When the worms come into contact they bend at first, but eventually they fuse and reorganize to form a small network of the two main worms connected by a short crossbar. Since in this setup there is little slack available that can be used for the additional branch, the network eventually yields under the continuously increasing pulling forces. In our opinion, therefore, these simulations confirm our conclusion that entangled worms merge, followed by an efficient tension release through sliding of the junctions of the network.

Besides the above discussed network formation process, an entangled network of worms may also relax its tension by the scission of the worms bearing the heaviest loads. It is to be expected that the free energy barrier for fusion depends on the tension alone, and does not vary with the length of the two entangled worms. The break-up process, on the other hand, combines a tension dependence with a geometrical contribution that makes long worms, or long segments between consecutive entanglement points, more prone to rupture than shorter worms. The prevailing conditions in the solvent, i.e., average micellar length and the concentration of entanglements, will then determine which relaxation process dominates. We never observed rupture of a worm in our simulations of entanglement points, although this did occur when linear worms of similar lengths were exposed to five times higher tensions than those needed for fusion. After fusion, the fast relaxation of the 2Y-structure makes it unlikely to build up enough tension for any of the worms in the network to rupture. By hardening the head-head interaction potential, which both impedes fusion and makes the worms stiffer, Yamamoto and Hyodo[33] caused their worms to break up repeatedly. The ruptures occurred specifically at the entanglement point, suggesting that it is triggered by the high local curvature rather then by tension itself, and hence that the length of the worm is unimportant in this process (see also Chapter 8, Sections 8.2.2 and 8.3.1.1).

In summary, the simulated process of micellar fusion and network formation is a viable mechanism for stress relaxation in sheared solutions, in competition with scission of tensile worms. Fusion at entanglement points offers a much more effective route to forming Y-junctions than the merging of a free end cap with a nearby central bond. The junctions in a network slide easily; under shear or tension forces a worm may grow in length by consuming a side branch, or by sliding the side branch to the end of the original worm. This may explain the observed increase in micellar lengths under shear.[40]

In atomistic and coarse-grained simulations alike, it is not uncommon that the amphiphilic particles are a minority, with most of the valuable computer time spent on the relatively uninteresting solvent particles. To remedy this situation, a number of solvent-free amphiphilic models have been proposed over the last years.[41–43] Because micelles and bilayers rely for their stability on the hydrophobic repulsion between solvent and tail particles, additional forces between the tail particles need to be introduced to prevent the aggregates from falling apart. Omitting the solvent particles will also eliminate hydrodynamic interactions between micelles, and volume constraints in the case of closed vesicles, and is therefore not always permitted. So far these models have mainly been used to study micellar self-assembly, bilayers, and vesicles, but it should be straightforward to extend them, by enlarging the head group relative to the tail, to yield stable wormlike micelles.

The data extracted from coarse-grained simulations of single and entangled worms provides the input for the mesoscopic level of description, which is discussed next.

3.5 DYNAMICS AND RHEOLOGY OF ENTANGLED WORMLIKE MICELLES

3.5.1 SIMULATING LARGE LENGTH AND TIME SCALES

The rich and complex rheology of wormlike micellar solutions is what makes them so interesting for practical applications. To realistically simulate the rheology, one would ideally use the atomistic model as described in Section 3.3, or a properly coarse-grained version as in Section 3.4. However, using such models and current-day computing power, it is impossible to determine the macroscopic rheology of a solution of entangled wormlike micelles. The reasons for this are twofold: (1) very large length scales combined with (2) very long time scales.

Focusing first on the length scales, it must be realized that the stress tensor is a collective property of a sufficiently large portion of fluid. In a molecular dynamics simulation, the instantaneous stress components are generally given by

$$\sigma_{\alpha\beta} = -\frac{1}{V}\left(\sum_{i=1}^{N} m_i v_{i\alpha} v_{j\beta} + \sum_{i=1}^{N-1}\sum_{j=i+1}^{N} r_{ij\alpha} F_{ij\beta}\right). \tag{3.33}$$

Here m_i is the mass and $v_{i\alpha}$ the α component of the velocity of particle i, $r_{ij\alpha}$ the α component of the vector from the position of particle j to particle i, $F_{ij\beta}$ the β component of the force exerted by particle j on particle i (here we have assumed a pairwise interacting and periodic system), and V the volume of the simulation box. In our case the volume V is sufficiently large if the simulation box contains at least several hundred wormlike micelles, providing a minimal representation of the very polydisperse (bulk) length distribution $c(L)$. Because the interesting rheological behavior occurs in entangled wormlike micellar solutions, the average length of a wormlike micelle must by definition exceed the entanglement length. The entanglement length depends on many factors, including the concentration of the solution and the bending stiffness of the wormlike micelle. It is typically $O(10–100)$ persistence lengths. This means that the simulation box must contain at least $O(10^4)$ persistence lengths. In Sections 3.3 and 3.4 we have found that one persistence length of wormlike micelle contains $O(10^3)$ surfactant molecules, leading to $O(10^7)$ surfactant molecules in total. In principle the solvent molecules need to be included as well, leading again to a more than tenfold increase of the number of molecules. Even when the surfactant molecules are coarse grained to a few beads, and the solvent effect can be treated in an implicit manner, a simulation of this magnitude will take an appreciable amount of computational time.

Next we focus on the time scales. Typical rheological experiments are performed at oscillation frequencies or shear rates between 10^{-2} and 10^2 or 10^3 s^{-1}, with corresponding inverse time scales. Conversely, most entangled wormlike micelles have a longest characteristic relaxation time τ ranging from 10^{-4} to 10^{3} s (although unentangled wormlike micelles can have smaller relaxation times). In order to simulate the rheological properties of entangled wormlike micelles, one must therefore be able to reach time scales of the order of at least milliseconds. For example, for the coarse-grained model of Section 3.4 a typical time step is 10^{-14} s, leading to a total of 10^{11} time steps. So, even when the simulation is sped up by coarse-graining surfactant molecules to a few beads, this still presents us with an unsurmountable amount of computational work.

3.5.2 THE FENE-C MODEL

By now it must have become clear to the reader that the only way forward is to coarse-grain even further, to a level where each unit represents several surfactant molecules. This will, of course, be at the cost of losing detailed information about the surfactant molecules. One important development in this direction is the generic FENE-C model. In FENE-C the wormlike micelles are represented by flexible chains of relatively hard spheres. Chains can grow by the addition of monomers at the chain ends, or by recombination with other chain ends. Conversely, chains can break if any of the bonds are stretched because of thermal fluctuations or tension. The FENE-C model was studied extensively by Kröger and others.[44–48] In the original model[44,45] solvent beads were included to account for solvent effects, but the solvent effect may be mimicked through Brownian dynamics as well.[46] The resulting worm length distribution is found to agree well with the theoretical (mean field) prediction $c(L) \propto \exp(-L/\bar{L})$, and shear thinning of the viscosity is observed, although not as strong as in experiments.[45,48]

In the FENE-C model, being a generic model, no reference is made to any specific real wormlike micellar system. This may be an advantage, since this allows the simulator the freedom to scale the simulation results onto experimental results. However, in our opinion great care must be taken if realistic and quantitative results for the dynamics and rheology of wormlike micelles are required. First, there are always multiple relevant length scales. Scaling one of them onto experimental values does not guarantee that the other length scales will be described correctly as well. For example, the persistence length of a typical wormlike micelle is several times its diameter, whereas in the flexible FENE-C model these are more or less the same. Second, the kinetics of break up and fusion of chain ends may not be as fast as predicted by the FENE-C model. In the original FENE-C model, recombination is relatively easy because chain ends can fuse instantly if their separation is smaller than some critical distance. In reality, before two chain ends can fuse, there may be specific demands on the conformations of the surfactants in the end caps, giving rise to a considerable free energy barrier. Chain recombination, like scission, may therefore be an activated process (see Chapter 10).[47,48]

These disadvantages may be alleviated by introducing a (strong) bending potential between the spherical beads and an additional radial interaction mimicking the activation barrier, or some Monte Carlo equivalent. Unfortunately, this quickly becomes computationally expensive, because many beads will be required to represent one persistence length of a realistic wormlike micelle, while the integration time step will still be limited by the relatively hard interactions at the scale of the beads.

3.5.3 THE MESOWORM MODEL

In this section we will describe a more recent coarse-grained model, called mesoworm (MESOscale WORMlike micelles).[49] The philosophy behind this model is that the material properties of individual wormlike micelles can in principle be measured from more detailed simulations or targeted experiments. Relevant material properties include the solvent viscosity and worm diameter, bending rigidity, compressibility, and scission and activation free energies for fusion and breaking. Allowing the model to have these parameters as input, the rheology can be predicted from realistic input, with as few assumptions as possible. This enables us to create a hierarchy of simulation models, ultimately linking the chemical details of the surfactants and other components of the solution to its macroscopic rheology.

Of course, some assumptions still have to be made. First, it is assumed that the stress is dominated by the network of wormlike micelles and that hydrodynamic interactions are relatively unimportant. This allows for a simple implicit treatment of the solvent effects, i.e., Brownian dynamics of rigid rods. The model therefore does not apply to dilute solutions, where the micelles are relatively short and hardly interact. Second, it is assumed that excluded volume interactions are unimportant. This allows us to treat wormlike micelles as chains of infinitely thin lines. The model should therefore not be used for very high concentrations either, where excluded volume effects may lead to spontaneous nematic ordering of the wormlike micelles. There is a large range of concentrations between dilute and concentrated where the model does apply. Indeed, the semidilute regime is of most practical interest, since in this regime the viscosity can be greatly enhanced with a relatively low amount of surfactant.

In the mesoworm model, each persistence length of wormlike micelle is represented by one unit (see Figure 3.4). This degree of coarse graining is as

(a)   (b)

FIGURE 3.4 (a) The wormlike micelle is represented by a string of thin rods, each of one persistence length. (b) The chain can break and fuse. Details of these processes are given in the main text. Reproduced from Ref. 49 with permission of IOP Publishing House.

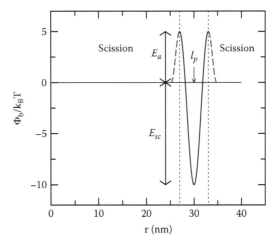

FIGURE 3.5 Interaction between two units in the MESOWORM model. The solid line is the potential well of depth E_{sc}, which is a function of the distance r between the mid-points of bonded units. The dashed line is the repulsive barrier of height E_a, which is a function of the distance x between the end caps of wormlike micellar chains (see main text).

large as possible to permit a large integration step and fewer particles, while it is still small enough to allow an accurate description of the overall conformation of the wormlike micelle. The units interact with each other through a potential of a form given in Figure 3.5. This potential contains a well of depth E_{sc}, the scission energy, and a barrier of height E_a, the (additional) activation energy associated with the fusion-recombination process. The well is centered around the persistence length l_p, where it has a second derivative equal to K_L/l_p, as required by the elastic response of a wormlike micelle of that length. The correct stiffness of the wormlike micelle is implemented by means of an angular potential between each bonded triplet. As soon as a bond is stretched or compressed too much, the bond is broken. At that moment, its potential energy is not zero but equal to E_a, the activation barrier associated with the fusion-breaking process. Conversely, before two chain ends are allowed to fuse, they must overcome this energy barrier. It is implemented as a repulsive potential which depends on the distance x between the *end caps* of the wormlike micelles. When the distance x is small enough, and the distance r between the midpoints is within the appropriate range, two chain ends are allowed to fuse.

The fusion energy barrier is purposely not a function of the distance between the midpoints of the units at the extremes of the wormlike micelles, because this would lead to far-too-long wormlike micelles when using realistic values of the scission energy E_{sc}. The average worm length is actually greatly overpredicted in the case of both the FENE-C model and mean-field theory. In the latter, a wormlike micelle is treated as a random walk on a lattice, with an energy penalty E_{sc} for each pair of chain ends. The random walk lattice model is usually justified by identifying each occupied lattice space with one persistence length of wormlike micelle. By variationally minimizing the free energy, under the constraint of fixed volume fraction ϕ, the theory then predicts an average length of $\bar{L} = c\phi^{1/2}\exp[E_{sc}/(2k_BT)]$, where the prefactor c is of order 1. Although the scaling with E_{sc}/T has been confirmed, the prefactor is hugely unrealistic. For example, experimental estimates of the scission energy of EHAC wormlike micelles vary between 25 and 50 k_BT.[50,51] For $\phi \approx 0.1$ mean-field theory then predicts an average worm length of $O(10^5-10^{10})$ persistence lengths, corresponding to a contour length of $O(10^{-3}-10^2)$ m. These estimates would make wormlike micelles truly giant! It is important to realize that wormlike micelles are relatively thin. By letting the fusion process take place within a small volume at the end cap positions, instead of a large volume associated with a sphere of diameter $O(l_p)$ around the midpoint of the extremal units, the average length of the wormlike micelles is brought down to realistic proportions.

In the model as described up to this point, there is no excluded volume between the worm segments. In other words, the wormlike micelles would be able to pass through each other like ghost chains, whereas the dominant stress contribution is expected to arise from entanglements between the wormlike micelles. To remedy this, the TWENTANGLEMENT technique is applied. In this technique, originally designed for polymer melt simulations,[25,52] an imminent bond crossing is detected and prevented by introducing a new coordinate (an entanglement point) at the crossing point. From that time onward, until the entanglement is removed again, the interaction between bonded units is a function of the path length measured via the entanglement points, instead of the usual distance between the units (see Figure 3.6). Entanglements are allowed to jump over the (central) position of a unit, and so can slide along the backbone of a wormlike micelle.

A snapshot of a simulation box created by the above method is given in Figure 3.7. By analyzing the fluctuations of the stress tensor in equilibrium simulations, we can obtain the (zero-)shear relaxation modulus $G(t)$:

$$G(t) = \frac{V}{k_BT}\langle\sigma_{xy}(t)\sigma_{xy}(0)\rangle. \qquad (3.34)$$

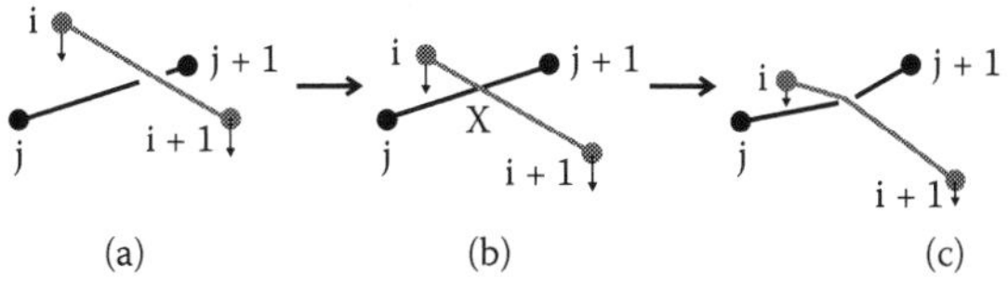

FIGURE 3.6 Basic function of the TWENTANGLEMENT algorithm.[52] (a) A bond crossing is imminent. (b) When two bonds cross, a new entanglement point is defined at X. (c) From then on, the bonded interaction is a function of the path length. Reproduced from Ref. 52 with permission of the American Institute of Physics.

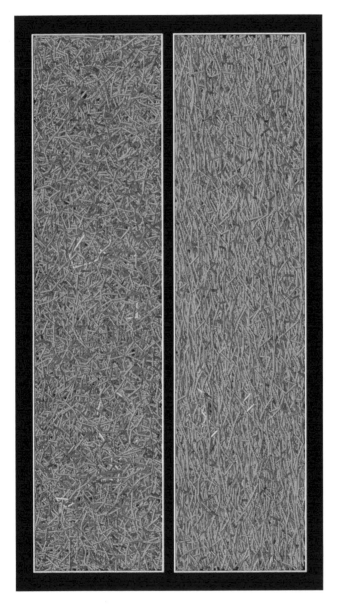

FIGURE 3.7 Snapshots of mesoworm simulation boxes in equilibrium (top) and at a shear rate of $10^5\,s^{-1}$ (bottom).

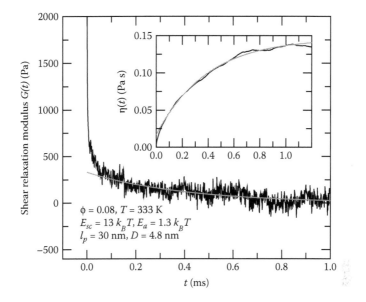

FIGURE 3.8 Main plot: The shear relaxation modulus $G(t)$, calculated from the stress-stress autocorrelation in an equilibrium mesoworm simulation. The gray line is a fit with a single exponential. Inset: The integral yields the zero-shear viscosity η_0 of the solution.

A typical result is given in Figure 3.8. Other properties characteristic of the linear rheology can be derived from $G(t)$. For example, the storage and loss moduli $G'(\omega)$ and $G''(\omega)$ are obtained through a Fourier transform, and the zero-shear viscosity is given by the infinite time integral, $\eta_0 = \int_0^\infty G(t)dt$.[37]

The nonlinear rheology of the wormlike micelles can be measured by applying sheared periodic boundary conditions[3] and analyzing the time dependence of different components of the stress tensor. For example, when a shear is suddenly applied to an initially quiescent solution of wormlike micelles, one often observes a characteristic overshoot in the transient shear stress. This is also observed in simulations, as in Figure 3.9.

3.6 CONCLUSION

We have shown that simulation can deliver a detailed and fundamental understanding of microscopic processes. Knowledge of these processes in wormlike micelles is important to understand their emergent dynamics and rheology. The main strength of simulations is that they enable us, at different levels of length and time, to verify or negate assumptions that are made in approximate theories. In our opinion this is the way forward, enabling a rational design of new viscoelastic materials based on wormlike micelles.

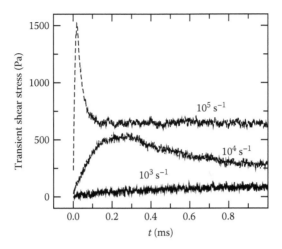

FIGURE 3.9 Transient shear stress response to a sudden shear flow. The applied shear rates are given in the figure.

REFERENCES

1. Israelachvili, J. N.; Mitchell, D. J.; Ninham, B. W. *J. Chem. Soc. Faraday Trans.* 1976, *72*, 1525.
2. Maitland, G. C. *Curr. Opin. Colloid Interface Sci.* 2000, *5*, 301.
3. Allen, M. P.; Tildesley, D. J. *Computer simulation of liquids;* Clarendon Press: Oxford, U.K., 1987.
4. Frenkel, D.; Smit, B. *Understanding molecular simulations. From algorithms to applications;* Academic Press: San Diego, CA, USA, 2002.
5. Links to common simulation packages can be found at http://en.wikipedia.org/wiki/Molecular_dynamics.
6. Links to common force fields can be found at http://en.wikipedia.org/wiki/Force_field_(chemistry).
7. Deutch, J. M.; Oppenheim, I. *J. Chem. Phys.* 1971, *54*, 3547.
8. Ermak, D. L.; McCammon, J. A. *J. Chem. Phys.* 1978, *69*, 1352.
9. Murphy, T. J.; Aguirre, J. L. *J. Chem. Phys.* 1972, *57*, 2098.
10. Hoogerbrugge, P. J.; Koelman, J. M. V. A. *Europhys. Lett.* 1992, *19*, 155.
11. Tao, Y.-G.; den Otter, W. K.; Padding, J. T.; Dhont, J. K. G.; Briels, W. J. *J. Chem. Phys.* 2005, *122*, 244903.
12. Hoover, W. G. *Phys. Rev. A* 1985, *31*, 1695.
13. Watanabe, K.; Ferrario, M.; Klein, M. L. *J. Phys. Chem.* 1988, *92*, 819.
14. Shelley, J. C.; Shelley, M. Y. *Curr. Opin. Coll. Int. Sci.* 2000, *5*, 101.
15. Mohanty, S.; Davis, H. T.; McCormick, A. V. *Langmuir* 2001, *17*, 7160.
16. Stephenson, B. C.; Beers, K.; Blankschtein, D. *Langmuir* 2006, *22*, 1500.
17. Maillet, J. B.; V., L.; Coveney, P. V. *Phys. Chem. Chem. Phys.* 1999, *1*, 5277.
18. Marrink, S. J.; Tieleman, D. P.; Mark, A. E. *J. Phys. Chem. B* 2000, *104*, 12165.
19. Tieleman, D. P.; Marrink, S. J.; Berendsen, H. J. C. *Biochim. Biophys. Acta* 1997, *1331*, 235.

20. Saiz, L.; Bandyopadhyay, S.; Klein, M. L. *Biosci. Rep.* 2002, *22*, 151.
21. den Otter, W. K.; Shkulipa, S. A.; Briels, W. J. *J. Chem. Phys.* 2003, *119*, 2363.
22. Boek, E. S.; den Otter, W. K.; Briels, W. J.; Iakovlev, D. *Phil. Trans. R. Soc. London A* 2004, *362*, 1625.
23. Magid, L. J. *J. Phys. Chem. B* 1998, *102*, 4064.
24. Smit, B.; Hilbers, P. A. J.; Esselink, K.; Rupert, L. A. M.; van Os, N. M.; Schlijper, A. G. *J. Phys. Chem.* 1991, *95*, 6361.
25. Padding, J. T.; Briels, W. J. *J. Chem. Phys.* 2002, *117*, 925.
26. Groot, R. D.; Rabone, K. L. *Biophys. J.* 2001, *81*, 725.
27. Marrink, S. J.; de Vries, A. H.; Mark, A. E. *J. Phys. Chem. B* 2004, *108*, 750.
28. Pool, R.; Bolhuis, P. G. *J. Phys. Chem. B* 2005, *109*, 6650.
29. Goetz, R.; Gompper, G.; Lipowsky, R. *Phys. Rev. Lett.* 1999, *82*, 221.
30. Marrink, S. J.; Mark, A. E. *J. Am. Chem. Soc.* 2003, *125*, 15233.
31. Goetz, R.; Lipowsky, R. *J. Chem. Phys.* 1998, *108*, 7397.
32. Maiti, P.; Lansac, Y.; Glaser, M. A.; Clark, N. A. *Langmuir* 2002, *18*, 1908.
33. Yamamoto, S.; Hyodo, S. *J. Chem. Phys.* 2005, *122*, 204907.
34. Egelhaaf, S. U.; Schurtenberger, P. *Phys. Rev. Lett.* 1999, *82*, 2804.
35. Gradzielski, M. *J. Phys.: Cond. Matter* 2003, *15*, R655.
36. Cates, M. E. *Macromolecules* 1987, *20*, 2289.
37. Doi, M.; Edwards, S. F. *The theory of polymer dynamics;* Clarendon Press: Oxford, U.K., 1998.
38. Briels, W. J.; Mulder, P.; den Otter, W. K. *J. Phys.: Cond. Matter* 2004, *16*, S3965.
39. Allen, M. P.; Masters, A. J. *Mol. Phys.* 1993, *79*, 277.
40. Berret, J.-F.; Gamez-Corrales, R.; Séréro, Y.; Molino, F.; Lindner, P. *Europhys. Lett.* 2001, *54*, 605.
41. Noguchi, H.; Takasu, M. *Phys. Rev. E* 2001, *64*, 041913.
42. Farago, O. *J. Chem. Phys.* 2003, *119*, 596.
43. Cooke, I. R.; Kremer, K.; Deserno, M. *Phys. Rev. E* 2005, *72*, 011506.
44. Kröger, M.; Makhloufi, R. *Phys. Rev. E* 1996, *53*, 2531.
45. Carl, W.; Makhloufi, R.; Kröger, M. *J. Phys. II* 1997, *7*, 931.
46. Kröger, M. *Physics Reports* 2004, *390*, 453.
47. Padding, J. T.; Boek, E. S. *Europhys. Lett.* 2004, *66*, 756.
48. Padding, J. T.; Boek, E. S. *Phys. Rev. E* 2004, *70*, 031502.
49. Padding, J. T.; Boek, E. S.; Briels, W. J. *J. Phys.: Cond. Matter* 2005, *17*, S3347.
50. Couillet, I.; Hughes, T.; Maitland, G.; Candau, F.; Candau, S. J. *Langmuir* 2004, *20*, 9541.
51. Raghavan, S. R.; Kaler, E. W. *Langmuir* 2001, *17*, 300.
52. Padding, J. T.; Briels, W. J. *J. Chem. Phys.* 2001, *115*, 2846.

4 Theoretical Rheology of Giant Micelles

Michael Cates and Suzanne Fielding

CONTENTS

4.1 INTRODUCTION

This chapter concerns rheology, which is the science of flow behavior. We shall address the theoretical rheology of giant micelles at two levels. The first (in Section 4.4) is microscopic modeling in which one seeks a mechanistic understanding of rheological behavior in terms of the explicit dynamics (primarily entanglement and reversible self-assembly) of the giant micelles themselves. This yields so-called constitutive equations, which relate the stress in a material to its deformation history. Solution of these equations for simple experimental flow protocols presents major insights into the fascinating flow properties of viscoelastic surfactant solutions, including near-Maxwellian behavior in the linear regime, and drastic shear-thinning at higher stresses. These successes mainly concern the strongly entangled region where the micellar solution is viscoelastic at rest; in this regime, strong shear thinning is usually seen. There are however a range of fascinating phenomena occurring at lower concentrations where the quiescent solution is almost inviscid, but becomes highly viscoelastic after a period of shearing. These will also be discussed (Section 4.4.7) although they remain, for the present, much less well understood.

Microscopic models of giant micelles under flow generally treat the micelles as structureless, flexible, polymer-like objects, albeit (crucially!) ones whose individual identities are not sustained indefinitely over time. This neglect of chemical detail follows a very successful precedent set in the field of polymer dynamics.[1,2] There, models that contain only a handful of static and dynamic parameters can explain almost all the observed features of polymeric flow behavior. Indeed, microscopic models of polymer rheology arguably represent one of the major intellectual triumphs of 20th century statistical physics.[3]

However, at least when extended to micelles,[4] these microscopic constitutive models remain too complicated to solve in general flows, particularly when flow instabilities are present. Therefore we also describe in Section 4.5 some purely macroscopic constitutive models, whose inspiration stems from the microscopic ones but which can go much further in addressing the complex nonlinear flow phenomena seen in giant micelles. These phenomena include for example rheochaos, in which a steady shear deformation gives chaotically varying stress or vice versa. Our discussion of macroscopic modeling will take us to the edge of current understanding of these exotic rheological phenomena. Prior to discussing rheology, we give in Section 4.2 a brief survey of the equilibrium statistical mechanics and kinetics of micellar self-assembly, addressing only those aspects needed for the subsequent discussion of rheology.

4.2 STATISTICAL PHYSICS OF EQUILIBRIUM MICELLES

Most of the thermodynamic modeling needed here can be addressed within mean-field theory (Sections 4.2.1–4.2.3), although more advanced treatments show subtleties that still await experimental clarification (Section 4.2.4). In Section 4.2.5 we turn to the kinetic question of how micelles exchange material with one another in equilibrium.

4.2.1 MEAN-FIELD THEORY: LIVING POLYMERS

In typical giant micellar systems the critical micelle concentration (cmc) is low — of order 10^{-4} molar for CTAB/KBr, for example. As concentration is raised above this, uniaxial elongation occurs and soon micelles become longer than their persistence length l_p (the length over which bending occurs[1]) and then resemble flexible polymers. Persistence lengths of order 10–20 nm are common, but much larger values are possible in charged systems.

As concentration is increased, viscoelasticity sets in at a concentration $\tilde{C}$ usually identified (but see Section 4.2.3) with an overlap concentration C^*. In a range above C^*, the wormlike micelles are semidilute[1] — overlapped and entangled at large distances, but well separated from one another at scales below ξ, the correlation length. In ordinary polymer solutions in good solvents, the behaviour at scales less than ξ is not mean-field-like but governed by a scaling theory.[1] We return to this in Section 4.2.4, but note that these scaling corrections become small when the persistence length of a micellar cylinder is much larger than its diameter.[1,2] Under these conditions a mean-field approach captures the main phenomena of interest, particularly in the regime of strong viscoelasticity at $C \geq \tilde{C}$.

The simplest mean-field theory[5,6] assumes that no branch points and no closed rings are present (rectified in Sections 4.2.2 and 4.2.3), and ascribes a free energy $E/2$ to each hemispherical endcap of a micelle relative to the free energy of the same amount of amphiphilic material residing in the cylindrical body. Denoting by $c(N)$ the number density of aggregates of N monomers, the mean-field free energy density obeys

$$\beta F = \sum_N c(N)[\ln c(N) + \beta E] + F_0(\phi) \tag{4.1}$$

Here $\beta = 1/k_B T$; the term in E counts two end-caps per chain, and the $c \ln c$ piece comes from the entropy of mixing of micelles of different lengths. Within a mean-field calculation, these are the *only* terms sensitive to the size distribution $c(N)$ of the micelles. All other contributions reside in $F_0(\phi)$ which depends only on total volume fraction ϕ; this obeys

$$\phi = v_0 C = v_0 \sum_N N c(N) \tag{4.2}$$

where v_0 is the molecular volume of the amphiphiles and C their total concentration.

Minimizing (4.1) at fixed ϕ gives an exponential size distribution

$$c(N) \propto \exp[-N/\bar{N}]; \quad \bar{N} \simeq \phi^{1/2} \exp[\beta E/2] \tag{4.3}$$

The exponential form in each case is a robust result of mean-field theory. The ϕ-dependence in the second equation is also robust, but only so long as parameters like E and v_0 are themselves independent of concentration. The formula for $\bar{N}$ absorbs into E various dependences on v_0, l_p, and a_0, where a_0 is the cross-sectional area of the micellar cylinders. So long as a_0 is constant, these results also control $c(L)$ and $\bar{L}$, where $L \propto N$ is the contour length of a micelle. In mean field, L controls the typical geometric size R of a micelle via $R^2 \simeq L l_p$.

We can now find the overlap concentration C^*, or overlap volume fraction $\phi^* = C^* v_0$. For a micelle of length $\bar{L}$ we have $R \simeq n^{1/2} l_p$ where $n = \bar{L}/l_p$ is the number of persistence length it contains; this obeys $n l_p a_0 / v_0 = \bar{N}$. The total volume of amphiphile within the region spanned by this micelle is $\bar{N} v_0$ so the volume fraction within it is $\phi \simeq \bar{N} v_0 / R^3$. At the threshold of overlap, this ϕ equates to the true value ϕ^*; eliminating $\bar{N}$ via (4.3) gives

$$C^* v_0 = \phi^* \simeq (a_0 / v_0^{1/3} l_p)^{6/5} e^{-\beta E/5} \tag{4.4}$$

For typical cases the dimensionless pre-exponential factor is smaller than unity, but nonetheless a fairly large E is required if ϕ^* is to be below, say, 1%. The regime of long, entangled micelles usually entails scission energies E of around 10–$20 k_B T$; in practice, experimental estimates of ϕ^* (best determined by light scattering) are often in the range 0.05–5%.[7]

The region around ϕ^* is where spectacular shear-thickening rheology occurs (see Section 4.4.7). In ionic micellar systems without excess of salt, the strong dependence of l_p, E, and other parameters on ϕ itself in this region means that the simple calculations leading to (4.3), and hence the estimate (4.4), are at their least reliable. More detailed theories, which treat electrostatic interactions explicitly, give a far stronger dependence of $\bar{L}$ on ϕ, a narrower size distribution for the micelles, and a higher ϕ^*.[8]

4.2.2 ROLE OF BRANCHING: LIVING NETWORKS

The above assumes no branching of micelles. While the general case of branched micelles is complicated,[9] things simplify considerably in the branching-dominated limit, when there are many branch points per end-cap. For branching via z-fold crosslinks one has, replacing (4.1), the following mean-field result:[9]

$$\beta F = \sum_N c(N)[\log c(N) + \beta E z^{-1}] + 2(z^{-1} - 1)C \ln(2C) + F_0(\phi) \qquad (4.5)$$

where C is as defined in (4.2), and $c(N)$ is now the concentration of network strands containing N amphiphiles. E is now the energy of a z-fold junction, not of an end-cap. The value of z most relevant to micelles is $z = 3$, since for a system whose optimal local packing is a cylinder, a three-fold junction costs less in packing energy than $z > 3$. (Low z is also favored entropically.[9])

Minimizing (4.5) to find the equilibrium strand length distribution, one finds this again to be exponential, with mean strand length $\bar{L} \sim \phi^{1-z/2} \exp[\beta E]$. This result applies whenever the geometric distance between crosslinks, $\Lambda \simeq (\bar{L} l_p)^{1/2}$ greatly exceeds the geometric mesh size ξ, which within mean-field theory obeys $\xi \sim a_0 / \phi l_p$. This situation of $\Lambda \gg \xi$ is called an unsaturated network[9] and arises at high enough concentrations ($\phi \gg \phi^{sat} \simeq v_0^{-1} \exp[-\beta E/(3-z/2)]$). For $\phi \leq \phi^{sat}$ one has a saturated network with $\Lambda \simeq \xi$.[9]

The rheology of living networks (see Section 4.4.6) should differ strongly from the unbranched micellar case. Such a regime has been identified in several systems, primarily cationic surfactants at relatively high ionic strength.[10,11,12] These accord with the expected trend for curvature packing energies: adding salt in these systems stabilizes negatively curved branch points relative to positively curved end-caps.[11]

4.2.3 ROLE OF LOOP CLOSURE: LIVING RINGS

We have assumed in (4.1) that rings do not arise. This assumption turns out to be satisfactory only when E is not too large, so that ϕ^* in (4.4) lies well above a certain volume fraction ϕ_r^{max}, defined below, which signifies a maximal role for ringlike micelles.

From (4.4), as $E \to \infty$, the overlap threshold ϕ^* for open micelles tends to zero. In this limit, there is formally just a single micelle of macroscopic length. This corresponds to an untenable sacrifice of translational entropy which is easily regained by ring formation. To study this, let us set $E \to \infty$ so that no open chains remain, but allows rings with concentration $c_r(N)$. Then, to replace (4.1), one has [13]

$$\beta F = \sum_N c_r(N)[\ln c_r(N) + \beta f_r(N)] + F_0(\phi) \qquad (4.6)$$

where $f_r(N) = -k_B T \ln(Z_r)$, and Z_r is the configurational free energy cost of ring closure. Put differently, $E - f_r(N)$ is the total free energy cost of hypothetically opening a ring, creating two new endcaps but gaining an entropy $k_B \ln Z_r$.

For Gaussian (mean-field-like) chains in three dimensions, it is easily shown that $Z_r = \lambda N^{-5/2}$,[1] where λ is a dimensionless combination (as yet unknown)[7] of a_0, v_0, l_p. Minimizing (4.6) at fixed ϕ then gives

$$c_r(N) = \lambda N^{-5/2} e^{\tilde{\mu}N} \tag{4.7}$$

where $\tilde{\mu}$ is a chemical-potential-like quantity. Interestingly, this size distribution for rings shows a condensation transition. That is, for $\tilde{\mu} > 0$ the volume fraction $\phi_r = v_0 \Sigma_N N c_r(N)$ is divergent, whereas for $\tilde{\mu} \leq 0$ it can apparently be no greater than

$$\phi_r^{max} = \lambda \sum_{N=N_{min}}^{\infty} N^{-3/2} \tag{4.8}$$

This apparent ceiling or limiting value $\phi_r^{max} < 1$ depends on N_{min}, the smallest number of amphiphiles that can make a ring-shaped micelle without prohibitive bending cost. This mathematical situation represents the following physical picture, valid in the $E \to \infty$ limit. For $\phi \leq \phi_r^{max}$ one has the power law distribution of ring sizes in (4.7), cut off at large N by an exponential multiplier (resulting from small negative $\tilde{\mu}$). For $\phi > \phi_r^{max}$, one has a pure power law distribution of rings, in which total volume fraction ϕ_r^{max} resides; plus an excess volume fraction $\phi - \phi_r^{max}$ which exists as a single giant ring of macroscopic length.

For any finite E, all rings with $N > \bar{N}$ as defined in (4.1) will break up into pieces (roughly of size $\bar{N}$). Indeed, if E is finite one can, within mean-field theory, simply add the chain and ring free energy contributions as

$$\beta F = \sum_N c(N)[\ln c(N) + \beta E] + \sum_N c_r(N)[\ln c_r(N) + \beta f_r(N)] + F_0(\phi) \tag{4.9}$$

From this one can prove that the singular behavior found above at ϕ_r^{max} is smoothed out.[14] Nonetheless, if E is large enough that the overlap threshold $\phi^*(E)$ obeying (4.4) falls below ϕ_r^{max} obeying (4.8), then this condensation transition of rings, though somewhat rounded, should still have experimental consequences. These should mainly affect a (roughly) factor-two window in concentration either side of ϕ_r^{max}. Because of uncertainty over the values of λ and N_{min} in (8), ϕ_r^{max} remains poorly characterized for giant micelles. So far there is relatively little (but some) [15,16] experimental evidence for a ring-dominated regime.

4.2.4 BEYOND MEAN-FIELD THEORY

As with conventional polymers,[1] short-range excluded-volume correlations should alter the various power law exponents that appear in equations such as (4.3) and (4.7). In practice, however, giant micellar systems can often be expected to lie in a messy crossover region between mean-field and the scaling theory. Here we briefly outline the scaling results.[7,9,17] First, consider a system with no branches or rings. In such a system, the excluded volume exponent $\nu \sim 0.588$ governs the non-Gaussian behavior of a self-avoiding chain; $R \sim L^{\nu}$.[1] This gives $\xi \sim \phi^{\nu/(1-\nu d)} \sim \phi^{-0.77}$ where $d = 3$ (the dimension of space). This leads to an altered scaling of the transient elastic modulus G_0 (defined in Section 4.4 below): $\beta G_0 \sim \xi^{-d} \sim \phi^{2.3}$ as opposed to ϕ^2 in mean field.[1] Second, one finds in place of (4.3)

$$c(L) \propto \exp[-L/\bar{L}]; \quad \bar{L} \simeq \phi^{y} \exp[\beta E / 2] \tag{4.10}$$

where $y = [1 + (\gamma - 1)/(\nu d - 1)]/2 \simeq 0.6$, with $\gamma \simeq 1.2$ another known polymer exponent.[1] Turning now to the case where branching dominates, the mean-field result for the mean network strand length $\bar{L} \sim \phi^{-1/2} \exp[\beta E]$ becomes $\bar{L} \sim \phi^{-\Delta} \exp[\beta E]$ with an exponent $\Delta \sim 0.74$; the expression for Δ in terms of standard polymer exponents is given in Ref. 9. Similarly the mean-field result $\phi^{sat} \sim e^{-2\beta E/3}$ becomes $\phi^{sat} \sim e^{-\beta E/y}$ with $y \sim 0.56$.[9] Finally, in any regime where rings dominate, even the chain swelling exponent ν is slightly reduced.[17] Near ϕ_r^{max} (which has the same meaning as before, but no longer obeys (4.8)) there is a power-law cascade of rings, controlled by a distribution similar to (4.7), but with an exponent somewhat larger than $5/2$.[17,18] Due to the difficulty in detecting any ring-dominated regime, these results for micelles remain mainly of academic interest.

4.2.5 REACTION KINETICS IN EQUILIBRIUM

A key ingredient into rheological modeling is the presence of reversible aggregation and disaggregation processes (or reactions), allowing micelles to exchange material. We will treat these only at the mean-field level. We also neglect branching and ring-formation in the first instance, and distinguish reactions that change the aggregation number of a particular micelle N by a small increment, $\Delta N \simeq 1$, from those which create changes ΔN of order N itself. The former reactions can alter a micelle's size, but as N increases, the time scale involved gets longer.[19] Unless the reaction rates for all reactions of the second ($\Delta N \simeq N$) type are extremely slow, the latter will dominate for giant micelles. From now on, we consider only these, of which there are three types: reversible scission, end interchange, and bond interchange, as shown in Figure 4.1.

In reversible scission, a chain of length L breaks spontaneously into two fragments of size L' and $L'' = L - L'$. In thermal equilibrium the reverse process (end-to-end fusion) happens with exactly equal frequency.[4] If, for simplicity,

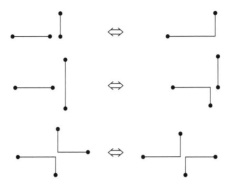

FIGURE 4.1 The three main types of micellar reaction: Top, reversible scission; middle, end interchange; bottom, bond interchange

we assume that the fusion rate of chains of lengths L and L' varies as the product of their concentrations, then the full mean-field kinetic equations (as detailed, e.g., in Ref. 4) involve only a single rate constant k_{rs}, which is the rate of scission per unit length of micelle. More relevant physically is

$$\tau_{rs} = (k_{rs}\bar{L})^{-1} \qquad (4.11)$$

which is the time taken for a chain of the mean length to break into two pieces by reversible scission. Note that the lifetime of a chain-end before recombination is also τ_{rs}.[4] Moreover, if $\bar{L}$ is perturbed by a temperature (or pressure) jump, it relaxes exponentially to equilibrium with a decay time $\tau_{rs}/2$.[20] Such experiments thus allow direct access to τ_{rs}.[21]

Turning to end interchange, this is the process where a reactive chain-end bites into another micelle, carrying away part of it (Figure 4.1). Assuming all ends to be equally reactive and/or that all points on all micelles are equally likely to be thus attacked, one finds once again a single relevant rate constant k_{ei}, with the lifetime of any individual chain-end now $1/k_{ei}\phi$. The lifetime of a micelle of the average length becomes[20]

$$\tau_{ei} = (4k_{ei}\phi)^{-1} \qquad (4.12)$$

In contrast to the reversible scission case, analysis of the full mean-field kinetic equations[20] shows that end interchange is *invisible* in T-jump: no relaxation whatever occurs by this mechanism. The root cause is that *end interchange conserves the total number of micelles*; if any disturbance is applied that perturbs the total chain number $\Sigma_N c(N)$, this will not fully relax until the time-scale τ_{rs} is attained, even if this is much larger than τ_{ei}.[20]

Finally we turn to the bond-interchange process[22] in which micelles transiently fuse to form a four-fold link before splitting again into differently connected components (Figure 4.1). This process, like end-interchange, conserves chain number. Bond interchange is far less effective than reversible scission or end-interchange in speeding up the disentanglement of micelles (see Section 4.4). In fact, although a breaking time $\tau_{bi} = (k_{bi}\bar{L}\phi)^{-1}$ can be defined, this enters the rheological models differently from τ_{rs} or τ_{bi} (Section 4.4.2). Bond interchange does however allow chains to effectively pass through one another by decay of the four-fold intermediate.[23] Such links are however prone to dissociate into threefold links, which decay by end interchange.

The reaction kinetics in *branched* micellar networks is less easy to cast in terms of simple mean-field equations. However, within such a network, alongside any bond-interchange reactions that are present, structural relaxation can still occur by reversible scission or end interchange of a section of the micellar network between junctions. Time-scales τ_{rs} or τ_{ei} can then be defined as the lifetime of a typical network strand before destruction by such a process. In the reversible scission case (4.11) still holds, now with $\bar{L}$ the mean strand length in the network.[10] Turning to the case where rings are present, the three reaction schemes of Figure 4.1 remain relevant. The chain number $\Sigma_N c(N)$, though not the ring number $\Sigma_N c_r(N)$, is still conserved by the two interchange processes. Whenever open chains are present, reversible scission is needed for them to reach full thermal equilibrium.[16]

As stated previously, the static mean-field theories given above (in Sections 4.2.1–4.2.3) take as their parameters E, l_p, a_0, v_0. Also relevant is an excluded volume parameter w.[1,2] (Within mean-field, this parameter only affects the purely ϕ-dependent term $F_0(\phi)$ in Equations (4.1), (4.5), and (4.6) and hence has no effect on the mean micellar length $\bar{L}$ or the size distribution $c(L)$.) All of the parameters E, l_p, a_0, w in principle can have explicit dependence on the volume fraction ϕ. This certainly occurs in ionic micellar systems at low added salt, where the ionic strength itself depends strongly on ϕ. Ion binding and similar effects can also be strongly temperature dependent. Similar remarks apply to the reaction rate constants k_{rs}, k_{ei}, and hence to their activation energies $E_A \equiv -\partial \ln k/\partial \beta$. The rheological consequences of these parameter variations are discussed in Section 4.4.5.

4.3 THEORETICAL RHEOLOGY

Rheology is the measurement and prediction of flow behavior. The basic experimental tool is a rheometer. Many rheometers use a Couette cell (two concentric cylinders, of radius r and $r + h$, with the inner one rotating). Others use a cone-plate cell where a rotating cone contacts a stationary plate at its apex, with opening angle θ. In the limit of small h/r or small θ, each results in a uniform stress in steady state; in each case, the shear stress can be measured from the torque.

4.3.1 STATISTICAL MECHANICS OF STRESS AND STRAIN

We shall use suffix notation, with roman indices and the usual summation convention, for vectors and tensors; letters $a \ldots w$ can therefore stand for any of the three cartesian directions x, y, z. Greek indices will be reserved for labels of other kinds. Consider a surface element of area dA with normal vector n_i. Denote by dF_i the force exerted on the interior of the surface element by what is outside. Writing $dS_i = n_i dA$, we have

$$dF_i = \sigma_{ij} dS_j \tag{4.13}$$

which defines the (symmetric) stress tensor σ_{ij}. The hydrostatic pressure is *defined* as $p = -\sigma_{ii}/3$; what matters in rheology is the (traceless) deviatoric stress $\sigma_{ij}^{dev} = \sigma_{ij} + p\delta_{ij}$. This includes all shear stresses, and also the two normal stress differences,

$$N_1 = \sigma_{xx} - \sigma_{yy}; \quad N_2 = \sigma_{yy} - \sigma_{zz} \tag{4.14}$$

The force $f_i^{\alpha\beta}$ exerted by some particle (identified below) α on another particle β depends on their relative coordinate $r_i^{\alpha\beta}$. But this pair of particles contributes to the force dF_i only if the surface element dS_i divides one particle from the other. The probability of this happening in volume V is $dS_i r_i^{\alpha\beta}/V$. (This is easiest seen for a cubic box of side L with a planar dividing surface of area $A = L^2$. The separation of the particles normal to the surface is $\ell = r_i^{\alpha\beta} n_i$, and the probability of their lying one either side of it is ℓ/L, which can be written as $A r_i^{\alpha\beta} n_i/V$.) The total force across a surface element dS_i is then $dF_i = -\sum_{\alpha\beta} dA(r_j^{\alpha\beta} n_j) f_i^{\alpha\beta}/V$. Using (4.13), this gives

$$\sigma_{ij} = -V^{-1} \sum_{\alpha\beta} r_i^{\alpha\beta} f_j^{\alpha\beta} = -\rho^2 V \langle r_i f_j \rangle \tag{4.15}$$

where the average is taken over pairs, and ρ is the mean particle density.

In applying this formula to polymers (or micelles), one could choose the individual monomers (or surfactant molecules) as the particles, and their covalent, van der Waals, and other interactions as the forces in (4.15). But so long as the force is suitably redefined as a coarse-grained quantity that includes entropic contributions, we can equally well consider a polymer chain as a sparse string of beads connected by springs. At this larger scale, $f_i^{\alpha\beta}$ has a universal and simple dependence on $r_i^{\alpha\beta}$, deriving from an "entropic potential" $U(r_i) = (3k_B T/2b^2) r_i r_i$, where $b^2 \equiv \langle r_i r_i \rangle$. This is a consequence of the well-known Gaussian distribution law for random walks, of whichthe polymer, at this level of description, is an example. The force now obeys

$$f_i^{\alpha\beta} = -dU\left(r_i^{\alpha\beta}\right)/dr_i^{\alpha\beta} = -\left(3k_B T/b^2\right) r_i^{\alpha\beta} \tag{4.16}$$

which gives, using (4.15), the polymeric contribution to the stress tensor:

$$\sigma_{ij}^{\text{pol}} = \frac{N_{\text{spr}}}{V} \frac{3k_B T}{b^2} \langle r_i r_j \rangle \tag{4.17}$$

Here the average is over the probability distribution $P(r_i)$ for the end-to-end vectors of our polymeric subchains; N_{spr}/V is their number per unit volume. There may also be a significant stress contribution from local viscous dissipation in the (Newtonian) solvent; this is $\sigma_{ij}^{\text{sol}} = \eta^{\text{sol}}(K_{ij} + K_{ji})$, where K_{ij} is the *velocity gradient tensor*, which we now define.

Consider a uniform deformation of a material to a strained from an unstrained state. The position vector r_i of a material point is transformed into r_i'; the *deformation tensor* E_{ij} is defined by $r_i' = E_{ij} r_j$. For small deformations, one can write this as $E_{ij} = \delta_{ij} + e_{ij}$ so that the displacement $u_i = r_i' - r_i$ obeys $u_i = e_{ij} r_j$. Alternatively, we may write this as $e_{ij} = \nabla_j u_i$. Consider now a time-dependent strain, for which $v_i \equiv \dot{u}_i(r_i)$ defines the fluid velocity. We *define* the velocity gradient tensor as $K_{ij} = \nabla_j v_i = \dot{e}_{ij}$. An important result, easily proved from this definition, is that $\dot{E}_{ij} = K_{ik} E_{kj}$.

For a simple shear flow at shear rate $\dot{\gamma}$ with a flow velocity along x that depends linearly on y, we have $v_i = \dot{\gamma} y \delta_{ix}$. The velocity gradient tensor is then $K_{ij} = \dot{\gamma} \delta_{ix} \delta_{jy}$. For a time-dependent shear rate $\dot{\gamma}(t)$, one finds $E_{ij}^{tt'} = \delta_{ij} + \gamma(t, t') \delta_{ix} \delta_{jy}$ where $E_{ij}^{tt'}$ is defined as the deformation tensor connecting vectors at time t to those at time t', and $\gamma(t, t') = \int_t^{t'} \dot{\gamma}(t'') dt''$ is the total strain between these two times.

4.3.2 LINEAR RHEOLOGY

Imagine an undeformed block of material which is suddenly subjected, at time t_1, to a small shear strain γ. The resulting deformation tensor is $E_{ij} = \delta_{ij} + \gamma \delta_{ix} \delta_{jy}$. Suppose we measure the corresponding stress tensor $\sigma_{ij}(t)$. Linearity requires that

$$\sigma_{yx} = \sigma_{xy} = G(t - t_1)\gamma \tag{4.18}$$

and that all other deviatoric components of σ_{ij} vanish by symmetry. (For example, $N_1(\gamma) = N_1(-\gamma)$, so that $N_1 = O(\gamma^2)$.) This defines the linear step-strain response function $G(t)$. This function is zero for $t < 0$ and at $t = 0$ jumps to an initial value controlled by microscopic degrees of freedom. There follows a very rapid decay to a more modest level, G_0 arising from mesoscopic (polymeric) degrees of freedom. On the time-scale of mesoscopic relaxations, which are responsible for viscoelasticity, $G(t)$ then falls further.

Now suppose we apply a time-dependent, but small, shear strain $\gamma(t)$. By linearity, we can decompose this into a series of infinitesimal steps of magnitude

$\dot{\gamma}(t')dt'$; the response to such a step is $d\sigma_{xy}(t) = G(t-t')\dot{\gamma}(t')dt'$. We may sum these incremental responses, giving

$$\sigma_{xy}(t) = \int_{-\infty}^{t} G(t-t')\dot{\gamma}(t')dt' \qquad (4.19)$$

This is an example of a *constitutive equation*. (The constitutive equation for nonlinear flows is far more complicated.) In steady shear $\dot{\gamma}(t)$ is constant; therefore from (4.19) one has $\sigma_{xy}(t) = \dot{\gamma}\int_{-\infty}^{t} G(t-t')dt'$. However, the definition of a fluid's linear viscosity (its zero-shear viscosity, η) is the ratio of shear stress to strain rate in a steady measurement when both are small; hence $\eta = \int_{0}^{\infty} G(t)dt$.

The case of an oscillatory flow is often studied. We write $\gamma(t) = \gamma_0 e^{i\omega t}$ (taking the real part whenever appropriate); substituting in (4.19) gives after trivial manipulation

$$\sigma_{xy}(t) = \gamma_0 e^{i\omega t} G^*(\omega) \qquad (4.20)$$

where $G^*(\omega) \equiv i\omega \int_{0}^{\infty} G(t)e^{-i\omega t}dt$. This is called the complex modulus, and is conventionally written $G^*(\omega) = G'(\omega) + iG''(\omega)$ where the real quantities G' and G'' are respectively the storage modulus and the loss modulus. Most polymeric fluids exhibit a longest relaxation time τ in the sense that for large enough t, the relaxation modulus $G(t)$ falls off asymptotically like $\exp[-t/\tau]$. In this case one has at low frequencies $G' \sim \omega^2$ and $G'' \sim \omega$. For polymer melts and concentrated solutions, as frequency is raised G' passes through a plateau whereas G'' starts to fall; at high frequencies both rise again. This is sketched in Figure 4.2 on double logarithmic axes. One can also study the steady-state flow response to an oscillatory stress. This defines a frequency-dependent complex compliance $J^*(\omega)$; however, within the linear response regime this is just the reciprocal of $G^*(\omega)$.

4.3.3 THE LINEAR MAXWELL MODEL

The simplest imaginable $G(t)$ takes the form $G(t) = G_0 \exp(-t/\tau_M)$ for all $t > 0$ and is called the linear Maxwell model. G_0 is a transient elastic modulus and τ_M is

FIGURE 4.2 Artist's impression of the viscoelastic spectrum for a typical polymeric material.

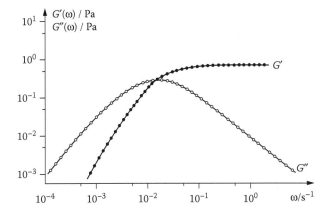

FIGURE 4.3 Viscoelastic spectrum for a system of entangled micelles:[27] arguably nature's closest approach to the linear Maxwell model, for which the peak in G'' is perfectly symmetric and G' crosses through this peak at the maximum. Figure courtesy H. Rehage.

called the Maxwell time; the viscosity is $\eta = G_0 \tau_M$. In nature, nothing exists that is quite as simple as the Maxwell model: but the low-frequency linear viscoelasticity of certain giant micellar systems is remarkably close to it (Figure 4.3). The Maxwell model has $G^*(\omega) = G_0 i\omega\tau_M/(1 + i\omega\tau_M)$ whose real and imaginary parts closely resemble Figure 4.3: a symmetric maximum in G'' on log-log through which G' passes as it rises towards a plateau. This is distinct from ordinary polymers, where the peak is lopsided (Figure 4.2). We return to this in Section 4.4 below.

4.3.4 Linear Viscoelasticity of Polymers: Tube Models

Figure 4.4 shows a flexible polymer. Dense polymers are like an entangled mass of spaghetti, lubricated by Brownian motion. The presence of other chains, shown schematically as circular obstacles in Figure 4.4, impedes the diffusion of any particular chain. Suppose for a moment that the ends of that chain are held fixed. In this case, the effect of the obstacles can be represented as a tube (also shown in Figure 4.4). Because it wraps around a random walk (namely the chain), the tube is also a random walk; its number of steps N_T and step-length b (comparable to the tube diameter) must obey the usual relation $\langle R^2 \rangle = N_T b^2$ where R is the end-to-end distance of both the tube and the chain. At present, there is no fundamental theory that can predict b; in what follows, it is a parameter. It is quite large, so that chains smaller than a few hundred monomers do not feel the tube at all.[2]

Suppose we take a dense polymer system and perform a sudden step-strain with shear strain γ. The chain will instantaneously deform with the applied strain, causing a drop in its configurational entropy. Quite rapidly, degrees of freedom

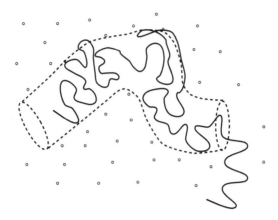

FIGURE 4.4 A polymer chain (heavy line). Surrounding chains present obstacles (circles) that the chain cannot cross. These can be modeled by a tube (dotted), from which this particular chain has partially emerged.

within the tube can relax by Brownian motion, leaving only deformation at the scale of the tube: the residual entropy change ΔS is that of the tube in which the chain resides. A calculation[2] of the entropy of deformed random walks gives a resulting free energy change

$$\Delta F = -T\Delta S = \frac{1}{2}G_0\gamma^2 \tag{4.21}$$

where $G_0 = 4k_B Tn/5$ is identified as the transient elastic modulus.

What happens next? The chain remains hemmed in by its neighbors and can diffuse only along the axis of the tube. The resulting curvilinear diffusion constant D_c is inversely proportional to chain length L.[1] Curvilinear diffusion allows a chain to escape through the ends of the tube. When it does so, the chain encounters new obstacles and, in effect, creates a new tube around itself. However, to a good approximation we may assume that this new tube, which is created at random *after* the original strain was applied, is undeformed.[2] The stored free energy then obeys $\Delta F(t) = G(t)\gamma^2/2$ where

$$G(t) = G_0\mu(t) \tag{4.22}$$

Here we identify $\mu(t)$ as the fraction of the original tube (created at time zero) which is still occupied, by any part of the chain, at time t. The problem of finding $\mu(t)$ can be recast[1] as the problem of finding the survival probability up to time t of a particle of diffusivity D_c which lives on a line segment $(0,L)$, with absorbing boundary conditions at each end; the particle is placed at random on the line segment at time zero. To understand this, choose a random segment of the initial tube and paint it red; then go into a frame where the *chain* is stationary and the tube is moving. The red tube segment, which started at a random place, diffuses relative to the chain and is lost when it meets a chain end.

The result of this calculation is:[2]

$$\mu(t) = \sum_{n=\text{odd}} \frac{8}{n^2\pi^2} \exp\left[-n^2 t/\tau_R\right] \tag{4.23}$$

where $\tau_R = L^2\pi^{-2}/D_c$. This parameter is called the reptation time. The calculated $\mu(t)$ is dominated by the slowest decaying term — hence it is not that far from the Maxwell model, though clearly different from it, and resembles the left part of Figure 4.2. (To understand the upturn at the right-hand side of that figure, you need to include intratube modes; see Ref. 2) The tube model gives quantitative interrelations between observable quantities, and the number of these relations significantly exceeds the number of free parameters in the theory — which can be chosen, in effect, as G_0 and a friction constant $\tilde{D}_c = D_c L$. The model predicts that $\eta = G_0 L^3/(12\tilde{D}_c) \sim L^3$ for long chains. The experiments lie closer to $\eta \sim L^{3.4}$ (at least for modest L) but the discrepancy can be attributed to intratube fluctuation modes and their effects on other chains; see Ref. 26.

4.3.5 NONLINEAR RHEOLOGY

Nonlinear rheology addresses the response of a system to finite or large stresses. In the absence of a superposition principle, such as the one as holds for linear response, the range of independent measurements is much wider.

In nonlinear step strain, a deformation $E_{ij} = \delta_{ij} + \gamma\delta_{ix}\delta_{jy}$ is suddenly applied at time t_1, just as in Section 4.3.2, but now γ need not be small. Analogous to (4.18) we define

$$\sigma_{xy} = G(t - t_1; \gamma)\gamma \tag{4.24}$$

where a factor of γ ensures that $G(t - t_1; 0) = G(t - t_1)$ (so the small-strain limit coincides with the linear modulus defined previously). A system is called factorable if $G(t - t_1; \gamma) = G(t - t_1)h(\gamma)$, but this is not the general case. Whereas at linear order all other deviatoric components of σ_{ij} vanished by symmetry, in the nonlinear regime one can expect finite normal stress differences N_1, N_2, as defined in (4.14). For micelles, these can be large.[27]

Another key experiment in the nonlinear shear regime is to measure the flow curve, that is, the relationship $\sigma(\dot{\gamma})$ in steady state. For a Newtonian fluid this is a straight line of slope η; upward curvature is called shear thickening and downward curvature shear thinning. Flow curves can also exhibit vertical or horizontal discontinuities: these are usually associated with an underlying instability to an inhomogeneous flow, to which we return in Section 4.5.

4.3.6 NONLINEAR RHEOLOGY OF POLYMERS

Imagine a dense polymer system to which a *finite* strain is suddenly applied. As previously discussed, the random-walk tube, which controls the slow degrees of freedom, is deformed. We may define the tube as a string of vectors bu_i^α (where α labels the tube segment) with the initial u_i^α random unit vectors. On deformation, $u_i^\alpha \to E_{ij}u_j^\alpha$ where one may show[2] that the average length of the vector has gone up: $\langle | E_{ij}u_j^\alpha | \rangle_\alpha \equiv \chi > 1$. This increase in the length of the tube is rapidly relaxed by a breathing mode,[2] killing off a fraction $1-1/\chi$ of the tube segments, and relaxing the magnitude, but not the direction, of the mean spring force in each tube segment. The result that then follows from (4.17) is

$$\sigma_{ij}(t > t_1) = \frac{3nk_BT}{\langle | E_{ij}u_j^\alpha | \rangle_\alpha} \left\langle \frac{E_{ik}u_k^\alpha E_{jl}u_l^\alpha}{| E_{im}u_m^\alpha |} \right\rangle_\alpha \mu(t - t_1) \qquad (4.25)$$

Here the final $\mu(t - t_1)$ is inserted on the grounds that, after retraction is over, the dynamics proceeds exactly as discussed previously for escape of a chain from a tube. This stress relaxation is of factorable form (now choosing $t_1 = 0$):

$$\sigma_{ij}(t) = 3nk_BTQ_{ij}(E_{mn})\mu(t) \qquad (4.26)$$

which defines a tensor Q_{ij} as a function of the step deformation E_{mn}. Expanding the result in γ for simple shear gives $Q_{xy} = 4\gamma/15 + 0(\gamma^2)$; this confirms the value of the transient modulus G_0 quoted after (4.21). In finite amplitude shear, Q_{ij} is sublinear in deformation: this is called strain-softening and the same physics is responsible for shear thinning in polymers under steady flow. Like many other predictions of the tube model, these ones are quantitative to 10 or 15 percent. Note that the factorability stems from the separation of time scales between slow (reptation) modes and the faster ones (breathing) causing retraction; close experimental examination shows that the factorization fails at short times, as it should.

Alongside shear thinning, polymeric fluids exhibit several exotic phenomena under strong flows.[28] A goal of theoretical rheology is to obtain for each material studied a *constitutive equation*: a functional relationship between the stress at time t and the deformation applied at all previous times. For the tube model in its simplest form (which invokes an independent alignment approximation or IAA) this is:[2]

$$\sigma_{ij}^{pol}(t) = G_0 \int_{-\infty}^{t} \dot{\mu}(t - t')Q_{ij}(E_{mn}^{tt'}) \, dt' \qquad (4.27)$$

where $Q_{ij}(E_{mn})$ is as defined in (4.26) and $E_{mn}^{tt'}$ is the deformation tensor between t and an earlier time t'. This is the deformation seen by tube segments that were created at time t'; $G_0 Q_{ij}$ gives the corresponding stress contribution. The factor $\dot{\mu}(t - t')$ (with $\mu(t)$ obeying (4.23)) is the probability that a tube segment, still alive at time t, was created at t'. The Doi–Edwards equation (4.27) has spawned advanced variants in which both IAA and other simplifications of the tube model have been improved upon; for a recent review see Ref. 29.

4.3.7 UPPER CONVECTED MAXWELL MODEL

Due to the complexity of (4.27), there is a temptation to start instead from simpler models, such as the dumbbell model, which in fact predated the tube model by many years. A polymer dumbbell is defined as two beads connected by a Gaussian spring. We forget now about entanglements, and represent each polymer by a single dumbbell, whose end-to-end vector is R_i. The force in the spring is taken as $f_i = -\lambda R_i$. In thermal equilibrium, it follows that $\langle R_i R_j \rangle_e = k_B T \delta_{ij} / \lambda$ and we can write (4.17) as $\sigma_{ij}^{\mathrm{pol}} = n_D \lambda \langle R_i R_j \rangle$, where $n_D = N_D / V$ is the number of dumbbells per unit volume. The dumbell model assumes that the two beads undergo independent diffusion subject to (a) the spring force, and (b) the advection of the beads by the fluid in which they are suspended. These ingredients can be combined to give a relatively simple equation of motion for $\sigma_{ij}^{\mathrm{pol}}$.[28] The result is

$$\frac{d}{dt}\sigma_{ij}^{\mathrm{pol}} = K_{il}\sigma_{lj}^{\mathrm{pol}} + \sigma_{il}^{\mathrm{pol}}K_{jl} + \tau^{-1}\left(G_0\delta_{ij} - \sigma_{ij}^{\mathrm{pol}}\right) \qquad (4.28)$$

where $\tau = \zeta/(4k_B T\lambda)$ is the relaxation time, and $G_0 = N_D\lambda/V$ is the transient modulus, of the system. This is a differential constitutive equation, which can also be cast into an integral form resembling (4.27); it is called the upper convected Maxwell model.[28]

The equations above consider only the polymeric contribution to the stress. To this can be added a standard, Newtonian contribution from the solvent (see Section 4.3.1)

$$\sigma_{ij} = \sigma_{ij}^{\mathrm{pol}} + \eta^{\mathrm{sol}}(K_{ij} + K_{ji}) \qquad (4.29)$$

which defines the so-called Oldroyd B fluid. This model is the most natural extension to nonlinear flows of the linear Maxwell model of Section 4.3.3, and so its adoption for macroscopic flow modeling in micellar systems, which are nearly Maxwellian in the linear regime, is highly appealing. However, this is not enough — in particular it cannot describe the spectacular shear-thinning behavior, and related flow instabilities, seen in these systems. The simplest tensorial model capable of this is called the Johnson–Segalman model, which will be presented in Section 4.5; it reduces to Oldroyd B in a certain limit, but has additional parameters allowing a much closer approach to micellar rheology.

4.4 MICROSCOPIC CONSTITUTIVE MODELING

In 1987 Cates[4] proposed an extension of the tube model that incorporates micellar reactions, giving a constitutive model for giant micelles. We review the model (Section 4.4.2), outline its main rheological predictions (Section 4.4.4), and briefly overview the extent to which these hold experimentally. There follows a discussion of ionicity effects and branching in entangled micelles (Section 4.4.6). Attempts to model shear-thickening phenomena around $\tilde{\phi}$ are outlined in Section 4.4.7, and in Section 4.4.8 we address structural memory effects.

4.4.1 SLOW REACTION LIMIT

Consider first a system of (linear, unbranched) giant micelles for which the kinetic time scales $\tau_{rs}, \tau_{ei}, \tau_{bi}$ are very long. The equilibrium size distribution obeys (4.3) but the system is otherwise equivalent to a set of unbreakable polymers; the identity of any individual chain is preserved on the time scale of its stress relaxation. Hence, for the purposes of calculating the stress relaxation function $\mu(t)$ defined in (4.22), one has a pure polymer problem. The simplest approach then is to write $\mu(t)$ as a length-weighted average:[4, 7]

$$\mu(t) = \sum_{L} Lc(L)\mu_L(t) / \sum_{L} Lc(L) \qquad (4.30)$$

Here $\mu_L(t)$ is the function defined in (4.23) appropriate to the given chain length L, which controls the reptation time τ_R in that expression via $\pi^2 \tau_R = L^3 / \tilde{D}_c$. (Recall that $\tilde{D}_c$, the curvilinear diffusivity, is L-independent, though it does depend on ϕ, l_p, a_0 and the solvent viscosity η^{sol} which controls the local drag on a chain.) An estimate of (4.30) gives[4]

$$\mu(t) \simeq \exp\left[-A(t/\tau_R)^{1/4}\right] \qquad (4.31)$$

This has a characteristic relaxation time given by the reptation time for a chain of the average length (we abbreviate $\tau_R \equiv \tau_R(\bar{L})$) but, in contrast to the result (4.23) for monodisperse chains, it represents an *extremely* nonexponential decay. This crude result[4] could be improved by applying modern dynamic dilution concepts.[26] Nonetheless, a strongly nonexponential relaxation would remain.[26] The experimental observation of a near-monoexponential relaxation (e.g., Figure 4.3) in many viscoelastic micellar systems thus proves the presence of a different relaxation mechanism from simple reptation.[7]

4.4.2 REPTATION–REACTION MODEL

We now define τ_b, the mean breaking time for a micelle, as the lesser of τ_{rs} in (4.11) for reversible scission, or τ_{ei} in (4.12) for end interchange. (Bond interchange

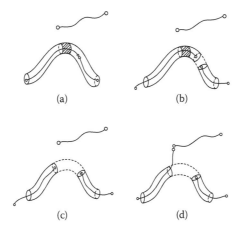

(a)   (b)

(c)   (d)

FIGURE 4.5 Stress relaxation in the reptation–reaction model with reversible scission.

is dealt with separately below.) This is the lifetime of a chain before breaking, and of an end before recombination. We also define $\zeta = \tau_b / \tau_R$, the ratio of breaking and reptation times. Ref. 4 proposes that, when ζ is small, the dominant mode of stress relaxation is as shown in Figure 4.5. The stress relaxation function $\mu(t)$ is, just as for unbreakable chains (Section 4.3.4), the probability that a randomly chosen tube segment, present at time zero, survives to time t without a chain end passing through it. However, the original chain ends do not survive long enough for ordinary reptation to occur; instead, each tube segment has to wait for a break to occur close enough to it, that the new chain end can pass through the given tube segment before disappearing again. The distance l an end can move by reptation during its lifetime τ_b obeys $\tilde{D}_c(\bar{L})l^2 \simeq \tau_b$; hence $(l/\bar{L})^2 \simeq \tau_b / \tau_R$. The waiting time τ for a new end to appear within l is $\tau_b \bar{L}/l$. This gives, for $\zeta \ll 1$, a mean stress relaxation time

$$\tau \simeq (\tau_b \tau_R)^{1/2} \qquad (4.32)$$

Moreover, if we also define $\bar{\zeta} = \tau_b / \tau$, then in the limit $\bar{\zeta} \ll 1$ of rapid breaking micellar lengths and identities change of order $\bar{\zeta}^{-1}$ times during stress relaxation. There is then no dispersion in relaxation rates and accordingly, in this limit, the resulting relaxation function $\mu(t) = \exp[-t/\tau]$ is a purely Maxwellian, monoexponential decay.[4]

For modest $\bar{\zeta}$, deviations from the Maxwellian form are expected; these have been studied numerically (see next section). Note also that as ζ falls below $1/N_T$, where N_T is the number of tube segments on the average chain, there is a crossover to a new regime, where the dominant motion of a chain end during its lifetime is not curvilinear diffusion, but breathing (see Section 4.3.6). This gives instead of (4.32)[4]

$$\tau \simeq \tau_b^{3/4} \tau_R^{1/4} N_T^{1/4} \qquad (4.33)$$

The deviations from a pure Maxwellian relaxation spectrum in this regime have also been studied;[35] they include a high-frequency turnup in $G'(\omega), G''(\omega)$ (compare Figure 4.2).

So far, we ignored bond interchange. This does not create or destroy chain ends (see Section 4.2.5) and so is rather ineffective at causing stress relaxation. Nonetheless, enhancement of relaxation does occur. As shown in Ref. 22 the result is to replace (4.32) with $\tau = \tau_R^{2/3} \tau_b^{1/3}$ There is a second effect, discussed already in Section 4.2.5, which is the evaporation of the tube caused by those bond interchange processes whose effect is to pass one chain through another. Closer inspection shows that this does not affect the regime just described, but does alter the analog of (4.33). For details, see Ref. 22.

4.4.3 CONSTITUTIVE EQUATION FOR GIANT MICELLES

A nonlinear constitutive equation for the reptation–reaction model was first worked out in Ref. 31. We assume $N_T^{-1} \ll \zeta \ll 1$ so that the linear response behavior is Maxwellian with relaxation time obeying (4.32), and all tube segments are governed by the same relaxation dynamics. We also assume that the rates of micellar reactions are unperturbed by shear.[31] The resulting constitutive equation gives for the deviatoric polymer stress:

$$\sigma_{ij}^{pol,dev} = \frac{15}{4}[W_{ij} - \delta_{ij}/3] \tag{4.34}$$

$$W_{ij}(t) = \int_{-\infty}^{t} B(v(t')) \exp\left[-\int_{t'}^{t} D(v(t''))\right] \tilde{Q}_{ij}(E_{mn}^{tt'}) dt' \tag{4.35}$$

$$\tilde{Q}_{ij}(E_{mn}) = \left\langle \frac{E_{ik} u_k E_{jl} u_l}{|E_{im} u_m|} \right\rangle_0 \tag{4.36}$$

In (4.34), $W_{ij}(t) = \langle u_i u_j \rangle$; this equation is the deviatoric part of (4.17). In (4.35), B and D are birth and death rates for tube segments. These are well approximated for $v > 0$ by $D = 1/\tau + v$, $B = 1/\tau$; and for $v < 0$ by $D = 1/\tau$, $B = 1/\tau - v$. Here $v(t) \equiv W_{ij}(t) K_{ij}(t)$ is the rate of destruction of tube segments by retraction (the same process as outlined in Section 4.3.5). In the absence of flow ($v = 0$), the birth and death rates are both $1/\tau$, the lifetime of a tube segment in the reptation–reaction process of Figure 4.5.

The physical content of (4.35) is that the stress in the system at any time t is found by integrating over past times t' the creation rate $B(v(t'))$ for tube segments, times an exponential factor which is the survival probablity of these segments up to time t, times $\tilde{Q}_{ij}$ which is the stress contribution of such a surviving segment (allowing for both its elongation and orientation by the intervening deformation $E_{mn}^{tt'}$). The quantity $\tilde{Q}_{ij}$ is in turn calculated in (4.36) where the average is over an isotropic distribution of initial tangent vectors u_i. This is a close relative (but not identical) to Q_{ij} defined for unbreakable chains in (4.27). In the linear viscoelastic limit, Equations (4.34)–(4.36) reduce to the linear Maxwell model, which obeys (4.19) with $G(t) = G_0 \exp[-t/\tau]$. Accordingly all linear viscoelastic quantities reduce to those of a pure Maxwellian fluid with τ obeying (4.32). However, the full constitutive model is, in both structure and content, quite unlike (4.28), which would be to an empirical rheologist the "natural" nonlinear extension of such a Maxwell model.

4.4.4 PRIMARY RHEOLOGICAL PREDICTIONS

We now discuss the model's primary predictions, which do not require much input in terms of how parameters such as E, k_{rs}, l_p depend on concentration or temperature. Its secondary predictions, which do require this information, are discussed in Section 4.4.5.

4.4.4.1 Linear Spectra: Cole–Cole Plots

The reptation–reaction model predicts (rather than assumes, as Maxwell did) that $\mu(t)$ approaches an exponential form in the fast breaking limit. Such behavior has by now been reported dozens of times in the literature (see, e.g., Refs. 27, 32, 33). This stems from the rapid averaging of tube-segment statistics on time scale $\tau_b = \zeta \tau$.

By numerical methods,[4,34,35] the model can also predict systematic deviations from the Maxwellian form for ζ of order unity. These predictions offer additional tests of the model, particularly in cases where τ_b can be estimated independently.[21] Such deviations are rendered most visible in the Cole–Cole representation[21,34] whereby $G''(\omega)$ is plotted against $G'(\omega)$; for exponential $\mu(t)$ the result should be a perfect semicircle. For systems well within the rapid-breaking regime, this has been confirmed in a large number of cases (see, e.g., Refs. 7, 21, 23, 33, 36, 37). Significant deviations are predicted[34] at $\bar{\zeta} \geq 0.4$. In at least one case, where τ_b is independently determined by temperature jump on the same system, a fit between the experimental and theoretical Cole–Cole plots gives good agreement on the value of $\bar{\zeta} = \tau_b/\tau$.[21] In more recent measurements, the breaking times determined by T-jump at low volume fractions were contiguous with those measured by rheology at higher ϕ;[38] see Figure 4.6. In cases where the deviation from the semicircle has a region of negative slope prior to an upturn (as in Figure 4.6), the location of the minimum can be used[35] to estimate $\bar{L}$. This was done for the system of Ref. 21, and a physically reasonable trend for $\bar{L}(\phi)$ found.[35]

In summary, the linear viscoelastic behavior of entangled micellar systems, across a wide range of different chemical types (see, e.g., Ref. 32) shows a regime of strongly Maxwellian relaxation in accord with the reptation–reaction model. The leading shape corrections to the relaxation spectrum as one departs this regime are also well accounted for, in many systems, by that model. These

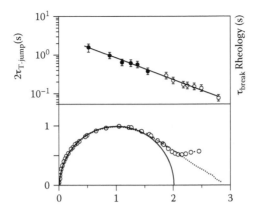

FIGURE 4.6 Upper panel: comparison of τ_{rs} measured by T-jump (●) and rheology (○) in micellar system. Lower panel: fit of Cole–Cole data (open symbols) to numerical model of Ref. 34 (dots). Figures courtesy C. Oelschlaeger.[39]

statements involve no knowledge about how the kinetic and structural inputs to the model vary with concentration, ionic strength, or temperature.

4.4.4.2 Nonlinear Rheology: Shear Banding

The main arena for comparing nonlinear predictions of the microscopic model with experiments on micelles has involved steady flow. Steady flow, if homogeneous, is fully characterized by the "flow curve" $\sigma(\dot{\gamma})$, which relates shear stress to strain rate in steady state, and the normal stress difference curves $N_1(\dot{\gamma})$ and $N_2(\dot{\gamma})$. Until the early 1990s, studies of the flow curve for giant micelles simply *assumed* homogeneity of the flow. (Checking for this has since become much easier with a variety of modern techniques.[24,25,40–43]) A very striking observation in a CPySal/NaSal system (compare Figure 4.3)[27] was that above a certain strain rate $\dot{\gamma}_p$, the shear stress σ attains a plateau value $\sigma = \sigma_p$, remaining at this level for at least two decades in $\dot{\gamma} \geq \dot{\gamma}_p$. At the same time the normal stress difference N_1 continues to increase almost linearly. This represents shear thinning of a quite drastic kind. It is used in technologies such as shampoos, allowing a highly viscous liquid to be pumped or squeezed out of the bottle through a narrow nozzle; in a non-shear-thinning fluid of equal viscosity the bottle would certainly break first.[44]

An explanation came in Ref. 45 where the reptation–reaction constitutive Equations (4.34)–(4.36) were solved in steady state (Figure 4.7). It was found that the shear stress σ has, as a function of $\dot{\gamma}$, a maximum at $(\dot{\gamma}, \sigma) = (2.6/\tau, 0.67G_0)$. Such a nonmonotonic flow curve is known to be unstable;[45–47] but a steady flow can often be recovered by developing shear bands: layers of fluid with unequal strain rate but equal stress, their layer normals in the velocity gradient direction.[45] In this way, the decreasing part of the flow curve is bypassed by coexistence

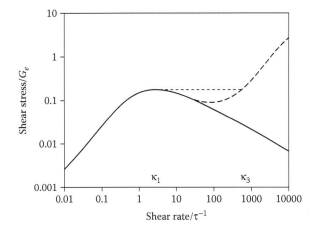

FIGURE 4.7 Flow curves for reptation–reaction model: solid line, by solution of (4.34–4.36); dashed line, with additional quasi-Newtonian stress calculated (with one fit parameter) as per Ref. 51; dotted line, top-jumping shear-banded solution. Figure courtesy N. Spenley.[46]

between two bands, one at low $\dot{\gamma} = \dot{\gamma}_1$ and one at high $\dot{\gamma}_2$, each of which is on an increasing part of the flow curve. Assuming the nature of the coexisting states does not vary as their amounts change (an assumption that ignores coupling to concentration fields, see Section 4.5), this gives a horizontal stress plateau as observed in Ref. 27. The value of this plateau could be reproduced by assuming top jumping, in which one of the coexisting states is at the maximum of the flow curve described previously (giving $\sigma_p = 0.67G_0$). However, it is now known that the mechanism for selecting coexisting stresses is more complicated than this.[48–50] The theory of Ref. 45 further assumes that the high shear branch (which is at $\dot{\gamma}_2 \geq 1000 \text{ s}^{-1}$) is not purely Newtonian, but contain highly aligned micelles that give a very large normal stress difference N_1 while maintaining a small shear stress σ.[51] This theory accounts for the observation in Ref. 27 that N_1 continues to increase almost linearly with $\dot{\gamma}$ throughout the shear-banding plateau at constant $\sigma(\dot{\gamma}) = \sigma_p$.

Since its first prediction in Ref. 45, the evidence for shear banding in viscoelastic micellar solutions has become overwhelming.[24,40–43,52] Some authors have preferred interpretations involving a shear-induced phase transition to a nematic state.[50,52] However, this is not excluded by the above arguments, which make negotiable assumptions about the physics of the high-shear branch at $\dot{\gamma} = \dot{\gamma}_2$. Even more complicated identifications of the high-shear branch are not ruled out; these might include a long-lived gel phase similar to that induced by flow in the shear-thickening region (see Section 4.4.7). In all cases, there can also be a concentration differential between the two bands; this is not addressed in the reptation–reaction model, and causes a ramp, rather than a plateau, in shear stress (Section 4.5.2). Such couplings to nematic and/or concentration fields take us beyond the limit of practical calculability within such a model; these phenomena can instead be addressed with the macroscopic approaches of Section 4.5.

One might ask whether such banding phenomena can be expected in ordinary, unbreakable polymers. Though seemingly robust for micelles, there is very limited (and perhaps controversial) evidence for shear banding in polymers.[54] This is, however, absent in even moderately *polydisperse* polymers; hence the reptation–reaction model includes among its primary predictions that the shear banding instability, seen in the Maxwellian regime, should disappear as one crosses into the unbreakable chain limit for $\zeta = \tau_b / \tau_R \gg 1$.

4.4.5 SECONDARY PREDICTIONS: CONCENTRATION AND TEMPERATURE DEPENDENCE

If E, k_{rs} (or k_{ei}, k_{bi}, as applicable), and l_p are independent of concentration, a number of secondary rheological predictions can be obtained for the concentration dependence of τ (and hence viscosity $\eta = G_0 \tau$) and related quantities. However, these depend on: (a) whether the scaling or mean-field theory applies; (b) whether reversible-scission, end interchange, or bond-interchange dominates; (c) whether one is in the pure reptation–reaction regime governed by (4.32), or in the breathing-reaction regime obeying (4.33). With these caveats, we present in Table 4.1 best estimate scaling laws for an exponent g relating the Maxwell time to concentration ($\tau \sim \phi^g$) for various different regimes.[22] These results use a blend of mean-field and scaling theory, but would not be very different if mean-field estimates were applied consistently instead. To convert from τ to the viscosity, one multiplies by $G_0 \sim \phi^{2.3}$.

We note that g is always positive; in no regime does τ fall with ϕ. Observation of *small* positive exponents is not disproof of the model, but might be evidence that a system is in the breathing regime with interchange kinetics. Third, even if parameters such as $k_{rs,ei,bi}$, E, l_p are ϕ-independent as assumed, there can be crossovers between at least six different regimes (Table 4.1), hindering experimental verification of simple power laws of any kind. Although perhaps more informative in being broken than obeyed, the scaling laws in Table 4.1 do give reasonable agreement with some of the simpler experimental systems (e.g., Refs. 7, 36, 55).

One can make further interesting predictions by assuming that E and also the relevant activation energy E_A are T-independent. Since E and E_A are likely to be

TABLE 4.1
The Exponent g (Where $\tau \sim \phi^g$ and $\eta \sim \phi^{g+2.3}$) for Various Regimes

Reaction	$\zeta \gg 1$	$1 \gg \zeta \gg N_T^{-1}$	$N_T^{-1} \ll \zeta$
reversible scission	3.4	1.4	0.9
end interchange	3.4	1.2	0.6
bond interchange	3.4	1.7	0.3

much bigger than the Arrhenius energy E_S for the solvent viscosity (which controls the curvilinear friction parameter $\tilde{D}_c$), Equation (4.32) for reversible scission (say) predicts that $\tau \propto \exp[\beta E']$ with $E' \simeq (E_A + E)/2$. Since E_A is measurable in temperature jump, this allows a sanity check on the equilibrium scission energy E; reasonable values (around $20k_BT$) have been obtained in this way, as have reasonable trends for dependence on ionic strength and other factors.[7,38,56] In one recent study,[56] strong evidence is given for the switching off of the reversible scission process at low ionic strength. At low salt, the micellar breaking time τ_b is then controlled by an interchange process, invisible to T-jump, and the activation energy comparison with T-jump data described above is inapplicable.

4.4.6 ROLE OF BRANCHING: IONICITY EFFECTS

In ionic micellar systems, and also zwitterionic ones, the overall ionic strength and/or the degree of specific counterion binding can strongly influence E, E_A, l_p, and other parameters. While naive application of Table 4.1 is then precluded, the trends can be understood, at least *a posteriori*, within the physical precepts of the reptation–reaction picture, so long as branching is allowed for. Much work along these lines is reviewed in Ref. 32.

A crucial theoretical idea is that micellar branch points are *labile*; they are always free to slide along the length of a micelle. This is quite different from crosslinks in conventional polymers, as was first recognized by Lequeux,[10] who showed that, if branch points are present in an entangled micellar network, the curvilinear diffusion constant of a chain end is $D_c = \tilde{D}_c/\bar{L}$, where $\bar{L}$ now denotes (roughly speaking) the distance to the nearest other chain end *or junction point* in the network. The main rheological predictions of the reptation-reaction model can be retained, so long as $\bar{L}$ carries this new interpretation. Once branching is widespread, as shown in Section 4.2.2, one then has $\bar{L} \sim \phi^{-1/2}$ (within a mean-field picture); for rheological purposes, the system behaves *as if* micelles were becoming shorter with concentration. This can lead to an exponent g that, for reversible scission reactions, is barely larger than zero ($\tau \sim \phi^g = \phi^{0.15}$ in a scaling picture).[38] For interchange reactions g can even be somewhat negative (contrast Table 4.1).

In a careful recent study,[38,39] Oelschlaeger et al. correlate surfactant hydrophobicity, salt, and counterion binding efficiency with the observed g values. These authors find evidence that the unbranched reptation–reaction model is applicable at relatively low ionic strength, high hydrophobicity, and low counterion binding efficiency. They then argue that the model of Lequeux[10] explains the falling g values seen when ionic strength is increased, surfactant hydrophobicity decreased, or counterion binding increased.[32]

4.4.7 SHEAR THICKENING: A ROLE FOR RINGS?

Shear thickening[56–61] is seen in a window of volume fraction around the onset of viscoelasticity $\phi \simeq \tilde{\phi}$. An initially near-inviscid system is found, after a period of prolonged shearing above a critical shear rate $\dot{\gamma}_c$, to convert into a much more viscous state. In some cases the viscosity increase is modest, and the state relaxes quite rapidly to the previous one when shearing ceases.[57] In other cases, the new phase is a long-lived gel which may show shear banding.[60] There is as yet no consensus among theorists as to the mechanisms involved. Model-building attempts have been made from time to time, based for example on shear-induced aggregation or polymerization of rodlike micelles.[62] There are also models that couple a gelation transition to shear bands,[63,64] and some that ascribe the shear thickening directly to ionic or electrokinetic phenomena.[65]

In one speculative approach, shear thickening stems from the presence of micellar rings.[16] As discussed in Section 4.2.3, the expected influence of micellar rings is maximal around ϕ_r^{max} at which a cascade of rings crosses over to a semidilute solution of open chains. (The reptation–reaction model assumes strong entanglement and hence requires $\phi \gg \phi_r^{max}$.) In the putative cascade-of-rings phase, governed in mean field by (4.7), it is an open question whether or not the rings interlink so as to form a percolating linked network.[16] Suppose for now that the rings are indeed linked for $\phi \simeq \phi_r^{max}$. If so, then were micellar kinetics suddenly to be switched off, creating a set of interlocked dead rings, the system would be a gel (with some modulus G_r, assumed small), indefinitely resisting attempts to impose a steady shear flow. Restoring a finite delinking time τ_l, this gel becomes a viscoelastic fluid of viscosity $\eta_r = G_r \tau_l$. It is possible that, if G_r is small enough, η_r remains comparable to that of the solvent; the sample is only marginally viscoelastic and in that sense would be considered to have $\phi \simeq \tilde{\phi}$ (and identified as $\phi \simeq \phi^*$ if an assumption of linear chains were made). Notice now that the viscoelastic linked-ring fluid does not in fact require complete percolation of linked rings; it only requires that linked structures extend far enough that their configurational relaxation times exceed τ_l.

If such a linked-ring fluid exists around $\phi \simeq \tilde{\phi}$, any shear rate $\dot{\gamma} \geq \tau_l^{-1}$ will cause strongly nonlinear effects. The orientating effect of elastic strains (of order $\dot{\gamma}\tau_l$) on the linked rings will alter the reaction rates. For example, pulling two inter-locked rings in opposite directions could promote bond interchange at their contact point, shifting the mean ring size upwards. This polymerization tendency could well cause shear thickening, as could tension-induced chain scission, by pushing the equilibrium in (4.9) toward polymerization. This scenario, though speculative, does explain several observed features of the shear-thickening pro-cess, including the very low shear-rate threshold for the transition, its long latency time, and a possible geometry dependence of this time.[16] Any ring-dominated regime requires suppression of open chains in the quiescent state and hence large E. Factors favoring large E include raising counterion lipophilicity and using gemini surfactants; such factors do seem broadly to cause a reduction in $\dot{\gamma}_c$ and enhancement of the viscosity jump.

4.4.8 Structural Memory Effects

Structural memory is the presence in a system of internal degrees of freedom, other than the stress, which relax on a time scale τ_s that are at least comparable to the stress relaxation time τ itself (and in some cases vastly longer). For example, one can find among micellar systems some instances where τ itself is at most a few seconds; but the value one measures for τ in repeat experiments depends on sample history over, say, the preceding 24 hours. Though hinted at anecdotally from the earliest days of the subject,[33,57] it is only very recently that structural memory effects in micellar systems have been studied in detail.[56,61]

Among effects observed in Refs. 56 and 61 are the following. In some shear-thickening micellar systems subjected to steady shearing, there is an initial latency time τ_{lat} (seconds or minutes) for the thickening to occur. After this, however, the stress level continues to adjust slowly over time scales τ_s of order hours or days before finally achieving a steady state. If shearing is stopped, the stress relaxes quite rapidly but the *memory of having been sheared* persists for times of order τ_s: if shearing is resumed within this period, a quite different τ_{lat} is measured. Moreover the stress level immediately after latency is closer to the ultimate steady-state value, and almost identical to it if the switch-off period has been short compared to τ_s. Finally, the latency time can also be raised or lowered by a prior incubation at elevated or reduced temperature,[56,61] even in a sample that has never been sheared.

These phenomena point to a robust structural property, perturbed by shear but also by temperature, as the carrier of structural memory in micellar systems. One such property immediately springs to mind, namely the micelle size distribution. However, as discussed in Section 4.2.5, this relaxes rapidly to its equilibrium form (in mean field, this is the usual $c(L) \propto \exp[-L/\bar{L}]$) when any kind of micellar reaction is present. On the other hand, as also discussed in Section 4.2.5, the relaxation of the mean micelle length $\bar{L}$ is contingent on the presence of reversible scission reactions. This is because interchange reactions conserve the chain number $\Sigma_L c(L)$; and, given the fixed shape of the distribution, $\bar{L}$ can only change if the chain number does so. As emphasized in Section 4.2.5, chain number is conserved by interchange reactions, even when rings are present; but ring number is not itself conserved.

The structural memory effects reported in Refs. 56 and 61 close to the shear-thickening transition can be explained in outline if one assumes that (a) both rings and open chains are present and (b) reversible scission reactions are very slow. The slow relaxation time is $\tau_s = \tau_{rs}$, the time scale for the chain number to reach equilibrium. Stress relaxation is not slow, since the faster rate for end-interchange or bond-interchange reactions will dominate in τ_b, allowing fast relaxation of all quantities other than $\bar{L}$. A system which is sheared or thermally treated for a time long compared to τ_{rs} will acquire a steady-state chain number appropriate to those conditions; if conditions are now changed, it will take a time of order τ_{rs} to relax to the new value. Among things that can vary with chain number are, of course, the latency time in the thickening transition; and also the Maxwell time for stress relaxation in the quiescent state.

The above scenario could hold equally well for giant micelles in the fully entangled regime. For a system showing slow reversible scission in T-jump,[38,56] it would be interesting to look for the effects of nonlinear flow, and also thermal pretreatment, on the Maxwell time. If the nonlinear flow creates a shear-banding region, this might reveal whether the average chain length in the high-shear band is significantly perturbed by the flow. A more radical speculation is that at least in some cases, when a shear-induced gel phase that forms below $\tilde{\phi}$ persists after stress is removed (and does not retain nematic order),[52] this state is in disequilibrium *solely* through having an enhanced $\bar{L}$ (of lifetime $\tau_s = \tau_{rs}$). If so, the shear-induced gel state is just another instance of the entangled regime of giant micelles, to which the reptation–reaction model can be applied. An open issue would be how $\bar{L}$ acquires its nonequilibrium value during the induction period of the shear-thickening transition.

4.5 MACROSCOPIC CONSTITUTIVE MODELING

So far we have discussed microscopic constitutive modeling, which aims to predict rheology from an understanding of the microscopic dynamics of giant micelles. Pursuing this approach further, particularly to address nonstationary shear-banded flows, becomes prohibitively complicated. To make progress, we now turn to macroscopic constitutive modeling.

Before discussing individual models, we sketch the basic features that are common to all of them. The stress σ_{ij} is taken to comprise additive contributions from the micelles and from a Newtonian solvent (as per (4.29)):

$$\sigma_{ij} = \sigma_{ij}^{pol} + \eta^{sol}(K_{ij} + K_{ji}) \qquad (4.37)$$

Neglecting inertia (for small Reynolds number), the stress obeys the force balance equation

$$\nabla_i \sigma_{ij} - \nabla_j p = \nabla_i \sigma_{ij}^{pol} + \eta^{sol}\nabla^2 v_j - \nabla_j p = 0 \qquad (4.38)$$

Here p is an isotropic pressure, which maintains fluid incompressibility:

$$\nabla_i v_i = 0 \qquad (4.39)$$

The viscoelastic stress σ_{ij}^{pol} is then written as a function of some underlying microstructural quantities, whose identities vary according to the system and regime of interest. Common choices include the volume fraction ϕ and molecular deformation W_{ij} of the polymeric component; the orientation tensor Q_{ij} in nematics; the micellar length distribution $P(L)$, etc.:

$$\sigma_{ij}^{pol} = \sigma_{ij}^{pol}(W_{ij},\phi,...) \qquad (4.40)$$

Among these microstructural variables we distinguish fast from slow variables. The former quickly relax to local steady-state values determined by the latter, whereas each slow variable requires its own equation of motion. Formally, the slowest variables are the hydrodynamic ones, which relax at a vanishing rate $\omega \propto k^{\alpha}$ ($\alpha > 0$) for small wavenumbers $k \to 0$. In viscoelastic solutions however, variables that are not strictly hydrodynamic nonetheless relax very slowly, and must be included in the above list.[66] An example is the molecular deformation W_{ij} governed by the Maxwell time τ (often seconds or minutes).

As discussed in Section 4.4.8, many micellar systems show a pronounced structural memory, with degrees of freedom that relax on a time scale greater than the stress relaxation time τ. One candidate for this is the mean micellar length $\bar{L}$, with a time scale $\tau_{rs} \gg \tau$; another is the micellar volume fraction ϕ. Consider, then, a scenario in which the relevant dynamical variables are W_{ij}, ϕ and $\bar{L}$, with respective relative relaxation time scales $\tau \ll \Lambda^2/D \ll \tau_{rs}$ (for a micellar diffusion coefficient D and sample size Λ). The concentration evolves with conserved dynamics of the form

$$D_t \phi = \nabla M \nabla \mu (\nabla_n v_m, W_{nm}, \phi) \tag{4.41}$$

with a mobility M, proportional to D. The derivative $D_t \equiv \partial_t + v_l \nabla_l$ denotes rate of change in a fluid element convected with the flow field v_i. The direct counterpart of (4.41) for nonconserved quantities such as W_{ij} and $\bar{L}$ is an equation of the form

$$D_t W_{ij} = \frac{1}{\tau} G_{ij}(\nabla_n v_m, W_{nm}, \phi) \tag{4.42}$$

Sometimes, however, this differential structure is replaced by an integral form

$$\bar{L} = \int_{-\infty}^{t} dt' M(t - t') g(t') \tag{4.43}$$

where g depends on time t' via all relevant quantities $W_{ij}(t')$, etc.

The basic structure outlined above encompasses the microscopic models discussed earlier. For example, the reptation–reaction model has (4.34) for (4.40); (4.35) to (4.36) which can be cast into the form of (4.42); and assumes the concentration to remain uniform (so that (4.41) is suppressed, with ϕ constant in (4.40)). In the macroscopic approach to constitutive modeling, one instead arrives at equations of the form (4.40) to (4.42) by ansatz, or by an exact description of a simplified system (such as the dumbbell model of Section 4.3.7). The crucial advantage of the macroscopic approach is that it allows coupling between the flow and microstructural quantities such as ϕ, Q_{ij}, and $P(L)$ to be incorporated in a simple way. Such coupling is common in shear-banding systems such as wormlike micelles, but often too complicated to capture in microscopic models. Another ingredient, almost always absent from microscopic models (though see Ref. 67), is that operators such as G in (4.42) should in principle contain spatially nonlocal terms, which are needed to correctly describe the structure of spatially inhomogeneous (i.e., banded) flows.[48]

4.5.1 JOHNSON-SEGALMAN MODELS FOR SHEAR THINNING

Within the framework just described, we now discuss some specific models of shear thinning. These are designed to reproduce, at the level of macroscopic modeling, the nonmonotonic constitutive curve of the microscopic reptation–reaction model (Section 4.4.4), for which homogeneous flow is unstable with respect to the formation of shear bands. From now on, we reserve the term "constitutive curve" for the underlying nonmonotonic relation between stress and strain rate in homogeneous flow, and "steady-state flow curve" for the actual stress/strain-rate relation measured in an experiment, where flow might be heterogeneous.

The most widely used model was originally devised by Johnson and Segalman,[68] and later extended by Olmsted et al.[69] to include the spatially nonlocal terms needed to describe the structure of the interface between the bands. Force balance and incompressibility are given by (4.38) and (4.39). The viscoelastic stress of (4.40) is assumed to depend linearly on the molecular deformation tensor W_{ij}, and on the volume fraction ϕ via a modulus G:

$$\sigma_{ij}^{pol} = G(\phi)W_{ij} \qquad (4.44)$$

The deformation tensor W_{ij} obeys diffusive Johnson–Segalman (dJS) dynamics as follows:

$$D_t W_{ij} = a(D_{il}W_{lj} + W_{il}D_{lj}) + (W_{il}\Omega_{lj} - \Omega_{il}W_{lj}) + 2D_{ij} - \frac{1}{\tau(\phi)}W_{ij} + D\nabla^2 W_{ij} \qquad (4.45)$$

in which D_{ij} and Ω_{ij} are respectively the symmetric and antisymmetric parts of the velocity gradient tensor $K_{ij} = \nabla_i v_j$. For the moment we assume that ϕ is uniform so that $G(\phi)$ and $\tau(\phi)$ are constants in space and time for any sample.

Setting $a = 1$ and $D = 0$ in (4.45) we recover Oldroyd B dynamics, as derived in Section 4.3.7 by considering an ensemble of relaxing dumbbells. The resulting constitutive curve $\sigma_{xy}(\dot\gamma) = G\dot\gamma\tau + \eta^{sol}\dot\gamma$ fails to capture the dramatic shear thinning of wormlike micelles. To allow for shear thinning, the JS model invokes a slip parameter a, with $|a| \leq 1$. When $|a| < 1$, the dumbbells no longer deform affinely, but slip relative to the flow field. The resulting constitutive curve is then

$$\sigma_{xy}(\dot\gamma) = \frac{G\dot\gamma\tau}{1 + (1 - a^2)\dot\gamma^2\tau^2} + \eta^{sol}\dot\gamma \qquad (4.46)$$

The viscoelastic (first) term now shear-thins dramatically and is nonmonotonic, increasing as $G\dot\gamma\tau$ for small $\dot\gamma$ before decreasing towards zero at higher shear

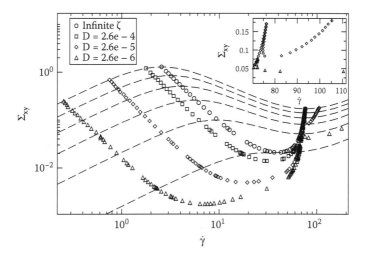

FIGURE 4.8 Dashed lines: constitutive curves for the dJS model with $G \sim \phi^{2.2}$, $\tau \sim \phi^{1.1}$, and various ϕ. Circles: limits of linear stability in the dJS model. Squares: corresponding limits for the full dJSϕ model with a realistic micellar diffusion coefficient. (Diamonds and triangles are for an artificially reduced D.) Inset: zoom on large $\dot{\gamma}$. Figure from Ref. 70. © 2003 by the American Physical Society.

rates. In contrast, the Newtonian solvent stress always increases with $\dot{\gamma}$. The overall shape of the constitutive curve thus depends on the relative strength of these contributions. For $\eta^{sol} > G\tau/8$ the solvent dominates, restoring monotonicity (bottom dashed curve in Figure 4.8). For lower η^{sol} the negative slope survives over some range of shear rates (upper dashed curves). With $G \sim \phi^{2.2}$ and $\tau(\phi) \sim \phi^{1.1}$,[22] one obtains the full family of dashed curves $\sigma_{xy}(\dot{\gamma}, \phi)$ of Figure 4.8. [70]

4.5.1.1 Concentration Coupling

So far, we have assumed the micellar concentration to remain spatially uniform. Generically, however, one expects concentration fluctuations to be important in sheared multi-component solutions when different species have widely separated relaxation times,[71–74] as seen experimentally in Refs. 75–78. This was first explained by Helfand and Fredrickson in the context of polymer solutions,[79] as follows. Under shear, parts of a stretched polymer chain (or micelle) in regions of low viscosity will, on relaxing to equilibrium, move further than parts mired in regions of high viscosity and concentration. A relaxing chain thus on average moves toward the high concentration region. This provides a positive feedback whereby chains migrate up their own concentration gradient, leading to flow-enhanced concentration fluctuations.

In a remarkable paper, Schmitt et al.[80] outlined the implications of this feedback mechanism for the shear-banding transition. They predicted an enhanced tendency to form bands, together with the existence of a concentration difference between the bands in steady state. They further noted that this difference would lead to a slight upward ramp in the stress "plateau" of the steady-state flow curve. Subsequently, strongly enhanced concentration fluctuations were seen in the early-time kinetics of shear-band formation.[81] Observations of a ramping stress plateau are now widespread; for example, see Ref. 52. In Ref. 70, therefore, one of us proposed an extension to the dJS model, by combining the constitutive equation (4.45) with a two-fluid model for flow-concentration coupling.

The basic assumption of this two-fluid approach[71–74] is a separate force-balance for the micelles (velocity v_i^m, volume fraction ϕ) and solvent (velocity v_i^s, volume fraction $1-\phi$) in any fluid element. Any relative velocity $v_i^{rel} = v_i^m - v_i^s$ (implicitly assumed zero in the ordinary dJS model) can then give rise to concentration fluctuations. The forces and stresses acting on the micelles are assumed as follows: (i) the usual viscoelastic stress $G(\phi)W_{ij}$; (ii) an osmotic force $\phi\nabla_i\delta F/\delta\phi$ derived from a free energy F, leading to conventional micellar diffusion; (iii) a drag force $\zeta(\phi)v_i^{rel}$ impeding motion relative to the solvent with a drag coefficient ζ; (iv) an additional Newtonian stress $2\phi\eta^{pol}D_{ij}^{m0}$ due to fast micellar relaxation processes such as Rouse modes, where D_{ij}^{m0} is the symmetric traceless part of the micellar strain rate tensor; and (v) a hydrostatic pressure. The solvent experiences the usual Newtonian viscous stress; a drag force (equal and opposite to that on the micelles) and a hydrostatic pressure. The constitutive curves of this dJSϕ model are the same as for the original dJS model (Figure 4.8). The relevance of the new coupling is that any *heterogeneity* in the flow variables now affects the concentration field, and vice versa. As we will show below, this enhances the tendency to form shear bands and leads to a concentration difference between the bands in steady state (Section 4.5.2), as predicted in Ref. 80.

4.5.1.2 A Simplified Scalar Model

The dJS model is the simplest *tensorial* model to capture the negatively sloped constitutive curve of the full reptation-reaction model. An even simpler model[82] neglects normal stresses and considers only the shear stress $\sigma = \sigma_{xy}$ and the shear component $\dot\gamma = \nabla_y v_x$ of the strain rate tensor. It further equates both τ and G to unity, so that $\sigma_{xy}^{pol} = W_{xy}$. The force balance and constitutive equations are then

$$\sigma = \sigma^{pol} + \eta^{sol}\dot\gamma \qquad (4.47)$$

$$\tau\partial_t\sigma^{pol} = -\sigma^{pol} + g(\dot\gamma\tau) + l^2\nabla^2\sigma^{pol} \qquad (4.48)$$

in which the choice $g(x) \equiv x/(1+x^2)$ is made; this recovers a JS-like constitutive curve.

4.5.2 Steady Shear Bands

For an applied shear rate in the regime of decreasing stress, steady homogeneous flow is unstable.[83] This can readily be confirmed by an explicit linear stability analysis of either the dJS model or the simpler scalar model of (4.47) and (4.48).[84] Such an

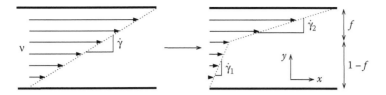

FIGURE 4.9 Left: homogeneous flow. Right: shear-banded flow.

analysis also shows that coupling to a concentration field can increase the tendency toward instability (Figure 4.8).

For a constitutive curve of the shape shown in Figure 4.8, this instability triggers formation of two bands of shear rates $\dot{\gamma}_1$ and $\dot{\gamma}_2$, one on each of the stable branches, with a flat interface between bands whose normal is oriented in the flow-gradient direction y (Figure 4.9). The relative volume fractions ($f, 1-f$) of the bands arrange themselves to match the spatially averaged shear rate $\dot{\gamma}$ imposed on the cell as a whole. (It is this averaged quantity that now appears on the abscissa of the experimental flow curve, $\sigma(\dot{\gamma})$.) As $\dot{\gamma}$ increases, the width f of the high shear band increases at the expense of the low shear band. Force balance demands that the shear stress σ_{xy} is common to both the bands. Assuming that the nature of each band does not vary as their amounts change (i.e., neglecting concentration coupling), this gives a plateau in the observed flow curve $\sigma(\dot{\gamma})$.

The scenario just described, first proposed for micelles in Ref. 45, was confirmed by explicit numerical calculation on the dJS model.[85] The resulting steady state flow curve indeed comprises two homogeneous branches ($\dot{\gamma} < \dot{\gamma}_1$ and $\dot{\gamma} > \dot{\gamma}_2$) connected by a plateau across the banding regime $\dot{\gamma}_1 < \dot{\gamma} < \dot{\gamma}_2$. A typical flow profile in the banding regime is shown in Figure 4.10. Note the

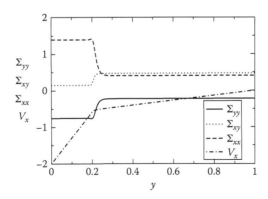

FIGURE 4.10 Shear-banded profile predicted by evolving the dJS equations in one spatial dimension, y. The imposed shear rate $\dot{\gamma} = 2.0$ lies toward the left of the stress plateau. $G = \tau = 1$. Figure from Ref. 85. © 2005 by the American Physical Society.

smooth variation across the interface, which has a width $l \propto \sqrt{D\tau}$ set by the nonlocal term in (4.45). This term confers a smooth interface and a robust, reproducibly selected value σ_{sel} of the plateau stress.[48]

In recent years, the experimental evidence for shear banding in wormlike micelles has become overwhelming.[24,40–43,52] Reports of kinks, plateaus and non-monotonicity in the flow curve are now widespread, while spatially resolved NMR[40–42] and birefringence[43] data provide direct evidence for banding in both shear rate and microstructure. As noted previously, in some cases, the associated stress plateau is not perfectly flat, but ramps upwards from left to right. In a cylindrical Couette geometry, there will always be a small slope caused by a slight stress gradient (absent in the planar case of Figure 4.9) causing the high shear band always to reside next to the inner cylinder. As this band expands outwards with increasing applied shear rate into regions of lower stress, the overall torque must increase to ensure that the interface between the bands stays at the selected stress σ_{sel}. An alternative explanation of the upward slope, independent of cell geometry, is coupling between flow and concentration.[80] If a concentration difference exists between the bands, the properties of each band must change as the applied shear rate is tracked through the coexistence regime, because material is redistributed between them as the high-shear band expands to fill the gap. This was confirmed by one of us in Ref. 53 by a numerical study of the dJSϕ model (Figure 4.11). The concentration and shear rate in each phase now define a family of tie lines, giving a nonequilibrium phase diagram.[53]

Beyond the Johnson–Segalman model, shear banding has also been studied in the Doi model of shear-thinning rigid rods.[86–88] In this case, the relevant

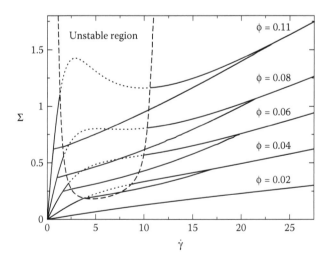

FIGURE 4.11 Steady-state flow curves in the dJSϕ model for different values of the average concentration ϕ. Concentration coupling now confers an upward slope in the banding regime. Figure from Ref. 53. Reprinted by permission.

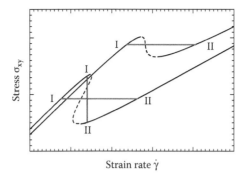

FIGURE 4.12 Stress–strain-rate curves for the Doi model with different excluded volume parameters. The dashed line segments are unstable. The straight lines indicate possible coexistence between branches I and II under conditions of common stress (horizontal lines) or strain rate (vertical line). Figure from Ref. 88. © 1999 by the American Physical Society.

microstructural variables in (4.40) are the nematic order parameter Q_{ij} and the concentration of rods ϕ. While this approach obviously ignores any effects of micellar flexibility, it takes a first step to incorporating orientational ordering, ignored by dJSϕ and likely to be important in concentrated micellar solutions close to an underlying isotropic-nematic transition. Depending on parameter values, the constitutive curve for homogeneous flow can now adopt either of the shapes in Figure 4.12. In both cases, the two stable branches correspond to a low-shear isotropic band (branch I) and a flow-induced paranematic phase (branch II). For concentrations inside the zero-shear biphasic regime, branch II touches down to the origin to form the zero-shear nematic phase. Coupling between concentration and flow arises because more strongly aligned rods in the high-shear band can pack more closely together, giving a higher concentration. In contrast, the Helfand–Fredrickson coupling in the dJSϕ model gives a less concentrated high-shear band.

The shape of the lower constitutive curve in Figure 4.12 opens up a new possibility, shown by the vertical line: that shear bands can coexist at a common shear rate with a different value of the stress in each band. In a Couette cell, this corresponds to bands stacked with layer normals in the vorticity direction, called vorticity banding. In contrast, the dJS model supports only gradient banding at a common stress, with the normal to the banding interface in the flow gradient direction (Figure 4.9).

In this section, we have explored steady shear-banded states within models whose underlying constitutive curve comprises two stable branches separated by an unstable region of negative slope. In Sections 4.5.5 and 4.5.6 below, we discuss the effects of (i) higher dimensionality and (ii) more exotic constitutive curves. As we will see, either can give rise to unsteady, chaotic shear bands. First, however, we introduce some models of shear thickening.

4.5.3 Simple Models for Shear Thickening

As discussed in Section 4.4.7 above, a window of shear thickening is seen in many micellar solutions, for volume fractions around the onset of viscoelasticity. [56–61] After a period of prolonged shearing, an initially inviscid fluid undergoes a transition to a notably more viscous state or even a long-lived gel, with shear banding often implicated in its formation.[60] Several features of this can be captured within a macroscopic approach[63,64] that couples flow to a generalised gelation transition by allowing a mixture of two species A (sol; concentration $\phi_A = 1 - \phi$) and B (gel; concentration $\phi_B = \phi$) to slowly interconvert under the influence of shear:

$$\partial_t \phi = R(\phi, \dot{\gamma}) + D\nabla^2 \phi \qquad (4.49)$$

$$R(\phi, \dot{\gamma}) \equiv |\dot{\gamma}|(1-\phi)\phi^2 - k\phi \qquad (4.50)$$

In the absence of shear, $\phi = \phi_B$ relaxes to zero, leaving pure A. In this way, B is identified as the shear-induced phase (gel). The model of Ref. 63 ignores normal stresses and considers an additive shear stress $\sigma = \sigma_A + \sigma_B$, with both contributions presumed Newtonian: $\sigma = [(1-\phi)\eta_A + \phi\eta_B]\dot{\gamma}$. Depending on the ratio $\eta_B/\eta_A \equiv c$, the model can capture either shear thinning or shear thickening. Here we focus on thickening, $c > 1$, for which the underlying constitutive curve is shown in Figure 4.13. As discussed in Ref. 63, for an applied shear rate $\dot{\gamma} = \Gamma$ in the nonmonotonic regime, the system can in principle choose between homogeneous states on branches 1 or 3 (circles in Figure 4.13) or it can gradient-band between these branches at a selected stress σ^* (horizontal line). But when the model equations are evolved numerically at imposed shear rate, gradient banding

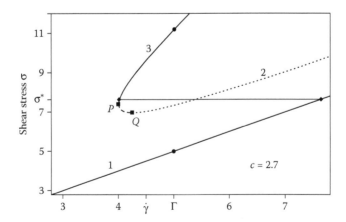

FIGURE 4.13 Shear-thickening constitutive curve from Ref. 63. Reprinted by permission.

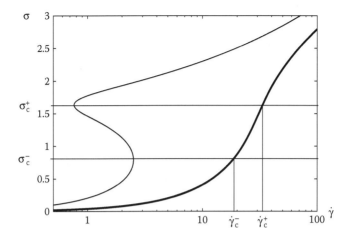

FIGURE 4.14 Bare (light line) and final (heavy) constitutive curve in the model of Ref. 89. The region of instability $\sigma_c^- < \sigma < \sigma_c^+$ is shown. © 2002 by the American Physical Society.

is not seen. The system instead always chooses homogeneous flow: on branch 1 below a critical shear rate $\dot{\gamma} = \dot{\gamma}^*$ (Figure 4.13), and on branch 3 for $\dot{\gamma} > \dot{\gamma}^*$. The sharp vertical jump between branches 1 and 3 at $\dot{\gamma}^*$ leaves a range of stresses that is unattainable under conditions of controlled shear rate. This range is instead accessed by controlling the stress, and marks a regime of vorticity banding.

A different model of shear thickening was devised by Cates, Head, and Ajdari (CHA) in Ref. 89. This has an unspecified, slowly evolving structural variable, but now with explicit (though still relatively fast) dynamics for the stress. While the instantaneous constitutive curve following fast stress equilibration at fixed structure is nonmonotonic (Figure 4.14), the slow structural evolution restores monotonicity in the true long-term constitutive curve. The short-term tendency to form shear bands is thus frustrated by the long-term structural evolution, leading to instability. We return to this model in Section 4.5.5.

We have seen that the microscopic mechanisms involved in shear-banding models differ from case to case, and often remain poorly understood. Nonetheless, the macroscopic phenomenology has many universal features, including kinks, plateaus. or nonmonotonicity in the flow curve. Reference 90 classifies all possible curves, and explains how a nonequilibrium phase diagram can be reconstructed from them.

4.5.4 OTHER RELATED MODELS

Here we list briefly some other approaches to modelling the rheology of wormlike micelles. In an article this length we have no room to discuss all of the many nonlinear rheological models that were developed without micellar systems specifically in mind. Of these, alongside the JS-type models described extensively above, we mention here only the Giesekus model[91] of shear-thinning polymeric fluids, based on the concept of a deformation dependent tensorial mobility. This was first applied to micelles in Ref. 30; it shows a plateau in the constitutive curve, but lacks the nonmonotonicity required to give a true banding instability. Among micelle-inspired models, Manero and co-workers developed a simple model of shear banding that combine the codeformational Maxwell constitutive equation with a kinetic equation accounting for the breaking and reformation of micelles, to give a nonmonotonic constitutive curve;[92] this is described elsewhere in the current volume. Goveas and Pine[93] developed a simple phenomenological model of shear thickening, based on a stress induced gelation transition. They predicted a re-entrant region in the flow curve, only accessible under controlled stress. Dhont and Briels[67] developed an expression for the stress tensor in inhomogeneous suspensions of rigid rods, in terms of the flow velocity and the probability density function for the position and orientation of a rod. By explicitly allowing for large spatial gradients in the shear rate, concentration, and orientational order parameter, this approach could potentially be applied to shear-banding systems. Dhont also introduced a phenomenological model incorporating spatial gradients into the standard expression for the shear stress, to describe shear banding.[94] Porte, Berret, and Harden[52] modeled shear banding by considering an effective non-equilibrium potential for a viscoelastic material under shear. Within this approach, banding is a manifestation of an underlying structural transition such as an underlying nematic/isotropic transition.[52]

4.5.5 Temporal Instability

The studies discussed so far capture the basic tendency of wormlike micelles to undergo a transition to shear-banded flow. However they fail to address recent reports that the constitutive response to steady mechanical driving is intrinsically unsteady in some regimes. In such cases, the stress response to a constant applied strain rate (or vice versa) fails to settle to a constant value. Instead it shows sustained periodic oscillations[95–99] or erratic behavior suggestive of low-dimensional chaos.[42,100–102] Such *long-time* unsteadiness is distinct from the *early-time* instability accessed by a linear stability analysis;[84] the latter merely provides the initial trigger for banding, of some sort, to occur.

Hydrodynamic instabilities have long been studied in simple liquids, where they stem from the inertial term ($\rho v_i \nabla_i v_j$) in the Navier–Stokes equation. For giant micelles, however, this term is usually negligible; the observed complexity stems from nonlinearity in the micellar constitutive equation. The term "rheochaos"[89] has been coined to describe this behavior. Irregular signals have also been reported in other complex fluids.[103,104] In many cases, evolution of the microstructure in concert with the rheological signal has been observed via birefringence imaging,[96,105] light scattering,[103] or spatially resolved NMR.[102]

A crucial question is whether these instabilities are spatiotemporal or purely temporal in character. In wormlike micelles they most often arise close to the banding regime, suggesting the spatiotemporal evolution of a heterogeneous (e.g., banded) state. Indeed, early optical experiments on wormlike micelles showed a temporally oscillating state comprising spatially alternating turbid and clear bands.[96,97] More recent measurements have unambiguously revealed fluctuating shear bands to be present.[102] However, spatial observations have been made in just a few of these cases, so that in others the question remains open. For example, a nearby banding instability could feasibly play a role in triggering temporal rheochaos, but with banding itself narrowly averted such that the system stays homogeneous, as discussed in Ref. 89. For systems close to the nematic transition, another possibility is the purely temporal director chaos captured theoretically in models of nematodynamics.[106,107] In this section, we therefore discuss purely temporal instability in spatially homogeneous models, before proceeding to the full spatiotemporal case in Section 4.5.6. We address thinning and thickening systems in turn.

4.5.5.1 Shear Thinning: Rigid Rods

Models of rigid rods have been widely studied in the context of liquid crystalline polymers. They capture an isotropic-to-nematic (I-N) transition, and make predictions for director dynamics in the nematic phase under shear. When applied to wormlike micelles, such models obviously ignore any effects of micellar flexibility, as noted above. However they do take a first step to incorporating orientational ordering, likely to be an important feature in the high-shear band of concentrated systems close to the nematic transition. Indeed, the interplay of shear banding with the I-N transition was studied in detail (for rigid rods) in. Ref. 88. However the precise link with wormlike micellar rheology remains to be established, so we will discuss such models only briefly. A recent review can be found in Ref. 108.

The studies in question, following Hess[109] and Doi,[86] consider a population of rods, with orientation vectors u_i chosen from a distribution $\psi(u_i,t)$. This is assumed spatially homogeneous. Taking account of macroscopic flow, excluded volume effects, and thermal agitation, the evolution of ψ is specified via a Fokker–Plank equation. To solve this numerically, one must first project it onto a finite number of degrees of freedom. One method is to expand ψ in a truncated set of spherical harmonics, giving a set of coupled ordinary differential equations for the time-dependent expansion coefficients.[110] An alternative is to construct the second-order orientation tensor $Q_{ij} \propto \langle u_i u_j \rangle_\psi$, and project the dynamics onto it via a closure approximation[111] to get an evolution equation for $Q_{ij}(t)$. Studies that confine the director to lie in the flow/flow-gradient plane then predict a sequence of transitions from "tumbling" through "wagging" to "flow-aligning" with increasing shear rate,[110] for suitable values of a "tumbling parameter" λ. (For other values of λ, flow alignment occurs at all shear rates.) In the tumbling and wagging regimes the director executes periodic motion in the flow/flow-gradient plane. Studies generalized to allow out-of-plane director components[112] predict richer dynamics, including new periodic regimes as well as chaos.[106, 113] Both intermittency and period-doubling routes to chaos are seen.

4.5.5.2   Shear Thickening: CHA Model

As noted in Section 4.5.3 above, a scalar model (CHA model) of temporal instability in shear thickening was devised in Ref. 89. The dynamics are defined as follows:

$$\dot{\sigma} = \dot{\gamma} - R(\sigma_1) - \lambda\sigma_2 \tag{4.51}$$

with the structural evolution modelled by "retarded stresses"

$$\sigma_i(t) = \int_{-\infty}^{t} M_i(t-t')\sigma(t')dt' \quad \text{for} \quad i = 1, 2 \tag{4.52}$$

The $M_i(t)$ are memory kernels, each having an integral of unity. In the absence of relaxation, the first term on the right-hand side of (4.51) causes the stress to increase linearly with straining (a Hookean solid with a spring constant of unity). The second and third terms respectively capture nonlinear and linear stress relaxation. In the simplest version of the model, M_1 is chosen to be a delta function such that $\sigma_1(t) = \sigma(t)$ is unretarded, and M_2 is chosen to be exponential, $M_2(t) = \tau_2^{-1}\exp(-t/\tau_2)$. In this case, the system can be rewritten as two coupled differential equations in the stress σ and structural variable $m \equiv \sigma_2$:

$$\dot{\sigma} = \dot{\gamma} - R(\sigma) - \lambda m; \qquad \tau_2\dot{m} = -(m - \sigma) \tag{4.53}$$

In steady state at a given applied shear rate $\dot{\gamma}$, we find the relation

$$\dot{\gamma} = R(\sigma) + \lambda\sigma \tag{4.54}$$

When inverted, this defines the constitutive curve $\sigma(\dot{\gamma})$ of Figure 4.14. Its two components $R(\sigma)$ and $\lambda\sigma$ stem respectively from rapid nonlinear stress relaxation on a time scale $t \approx R^{-1} = O(1)$ and retarded linear relaxation on a time scale $t = \tau_2 \gg 1$. Thus $R(\sigma)$ represents an "instantaneous" constitutive relation, describing the relaxation of stress at fixed structure. The much slower structural relaxation eventually recovers the full curve $R(\sigma) + \lambda\sigma$. The interesting case arises when $R(\sigma) + \lambda\sigma$ is monotonic but $R(\sigma)$ is not (Figure 4.14). The system then exhibits a shear banding instability, at short times, in the regime where $R'(\sigma) < 0$. If the linear contribution $\lambda\sigma$ is sufficiently retarded (τ_2 large), it fails to overcome this instability, despite the monotonic constitutive curve. Accordingly, the long-term dynamics of the model remain unsteady in a region $\sigma_c^- < \sigma < \sigma_c^+$. The dynamical system defined by (4.53) undergoes a Hopf bifurcation at σ_c^+ and σ_c^- (Figure 4.14), signifying the onset of finite frequency sinusoidal oscillations with an amplitude varying as $|\dot{\gamma} - \dot{\gamma}_c|^{1/2} a$. Chaos requires a phase space of dimensionality at least 3, and so cannot occur in the dynamical system (4.53). Without invoking flow inhomogeneity (which gives infinite dimensionality), sufficient dimensions can be achieved by assuming $\sigma_1 \equiv n$ to be retarded as well as σ_2, with $M_1(t) = \tau_1^{-1}\exp(-t/\tau_1)$. In harmony with the simpler version of the model, one takes $\tau_1 \lesssim 1 \ll \tau_2$ and considers the situation where monotonicity of the constitutive curve is restored only via the more retarded relaxation term. In the unstable regime, one now finds a period-doubling cascade leading to temporal chaos.

4.5.6 SPATIOTEMPORAL INSTABILITY; RHEOCHAOS

Above we have discussed the unsteady rheological response of models with purely temporal dynamics. As noted above, however, reports of unsteady dynamics in wormlike micelles are most common close to or inside the shear banding regime. In such cases, a *spatio* temporal description is essential, to allow for an evolving state that is heterogeneous (e.g., banded) at any instant. Indeed, recent experiments have unambiguously revealed fluctuating shear bands in wormlike micelles.[102] We now turn to models of spatiotemporal dynamics, considering thinning and thickening systems in turn.

4.5.6.1 Shear Thinning: A One-Dimensional Model

As explained previously, homogeneous flow on the negatively sloping branch of the constitutive curve is unstable with respect to the formation of shear bands.[83] Above we further saw that coupling between the flow and an auxiliary variable such as concentration can enhance this instability, causing it to extend into regions of positive constitutive slope (Figure 4.8). In Ref. 114, one of us exploited this fact to construct a model in which the high shear band is itself destabilised, leading to unsteady banding dynamics.

The model is defined as follows. We work in one spatial dimension, the flow-gradient direction y, with a velocity $v_i = v(y)\delta_{ix}$ and shear rate $\dot{\gamma}(y) = \partial_y v(y)$. Normal stresses are neglected, and the total shear stress is assumed to comprise additive viscoelastic and Newtonian components. At zero Reynolds number, $\sigma(t)$ must be uniform across the gap:

$$\sigma(t) = \sigma^{\text{pol}}(y, t) + \eta^{\text{sol}}\dot{\gamma}(y, t) \tag{4.55}$$

For the dynamics of the viscoelastic component, we use the scalar model of Section 4.5.1

$$\partial_t\sigma^{\text{pol}} = -\frac{\sigma^{\text{pol}}}{\tau(n)} + \frac{g[\dot{\gamma}\tau(n)]}{\tau(n)} + D\partial_y^2\sigma \tag{4.56}$$

with a relaxation time τ that now depends on a structural variable $n = n(y, t)$, according to $\tau(n) = \tau_0 (n/n_0)^{\alpha}$. As before, $g(x) = x/(1 + x^2)$ is chosen to ensure a region of negative constitutive slope. The auxiliary variable n is taken to represent a nonconserved quantity. For definiteness it is identified as the mean micellar length (previously denoted by $\bar{L}$) although there could be other candidates for its interpretation. Coupling of n to the flow is completed by assuming that it evolves with its own relaxation time τ_n, distinct from $\tau(n)$:

$$\partial_t n = -\frac{n}{\tau_n} + \frac{N(\dot{\gamma}\tau_n)}{\tau_n} \qquad (4.57)$$

Here the coupling term $N(x) = n_0/(1 + x^{\beta})$ represents (say) shear-induced scission.

As in the dJS model, this model shows the familiar banding instability when the slope of the flow curve is negative, at intermediate shear rates. It also shows a new instability not seen in dJS. This destabilizes the high shear branch to a degree that depends on τ_n, eventually terminating in a Hopf bifurcation at high $\dot{\gamma}$. Allowing for spatiotemporal variation, for small τ_n we find stable shear bands. For higher τ_n, the high-shear band is unstable, leading to unsteady dynamics of the banded state. This is explored in Figure 4.15. Several regimes are evident. At low applied shear rate a thin pulse or band of locally high shear ricochets back and forth across the cell. (A thin fluctuating high shear band, away from the rheometer wall, was seen experimentally in Ref. 42.) At larger shear rates, we find two or more such pulses. For two pulses (not shown), we typically find a periodically repeating state with the pulses alternately bouncing off each other (mid-cell) and the cell walls. Once three pulses are present (e.g., $\dot{\gamma} = 7.0$), periodicity gives way to chaotic behavior. At still higher mean shear rate, $\dot{\gamma} = 19.2$, we find regular oscillations of spatially extended bands pinned at a stationary defect. The local shear rates span both the low and high shear constitutive branches. (Oscillating vorticity bands were seen experimentally.[105,115]) For the intermediate value $\dot{\gamma} = 11.35$ we find intermittency between patterns resembling those for $\dot{\gamma} = 7.0$ and $\dot{\gamma} = 19.2$. Finally for $\dot{\gamma} = 23.0, 31.0$ we find oscillating bands separated by *moving* defects, with the flow now governed only by the high-shear constitutive branch: in each band, the shear rate cycles round the periodic orbit of the local model. For different τ_n we find many other interesting phenomena.[114] For example, for weaker instability ($\tau_n = 0.13$) at low applied shear rates we see a high-shear band that pulsates in width while adhering to the rheometer wall (compare Refs. 41 and 116), or meanders about the cell.

4.5.6.2 Shear Thinning: Higher Dimensional Model

In Section 4.5.1 we introduced the diffusive Johnson–Segalman (dJS) model and in Section 4.5.2 we discussed its predictions for steady shear-banded states. Those calculations were restricted to 1D variations only (in the flow gradient direction y) and implicitly assumed that the interface remains perfectly flat at all times.

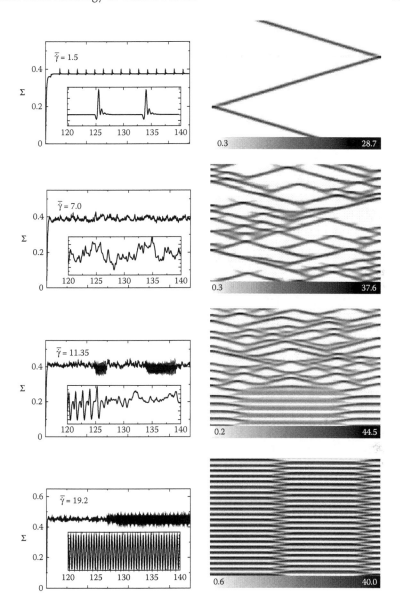

FIGURE 4.15 Right: space–time plots for shear rate evolution for $\tau_n = 0.145, D = 0.0016$ in the 1D model. Space coordinate $0 < y < 1$ runs left to right and time $120 < t < 140$ runs bottom to top; the shading denotes local shear rate (as per gray scale bar at bottom). Each horizontal slice through the space–time plot identifies regions of high shear rate at a certain time; a slice higher up the plot shows the system at a later time. Hence a shear band moving across the gap with velocity v shows up as a gray stripe of slope v. Left: corresponding stress vs. time. The quantity $\bar{\dot{\gamma}}$ is denoted $\dot{\gamma}$ in the main text. Figures adapted from Ref. 114. © 2004 by the American Physical Society.

Recent experiments suggest more complex interfacial dynamics[102] however, and a natural question is whether these 1D states remain stable in higher dimensions. This question was recently addressed in a study comprising, first, a linear stability analysis of an initially 1D banded state with respect to small interfacial undulations;[85] and, subsequently, a numerical study of nonlinear interfacial dynamics in the flow/flow-gradient plane.[117]

Linear stability analysis: As discussed in Section 4.5.1, a 1D flow-gradient calculation of planar shear within the dJS model predicts a steady state flow curve resembling Figure 4.7, with a corresponding banded profile shown in Figure 4.10. To study the linear stability of this 1D "base" profile with respect to small fluctuations with wavevector in the (xz) plane of the interface, we linearize the model for small perturbations (lower case) about the (upper case) base profile: $\tilde{\Phi}(x,y,z,t) = \Phi(y) + \phi_q(y)\exp(\omega_q t + iq_x x + iq_z z)$. The vector Φ comprises all components $\Phi = (W_{\alpha\beta}, V_\alpha)$, the pressure being eliminated by incompressibility. This gives a linear eigenvalue equation with an operator L and eigenfunction $\phi_q(y)$: We are interested here only in the resulting eigenvalue $\omega_{max}(\mathbf{q})$ with the largest real part, $\Re\omega_{max}(\mathbf{q})$: in particular, whether it is stable, $\Re\omega_{max} < 0$, or unstable, $\Re\omega_{max} > 0$. A dispersion relation $\Re\omega_{max}(q_x, q_z = 0)$ for wavevectors confined to the direction of the unperturbed flow is shown in Figure 4.16. At any q_x, $\Re\omega_{max}$ increases with decreasing values of the parameter $l \equiv \sqrt{D\tau}$, which sets the width of the interface between the bands. For small enough l the dispersion relation is positive over a range of wavevectors, rendering the 1D profile unstable. This applies to shear rates across most of the stress plateau, and suggests that, for

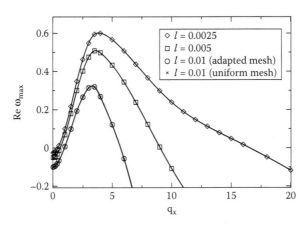

FIGURE 4.16 Real part of the eigenvalue of the most unstable mode. $a = 0.3$, $\eta = 0.05$, $\dot{\gamma} = 2.0$, Reynolds number $\rho/\eta = 0$. The data for $l = 0.01$ correspond to the base profile in Figure 4.10. Symbols: data. Solid lines: cubic splines. Figure from Ref. 85 © 2005 by the American Physical Society.

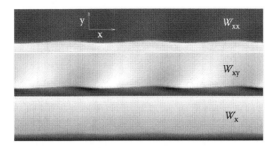

FIGURE 4.17 Greyscale of order parameters for traveling wave in the (x, y) plane for $l = 0.015$, $\Lambda_x = 6$, and upper wall velocity $V \equiv \dot{\gamma}\Lambda_y = 2$ to the right. Figure from Ref. 117. © 2006 by the American Physical Society.

realistic experimental parameters, the entire stress plateau region will be unstable to perturbations away from a flat interface between shear bands.[85]

Nonlinear interfacial dynamics: As soon as the undulations attain a finite amplitude, nonlinear effects become important. The linear calculation then breaks down, and so is unable to predict the ultimate behavior of the system at long times. To access this, one must perform a full nonlinear study of the dJS model in the flow/flow-gradient (xy) plane.[117] At large values of l, for which the dispersion relation of the linear analysis is negative at all wavevectors, the 1D base profile remains stable as expected: the interface stays flat at all times. For smaller values of l, the dispersion relation is positive over a window of q_x (Figure 4.16); at fixed l, the number of linearly unstable modes (satisfying periodic boundary conditions in the flow direction) increases with the system size Λ_x. For small l/Λ_x, just inside the unstable regime, the ultimate attractor comprises a traveling wave (Figure 4.17). The wall-averaged shear stress is constant in time, with a value $\bar{W}_{xy,ss}$ that depends on l and Λ_x and is slightly higher than the selected value W_{xy}^{sel} of the 1D calculation. For l/Λ_x values deeper inside the unstable regime, we see a new regime in which the traveling wave now periodically "ripples" (Figure 4.18). The corresponding wall-averaged stress $\bar{W}_{xy}$ is periodic in time, with variations of the order of 1 percent, and an average value larger than the 1D selected stress W_{xy}^{sel}. The interface height $h(x,t)$ is shown as a white line in Figure 4.18d.

4.5.6.3 Shear Thickening

Aradian and Cates recently extended the CHA model of Section 4.5.5 to allow for spatial modulation of the stress along the vorticity direction:[118,119] $\sigma(t) \rightarrow \sigma(z, t)$. To allow for the interfaces that are now expected to arise, they further incorporated a spatial gradient term, $D\nabla^2\sigma$. (Recall the discussion of

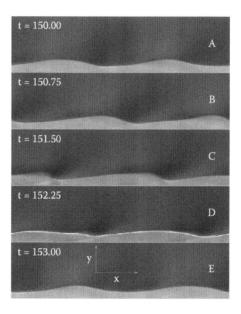

FIGURE 4.18 Rippling wave at $l = 0.005, \Lambda_x = 4, \dot{\gamma} = 2$. Grayscale of $W_{xx}(x,y)$. Upper wall moves to the right. White line in D): interface height. Figure from Ref. 117. © 2006 by the American Physical Society.

Section 4.5.2.) For simplicity, they focused on the version in which only σ_2 is retarded, extending (4.53) as follows:

$$\dot{\sigma}(z,t) = \dot{\gamma}(t) - R(\sigma) - \lambda m + D\nabla^2\sigma \qquad (4.58)$$

$$\dot{m}(z,t) = -\frac{m - \sigma}{\tau_s} \qquad (4.59)$$

As before, the instantaneous nonlinear relaxation term is chosen as

$$R(\sigma) = a\sigma - b\sigma^2 + c\sigma^2 \qquad (4.60)$$

giving in linear response the Maxwell time $\tau_M = 1/a$. Note that the shear rate $\dot{\gamma}(t)$ in (4.58) is now *uniform*: the moving wall of the rotor imposes the same velocity for all heights z, so that $\dot{\gamma}(z, t) = \dot{\gamma}(t)$ only. The model is studied under an imposed value of the spatial mean of the stress $\langle\sigma\rangle$. The two main control parameters are taken to be $\langle\sigma\rangle$ and the ratio τ_s/τ_M. Depending on the values of these parameters, the model shows a rich variety of oscillatory and chaotic banding dynamics (Figure 4.19), as summarized in the phase diagram of Figure 4.20.

In the regime where the structural evolution is much slower than the stress relaxation (marked O in Figure 4.20), the model predicts oscillating shear bands. Varying the imposed stress $\langle\sigma\rangle$ along any horizontal line of fixed τ_s/τ_M in this

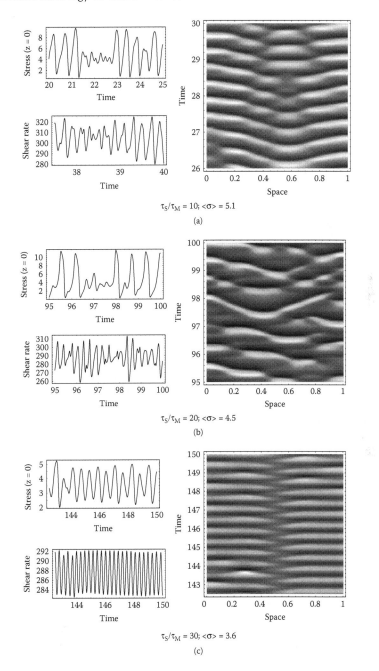

FIGURE 4.19 Various types of spatiotemporal rheochaos observed in the shear thickening model of Ref. 118. Left: time series for a local shear rate and for stress. Right: space–time plots (as explained in Figure 4.15). Figures from Ref. 119. © 2006 by the American Physical Society.

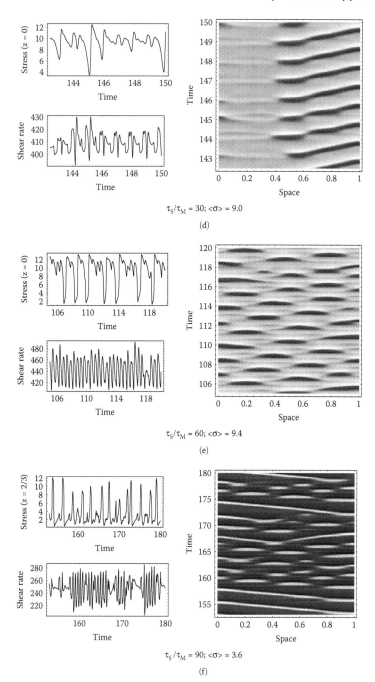

FIGURE 4.19 (Continued).

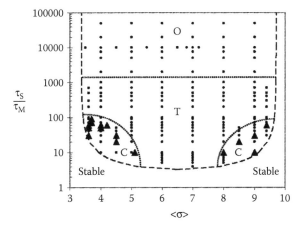

FIGURE 4.20 Nonequilibrium phase diagram of the shear-thickening model of Ref. 119 when τ_s and $\langle\sigma\rangle$ are varied. Black triangle: chaotic points, black circle: periodic point. Three main regimes are observed: (O) oscillating shear bands, (T) traveling shear bands, (C) chaotic regions. The outer dashed line is the linear stability limit. Figure from Ref. 119. © 2006 by the American Physical Society.

regime, the associated waveforms range from simple to very complex. Near the middle of the line one sees simple flip-flopping bands, with the cell divided equally between a high-stress and a low-stress band, the identities of which repeatedly switch with a period of order the structural time τ_s. Moving slightly off-center along the line, the interface between the bands now adopts a zig-zagging motion, superposed on the flip-flopping motion just described. The regime $10 \leq \tau_s/\tau_M \leq 10^3$ marked T in Figure 4.20 shows periodic nucleation of shear bands, which then cross the cell with roughly constant velocity. In the two disconnected pockets marked C, the complex oscillations of regimes O and T give way to true rheochaos, characterized by a positive Lyapunov exponent. This model also admits an interesting low-mode truncation in which only the lowest two nonhomogeneous Fourier modes are retained for each of $\dot{\gamma}$ and n, giving 4 modes in all.[119] The basic structure of the phase diagram of Figure 4.20 was found to be preserved by this truncation, showing that rheochaos is robust within the model, and is not dependent on the presence of sharp interfaces between the bands. Within this truncation, the chaos in the pockets C was found to set in via a classical period-doubling scenario.

4.6 CONCLUSION

Above we have reviewed theoretical modeling efforts that address the rheology of giant micelles. In the well-entangled regime, the linear viscoelastic spectra, often close to pure-Maxwell in character, are well explained by the reptation–reaction model which couples micellar kinetics to the tube dynamics of entangled objects. For the concentration dependence and ionic strength dependence of the Maxwell time, the model has more mixed success but, when branching is allowed for, is capable of accounting for the main trends. In the nonlinear regime, shear banding is widely observed, and is predicted by the reptation–reaction model. To address the rather complex nature and dynamics of shear-banded states, one must turn to macroscopic models which couple features of the reptation–reaction approach to spatially varying order parameters. These models currently offer quite a good account of a wide range of phenomena, involving unsteady or chaotic banded flow, observed in recent experiments. Current understanding is less good for systems close to the onset of viscoelasticity. Here, drastic shear thickening can be seen, often accompanied by shear banding and/or chaotic flows, with complex sample-history dependences in some cases. Microscopic modeling in this regime (for instance involving micellar rings) remains speculative, and offers limited guidance to formulating macroscopic models. Nonetheless, just as in the well-entangled, shear-thinning regime, models in which the local Maxwell time is coupled to slowly evolving structural variables (such as the mean micellar chain length) offer a promising way forward.

REFERENCES

1. de Gennes, P.-G. *Scaling Concepts in Polymer Physics*, Cornell University Press, Ithaca, 1979.
2. Doi, M., Edwards, S. F. *The Theory of Polymer Dynamics*, Clarendon Press, Oxford, 1986.
3. de Gennes, P.-G. *Soft Matter (Nobel Lecture)*, *Rev. Mod. Phys.* 1992, *64*, 645.
4. Cates, M. E. *Macromolecules* 1987, *20*, 2289.
5. Scott, R. L. *J. Phys. Chem.* 1965, *69*, 261.
6. Mukerjee, P. *J. Phys. Chem.* 1972, *76*, 565.
7. Cates, M. E., Candau, S. J. *J. Phys. Cond. Matt.* 1990, *2*, 6869.
8. MacKintosh, F. C., Safran, S. A., Pincus, P. A. *Europhys. Lett.* 1990, *12*, 697; Porte, G., Marignan, J., Bassereau, P., May, R. *J. Physique Paris* 1988, *49*, 511; Odijk, T. *J. Phys. Chem.* 1989, *93*, 3888.
9. Drye, T. J., Cates, M. E. *J. Chem. Phys.* 1992, *96*, 1367.
10. Lequeux, F. *Europhys. Lett.* 1992, *19*, 675.
11. Khatory, A., Kern, F., Lequeux, F., Appell, J., Porte, G., Morie, N., Ott, A., Urbach, W., *Langmuir* 1993, *9*, 933.
12. Appell, J., Porte, G. *Europhys. Lett.* 1990, 12, 185; Appell, J., Porte, G., Khatory, A., Kern, F., Candau, S. J. *J. Physique Paris II* 1992, *2*, 1045.
13. Petscheck, R. G., Pfeuty, P., Wheeler, J. C. *Phys. Rev. A* 1986, 34, 2391; Cordery, R. *Phys. Rev. Lett.* 1981, *47*, 457.
14. Jacobson, H., Stockmayer, W. H. *J. Chem. Phys.* 1950, 18, 1600; Porte, G. *J. Phys. Chem.* 1983, *87*, 3541.
15. In, M., Aguerre-Chariol, O., Zana, R. *J. Phys. Chem. B* 1999, *103*, 7747; Bernheim-Groswasser, A., Zana, R., Talmon, Y. *J. Phys. Chem. B* 2000, *104*, 4005.
16. Cates, M. E., Candau, S. J. *Europhys. Lett.* 2001, *55*, 887.
17. Cates, M. E. *J. Physique Paris* 1988, *49*, 1593.

18. Cates, M. E. *J. Physique Paris Lett.* 1985, *46*, 1059.
19. Marques, C. M., Turner, M. S., Cates, M. E. *J. Chem. Phys.* 1993, *99*, 7260.
20. Turner, M. S., Cates, M. E. *J. Physique Paris* 1990, 51, 307.
21. Kern, F., Lemarechal, P., Candau, S. J., Cates, M. E. *Langmuir* 1992, *8*, 437.
22. Turner, M. S., Marques, C. M., Cates, M. E. *Langmuir* 1993, *9*, 695.
23. Shikata, T., Hirata, H., Kotaka, T. *Langmuir* 1987, 3, 1081; *Langmuir* 1988, *4*, 354; *Langmuir* 1989, 5, 398; Shikata, T., Hirata, H., Takatori, E., Osaki, K. *J. Non-Newtonian Fluid Mech.* 1988, *28*, 171.
24. Pine, D. J. In *Soft and Fragile Matter: Nonequilibrium Dynamics, Metastability and Flow*, Cates, M. E., Evans, M. R., Eds., IOP Publishing, Bristol, 2000.
25. Callaghan, P. T. *Repts. Prog. Phys.* 1999, *62*, 599.
26. Milner, S. T., McLeish, T. C. B., Likhtman, A. E. *J. Rheol.* 2001, *45*, 539; Milner, S. T., McLeish, T. C. B. *Phys. Rev. Lett.* 1998, *81*, 725.
27. Rehage, H., Hoffmann, H. *Mol. Phys.* 1991, *74*, 933.
28. Larson, R. G. *The Structure and Dynamics of Complex Fluids*, Clarendon Press, Oxford, 1999.
29. McLeish, T. C. B. *Adv. in Phys.* 2002, *51*, 1379.
30. Holz, T., Fischer, P., Rehage, H. *J. Non-Newtonian Fluid Mech.* 1999, *88*, 133; Fischer, P., Rehage, H. *Rheol. Acta* 1997, *36*, 13.
31. Cates, M. E. *J. Phys. Chem.* 1990, *94*, 371.
32. Magid, L. J. *J. Phys. Chem. B* 1998, *102*, 4064.
33. Hoffmann, H., Loebl, M., Rehage, H., Wunderlich, I. *Tenside Detergents* 1986, *22*, 290; Rehage, H., Hoffmann, H. *J. Phys. Chem.* 1988, *92*, 4712; Hoffmann, H., Ebert, G. *Angew. Chemie* 1988, *27*, 902; Hoffmann, H., Platz, G., Rehage, H., Schorr, W., Ulbricht, W. *Ber. Bunsen-Ges. Phys. Chem. Chem. Phys.*, 1981, *85*, 255.
34. Turner, M. S., Cates, M. E. *Langmuir* 1991, *7*, 1590.
35. Granek, R., Cates, M. E. *J. Chem. Phys.* 1992, *96*, 4758; Granek, R. *Langmuir* 1994, *10*, 1627.
36. Berret, J. F., Appell, J., Porte, G. *Langmuir* 1993, *9*, 2851.
37. Kern, F., Zana, R., Candau, S. J. *Langmuir* 1991, *7*, 1344.
38. Oelschlaeger, C., Waton, G., Candau, S. J. *Langmuir* 2003, *19*, 10495.
39. Oelschlaeger, C. Ph.D. Thesis, University of Strasbourg, 2003.
40. Mair, R. W., Callaghan, P. T. *Europhys. Lett.* 1996, *36*, 241; *J. Rheol.* 1997, *41*, 901; Fischer, E., Callaghan, P. T. *Phys. Rev. E* 2001, *64*, 011501.
41. Britton, M. M., Callaghan, P. T. *Eur. Phys. J. B* 1999, *7*, 237.
42. Holmes, W. M., Lopez-Gonzales, M. R., Callaghan, P. T. *Europhys. Lett.* 2003, *64*, 274.
43. Decruppe, J.P., Cressely, R., Makhloufi, R., Cappalaere, E. *Colloid Polym. Sci.* 1995, *273*, 346; Makhloufi, R., Decruupe, J. P., Aitali, A., Cressely, R. *Europhys. Lett.* 1995, *32*, 253; Lerouge, S., Decruppe, J. P., Humbert, C. *Phys. Rev. Lett.* 1998, *81*, 5457.
44. Cates, M. E. In *Structure and Flow of Surfactant Solutions*, Herb, C. A., Prud'homme, R. K., Eds., ACS, Washington 1994 (Symp. Ser. 578).
45. Spenley, N. A., Cates, M. E., McLeish, T. C. B. *Phys. Rev. Lett.* 1993, *71*, 939.
46. Spenley, N. A. Ph.D Thesis, University of Cambridge, 1994.
47. Olmsted, P. D. *Curr. Opinion in Colloid Interface Sci.* 1999, *4*, 95.
48. Lu, C. Y. D, Olmsted, P. D., Ball R. C. *Phys. Rev. Lett.* 2000, *84*, 642.
49. Grand, C., Arrault, J., Cates, M. E. *J. Physique Paris II* 1997, *7*, 1071.

50. Berret, J. F., Porte, G., Decruppe, J. P. *Phys. Rev. E.* 1997, *55*, 1668.
51. Cates, M. E., McLeish, T. C. B., Marrucci, G. *Europhys. Lett.* 1993, *21*, 451.
52. Berret, J. F., Roux, D. C., Porte, G., Lindner, P. *Europhys. Lett.* 1994, *25*, 521; Porte, G., Berret, J. F., Harden, J. L. *J. Physique Paris II* 1997, *7*, 459.
53. Fielding, S. M., Olmsted, P. D. *Eur. Phys. J E* 2003, *11*, 65.
54. Tapiada, P., Wang, S. Q. *Phys. Rev. Lett.* 2003, 91, 198301; *Macromolecules* 2004, *37*, 9083.
55. Candau, S. J., Hirsch, E., Zanan R., Delsanti, M. *Langmuir* 1989, *5*, 1525.
56. Oelschlaeger, C., Waton, G., Buhler, E., Candau, S. J., Cates, M. E. *Langmuir* 2002, *18*, 30276; Oelschlaeger, C., Waton, G., Candau, S. J., Cates, M. E. *Langmuir* 2002, *18*, 7265.
57. Rehage, H., Wunderlich, I., Hoffmann, H. *Prog. Colloid Polym. Sci.* 1986, *72*, 11; Hoffmann, H., Rehage, H., Wunderlich, I. *Rheol. Acta* 1987, *26*, 532; Rehage, H., Hoffmann, H. *Rheol. Acta*, 1982, *21*, 561; Hoffmann, H., Rasucher, A., Hoffmann, H. *Ber. Bunsen-Ges. Phys. Chem. Chem. Phys.* 1991, *95*, 153.
58. Hu, Y. T., Matthys, E. F. *J. Rheol.* 1997, *41*, 151.
59. Oda, R., Panizza, P., Schmutz, M., Lequeux, F. *Langmuir* 1997, *13*, 6407.
60. Boltenhagen, P., Hu, Y. T., Matthys, E. F., Pine, D. J. *Phys. Rev. Lett.* 1997, *79*, 2369; *Europhys. Lett.* 1997, *38*, 389.
61. Berret, J. F., Gamez-Corrales, R., Lerouge, S., Decruppe, J. P. *Eur. Phys. J. E* 2000, *2*, 343.
62. Cates, M. E., Turner, M. S. *Europhys. Lett.* 1990, *7*, 681; Wang, S. Q., Gelbart, W., Ben-Shaul, A. *J. Phys. Chem.* 1990, *94*, 2219.
63. Goveas, J. L., Olmsted, P. D. *Eur. Phys. J. E* 2001, *6*, 79.
64. Picard, G., Ajdari, A., Bocquet, L., Lequeux, F. *Phys. Rev. E* 2002, *66*, 051501; Ajdari, A. *Phys. Rev. E* 1998, *58*, 6294.
65. Barentin, C., Liu, A. J. *Europhys. Lett.* 2001, *55*, 432.
66. Milner, S. T. *Phys. Rev. E* 1993, *48*, 3674.
67. Dhont, J. K., Briels, W. J. *J. Chem. Phys.* 2003, *118*, 1466.
68. Johnson, M., Segalman, D. *J. Non-Newtonian Fluid Mech.* 1977, *2*, 255.
69. Olmsted, P. D., Radulescu, O., Lu, C.-Y. D. *J. Rheology* 2000, *44*, 257.
70. Fielding, S. M., Olmsted, P. D. *Phys. Rev. Lett.* 2003, *90*, 224501.
71. Brochard, F., de Gennes, P.-G. *Macromolecules* 1977, *10*, 1157.
72. Milner, S. T. *Phys. Rev. Lett.* 1991, *66*, 1477.
73. de Gennes, P.-G. *Macromolecules* 1976, *9*, 587.
74. Brochard, F. *J. Physique (Paris)* 1983, *44*, 39.
75. Wu, X. L., Pine, D. J., Dixon, P. K. *Phys. Rev. Lett.* 1991, *66*, 2408.
76. Gerard, H., Higgins, J. S., Clarke, N. *Macromolecules* 1999, *32*, 5411.
77. Wheeler, E., Izu, P., Fuller, G. G. *Rheol. Acta* 1996, *35*, 139.
78. Kadoma, I. A., van Egmond, J. W. *Langmuir* 1997, *13*, 4551.
79. Helfand W., Fredrickson, G. H. *Phys. Rev. Lett.* 1989, *62*, 2468.
80. Schmitt, V., Marques, C. M., F. Lequeux, F. *Phys. Rev. E* 1995, *52*, 4009.
81. Decruppe, J. P., Lerouge, S., Berret, J.-F. *Phys. Rev. E* 1001, *63*, 022501.
82. Spenley, N. A., Yuan, X.-F., Cates, M. E. *J. Physique (Paris) II* 1996, *6*, 551.
83. Yerushalmi, J., Katz, S., Shinnar, R. *Chem. Eng. Sci.* 1970, *25*, 1891.
84. Fielding, S. M., Olmsted, P. D. *Phys. Rev. E* 2003, *68*, 036313.
85. Fielding, S. M. *Phys. Rev. Lett.* 2005, *95*, 134501.
86. Doi, M. *J. Polym. Sci: Polym. Phys.* 1981, *19*, 229.
87. Kuzuu, N., Doi, M. *J. Phys. Soc. Jap.* 1983, *52*, 3486.

88. Olmsted, P. D., Lu, C.-Y. D. *Phys. Rev. E* 1999, *60*, 4397.
89. Cates, M. E., Head, D. A., Ajdari, A. *Phys. Rev. E* 2002, *66*, 025202.
90. Olmsted, P. D. *Europhys. Lett.* 1999, *48*, 339.
91. Giesekus, H. *J. Non-Newtonian Fluid Mech.* 1982, *11*, 69.
92. Escalante, J. I., Macias, E. R., Bautista, F., Perez-Lopez, J. H., Soltero, J. F. A., Puig, J. E., Manero, O. *Langmuir* 2003, *19*, 6620; Bautista, F., Soltero, J. F. A., Perez-Lopez, J. H., Puig, J. E., Manero, O. *J. Non-Newtonian Fluid Mech.* 2000, *94*, 57.
93. Goveas, J. L., Pine, D. J. *Europhys. Lett.* 1999, *48*, 706.
94. Dhont, J. K. *Phys. Rev. E* 1999, *60*, 4534.
95. Hu, Y. T., Boltenhagen, P., Matthys, E., Pine, D. J. *J. Rheology* 1998, *42*, 1209.
96. Wheeler, E. K., Fischer, P., Fuller G. G. *J. Non-Newtonian Fluid Mech.* 1998, *75*, 193.
97. Fischer, E., Callaghan, P. T. *Europhys. Lett.* 2000, *50*, 803.
98. Fischer, E., Callaghan, P. T., Heatley, F., Scott, J. E. *J. Mol. Struct.* 2002, *602*, 303.
99. Herle, V., Fischer, P., Windhab, E. J. *Langmuir* 2005, *21*, 9051.
100. Bandyopadhyay, R., Basappa, G., Sood, A. K. *Phys. Rev. Lett.* 2000, *84*, 2022.
101. Bandyopadhyay, R., Sood, A. K. *Europhys. Lett.* 2001, *56*, 447.
102. Lopez-Gonzalez, M. R., Holmes, W. M., Callaghan, P. T., Photinos, P. J. *Phys. Rev. Lett.* 2004, *93*, 268302.
103. Salmon, J. B., Colin, A., Roux, D. *Phys. Rev. E* 2002, *66*, 031505.
104. Lootens, D., Van Damme, H., Hebraud, P. *Phys. Rev. Lett.* 2003, *90*, 178301.
105. Hilliou, L., Vlassopoulos, D. *Ind. Eng. Chem. Res.* 2002, *41*, 6246.
106. Grosso, M., Keunings, R., Crescitelli, S., Maffettone, P. L. *Phys. Rev. Lett.* 2001, *86*, 3184.
107. Rienacker, G., Kroger, A., S Hess, S. *Physica A* 2002, *315*, 537.
108. Das, M., Bandyopadhyay, R., Chakrabarti, B., Ramaswamy, S., Dasgupta, C., Sood, A. K. In *Molecular Gels*, Eds., Terech, P., and Weiss, R. G., Kluwer, 2006.
109. Hess, S., *Z. Naturforsch.* 1976, *31a*, 1507.
110. Larson, R. G. *Macromolecules* 1990, *23*, 3983.
111. Feng, J., Chaubal, C. V., Leal, L. G. *J. Rheology* 1998, *42*, 1095.
112. Larson, R. G., Ottinger, H. C. *Macromolecules* 1991, *24*, 6270.
113. Rienacker, G., Kroger, M., Hess, S. *Phys. Rev. E* 2002, *66*, 040702.
114. Fielding, S. M., and Olmsted, P. D., *Phys. Rev. Lett.* 2004, *92*, 084502.
115. Fischer, P., Wheeler, E. K., Fuller, G. G. *Rheol. Acta* 2002, *41*, 35.
116. Hu, Y. T., Boltenhagen, P., Pine, D. J. *J. Rheology* 1998, *42*, 1185.
117. Fielding, S. M., Olmsted, P. D. *Phys. Rev. Lett.* 2006, *96*, 104502.
118. Aradian, A., Cates, M. E. *Europhys. Lett.* 2005, *70*, 397.
119. Aradian, A., Cates, M. E. *Phys. Rev. E* 2006, *73*, 041508.

5 *Seeing* Giant Micelles by Cryogenic-Temperature Transmission Electron Microscopy (Cryo-TEM)

Yeshayahu (Ishi) Talmon

CONTENTS

5.1 INTRODUCTION

To fully characterize liquid, semiliquid, gel, or solid systems requires direct, supramolecular-level information, that is, images, which show how molecules assemble to form aggregates of various sizes and shapes. Cryogenic-temperature transmission electron microscopy (cryo-TEM) is an excellent tool to obtain direct imaging of liquid or semiliquid specimens, thermally fixed into a vitreous or quasi-solid state. Sample preparation for cryo-TEM makes it possible to preserve the nanostructure of the system in its native state while making it compatible with the stringent requirement of the microscope. Because cryo-TEM provides high-resolution direct images of the nanostructures and microstructures in the system, it can elucidate the nature of the basic building blocks that make up the systems, covering a wide range of length scales, from a few nanometers to several microns. Quite often many different types of assemblies coexist in the examined systems. Cryo-TEM makes it possible to observe them all, even those that are difficult to tell apart and identify by other, so-called indirect methods, such as scattering techniques. The interpretation of cryo-TEM images is usually quite straightforward and not model dependent, whereas data interpretation in "indirect methods" is model dependent and is complicated when the system contains more than one type of aggregate or a broad size distribution.

In the case of micellar systems, including "giant micelles," cryo-TEM is most useful to image the range of the nanostructures present in those systems, for example, threadlike micelles (TLMs, also referred to as wormlike micelles, WLMs) coexisting with spheroidal micelles (SMs), branched TLMs, and TLMs that form loops or lassolike micelles. In some cases, reliable direct images provide the only way to prove a suggested or a theoretically predicted model. That was demonstrated, for example, in the case of the theoretically predicted branched micelles,[1] and the shape of the end-caps of threadlike micelles.[2,3]

It should be emphasized here that while micrographs are most useful, cryo-TEM is not a strictly quantitative technique. It is the technique of choice to determine the structural building blocks of complex fluid systems, but the quantitative data should be usually provided by other techniques, such as small-angle x-ray scattering (SAXS) and small-angle neutron scattering (SANS) (see Chapter 6), or nuclear magnetic resonance (NMR). Another advantage of these scattering techniques is that they probe the bulk of the system, not just a small sample of it; they thus provide a real statistical average. However, in a very heterogeneous system, such an average may be difficult to interpret. In addition, these techniques are "model dependent," not "observer dependent." Based on that, the best experimental approach is to apply cryo-TEM to study the nature of the nano-building blocks of the system, use that information to construct a physical model to be applied to interpret data from the above mentioned "indirect techniques," and then check whether the two sets of results from, say, scattering and TEM agree, to rule out possible artifacts.

The reader will find next a description of the basics of cryo-TEM, especially as it is applied to the study of "giant micelles." That is followed by a review of the applications of the technique in the study of the systems in the focus of this book. The interested reader will find more details about the technique and its application to other systems in Talmon[4] and Danino.[5]

5.2 BASICS OF DIRECT-IMAGING CRYO-TEM

The term cryo-TEM covers the two main techniques that involve fast cooling of the specimen as part of its preparation before it is examined in the TEM. The more widely used technique is "direct-imaging cryo-TEM," which involves fast cooling of a thin specimen and its transfer at cryogenic temperature into the TEM, where it is maintained and observed at cryogenic temperature. Alternatively, a larger sample of the system is quickly cooled, the frozen sample is fractured, and a thin metal-carbon replica is prepared of the fracture surface. Then the sample is melted away, the replica is cleaned and dried and observed in the TEM at room temperature. That is called "freeze-fracture-replication cryo-TEM," or, in short, FFR.

This section describes in some detail the basics of direct-imaging cryo-TEM. FFR is described in the next section of the chapter. Some of the considerations involved in cryo-TEM do apply to both classes and are described in the current section.

Samples that contain high concentrations of liquids must be made compatible with the TEM. It is necessary to lower the vapor pressure of these systems to make them compatible with the high vacuum in the microscope column, typically better than 10^{-6} Pa. Also, any supramolecular motion must be stopped to prevent blurring of the recorded image. Of course, all TEM specimens must be thin, not thicker than about 300 nm, for the usual accelerating voltage of 120 kV. Thicker specimens give rise to inelastic electron scattering that deteriorates image quality. However, inelastically scattered electrons may be filtered out in those electron microscopes that are equipped with an in-column or post-column energy filter.

The reduction of vapor pressure and arresting supramolecular motion is called "fixation," that can be either chemical or physical (thermal). Chemical fixation involves addition of a chemical substance alien to the sample. Because nano-structured liquids, particularly surfactant-based systems, are very sensitive to changes in composition, addition of compounds such as a stain (a substance that enhances contrast) or fixative, followed in some cases by a chemical reaction between the fixative and the specimen, and often by drying of sample, may alter the original nanostructures in the system. Thus, chemical fixation is unacceptable (in most cases) for the study of nanostructured liquids. Hence, the method of choice is thermal fixation, that is, ultrafast cooling of the liquid specimens into a vitrified or quasi-solid state. This is achieved by rapidly plunging the specimen into a suitable cryogen. Because thermal diffusivities are larger than mass diffusivities, thermal fixation is much more rapid than chemical fixation, and, of course, eliminates the addition of an alien compound to the system.

The cooling rate needed for vitrification of water is on the order of 10^5 K/s, as estimated theoretically.[6] The cooling rate during vitrification was measured experimentally in an actual specimen preparation setup,[7] and was indeed found to be the order of 10^5 K/s. When cooling is too slow, crystalline ice, hexagonal or cubic, forms in aqueous systems. In nonaqueous systems, various other crystalline matrices may form in the cooled specimen. Crystalline matrix causes optical artifacts, mechanical damage to the microstructure, and redistribution of solutes. Solutes are often expelled from the growing ice lattice, and are deposited either in the crystal grains or, often, at grain boundaries.

The high cooling rates needed for vitrification require very large surface-area-to-volume ratio. To achieve that, the geometry of choice is a thin film. The limited penetration power of even high-energy electrons also requires thin films (up to 300 nm thick, as stated above). High-resolution imaging requires thinner samples. In practice most direct-imaging vitrified specimens display a wide thickness range. While microscopes operating at 200, 300, and 400 kV are capable of imaging specimens thicker than specified above, image interpretation becomes increasingly more difficult with specimen thickness. It is the high depth of field of the TEM that leads to superposition of information from many layers of thick specimens, all projected onto the plane of the detector. In FFR, samples are usually much thicker, thus cooling rates are slower and vitrification is rarely achieved.

The cryogen needed to successfully vitrify the specimen has to be at a low temperature, and well below its boiling point to avoid formation of a gas film around the specimen during quenching; such a gas film acts as a thermal insulator (the Leidenfrost effect). A good cryogen also has high thermal conductivity. Liquid nitrogen is a poor cryogen for most applications (see exceptions below), because of the narrow temperature range between its freezing and boiling. In contrast, liquid ethane, cooled to its freezing point (−183°C) by liquid nitrogen, is an excellent cryogen (its normal boiling point is about 100 K higher).

Another important issue in cryo-TEM preparation is the preservation of the nanostructure at precise conditions, namely, temperature and concentration. This cannot be achieved unless the specimen is prepared in a controlled environment of the desired temperature and atmosphere that prevents loss of volatiles, for example, water vapor. This requires a so-called controlled-environment vitrification system (CEVS). Several models are available, especially the relatively simple, but very reliable system developed by Bellare et al.[8] and modified by Talmon and co-workers over the years,[7,9] and the automatic "Vitrobot" of the FEI company that was developed by Frederik and co-workers.[10] The former CEVS can be used from −10 to +70°C and with various saturated or unsaturated atmospheres. It can be used for the preparation of FFR specimens, too.

Cryo-TEM specimen preparation is performed inside the CEVS, where the atmosphere is closed and controlled from the outside. A small drop, typically 3 to 5 μL, of a pre-equilibrated system is pipetted onto a perforated carbon film, supported on a TEM copper grid, held by tweezers and mounted on a spring-loaded plunger. Most of the liquid is blotted away by a piece of filter paper wrapped around a metal strip, leaving thin liquid films supported on the hole edges of the perforated carbon film. After blotting, the plunging mechanism is activated, a trap door opens simultaneously, the specimen is driven into the cryogen and is vitrified. Finally, the vitrified sample is transferred under liquid nitrogen to the working station of a cooling holder, where it is loaded into the special holder and transferred into the microscope. In some cases, a bare grid (i.e., a microscope grid not covered by a perforated film) is used. More technical details can be found elsewhere.[4,8]

The blotting step is very important for successful sample preparation. It may be performed in a number of ways. The simplest is wicking most of the liquid by simply touching the filter paper to the edge of the grid carrying the drop. Viscoelastic fluids, for example, TLM solutions, require blotting with a shearing or smearing action. That temporarily reduces the viscosity of a shear-thinning liquid, allowing the formation of a thin enough liquid film on the support. Another option is to press two pieces of blotting paper on the two sides of the specimen; that can be performed either manually or, as in the case of the Vitrobot, automatically. That mode usually produces more uniform films. The blotting process and the confinement of the liquid in a thin specimen may introduce artifacts one should be aware of (see below). In addition to changes of the nature of the nanostructure, distortions of large objects, and alignment of slender "one-dimensional" (rods or threads) or large "two-dimensional" (sheets) objects may take place.

After blotting, while the thin specimen is still liquid, it may be kept in the controlled environment of the CEVS for some time. In the case of many self-aggregating systems such as TLMs, this allows the specimen to relax after shear and elongation it may undergo during blotting.[11,12] The specimen may also undergo different processes directly on the grid, including chemical or physical reactions induced by different triggers such as fast heating[7,9] or cooling,[13] pH jumps,[14] or gelation. Those processes may be stopped at any intermediate stage by plunging the specimen into the cryogen. The experiment may be repeated a number of times, each time allowing the process to proceed further towards completion. That way one obtains a sequence of vitrified specimens that gives time sectioning of the process. This is called "time-resolved cryo-TEM." Several variations of the CEVS have been built to facilitate such experiments.[7,9,15] In the case of self-aggregating systems, when one is interested in high-viscosity phases that cannot be directly made into a thin liquid film, on-the-grid cooling (or, in rare cases, heating) is a convenient way to produce high-viscosity phases on the grid, starting with a low-viscosity precursor.

There is an increasing interest in self-aggregation, including giant micelle formation, in nonaqueous systems. While direct-imaging cryo-TEM can be extended to nonaqueous systems, liquid ethane, the cryogen of choice, cannot be used in many cases because it is a good solvent for many nonpolar liquids. However, a good number of systems, such as those of branched hydrocarbons, aromatics, or systems that contain glycerides, do not crystallize readily upon cooling, and thus can be vitrified even in liquid nitrogen.[16] That is also true for aqueous systems containing sufficiently high concentrations of glycols (> 20%).[17]

5.3   FREEZE-FRACTURE-REPLICATION

As described above, freeze-fracture-replication (FFR) is an indirect route to cryo-TEM. As in direct-imaging cryo-TEM, FFR involves fast cooling of the specimen in an appropriate cryogen. However, the specimen is larger than that of the former technique, thus in most cases vitrification is not accomplished, and freezing artifacts should to be taken into consideration. The next step of FFR involves fracturing the frozen specimen, followed by the preparation of a metal-carbon replica of the fracture surface by vapor deposition. Fracturing and replication are performed in a vacuum chamber, while the specimen is kept at cryogenic temperature, typically below −150°C. First, a heavy-metal (typically platinum or a mixture of platinum and carbon) is deposited, usually at an angle of 45° to the horizon. This enhances contrast by a "shadowing" effect. A smaller angle is needed to bring out finer details. Then a carbon layer is deposited perpendicularly to the specimen to enhance the mechanical stability of the replica. In modern equipment the sources for the deposition of the metal and carbon are electron guns bombarding a suitable target. An additional preparation option before replication is "etching," namely, removal of high vapor-pressure components. This requires warming up the specimen for a few minutes to about $-100°C$ in the case of aqueous systems, a temperature at which the vapor pressure of ice becomes sufficiently high to give an appreciable rate of sublimation. After replication, the sample is melted, the replica is cleaned, collected on a TEM copper grid, dried, and imaged in the TEM at room temperature. In most laboratories that perform FFR, the entire process of fracturing the specimen and replication is carried out in commercially available systems. Fast cooling may be carried out in the CEVS to allow quenching from given, well controlled, conditions.[18] FFR is most useful when relatively high-resolution images are needed, but direct-imaging cryo-TEM is not practicable, for example, in the case of high-viscosity systems, or systems containing large particles that cannot be accommodated in the thin specimens of direct-imaging cryo-TEM. Of course, fine details or fine particles can be imaged by the technique. In fact, one early success in imaging a network structure of molecular organogel was achieved by applying the FFR technique in the study of a steroid/cyclohexane physical gel by Wade et al.[19] While the technique is excellent to complement direct-imaging cryo-TEM,[20] it has lost popularity in the last two decades and, regrettably, is used in only a few research laboratories. The reasons are that the technique is labor intensive, it relies more than many other techniques on the technical skills of the user, and it requires complex and quite expensive equipment. In recent years the commercial FFR units on the market need modifications before they can be successfully used. The gradual disappearance of the techniques is quite unfortunate, because FFR-TEM is a most useful imaging technique.

5.4 LIMITATIONS, PRECAUTIONS, AND ARTIFACTS

While cryo-TEM, both in the direct imaging and the FFR modes, is probably the most reliable technique to obtain direct, nanometric-resolution images of liquid systems, it cannot always be applied, and, despite all the precautions, the user has to be aware of possible artifacts. The experienced user knows when the technique can be applied and is able to distinguish between real nanostructures and artifacts. Sometimes the distinction is not easy, and data from other techniques should be compared to the physical model emerging from the micrographs.

It has been already mentioned that high-viscosity systems cannot be made into thin direct-imaging cryo-TEM specimens, but can be made into FFR cryo-TEM replicas. It is worthwhile restating here that shear-thinning viscoelastic liquids can be made into thin liquid films, even if their zero-shear viscosity is high. Care must be taken to allow sufficient relaxation time on the grid, prior to vitrification, to avoid flow-induced structures. Knowing the relaxation time of the system is most useful in such cases.

Occasionally, one works close to a phase boundary. In that case chemical-potential control of the gas phase in the CEVS becomes very important. If possible, one should saturate the atmosphere in the CEVS with the same kind

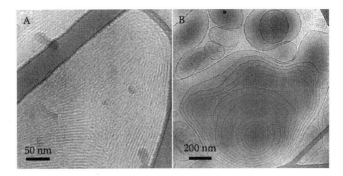

FIGURE 5.1 Vitrified specimens of 90 mM $C_{14}DMAO$, 10 mM SDS, and 200 mM heptanol in water. (A) Specimen prepared in an environment lacking the alcohol, after 90 seconds of relaxation following blotting, showing TLMs, aligned by the flow during blotting. (B) The same system prepared by the same protocol, but in a properly saturated atmosphere, showing multilamellar liposomes.

of solution as the one used to prepare specimens. In many cases, when only surfactants or polymers at low concentrations in water are used, the chamber may be saturated with pure water. However, if other volatiles are present in the mix, such as a low molecular weight cosurfactant, failing to use the actual studied solution for saturation may lead to artifacts. An example is shown in Figure 5.1 of the system made of 90 mM $C_{14}DMAO$ (tetradecyl N,N-di-methylammonium N-oxide), 10 mM sodium dodecylsulfate (SDS), and 200 mM heptanol in water, whereas in Figure 5.1A we see a sample prepared in an environment lacking the alcohol, after 90 seconds of relaxation following blotting, and Figure 5.1B shows the same system prepared by the same protocol in a properly saturated atmosphere. The TLMs that are formed in this case are an artifact; the vesicles imaged in the properly executed experiment are the real nanostructures in this system. Note another artifact in this case: the alignment of the TLMs in Figure 5.1A, a result of flow during the blotting stage of the specimen preparation.

Contrast is an important factor in any TEM technique. In cryo-TEM this is especially crucial, as in most cases inherent contrast in the systems studied is low. The contrast becomes even lower when the continuous phase is not pure water, but a mixture of water and relatively high percentage of organic materials such as ethylene glycol or glycerol. At high concentration of glycerol (about 30 wt% and higher), TLMs become practically invisible, as contrast becomes close to zero. Contrast is also quite poor in organic solvents, although in many cases reasonable images can be recorded. In some organic solvents contrast is reversed, that is, the continuous phase is more optically dense than the nano-aggregate. An example is given in Figure 5.2 showing TLMs of 1% polystyrene-poly(isobuty-lene) diblock copolymer (molecular weight of 13000 : 79000) in a 1:1 mixture of dibutyl and diethyl phthalates.[21] Note the background of the vitrified phthalates

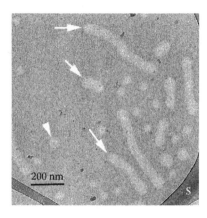

FIGURE 5.2 Cryo-TEM image of 1% polystyrene-poly(isobutylene) diblock copolymer (molecular weight of 13000 : 79000) in a 1:1 mixture of dibutyl and diethyl phthalates.[21] Note swollen end-caps of the TLMs (arrows), and the coexistence of TLMs with spheroidal micelles (arrowheads).

is more optically dense than the micelles. An interesting feature here are the swollen endcaps of the TLMs (arrows), and the coexistence of TLMs with spheroidal micelles (arrowheads). With increasing acceleration voltage in the TEM micrograph, contrast is reduced. For imaging soft materials we prefer to work at the moderately low acceleration voltage of 120 kV, whereas for high resolution TEM of "hard materials," higher acceleration voltage of up to 400 kV usually gives better results.

To overcome the poor contrast in vitrified cryo-TEM specimens, we almost always use "phase contrast," which is equivalent in principle to phase contrast in light microscopy, namely, enhancing contrast by converting phase differences to amplitude differences. In the electron microscope, phase contrast is formed by defocusing the microscope objective lens. This must be applied with care to avoid loss of resolution and introduction of imaging artifacts.

Normally, we refrain from enhancing contrast by staining, that is, adding a high-optical-density material to the system, in the hope that it will selectively attach to certain domains in the specimen; this is similar to biological specimens staining in light microscopy. Because most electron microscopy stains are either salts or acids of heavy elements, their addition to the system may alter its nanostructure. However, in some cases one can change the system slightly, with little effect on the nanostructure, in order to image the desired structures. In the case of ionic surfactants, one can often safely replace the light sodium ions with the much heavier cesium ions, or chloride ions with bromide ones. Of course, one should follow closely any possible structural changes caused by that substitution. An example where contrast was enhanced by replacing Na^+ by Cs^+ is given by Wittemann et al.[22] in the case of polymer brushes attached to polystyrene latex, a situation not unlike that of TLMs. Another example is shown in Figure 5.3 for

FIGURE 5.3 Nanotubes or tubular TLMs formed by complexation of 9 mM ionic block copolymer Eusolex with 1 mM cetyltrimethylammonium bromide (CTAB).[23]

tubular TLMs (essentially a nanotube) formed by complexation of the ionic block copolymer Eusolex (9 mM solution) with cetyltrimethylammonium bromide (CTAB 1mM). Note the fine structure, each wall exhibits two fine lines, visible thanks to the enhanced contrast of the Br⁻ ions. Those lines are most probably domains rich in the bromine anions.[23]

Another limitation for imaging, a possible source of artifacts, but occasionally a means to enhance contrast is electron-beam radiation damage. Organic compounds in vitrified aqueous or nonaqueous matrices are very susceptible to the electron beam.[24,25] Thus, all images must be recorded at low-dose conditions, namely at an exposure of just a few electrons per Å². Higher exposures to the electron beam may give rise to destruction of finer details, as demonstrated in Figure 5.4, to formation of new artifactual structures, or, as been shown in some solvents, to loss or reversal of contrast.[25] The threadlike micelles shown in Figure 5.4 were recorded by direct imaging cryo-TEM in an aqueous solution of 5 mM cetyltrimethylammonium chloride (CTAC) with 3 mM Na-p-iodobenzoate. In comparing the low-dose image (Figure 5.4A) with the high-dose image

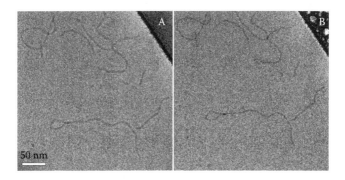

FIGURE 5.4 Threadlike micelles in a cryo-TEM image of an aqueous solution of 5 mM cetyltrimethylammonium chloride (CTAC) with 3 mM Na-p-iodobenzoate. Compare the low-dose image (A) with the high-dose image (B) and note the overall loss of material and fine detail. Note also the "lasso-type" TLM seen in this field of view.

(Figure 5.4B) note the overall loss of material and fine detail. Note also the "lasso-type" TLM seen in this field of view. Electron-beam radiation damage is certainly a problem in direct-imaging cryo-TEM. However, because radiolysis takes place preferentially at the organic material–vitrified water interface, careful exposure by the electron beam can "develop" contrast in areas where it is minimal.[26]

Finally, one should bear in mind that TEM is not strictly a quantitative technique (electron diffraction and lattice-imaging are exceptions). Image magnification depends rather strongly on the position of the imaged area on the optical axis of the microscope. Slight deviations from the ideal position (the "eucentric plan") give rise to changes in magnification that can be as high as ±10%; that is coupled by errors due to calibration and electromagnetic lens hysteresis. Thus measurement taken from electron micrographs should always be backed by data from more quantitative techniques, such as x-ray or neutron scattering. However, a reliable physical model to interpret the data for the latter techniques is to be based on the microscopy.

The apparent concentration one observes in cryo-TEM images may be misleading for two reasons. One is the redistribution of suspended aggregates in the liquid. Particles tend to move from thinner to thicker areas; in fact this is more so for larger particles, thus one observes also size segregation, with larger particles in larger number in the thicker areas while smaller one are left in the thinner areas. The second reason is the high depth-of-field mentioned above. It causes focused images of objects in different depths in the specimen to be simultaneously projected on the detector, thus giving the impression of higher concentration than the real one.

5.5 ADDITIONAL RECENT EXAMPLES

Several examples of the application of cryo-TEM in the study of giant micelles have been given above. The additional examples presented here are to illustrate the strength of the technique, and to serve as a guide to typical features often observed in images of "giant micelles." The reader may also find examples of the application of cryo-TEM in the study of various TLM systems in several chapters in this volume, for example, Chapters 7, 12, 14, and 16.

Figure 5.5 shows relatively high-magnification images of TLMs of the commercial cationic surfactant Habon G, which we studied some time ago.[5] White arrows in Figure 5.5A point to branching points, while white arrowheads point to overlap of two TLMs. The distinction between the two is made simple by paying close attention to the difference in appearance: while the branch has uniform optical density, the area of overlap is darker than the single TLM. Fresnel fringes (black arrowheads), an optical effect of interference between electrons traversing areas of high electron density (TLM) and lower density (vitrified ice), may also help distinguish between branches and points of overlap. The appearance of the Fresnel fringes, as well as the appearance of the entire images (contrast and resolution) depend on the amount of defocus of the TEM objective lens. Note that Figure 5.5B, showing the same system of Figure 5.5A, looks more grainy,

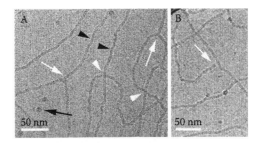

FIGURE 5.5 Cryo-TEM images of 0.1% commercial quaternary ammonium surfactant Habon G in water quenched from 20°C.[5] (A) White arrows point to branching points, while white arrowheads point to overlap of two TLMs. Fresnel fringes (black arrowheads) may also help distinguish between branches and points of overlap. Black arrow indicates a frost particle. (B) White arrow shapes a TLM end-cap, which is *not* hemispherical. Note that the image in A is more underfocused than the one in B.

and the Fresnel fringes look wider, the result of more defocus of the TEM objective lens. While the images in Figure 5.5 are quite clean, some small specks of frost (black arrow in Figure 5.5A) are visible. This is the result of water condensation on the cold specimen surface during specimen preparation and transfer. Good equipment and operator proficiency keep the amount of such contamination to a minimum.

Figure 5.5B shows a TLM tip (white arrow). In systems of very long TLMs such ends are difficult to locate in the image, especially in the case of relatively concentrated solutions where much overlap of the micelles occurs. Note the swollen so-called "end-cap" of the micelle, which does not form a hemisphere, as is quite often depicted in schematics of TLMs. Cryo-TEM has proven that the end caps are swollen, as shown here and in previous publications.[2,3] The same phenomenon can be seen in micelles in nonaqueous systems, as demonstrated in Figure 5.2 above.

The complementarity of direct-imaging cryo-TEM and freeze-fracture replication cryo-TEM (FFR) is demonstrated in Figure 5.6. It shows the evolution of the micellar structures observed in the mesoporous material SBA-15. The reaction mixture was sampled at different times after the addition of tetramethoxyorthosilane (TMOS) to an acidic (HCl) solution of Pluronic P123 held at 35°C.[20] Figure 5.6A is a direct-imaging cryo-TEM micrograph of the system after a short reaction time of 5'13". The structure is that of spheroidal micelles. Due to the relatively high concentration, the projection of several layers of micelles on the TEM detector are observed and it is difficult to make out the individual micelles. After 14'05" the same technique shows TLMs (arrows in Figure 5.6B). In some areas of the micrograph we note the micelles were aligned by the flow during specimen preparation, while in other areas one sees the projection of randomly oriented TLMs. This is quite common in images of TLMs, especially at relatively high concentration. Freeze-fracture-replication of the system at about the same

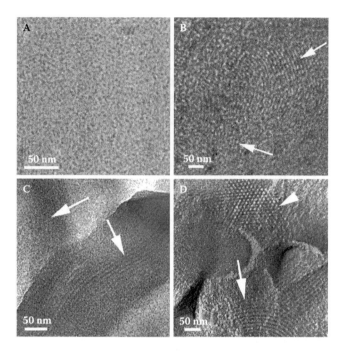

FIGURE 5.6 Evolution of nanostructure, as shown by cryo-TEM images of the SBA-15 reaction in HCl, vitrified from 35°C. (A) At t = 5'13" only spheroidal micelles are seen. (B) t = 14'05"; arrows show the TLMs. (C) Freeze-fracture-replication of the same system at t = 16'27"; arrows point to TLMs. (D) Same at t = 21'42". The hexagonal arrangement of the TLMs in this liquid crystalline phase is seen along their long axes (arrowhead), and the view perpendicular to the long axes (arrow).[20] Reproduced from Ref. 20 with permission of the American Chemical Society.

time of 16'27" (Figure 5.6C), also shows the TLMs (arrows). It is reassuring that the two techniques show the same nanostructures! At longer reaction times the system becomes too viscous because of liquid crystal formation, and it is impossible to image it by direct-imaging cryo-TEM; only FFR can be used. At 21'42" (Figure 5.6D) FFR reveals the arrangement of the TLMs into a hexagonal liquid crystalline phase. Note the hexagonal arrangement of the micelles seen along their long axes (arrowhead), and the view perpendicular to the long axis (arrow). While the spacings at 21'42" are very uniform, the nonuniform spacings between the TLMs in Figure 5.6C (16'27") indicate that those TLMs are, indeed, free.

In many surfactant-polymer systems threadlike micelles are formed. Quite often those TLMs arrange themselves into hexagonal liquid phases. Those complexes may be solubilized by excess surfactant to form free micelles. The evolution of phases, including TLMs in the system of poly(diallyldimethylammoniumchloride) (PDAC) and sodium dodecyl sulfate (SDS) was studies by

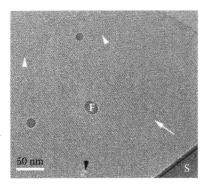

FIGURE 5.7 Vitrified specimen of the hexagonal liquid crystalline phase formed when CTAB is mixed at a charge ratio of 1:2 with 0.1 wt% PAAS. White arrow indicates the structure imaged perpendicularly to the long micelle axis; white arrowheads point to domains imaged parallel to the long axis. The black arrowhead indicates an area damaged by the beam. "F" is a frost particle. "S" is the edge of the support film.

Nizri et al.[27] In the reverse case of an anionic polyelectrolyte, poly(acrylic acid) sodium salt (PAAS), and various cationic surfactants, alkyltrimethylammonium bromides, the system has even a stronger propensity to form very long TLMs, ordered hexagonal arrays and nanoparticles of the hexagonal phase, depending on the surfactant hydrocarbon chain length and the polymer cation to surfactant anion ratio.[28] An example from this work is given in Figure 5.7, showing a very well developed array of the hexagonal phase made of the complexed TLMs. Some areas show hexagonal order (white arrowheads; the hexagonal phase imaged parallel to the TLMs long axes), while others are made of arrays of lines (white arrow; the phase imaged perpendicular to the TLMs). The order is not unlike what one observes by TEM of "hard-materials" crystals, but here we image a liquid crystalline phase in vitrified liquid state. The typical sizes here are nanometers, while in "hard" crystals the sizes are typically on the order of several Ångströms. These liquid crystalline phases are very electron-beam radiation-sensitive. Such images are recorded at very low electron exposures of not more than 15 e⁻/Å², but some damage is already visible (black arrowhead).

Finally, in Figure 5.8 we see two direct cryo-TEM images of inverse TLMs of 0.38 wt% soybean lecithin in isooctane, vitrified in liquid nitrogen. The micelles are swollen by a small amount of water (0.02 wt%) that makes the packing of the lecithin in this form possible. These images, taken from an on-going project, show how the inverse micelles arrange themselves within the holes of the support film (Figure 5.8A). Their arrangement into a rather curious patterns, especially at the hole edge, as shown in Figure 5.8B, is, of course, an artifact of sample preparation. However that such images had not been seen before in other TLMs is especially intriguing, and is being further investigated.

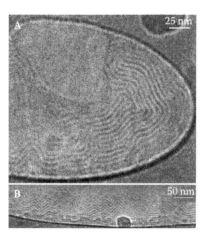

FIGURE 5.8 Two direct-imaging cryo-TEM images of inverse TLMs of 0.38 wt% soybean lecithin in isooctane, vitrified in liquid nitrogen. The micelles are swollen by a small amount of water (0.02 wt%).

5.6   CONCLUSIONS

Because of the complexity of giant micelle systems, namely, coexistence of TLM with spheroidal micelles, size distribution of TLM, branching and network formation, cryo-TEM is the best technique to capture directly all those features. The technique is mature now, with much of the basic physics behind it well understood, many of the technical difficulties solved, and a large number of possible artifacts well documented and avoided. Direct-imaging cryo-TEM is the most suitable cryo-technique for the study of giant micelles, but freeze-fracture-replication has been shown to be a useful complementary technique, especially for more viscous, gel-like systems. However cryo-microscopy requires expensive, sophisticated, and complicated equipment. Its applications require expertise, experience and technical dexterity, thus colloid science cryo-TEM is applied in just a few research centers. But the advent of nano- and bio-sciences that require the information the technique can afford, will no doubt make it much more available and accessible in the coming decade.

ACKNOWLEDGMENTS

My work on the cryo-TEM of threadlike micelles started back in 1985 in collaboration with Ted Davis and Skip Scriven of the University of Minnesota (my former Ph.D. advisors). I recorded the very first images of TLMs while on sabbatical in the laboratory of Wah Chiu, then at the University of Arizona, Tucson. Many other people have taken part in joint TLM related research projects ever since. Here I will just name a few who are represented in this chapter by images taken as part of those projects: Jack Zakin of the State University of Ohio,

Shlomo Magdassi of the Hebrew University of Jerusalem, Yachin Cohen of the Technion, Tim Lodge of the University of Minnesota, Heinz Hoffmann of the University of Bayreuth, Daniella Goldfarb of the Weizmann Institute, and Raoul Zana of the CNRS, Strasbourg, with whom we first identified branching in TLMs. Special thanks go to my former students, who are also represented here, and who are now faculty members, Dganit Danino of the Technion, Anne Bernheim-Groswasser of the Ben Gurion University, and to my current graduate student Alona Makarsky. The images of Figure 5.8 were taken by an undergraduate student working in my group, Mr. Ithai Lomholt-Levy. Last but not least I would like to express my gratitude to the permanent staff of my research group: Ellina Kesselman, Judith Schmidt, and Berta Shdemati.

REFERENCES

1. Danino, D., Talmon, Y., Levy, H., Beinert, G., Zana, R. *Science* 1995, *269*, 1420.
2. Bernheim-Groswasser, A., Zana, R., Talmon, Y. *J. Phys. Chem. B* 2000, *104*, 4005.
3. Zheng, Y., Won, Y.Y., Bates, F.S., Davis, H.T., Scriven, L.E., Talmon, Y. *J. Phys. Chem. B* 1999, *103*, 10331.
4. Talmon Y. In *Modern Characterization Methods of Surfactant Systems.* B.P. Binks, Ed., Marcel Dekker: New York, 1999, Chapter 5, p. 147.
5. Danino, D., Bernheim-Groswasser, A., Talmon, Y. *Colloid Surf. A* 2001, *183*, 113.
6. Uhlmann, D.R. *J. Non-Cryst. Solids* 1972, *7*, 337.
7. Siegel, D.P., Green, W.J., Talmon, Y. *Biophys. J.* 1994, *66*, 402.
8. Bellare, J.R., Davis, H.T., Scriven, L.E., Talmon, Y. *J. Electron Microsc. Techn.* 1988, *10*, 87.
9. Chestnut, M.H., Siegel, D.P., Burns, J.L., Talmon, Y. *Microsc. Res. Techn.* 1992, *20*, 95.
10. http://www.vitrobot.com/.
11. Danino, D., Talmon, Y., Zana, R. *Colloid Surf. A* 2000, *169*, 67.
12. Zheng, Y., Lin Z., Zakin, J.L., Talmon, Y., Davis, H.T. Scriven, L.E. *J. Phys. Chem. B* 2000, *104*, 5263.
13. Schmidt, J., Eger, S., Talmon Y. Manuscript in preparation.
14. Talmon, Y., Burns, J.L., Chestnut, M.H., Siegel, D.P. *J. Electron Microsc. Techn.* 1990, *14*, 6.
15. Fink, Y., Talmon, Y. In *Proc. 13th Int'l Congress on Electron Microsc.* 1994, 1: p. 37.
16. Danino, D., Gupta, R., Satyavolu, J., Talmon, Y. *J. Colloid Interface Sci.* 2002, *249*, 180.
17. Zhang, Y., Schmidt, J., Talmon, Y., Zakin, J. *J. Colloid Interface Sci.* 2005, *286*, 696.
18. Burns, J.L., Talmon, Y. *J. Electron Microsc. Techn.* 1988, *10*, 113.
19. Wade, R.H., P. Terech, P., Hewat, E.A., Ramasseul, R. Volino, F. *J. Colloid Interface Sci.* 1986, *114*, 442.
20. Ruthstein, S., Schmidt, J., Kesselman, E., Talmon, Y., Goldfarb, D. *J. Am. Chem. Soc.* 2006, *128*, 3366.
21. Bang, J., Jain, S., Li, Z., Lodge. T.P, Pedersen, J.S., Kesselman, E., Talmon, Y. *Macromolecules* 2006, *39*, 1199.

22. Wittemann, A., Drechsler, M., Talmon, Y., Ballauff, M. *J. Am. Chem. Soc.,* 2005 *127*, 9688.

23. Hoffmann, H., Schmidt, J., Talmon, Y. Manuscript in preparation.

24. Talmon, Y., Adrian, M., Dubochet, J. *J. Microscopy* 1986, *141*, 375.

25. Kesselman, E., Talmon, Y., Bang, B., Abbas, S., Li, Z., Lodge, T.P. *Macromolecules* 2005, *38*, 6779.

26. Mortensen, K., Talmon, Y. *Macromolecules* 1995, *28*, 8829.

27. Nizri, G., Magdassi, S., Schmidt, J., Cohen, Y., Talmon, Y. *Langmuir* 2004, *20*, 4380.

28. Nizri, G., Magdassi, S., Schmidt, J., Talmon, Y. Manuscript in preparation.

6 Scattering from Wormlike Micelles

Jan Skov Pedersen, Luigi Cannavacciuolo,
and Peter Schurtenberger

CONTENTS

6.1 INTRODUCTION

6.1.1 BACKGROUND

While micelles are frequently spherical, they can also exhibit a sphere-to-rod transition and even grow to giant flexible and polymer-like aggregates. This micellar growth can be induced by a reduction in the so-called spontaneous curvature H_0 due to a change in a "control parameter" such as temperature (in nonionic surfactant systems), ionic strength (in ionic surfactant systems), or cosurfactant concentration, which subsequently results in a characteristic transition in the particle morphology from spheres to cylinders to lamellae.[1,2] Another remarkable feature in such systems is the change in the micellar size distributions upon an increase of the surfactant

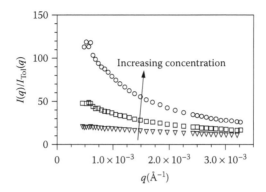

FIGURE 6.1 Typical examples of the q dependence of the normalized intensity $\Delta\langle I(q)\rangle/\langle I_{Tol}(q)\rangle$ for several concentrations at 25.5°C for $C_{16}EO_6$ in D_2O (see Ref. 3 for details): (triangles), $C = 2.21\times10^{-4}$ g/cm$_3$; (squares), $C = 4.69\times10^{-4}$ g/cm^3; (circles), $C = 1.1\times10^{-3}$ g/cm^3.

concentration, C. An example is given in Figure 6.1 for the nonionic surfactant $C_{16}EO_6$ in D_2O.[3] With increasing surfactant concentration the scattering intensity increases and a very strong angular dependence develops. The micellar size is now large enough to produce a measurable particle form factor, which permits, for example, determination of an apparent radius of gyration $R_{g, app}$.

These micelles have a locally cylindrical structure, and it is possible to find conditions where they grow dramatically with increasing surfactant concentration. Since they have some degree of flexibility, they attain conformations similar to those of polymers in solution. Therefore the micelles are sometimes called polymer-like or wormlike micelles. The micelles are also called equilibrium polymers or living polymers since their length depends on concentration as a consequence of the fact that they continuously break and recombine (see Chapters 4, 8, and 10).[4,5,6] Equilibrium polymers are transient structures with a relatively short lifetime, and this has consequences for their rheological and dynamic properties, which are much different from those of classical covalently bonded polymers.[7–12]

The analogy between classical polymers and polymer-like micelles is illustrated for scattering data from micelles at relatively low concentrations, that is, in the dilute regime (see Figure 6.2). Since the characteristic length scales of the micelles span from a few nanometers for the cross-section size to micrometers for the overall size, the scattering curves display variation over many orders of magnitude of scattering vectors q, where $q = 4\pi\sin\theta/\lambda$ (θ = scattering angle; λ = wavelength of the radiation). Due to this large span of length scales, the total scattering curve is often obtained as the combination of static light scattering (SLS) data and small-angle neutron scattering (SANS) data. The scattering curve usually levels off at low scattering vectors due to the finite size of the micelles relative to the length scale probed ($\sim 1/q$). The curve follows the Guinier behavior $I(q) \sim 1 - q^2\langle R_g^2\rangle/3$ for low $q^2\langle R_g^2\rangle \ll 1$, where $< R_g^2 >$ is the square average

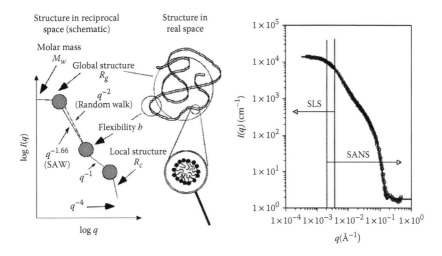

FIGURE 6.2 Schematic representation of a typical scattering curve from giant wormlike micelles obtained by combination of SLS and SANS. Shown are the different regimes in analogy to polymer theory, together with a schematic representation of the micellar structure in real space. This schematic representation can be compared with experimental data for polymer-like inverse micelles. The scattering intensity $I(q)$ versus the scattering vector q is shown for solutions of soybean lecithin in deuterated isooctane (water-to-lecithin molar ratio $w_o = 1.5$) at a volume fraction $\phi = 0.0021$. Data have been obtained from light (triangles) and neutron (circles) scattering experiments.[13]

radius of gyration of the chains. Beyond the Guinier region, the scattering crosses over to a power law behavior $q^{-\alpha}$, where $\alpha \approx 5/3$ due to the excluded-volume statistics of the micelles. At higher q, the scattering probes the local cylindrical structure and it crosses over to q^{-1}. However, this power law is often masked by the cross-section Guinier region that originates from the finite diameter of the micelles. Beyond this region the curve drops strongly and at the highest q it might follow the Porod law q^{-4} which occurs for objects with sharp interfaces. The Porod law might be masked by scattering from the constituting surfactant molecules.

Figure 6.2 also shows experimental scattering data for a microemulsion system of inverse micelles of lecithin in isooctane with small amounts of water.[13] In this system the length of the micelles can be tuned by the amount of water added. We see that the generic behavior of a polymer-like object is indeed qualitatively observed. Figure 6.2 can be seen as a starting point for the remainder of this chapter, where we will illustrate how polymer theory and a combination of scattering methods and computer simulations can be utilized to obtain a wealth of information about the static and dynamic properties of wormlike micelles.

Due to the analogy between polymer-like micelles and polymers, polymer-like micelles have frequently been used as model systems for polymers and polyelectroytes.[14] In turn, the application of theoretical concepts from polymer

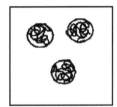

FIGURE 6.3 The different concentration regimes of polymer-like micelles. Left: dilute regime with well-separated micelles; center: at the overlap concentration C^*, interpenetration of the micelles sets in; right: the micelles forms a network for $C > C^*$.

physics can provide a deeper understanding of polymer-like aggregates. Modeling of the scattering at low concentration provides information on overall contour length and the related distribution function, persistence length, and cross-section structure. Such detailed modeling has become possible using the numerical scattering functions developed by Pedersen and Schurtenberger,[15] which are based on Monte Carlo simulations.[16]

At higher concentration the interpretation of scattering data is more complicated due to the presence of intermicellar interference effects. As for polymers, the polymer-like micelles have an overlap concentration C^* beyond which the micelles form an interpenetrated network (Figure 6.3). In this network, the system is homogeneous on a length scale larger than the distance between entanglement points between individual micelles, and a scattering experiment provides a correlation length rather than a micellar size. Similarly, the mass deduced from the experiment will be smaller than the micellar mass (cf. Figure 6.3), as intermicellar interactions will mask the concentration-induced growth of the micelles. The effects of intermicellar interactions and micellar growth are illustrated in Figure 6.4, where examples of scattering data obtained with weakly charged wormlike micelles at different concentrations are shown. Experiments were performed with aqueous solutions of the nonionic surfactant $C_{16}EO_6$, doped with a small amount of an ionic surfactant with the same hydrophobic chain length, but an ionic head group (see Ref. 3 for details). Shown are data at a doping level of 3% and two different ionic strengths of 0.01 M NaCl, and 0.001 M NaCl, respectively. Figure 6.4 illustrates the two different regimes generally found in solutions of wormlike micelles, where the micelles either behave as neutral polymers (nonionic surfactants or high salt regime, Figure 6.4(a)), or as polyelectrolytes (ionic surfactants at low ionic strength, Figure 6.4(b)) that exhibit a pronounced structure factor peak typically found for polyelectrolytes at low ionic strength.

A comparison of Figure 6.2 and Figure 6.4 illustrates the importance of incorporation of intermicellar interaction effects in the analysis of scattering data. In the dilute range, we see the typical features of semiflexible polymer and polyelectrolyte chains with excluded-volume effects. At high salt content, the SANS

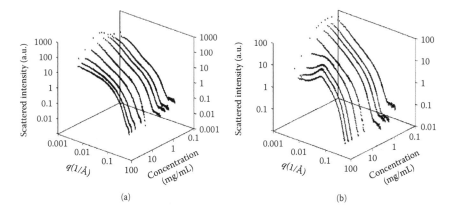

(a)   (b)

FIGURE 6.4 Concentration dependence of the normalized scattered intensity dσ(q)/dΩ from SANS as a function of the magnitude of the scattering vector q for wormlike micelles formed in aqueous solutions of the nonionic surfactant $C_{16}EO_6$, doped with a small amount of an ionic surfactant with the same hydrophobic chain length.[3] Shown are data at a doping level of 3% and ionic strengths (a) 0.01M NaCl and (b) 0.001M NaCl. The figure illustrates the effects of a combination of micellar growth and intermicellar interactions for polymer-like and polyelectrolyte-like behavior. Reproduced from Ref. 3 with permission of the American Chemical Society.

data closely resemble those obtained previously with uncharged polymer-like micelles, and we observe all the typical features of semiflexible polymers in the dilute and semidilute regimes. The micelles are highly flexible, and at low concentrations we clearly see the crossover between the different asymptotic regimes for flexible chains ($I(q) \sim q^{1.67}$) and rigid rods ($I(q) \sim q^1$). At concentration $C > C^*$, the data show the typical features of semidilute solutions of neutral polymers, with a strongly decreasing forward intensity and a concentration-dependent correlation length ξ_s that follows a power law of the form $\xi_s \sim C^x$ where $x \approx -0.75$. For the samples with low salt concentration the micelles are extremely stiff and the persistence length is comparable to the radius of gyration. Therefore we can hardly observe a flexible coil asymptotic regime and a corresponding crossover from the flexible coil to rigid rod asymptotic regimes. With increasing concentration, the forward intensity increases first due to the micellar growth and then decreases strongly in the semidilute regime at concentrations $C > C^*$. In the semidilute regime, we see the formation of a pronounced structure factor peak at a characteristic value q^*. From Figure 6.4 it becomes obvious that a quantitative determination of important parameters such as the overall contour length or micellar mass or the micellar flexibility requires an approach that combines concentration-induced micellar growth and intermicellar interactions.

Schurtenberger and co-workers[17,18] introduced the application of conformation space renormalization group theory[19] originally developed for semidilute polymer solutions for analyzing the measured apparent molar mass of the micelles. This self-consistent approach was originally applied in the analysis of static light scattering data and it explains the decrease in intensity above the overlap concentration and provides information on the growth law to higher concentration. The approach has been further developed in combination with Monte Carlo simulations,[20] so that the full q-dependence of the scattering curve can be analyzed at higher concentration.[21]

In this chapter we review the concepts and theories used in the analysis and interpretation of scattering experiments from giant polymer-like micelles and we provide examples of applications of the various approaches. First, some of the basic concepts of the thermodynamics of these systems will be given. Then the low concentration behavior of the neutral surfactant system is described and some details on the simulations and the derivation of scattering functions are provided. The necessary equations for analyzing experimental data are given and the results of some applications both to neutral and weakly charged systems are discussed. In addition, methods for determination of the cross-section structure by form-free methods are described.

In the section that follows, the influence of concentration effects in neutral systems on the micellar properties and the scattering is treated. The self-consistent approach by Schurtenberger and co-workers for obtaining the growth law from light scattering data is described. The derivation of scattering functions from Monte Carlo simulations on many chain systems is reviewed and empirical modifications for their application in the analysis of experimental data are described. The scattering from ionic systems is the last topic treated. These systems have attracted attention as model systems for polyelectrolytes. Monte Carlo simulations on such micellar systems have to employ a model that is more specifically adapted to the experimental system under consideration with respect to cross-section diameter, charge density, and salt concentration, and therefore the results are less general. We will review the results from simulations on relatively weakly charged micelles with varying concentrations of salts. For these systems, both single-chain[22] and many-chain[23] simulations have been performed. The results are compared with the experimental results for hexa-ethyleneglycol monohexadecyl ether, $C_{16}EO_6$, doped with ionic surfactant sodium 1-hexadecanesulfonate ($NaC_{16}SO_3$).[3] The chapter ends with a summary and an outlook.

Here, we focus on the systematic development of theoretical descriptions of scattering from giant wormlike micelles and on the most systematic applications of the theory to experimental systems. The theory covers self-consistent approaches that include, for example, concentration-induced growth, polydispersity, semiflexibility, excluded-volume interactions as well as interparticle interactions. The literature on investigations of wormlike micelles by scattering techniques is very large and it is beyond the scope of the present chapter to give a complete review of these applications. Although there are many applications, only a few of them are extensive and systematic. We have chosen to put emphasis on these applications since they best illustrate the development of the theory.

6.1.2 THERMODYNAMICS

Surfactants in solution exhibit a complex aggregation behavior as a result of a delicate balance of opposing forces.[24] Micellar solutions and microemulsions represent thermodynamically stable liquid dispersions containing surfactant aggregates, which can often be found in a large region of the phase diagram of two- or multicomponent surfactant systems.[25–27] The relation between microstructure and phase equilibria is an important aspect of surfactant systems. Several theoretical concepts based either on packing considerations of the surfactants in the aggregates or on the role of the bending elastic energy of the surfactant monolayer have provided a theoretical framework for a better understanding of these systems.[1,28–30] We can, for example, try to rationalize the morphology sequence and phase behavior of surfactant aggregates as being driven by the spontaneous curvature of the hydrophobic/hydrophilic interface, H_0. This quantity does not only depend on the space filling dimensions of the surfactant molecule, but may be tuned by various external factors such as the amount and nature of added electrolyte, the presence of other species in solution, the pH, or the temperature. For example, if the head groups are charged, they repel each other, which increases the effective head group area and favors the formation of small spherical micelles. The addition of electrolyte subsequently screens the electrostatic interactions, even more efficiently in the case of a strongly binding salt, which allow the head groups to approach each other closer and induces the formation of cylindrical structures. Similar arguments can be made for the known effect of temperature on nonionic surfactants of the C_mEO_n family, where temperature is known to change the spontaneous curvature of the surfactant due to the fact that water decreases its solvent quality with increasing temperature for the ethylene oxide head group. These ideas are developed further in Chapter 7 in regards to their implications for phase behavior.

Here we next focus on a brief summary of the thermodynamics of wormlike micelle formation and in particular on the question of the micellar size and flexibility as a function of various parameters such as concentration, temperature, or ionic strength. For surfactants forming cylindrical micelles, the free energy expression ε_N of a molecule in an aggregate is composed of one term related to the endcaps and another related to the central cylindrical part of the micelle. The free energy ε_∞ of a surfactant, located away from the endcaps, is independent of the length of the micelle. The endcap energy, $2E_C$, is distributed among all molecules, to give:

$$\varepsilon_N \approx \varepsilon_\infty + 2E_C/N \qquad (6.1)$$

In a mean field approach, the minimization of the total free energy of the micellar solution leads to a broad, exponential distribution of micellar sizes and to a mean aggregation number $\bar{N}$ equal to:

$$\bar{N} = C^{1/2} \exp(E_C/2k_BT) \qquad (6.2)$$

where T is the absolute temperature, and k_B the Boltzmann constant.

The number density of chains with aggregation number N is:

$$D(N) \propto \exp[-N/\bar{N}] \qquad (6.3)$$

If E_C is very high, cylindrical micelles can become extremely long (giant!) and attain almost macroscopic lengths. Equation (6.3) immediately also reveals the effect of temperature generally found for wormlike micelles, where increasing the temperature results in a drastic reduction of the average micellar size due to the Boltzman factor $\exp(E_C/2k_BT)$. This is of course different for the nonionic C_mEO_n surfactants, where the Boltzman factor in Equation (6.3) is often (over)compensated by the temperature-induced change in the spontaneous curvature, and where an increase of the temperature may even lead to an increase of the average micellar size at constant concentration.

The growth law $\bar{N} \propto C^{1/2}$ can be considered as a mean-field result and it is indirectly based on an assumption of low concentration or very weak interparticle interactions. The growth law changes to $\bar{N} \propto C^{0.6}$ in the semidilute regime,[4] where the intermicellar interactions become important. However, as we will point out below, the value of the growth exponent α in $\bar{N} \propto C^{\alpha}$ is a rather controversial topic, and the lack of agreement between the theoretical predictions and the experimental findings published so far clearly point out the need for further theoretical work on this issue. (See also Chapter 8, Section 8.4.1.)

For charged micelles, the situation is more complicated.[31,32] In this case, the scission energy has two contributions. It is now composed of the endcap energy that favors the micellar growth, and a repulsive contribution due to the charges along the micelle that favors micellar breakage. For charged systems, three different regimes have been discussed. In the dilute regime (regime i), where the Debye length is larger than the mean micelle size, the competition between endcap and repulsive energies results in a minimum in the free energy at some specific micellar size. This leads to micelles that are almost monodisperse, contrary to the broad distribution of lengths observed for neutral systems. Their length increases only very slowly with concentration, and one expects

$$\bar{N} \approx \frac{1}{l_B a v^2} \left[\frac{E_C}{k_B T} + \log\left(\frac{\phi}{\bar{N}}\right) \right] \qquad (6.4)$$

where v is the effective charge density per unit length, l_B the Bjerrum length, a the length of the surfactant molecule, and ϕ the surfactant volume fraction.

At higher concentration, the situation dramatically changes. We now reach the semidilute or growth regime (regime ii) in which the Debye length is smaller than the mean micelle size. This leads to a strongly enhanced micellar growth, and the micelles have a broad size distribution and their length is influenced by the charge of the micelles and the ionic strength of the solution:

$$\bar{N} \approx 2\phi^{1/2} \exp\left(\frac{E_C}{2k_B T} - \frac{v^2 a l_B}{2\phi^{1/2}} \right) \qquad (6.5)$$

The first term in the exponential favors growth, while the second term represents electrostatic repulsions, which favor the breaking of the micelles. Because of the explicit concentration dependence found in this second term, the micelles do not exhibit a simple power law dependence for the micellar size, but a strongly enhanced or accelerated growth as soon as the system enters this growth regime. The crossover between these two regimes is rather sharp and occurs when the Debye length is approximately equal to the size of the micelle. If no salt is added, it corresponds to the overlap volume fraction $\phi*$ for which the endcap energy E_C is equal to the repulsive energy:

$$\phi* \approx \left(k_B T l_B a v^2 / E_C \right)^2 \tag{6.6}$$

As the salt concentration is increased, the crossover volume fraction decreases. In the high salt limit (regime iii), the crossover is predicted to disappear and we expect to recover the simple power law behavior already derived for the neutral case. It is clear that the concentration-dependent micellar size distribution complicates the data analysis for small-angle scattering experiments considerably, as interaction effects and micellar growth will have to be taken into account. In the following sections we describe how this can be achieved based on the analogy to classical polymers and polyelectrolytes in a self-consistent way.

6.2 DILUTE SOLUTIONS: AN ANALOGY TO POLYMERS IN A GOOD SOLVENT

6.2.1 Monte Carlo Simulations

Modeling wormlike micelles (WLM) is by no means a simple task even if one just aims at reproducing equilibrium properties of uncharged structures in dilute systems. This is due to the very complex interactions acting between hydrophobic and hydrophilic parts of the surfactant molecules and the solvent. To make things tougher, WLM are living structures, that is, they can break and recombine among them (see Chapters 4, 8, and 10), giving rise to highly polydisperse systems. It is therefore clear that a "smart" coarse-grained model is necessary to overcome the complexity of these systems and make them tractable with the computer power presently available. Care should also be taken in modeling the solvent. Most of the time, one is not interested in the description of its physical behavior so its degrees of freedom are omitted from the model. Nevertheless, an explicit treatment of the solvent is sometimes required in order to end up with a realistic model. This is, for instance, the case of simulations of WLM in shear flow or under a flow field where accurate (long-range) hydrodynamic interactions have to be reproduced. In this case a reasonably realistic model has to include explicitly some millions of solvent particles. We stress that this kind of simulation is currently of extraordinary interest, because a large variety of phenomena are generated by the flow, most of them still poorly understood. A review on coarse-grained models for WLM is presented in Chapter 3, Section 3.4.

In what follows we will only consider simple models for a linear chain in which the sole properties we account for are the stiffness, the excluded volume, and the electrostatic charges on the chain. These are basically modified models of linear semiflexible polymers, where we start with a description of the structural properties of a WLM using a discrete representation of a so-called wormlike chain as introduced by Kratky and Porod.[37] Readers interested in more specific issues, such as self-assembly and nonequilibrium systems can see Chapter 3 and references therein. By comparing simulation results and experimental scattering measurements we will show that the physics included in the model is sufficient to reproduce the important structural properties of WLM.

Consider a chain comprising N_s beads connected by N_s-1 rigid bonds of length l_0. One way to include stiffness on the chain is to superimpose the condition that the angles between any two consecutive bonds are quenched at a common value θ, the so-called valence angle. In this way, the only possible movements of a bead c_n are the rigid rotations around the direction of the bond connecting beads c_{n-1} and c_{n-2} (see Figure 6.5). The stiffness is quantitatively expressed by a characteristic length b, the so-called Kuhn length or statistical segment length, which is, ultimately, a measure of the correlations between the bonds, or in simpler

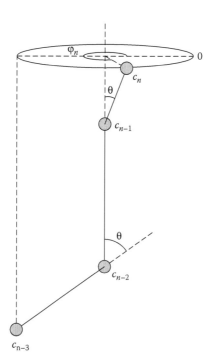

FIGURE 6.5 Backbone structure of the semiflexible chain model with fixed valence angle.

words, the length below which the chain can be considered stiff. Sometimes the persistence length $l_p = b/2$ is considered instead. Of course, b is related to the valence angle θ; indeed it can be shown to be:

$$b = l_0 \frac{1 + \cos\theta}{1 - \cos\theta} \tag{6.7}$$

We can now specify a WLM by assigning its contour length $L = N_s l_0$, cross-section radius R, and Kuhn length b. Typical values are $L \sim 10^4$ Å, $b \sim 200$ Å, and $2R \sim 20$ Å. It is, however, more convenient to express all lengths in units of the Kuhn length, that is, by the dimensionless quantities L/b and R/b. The next step towards a realistic WLM model is to extend the model from a discrete chain to a continuous one. From a theoretical point of view this can be achieved by letting $N_s \rightarrow \infty$, $l_0 \rightarrow 0$, and $\theta \rightarrow 0$ in such a way that L/b remains constant at its assigned value. In the simulations the procedure consists in performing many runs with L/b fixed but at different increasing values of N_s until the asymptotic regime is reached, and an appropriate extrapolation is possible.

Chain conformations where beads overlap may occur in this model. This is of course not the case in a real WLM in which different surfactant molecules cannot physically occupy the same volume. This is the so-called excluded-volume (EV) effect. The implementation of this constraint in our model requires some care. One way is to model the beads as hard spheres of radius R (see Figure 6.6). This is a reasonable approximation to the local cylindrical structure. However, neighboring spheres overlap for the large values of N_s required in the continuous limit procedure, regardless of the chain conformation. Therefore, in a simulation, the search for overlap should only concern spheres at distances larger than some

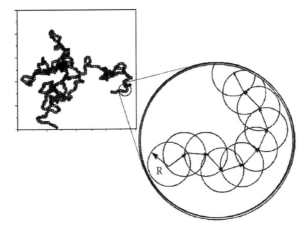

FIGURE 6.6 A chain configuration for the semiflexible models with spheres making up the excluded volume of the chain.

value, r_{min}, along the backbone. A reasonable choice has been shown[16] to be r_{min} = πR. This allows the chain to have a maximal bend of a semicircle before overlap occurs.

As opposed to Molecular Dynamics simulations which attempt a direct numerical integration of the equation of motion, the aim of Monte Carlo (MC) techniques is to generate trajectories in the phase space sampled from a given statistical ensemble. This is achieved by using random numbers to generate a sequence of random states of the system with the appropriate probability. These trajectories are eventually employed to calculate thermodynamic averages. A crucial trick that makes MC efficient is to do the sampling not completely at random but to privilege the regions of the phase space that give the most important contribution to the averages (importance sampling). This can be achieved by building up a Markov chain of states whose distribution function converges to the desired distribution of random states of the system. For a more complete description of MC techniques see Ref. 33.

6.2.2 SINGLE UNCHARGED CHAINS

For short chains (L/b~100) a simple self-avoiding walk algorithm can be used to generate independent chain configurations. More specifically, a chain is grown from scratch; if any overlap occurs, the configuration is rejected and a new attempt is performed. It is clear that as L/b increases this method becomes inefficient, since the number of rejections grows with L/b.

To improve efficiency, one can introduce moves to generate the Markov chain of states. The pivot algorithm[34] turns out to be very efficient with a typical fraction of accepted configurations of ≈ 0.8–0.9 in the MC importance sampling. Starting from a randomly chosen chain conformation, the pivot algorithm generates a new one by choosing at random a bond n on the chain and by performing a rigid rotation of the part of the chain on one side of the bond relative to the part on the other side. Thus, each of the beads n, $n + 1,..., N_s$ moves along some arch, and it turns out that even if only one angle has been changed the resulting chain conformation may be very different from the starting one. After a new configuration has been generated, a check for overlap has to be carried out. The zippering method[35] speeds up this process by excluding a priori from the check any configurations that, for geometrical reasons, cannot overlap.

Monte Carlo simulations have been performed with and without EV for chains of size ranging from $L/b = 0.25$ to 16 384.[16] Results shown in Figure 6.7 clearly demonstrate that for longer chains, the EV leads to expansion. A full quantitative analysis of this effect by determining the expansion factor of the chain is reported in Ref. 16.

One major success of the above model is that, despite its simplicity, it has provided accurate scattering function of semiflexible WLM, with and without excluded effects (see Figure 6.8), which could be numerically parameterized.[15] This has turned out to be an essential tool in fitting experimental scattering data, as no analytical expression for the scattering function of the Kratky-Porod (KP)

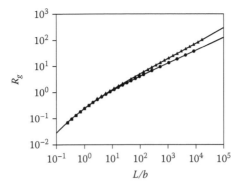

FIGURE 6.7 Radius of gyration of the semiflexible chain model.[16] Triangles are with excluded volume whereas spheres are without. Reproduced from Ref. 16 with permission of the American Physical Society.

model is available.[37] Figure 6.8 shows scattering functions calculated for the backbone points for chains with $L/b = 640$ both with and without excluded volume. At intermediate q, the scattering follows a power law with exponent -2 for the chains without excluded volume and with about $-5/3$ with excluded volume. At high angle where the locally stiff structure is probed, there is no influence of excluded volume, and the curve follows in both cases a power law with exponent -1.

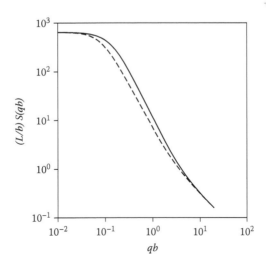

FIGURE 6.8 Scattering functions of semiflexible chains with $L/b = 640$.[15] The full curve is without excluded-volume effects and the broken curve is with excluded volume.

6.2.3 COMPARISON WITH EXPERIMENTAL DATA

The Monte Carlo simulations provide a numerical expression for the scattering functions of semiflexible self-avoiding chains. The form factor can formally be written as

$$P_L(q) = L^2 P_{KP}(q, L, b) \tag{6.8}$$

where L is the contour length of the chains and b is the Kuhn length, which is equivalent to the effective step length of the corresponding random walk in the limit $L \gg b$. As mentioned previously, the Kuhn length can also be expressed in terms of the persistence length l_p as $b = 2l_p$, where l_p (for chains without excluded volume[36]) is the $1/e$ decay length of the angular correlation function. The subscript KP on the form factor $P_{KP}(q, L, b)$ refers to the relation to the KP model.[37] The scattering function in Equation 6.8 is calculated only for the backbone points and it is thus, from a scattering perspective, calculated for infinitely thin chains.

In order to use the expression in the analysis of experimental data, a few additional points have to be considered. In most cases, the length of the micelles is significantly longer than the radius, $L \gg R$, and one can use a decoupling approximation for incorporating a cross-section part, $P_{CS}(q)$[38,39] into the expression. With this the scattering cross section for the (unrealistic) case of monodisperse micelles becomes

$$\frac{d\sigma}{d\Omega}(q) = nP_L(q)P_{CS}(q) \tag{6.9}$$

where n is the number density of particles.

For stiff micelles of length L one would use

$$P_L(q) = L^2 P_{rod}(q, L, b) \tag{6.10}$$

where

$$P_{rod}(q) = [2\mathrm{Si}(qL)/(qL) - 4\sin^2(qL/2)/(q^2L^2)] \tag{6.11}$$

is the scattering form factor of an infinitely thin rod. The function $\mathrm{Si}(x)$ is given by $\mathrm{Si}(x) = \int_0^x (\sin x / x)dx$.

The cross-section scattering functions for micelles with a circular cross section of radius R is:

$$P_{CS}(q) = \Delta\rho^2(\pi R^2)^2 \left(\frac{2J_1(qR)}{qR}\right)^2 \tag{6.12}$$

where $\Delta\rho$ is the difference in scattering length between the particle and solvent, J_1 is the first-order Bessel function, and the normalization is such that $P_{CS}(q = 0)$ $= \Delta\rho^2(\pi R^2)^2$. The contrast factor $\Delta\rho$ is (for surfactants in water) given by $\Delta\rho = \rho_{surf} - \rho_{solv}$, where ρ_{surf} is the scattering length density of the surfactant and ρ_{solv} that of the solvent. Since for water as solvent specific interactions in terms of hydration effects and water ordering might be important, it is crucial that the scattering densities are calculated from the apparent specific density of the surfactant and the density of the solvent. The term "apparent specific density" means the effective density of the molecule in the solution, when also the changes in the water in the vicinity of the molecules are ascribed to the molecule. The scattering length densities are calculated as the sum of all scattering lengths of the molecule, divided by the (effective) volume of the molecule calculated from the density. Since the densities and apparent specific densities can be determined accurately,[40,41,42] it is also possible to obtain accurate values for the scattering length densities.

For an elliptical cross section with axis $(R, \varepsilon R)$, the corresponding form factor is:

$$P_{CS}(q) = \Delta\rho^2 (\pi\varepsilon R^2)^2 \frac{2}{\pi} \int_0^{\pi/2} \left(\frac{2J_1(qr(R,\varepsilon,\theta))}{qr(R,\varepsilon,\theta)} \right)^2 d\theta \qquad (6.13)$$

with $r(R,\varepsilon,\theta) = R[\sin^2\theta + \varepsilon^2 \cos^2\theta]^{1/2}$.

For micelles, where the scattering densities of the hydrophilic and hydrophobic parts of the surfactant molecule are significantly different, a form factor of a core-shell cylindrical structure has to be applied. With a core radius R_{core} and an outer radius R_{out}, it is:

$$P_{CS}(q) = \left[\Delta\rho_{shell} \left(\pi R_{out}^2 \right) \frac{2J_1(qR_{out})}{qR_{out}} - (\Delta\rho_{shell} - \Delta\rho_{core})\left(\pi R_{out}^2 \right) \frac{2J_1(qR_{core})}{qR_{core}} \right]^2 \quad (6.14)$$

Molecular constraints and water penetration in the head group shell are quite easy to incorporate in the model as long as we consider the scattering length densities as being per unit length of the cylinder. The inclusion of water means that the volume of the head group region increases and that the scattering length density of it decreases. The number density of the micelles can be calculated as $n = (C - \text{cmc})/(NM_{surf})$, where M_{surf} is the mass of a surfactant molecule, and N is the aggregation number calculated from the volume of the micelle and of the surfactant. cmc is the critical micelle concentration of the surfactant.

As mentioned previously, the cylindrical micelles are usually very polydisperse and have a size distribution that is exponential-like:

$$D(L) = \exp(-L/L_n) \qquad (6.15)$$

where L_n is the number-average length of the micelles. The polydispersity has to be included in the longitudinal part of the form factor as:

$$\overline{P_L(q)} = \frac{\int_{2R}^{\infty} D(L)P_L(L)dL}{\int_{2R}^{\infty} D(L)dL} \qquad (6.16)$$

where the condition that $L > 2R$ has been introduced as a lower cutoff in the integrations. This expression for the average form factor has to be used in Equation (6.9) replacing $P_L(q)$. Usually polydispersity in the cross section of the micellar structure is neglected, as it is generally much smaller than that related to the length.

The number density of micelles is calculated as $n = (C - \text{cmc})/(N_n M_{surf})$, where N_n is the number-average aggregation number $N_n = \pi R^2 L_n / V_{surf}$. The number-average length is given by:

$$L_n = \frac{\int_{2R}^{\infty} D(L)LdL}{\int_{2R}^{\infty} D(L)dL} \qquad (6.17)$$

These expressions for the micellar form factor summarized above have been widely used for fitting experimental SANS and light scattering data from giant wormlike micelles at low concentration. In most cases the most interesting parameter to derive is the persistence length or Kuhn length describing the flexibility of the micelles, together with the average contour length of the micelles. However, due to the open coil-like structure of the micelles, the scattering data are influenced by intermicellar interaction effects even at low concentration. This is important when attempting to analyze scattering data as a function of concentration as shown for example in Figure 6.1 for the nonionic surfactant $C_{16}EO_6$. An increase in the scattering intensity at low scattering vectors is observed, which is due to the concentration-induced micellar growth according to $\bar{N} \propto C^\alpha$, where α is a growth exponent. At higher concentration, however, the scattering at low scattering vectors decreases[47] due to interference effects of scattering from different micelles. The onset of this is quite gradual, and is caused by the combination of the opposing effects of intermicellar interactions and concentration-induced growth. Both effects have to be incorporated in a proper analysis of the forward scattering, and this is discussed further in the next section.

Concentration or intermicellar interaction effects not only influence the low-q data and consequently the determination of the overall size or the aggregation number of the micelles. They also influence the corresponding scattering data at higher q values, and thus also the determination of parameters such as the persistence length. This is demonstrated in Figures 6.9 and 6.10, which summarize attempts to fit the scattering data from WLM at low concentrations using single-chain form factors, that is, neglecting any interference effects. Figure 6.9 shows an example for inverse lecithin micelles in isooctane with small amounts of water. The model gives an excellent fit to the data. The contour length of the micelles

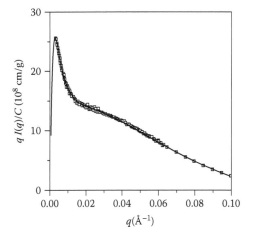

FIGURE 6.9 Experimental SANS data for inverse lecithin micelles in isooctane.[13] The micelles contain 1.5 water molecules per lecithin. The data are shown in a Holtzer plot ($qI(q)/C$ vs. q) to emphasize any deviation between data and model curve, which is a polydisperse single-chain scattering form factor with excluded-volume effects. The concentration is 2.09 mg/mL. Reproduced from Ref. 13 with permission of the American Physical Society.

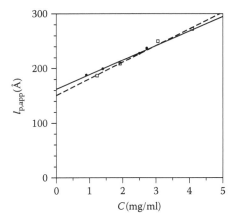

FIGURE 6.10 Apparent persistence length for inverse lecithin micelles in isooctane with small amounts of water.[13] Open squares are for 1.5 molecules of water per lecithin and filled circles are for 2.5 molecules of water per lecithin. The linear extrapolations gives a persistence length of $l_p = 150 - 160$ Å. Reproduced from Ref. 13 with permission of the American Physical Society.

derived in this way will, however, be somewhat smaller than the true ones, and it further turns out that there is a significant influence on the derived persistence length. The correct intrinsic value of the persistence length has to be found by extrapolation to zero concentration.[13] Figure 6.10 shows an example of this extrapolation procedure for the inverse lecithin micelles. One sees that even at very low concentration, there is a significant influence of the concentration on the determined value.

6.2.4 CROSS-SECTION STRUCTURE

In the modeling approach described in the previous section, a particular structure is assumed for the cross section of the micelle, and this structure is included by multiplying the longitudinal scattering function by the square of the Fourier transform of the radial profile. However, there exists also a free-form approach for structure determination that is based on inversion[43] and deconvolution techniques.[44] For cylindrical structures where the length is much longer than the cross-section diameter, the scattering cross section can be written at intermediate to high q as:[45,46]

$$\frac{d\sigma}{d\Omega}(q) = \frac{\pi}{q} 2\pi \int_0^\infty p_{cs}(r)J_0(qr)dr \qquad (6.18)$$

where J_0 is the zeroth-order Bessel function and $p_{cs}(r)$ is the cross-section pair distance distribution function:

$$p_{cs}(r) = C\Delta\rho_M^2 M_L\, r\left\langle \int_0^\infty \Delta\rho(\mathbf{r}')\Delta\rho(\mathbf{r}'+\mathbf{r})d\mathbf{r}' \right\rangle \qquad (6.19)$$

where $\Delta\rho_M$ is the excess scattering length density per unit mass, M_L is the mass per unit length, $\Delta\rho(\mathbf{r})$ is the normalized excess radial scattering length density distribution, and $\langle.\rangle$ means an orientational average. M_L is given by:

$$M_L = \int_0^\infty p_{cs}(r)dr\, /C\Delta\rho_M^2 \qquad (6.20)$$

The cross-section pair distance distribution function $p_{cs}(r)$ is written as a linear combination of appropriately chosen smooth functions and the coefficients determined using (linear) least-squares methods in combination with a smoothness constraint on the $p_{cs}(r)$. This method is known as indirect Fourier transformation.[43] When this function is obtained it is possible for centrosymmetric structures to go one step further and parameterize $\Delta\rho(r)$ as a linear combination of step functions and to determine the coefficients using (nonlinear) least-squares methods in combination with a smoothness constraint on $\Delta\rho(r)$. This approach is known as a convolutions square-root operation[44] or square-root deconvolution.[46]

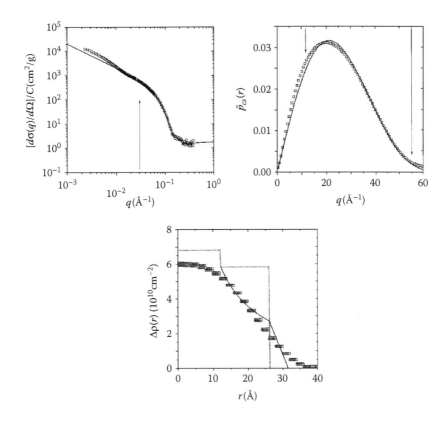

FIGURE 6.11 Local structure of polymer-like micelles formed by $C_{16}EO_6$ in D_2O at a surfactant concentration of 0.6 mg/mL. Upper left: Fit by indirect Fourier transformation (IFT) to the high-q part of the SANS data at 26°C.[47] Upper right: Distance distribution function $p_{cs}(r)$ as determined by IFT. The solid curve is a fit in the range indicated by arrows by a square-root deconvolution. Lower part: Radial excess scattering length density profile $\Delta\rho(r)$ obtained by deconvoluting $p_{cs}(r)$ from indirect Fourier transformation. The dotted lines show the expected profile of a simple geometrical model of two box functions representing the inner core (surfactant tail region) and outer shell (surfactant head group) without solvent penetration in the head group. The solid lines show the same profile with solvent penetration. Reproduced from Ref. 47 with permission of the American Chemical Society.

Figure 6.11 shows an example of an application of this approach. The data are SANS data for polymer-like micelles of $C_{16}EO_6$ in D_2O at a surfactant concentration of 0.6 mg/mL.[47] The indirect Fourier transformation (IFT) of the high-q part of the SANS data gives the distance distribution function $p_{cs}(r)$. The radial excess scattering length density profile $\Delta\rho(r)$ obtained by deconvoluting $p_{cs}(r)$ from indirect Fourier transformation is also shown. It displays a very gradual decay and is quite different from that of a homogeneous cross section. For comparison, the expected profile of a simple geometrical model of two box functions representing the inner core (surfactant tail region) and outer shell (surfactant head group) is also shown. The absolute value of the excess scattering density is in good agreement with the expected level in the hydrocarbon core. However, the decay at larger values of r is much smoother than predicted by the simple box model which assumes sharp interfaces, and also extends significantly beyond the value estimated from the average dimension of the surfactant tail and head group only. Much better agreement is found when a more realistic model that accounts for solvent penetration and a small degree of chain and head group extension instead of a sharp interface is used.

This approach has also been applied to contrast variation SANS data from inverse micelles of lecithin in deuterated cyclohexane with small amounts of water.[45] Contrast variation was done using, respectively, H_2O and D_2O and this leads to somewhat different SANS scattering curves, with a more pronounced secondary bump for the shell-like structure with D_2O (Figure 6.12). The pair distance distributions are also quite different. The function for D_2O is broader and flatter at the top. The differences are also reflected in the radial excess scattering length densities, which is shell-like for D_2O and more homogeneous for H_2O. The profiles agree very well at large distances where the density is due to the hydrocarbon tails of the lecithin. The decay in this region is quite broad and is a result of a significant solvent penetration in this region. For the profile with D_2O, the scattering length density of the water is close to zero and only the lecithin is observed. This profile shows significant diffuseness of the head group-water interface, which is due to water presence in the head group region and to some additional spread probably due to protrusion of some of the lecithin molecules. For the H_2O sample, the absolute value of the excess scattering length in the core of the particles is close to the expected value. Integrations of the profile for D_2O and of the difference of the two profiles provide values proportional to the mass of the lecithin and of the water, respectively. The ratio of the two masses is in close agreement with the mixing ratio of the sample, and this shows that all water is present in the cores of the micelles.

This example demonstrates that the application of model-free approaches that include square-root deconvolution methods to data from SANS measurements with cylindrical polymer-like micelles allows an accurate and detailed characterization of the local structure of these aggregates. For the case of inverse WLMs they have allowed to directly verify the previously postulated geometrical model of flexible tubular structures with a well-defined water core and a surfactant shell. Due to the fact that contrast variation experiments and data analysis have been performed on absolute scale, this approach has made it possible to quantitatively deduce information on properties such as the extension of the aqueous core, and the degree of water penetration into the head group region, and solvent penetration into the tail region.

Another detailed study was performed by von Berlepsch et al.,[48] who investigated aqueous 0.25 M NaCl solutions of rodlike micelles of the ionic surfactant sodium sulfopropyl-octadecylmaleate at 50°C by SANS. The scattering contrast

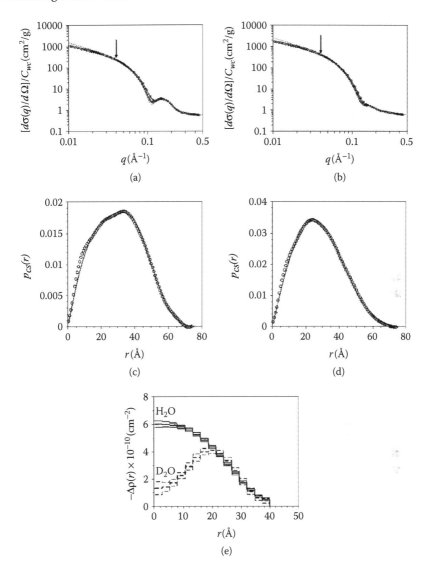

FIGURE 6.12 Cross-section structure of inverse lecithin micelles in deuterated cyclo-hexane with small amounts of water.[45] (a) SANS data with D_2O in the core. (b) SANS data with H_2O in the core. The curves are fits by the indirect Fourier transform (IFT) method. The solid curves correspond to the intensity smeared by the instrumental resolution, and the dotted line corresponds to the ideal intensity. Note that data obtained with different instrumental settings are slightly shifted due to resolution effects. (c) and (d) are the corresponding distance distribution functions $p_{cs}(r)$. Points are from IFT and solid curve is the fit by the square-root deconvolution method. (e) The radial cross-section excess scattering length density profiles. Reproduced from Ref. 45 with permission of the American Chemical Society.

was varied using different D_2O–H_2O mixtures. As the head group has a match point at higher D_2O fractions than the hydrocarbon chains, a Guinier analysis demonstrated that the effective size decreased as the D_2O fraction was lowered. More detailed analysis was done using two different approaches of which one was a modeling approach and the other was a free-form method using indirect Fourier transformation and deconvolution. The modeling approach used a core-shell model and polydispersity of the cross section, and also for the free-form approach, the deconvolution procedure was modified to include polydispersity.[49] The two analysis methods both gave a polydispersity of about 10% of the mean radius of 27 Å. This was taken as a sign that there are molecular protrusions of individual surfactant molecules into the surrounding water. There was good agreement between the core-shell and the free-form profile, although the head group region of the profile was not resolved by the free-form methods, which have a homogenous structure with a graded outer surface.

Iampietro et al.[50] used free-form methods for determining the cross-section structure of both vesicles and rodlike micellar structures formed in mixtures of the oppositely charged surfactants cetyltrimethylammonium bromide (CTAB) and sodium octylsulfate (SOS) with a composition of 21 mol % CTAB. Without salt, rodlike micelles are formed, and when electrostatics is screened by adding 5% NaBr, so that the spontaneous curvature is decreased, vesicles are formed. Two different contrasts were obtained using SANS and SAXS, respectively. In the SANS experiments, fully deuterated SOS was used and since the contrast of CTAB head group is very low in D_2O, the SANS gives the distribution of the C_{16} chains of CTAB. The combination with SAXS allows head group and tail regions to be separated. For the vesicle bilayers, a core radius of 9 Å and a head group region extending with a width of 9 Å were determined. The cross-section sizes of the rodlike micelles are slightly larger with a core radius of 10–11 Å and a head group region of 10 Å. It was also concluded that the head group region is rather diffuse due to molecular protrusions of individual surfactant molecules.

6.3 CONCENTRATION EFFECTS

6.3.1 ANALYSIS OF STATIC LIGHT SCATTERING DATA

Static light scattering data are easily normalized to absolute scale using a solvent such as toluene as a standard. The scattering contrast of an object is expressed in terms of the refractive index increments which can be obtained experimentally using a refractometer. With this, the intensity can be converted so that the forward scattering intensity is directly the weight-average molar mass of the objects in the solution, $I(q = 0) = M_w$.[8] However, this relation is only valid at low concentration, where there are no interparticle interference effects. At higher concentration, the relation becomes:

$$I(q = 0) = M_w S(0) \tag{6.21}$$

where $S(0)$ is the $q = 0$ value of the structure factor that describes the interparticle interference effects. Following Schurtenberger and co-workers,[5,17,18] $S(0)$ can be written as a function of a reduced concentration $X \approx C/C^*$, where C^* is the overlap concentration. Renormalization-group theory[19] for flexible polymers can then be used to relate X to the weight concentration C

$$X = A_2 C M_w / [9/16 - (\ln(M_w/M_n))/8] \qquad (6.22)$$

where A_2 is the osmotic second virial coefficient and M_n is the number average molar mass of the micelles. A_2 can be expressed in terms of the radius of gyration, the mass, and the interpenetration function Ψ:[51]

$$A_2 = 4\pi^{3/2}\left(R_g^3 / M_w^2\right)\Psi \qquad (6.23)$$

We can now exploit the analogy between wormlike micelles and classical polymers in a good solvent. One has the relation $R_g \propto M_w^{\nu}$ with $\nu = 0.588$ for excluded-volume chains, and therefore $A_2 = BM_w^{(3\nu - 2)}$, where B is a constant to be determined. B gives the dependence of the reduced concentration X on C. Schurtenberger and co-workers employed an explicit functional form of $S(0)$ which has been calculated using a renormalization-group method:[19]

$$S(0)^{-1} = 1 + \frac{1}{8}\left(9X - 2 + \frac{2\ln(1+X)}{X}\right)\exp\left\{\frac{1}{4}\left[\frac{1}{X} + \left(1 - \frac{1}{X^2}\right)\ln(1+X)\right]\right\} \qquad (6.24)$$

In contrast to classical polymers the average molar mass for wormlike micelles is not a constant, but depends also on the concentration, which results in an additional implicit concentration dependence of X and A_2. This can be incorporated using a growth law of the micelles according to $M_w = K(C - cmc)^\alpha$ with K and α being fit parameters. The cmc is usually known from other measurements.

For stiff particles or particles that are short compared to their persistence length, the micelles have to be described as stiff cylinders. In this case $S(0)$ can be taken from scaled particle theory:[52,53]

$$S(0) = \frac{(1 - \phi - E)^4}{[1 + 2(\phi + E)]^2 + 2D\ [1 + \phi + (5/4)E]} \qquad (6.25)$$

where $\phi = \pi R^2 L\, n$, $D = (1/2)\pi R L^2 n$, and $E = (4/3)\pi R^3 n$. ϕ is the volume fraction of the micelles and can be calculated from the concentration as $\phi = (C - cmc)/\rho$ using the specific density of the surfactant ρ. D and E are not directly related to macroscopic parameters and can only be calculated if knowledge of the radius of the micelles is available. It can either be determined from small-angle scattering or estimated from the length of the surfactant molecule. The mass of the micelles is again assumed to follow $M = K(C - cmc)^{\alpha}$, so that the density of particles can be calculated as $n = \phi/(M/\rho)$ and the length of the micelles is given as $L = M/(\pi R^2 \rho)$. With these expressions D and E can be calculated. Note that compared to the approach for the flexible chains the expression for rods includes one less fit parameter.

The expression for $S(0)$ for rods can only be expected to be valid in the case of monodisperse particles. For polymers in good solvents Equations (6.22)–(6.24) work for polydisperse systems as reported in the literature[54,55] and they have also been applied to the analysis of data from WLM.[71]

The growth exponents α determined from experimental data for the forward scattering are in general not in agreement with the mean-field values of 0.5–0.6. Data from inverse micelles of lecithin in cyclohexane[17] or isooctane[13] with small amounts of water were analyzed by the flexible chain model, and exponents of 1.1–1.2 were determined (see Figure 6.13). A similar exponent was found for nonionic $C_{16}EO_6$ micelles in water[47] (Figure 6.13), whereas ionic micelles of half-ionized[21] and fully ionized[56] tetradecyldimethylamine oxide (TDAO) in 0.1 M NaCl water solutions followed mean-field growth laws with $\alpha = 0.5$. For the fully ionized TDAO micelles, the micelles can be described as stiff rods. However, for comparison the data were analyzed both as stiff rods and flexible chains and for both approaches the data are in agreement with $\alpha = 0.5$ with similar quality of the fits. For aqueous solutions of the alkyl polyglucoside $C_{8/10}G_{1.5}$ (Glucopon

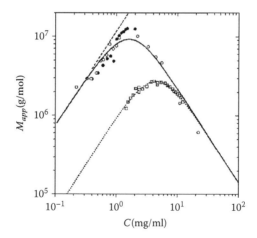

FIGURE 6.13 Apparent molar mass $M_{app} = M_w S(0)$ as a function of concentration C. Upper data (circles) are for $C_{16}EO_6$ micelles in D_2O (filled symbols Ref. 47; open symbols Ref. 71). Lower data (squares) are for lecithin inverse micelles in deuterated isooctane with a molar water-to-lecithin ratio of 1.5.[13] The curves are the fits to the data using the renormalization group theory approach with a growth exponent of $\alpha = 1.2$. The growth law without intermicellar interference effects is also shown for the $C_{16}EO_6$ micelles as the dashed line. Reproduced from Ref. 47 with permission of the American Chemical Society.

215CSUP) a transition from globular to giant wormlike micelles is induced by the addition of hexanol.[57] In this system $\alpha = 0.8$–1.2, with increasing values for increasing amounts of hexanol. For wormlike micelles of sodium dodecylsulphate (SDS) induced by addition of various concentrations of salt (NaBr),[58] the growth was in agreement with the mean-field value. For mixed micelles, the situation is usually much more complicated due to the possibility of having different cmc's of the surfactants as well as segregation effects. Note that for lecithin-bile salt mixed micelles, a *growth upon dilution* is found.[59] In general it is fair to say that our current understanding of the concentration dependence of the size distribution for polymer-like micelles is still not satisfactory, and the reason for the discrepancy between the theoretical descriptions and the experimental data remains to be resolved.

6.3.2 MONTE CARLO SIMULATIONS OF CONCENTRATION EFFECTS

In order to get quantitative information on the concentration effects in solutions of semiflexible chains with excluded-volume effects, Pedersen and Schurtenberger[20] applied Monte Carlo simulations on the same model as used previously for the single-chain simulations.[16] Extensive simulations as a function of chain length and concentration were performed for chains in a box with periodic boundary conditions. At concentrations higher than C^* the reptation algorithm[33] was preferred to the pivot because it is more efficient in generating non-overlapping configurations of the chains. Scattering functions for the single chains and full system were sampled, together with a series of other parameters. The full system scattering functions are displayed in Figure 6.14. The scattering is greatly depressed at the highest concentrations and approaches the individual chain scattering function at the lowest concentration.

The correlation length ξ and the forward scattering $S(0)$ were derived from the simulated scattering data and are shown in Figure 6.15 as a function of the reduced concentration as given by Equation (6.22). The data follow scaling behavior and are in very good agreement with the normalization group theory (RGT) calculations of Schäfer.[72] The agreement with the result of Ohta and Oono[19] for $S(0)$ given by Equation (6.24) was also investigated and it was found that the scaling at high reduced concentration was not in agreement with the expression. However, the simulations are reproduced by

$$S(0)^{-1} = 1 + \frac{1}{8}\left(9X - 2 + \frac{2\ln(1+X)}{X}\right)\exp\left\{\frac{1}{2.565}\left[\frac{1}{X} + \left(1 - \frac{1}{X^2}\right)\ln(1+X)\right]\right\}$$

$$(6.26)$$

The scattering data were also investigated in terms of a random-phase approximation (RPA)[60] and a polymer reference interaction model (PRISM)[61] type expression:

$$I(q) = P_{chain}(q)S(q) \qquad (6.27)$$

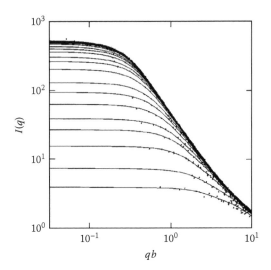

FIGURE 6.14 Scattering intensity $I(q)$ for $L/b = 90$ from Monte Carlo simulations.[20] Volume fractions of the chains are (from below) 0.075, 0.05, 0.03, 0.02, 0.015, 0.01, 0.007, 0.005, 0.003, ... The curves are from the parameterization described in the text. Reproduced from Ref. 20 with permission of EDP Sciences.

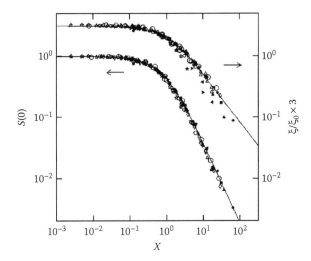

FIGURE 6.15 Forward scattering $S(0)$ and correlation length ξ as a function of the reduced concentration X derived from the simulation data[20] in Figure 6.14. ξ is multiplied by 3 to shift the data. Open symbol are from the simulations with $L/b = 3.2$ (small circle), 10.88 (triangle), 30 (diamond), 90 (star), and 270 (large circle). Filled symbols are experimental data from various polystyrene samples in the good solvent toluene. The full curves are the RGT results of Schäfer.[5,24,72] Reproduced from Ref. 20 with permission of EDP Sciences.

where $P_{chain}(q)$ is the single-chain scattering function and

$$S(q) = 1/[1 + \beta c(q) P_{chain}(q, L, b)] \qquad (6.28)$$

$c(q)$ is the direct correlation function between sites on the chains, and the parameter β is given by $\beta = [1 - S(0)]/S(0)$. The single-chain function $P_{chain}(q)$ depends on concentration and approaches the behavior of ideal chains without excluded volume at high concentration, due to the screening of these effects at high concentration. Since both $I(q)$ and $P_{chain}(q)$ were sampled, $c(q)$ could be calculated. It was found that the best approximation for it is the form factor of an infinitely thin rod:

$$c(q) = \left[2\text{Si}(qL_c)/(qL_c) - 4\sin^2(qL_c/2)/\left(q^2 L_c^{\;2}\right) \right] \qquad (6.29)$$

where L_c is the characteristic length of the function which has a magnitude between the overall size of the chains and the cross-section dimension of the micelle. There is not a simple length and concentration dependence of L_c and therefore it has to be determined in the fit to the data. Equations (6.27) through (6.29) have in fact been systematically tested with experimental data obtained for polystyrene in deuterated toluene by SANS, and excellent agreement was found.[20,62] The scattering functions obtained from the MC simulations fit the SANS data in the full measured range of scattering vectors, demonstrating agreement almost down to atomic level, and thus can be expected to work also for WLM, provided that the concentration dependence of the micellar size is taken into account.

6.3.3 ANALYSIS OF COMBINED SANS AND SLS DATA

The Monte Carlo simulations provide an empirical expression for the scattering function of semiflexible self-avoiding chains with concentration effects, which can be used for analysis of experimental data:

$$\frac{d\sigma}{d\Omega}(q) = n P_L(q) P_{CS}(q) S(q) \qquad (6.30)$$

where $S(q)$ is the structure factor describing the q dependence of the concentration effects. As mentioned above, a good approximation[20,62] for the structure factor is provided by an RPA/PRISM-type expression:

$$S(q) = 1/[1 + \beta c(q) P_{KP}(q, L, b)] \qquad (6.31)$$

The simulations were performed for monodisperse chains only, and we thus have to incorporate the intrinsic polydispersity of the wormlike micelles in an empirical way. This is done based on the experimental finding that for nonionic micelles at not too high concentration $C > C^*$, we observe a flat intensity at low q without indication of a peak. Therefore the behavior of $P_L(q)$ and the denominator of $S(q)$ have to be identical at low q. This can be obtained by replacing the form factor $P_{KP}(q)$ in the denominator of $S(q)$ with the L^2-weighted form factor.

We also neglect screening effects present at high concentration and use excluded-volume expressions in the form factors for $P_L(q)$ as well as in the structure factor.

The direct correlation function can be approximated by the form factor of an infinitely thin rod (Equation (6.29)). Alternatively, $c(q)$ can be approximated by the Fourier transform of the correlation hole around each site:[62]

$$c(q) = \frac{3[\sin(q2R') - q2R'\cos(q2R')]}{(q2R')^3} \tag{6.32}$$

This expression assumes that the size of the correlation hole is $2R'$. In practice R' is not equal to the radius of the micelle R, but it depends both on concentration and micellar size. Another simpler approximation[20] is to replace the product $c(q)P_{KP}(q)$ by the flexible chain equivalent of $P_{KP}(q)$.[63]

For very stiff micelles, an empirical expression for the structure factor is:[64]

$$S(q) = 1/[1 + \beta c(q)P_{rod}(q, L - 2R)] \tag{6.33}$$

where $P_{rod}(q)$ is the form factor of an infinitely thin rod (Equation (6.11)) and the function $c(q)$ is the direct correlation function between sites on the rod. $c(q)$ can be approximated by the Fourier transform of the correlation hole at each site (Equation (6.32)). In general, there is a concentration and length dependence of the size of the correlation hole $2R'$. However, the actual form of the dependence has not been established to our knowledge. The parameter β is again given by $\beta = [1 - S(0)]/S(0)$ where $S(0)$ is given by Equation (6.25). Polydispersity can also be taken into account by an empirical approach similar to that used for semiflexible chains. Also for rods, a flat intensity at low q without indications of a peak, is observed experimentally at not too high concentration. Therefore the form factor $P_{rod}(q)$ in the denominator of $S(q)$ has to be replaced by the corresponding L^2-weighted form factor.

When analyzing combined experimental SANS and SLS data, the SLS data extrapolated to $q = 0$ are usually analyzed first to establish the growth law and the concentration dependence of $S(0)$. In a next step the full q dependence is analyzed at each concentration with the mass calculated from the analysis of the $q = 0$ data. In order to calculate the form factor and q dependence of the structure factor, the relation between M_w and L_w has to be established. This can be done by determining the mass per unit length from the high-q part of the scattering data, for example, by indirect Fourier transformation of data on absolute scale (see Section 6.2.4). The mass of the micelles is also used for calculating the number density n of the micelles.

The first application of this analysis scheme, which incorporated corrections for concentration effects, were for inverse lecithin micelles in isooctane with small amounts of water.[13] In this work, modest concentrations were investigated and the concentration effects were included using a simple Gaussian chain form factor for $P_{chain}(q)$ in the structure factor (Equation (6.31)). This gave a reasonable reproduction of the data at moderate concentrations.

Wormlike micelles of sodium dodecylsulphate (SDS) at relatively high salt concentration (NaCl, 1–2 M) were studied using SANS by Magid and co-workers.[65] Due to the high salt concentration, the electrostatic interactions are to a large degree screened out and one can consider the micelles as neutral excluded-volume chains. The surfactant concentrations in the study were relatively low, however the intermicellar interaction effects were included in the analysis in order to allow reliable determination of the contour length and the persistence length. Therefore the structure factor at $q = 0$ was included in the analysis when obtaining the contour length, and the persistence length was calculated from an extrapolation to zero concentration. The micelles become more flexible as the ionic strength is increased and it was found that the electrostatic interactions still contribute to the persistence length even at 2 M NaCl. This was explained by the high surface charge density at the micellar head groups. Simple scaling relations for the total and the electrostatic part of the persistence lengths with salt concentration were absent.

In another publication, Magid and co-workers[66] reported a SANS study of mixed micelles of cetyltrimethylammonium 2,6-dichlorobenzoate (CTA26ClBz) and cetyltrimethylammonium chloride (CTACl) in water as a function of composition, and of surfactant and NaCl salt concentrations. At high salt concentrations of 1–2 M, wormlike micelles are present and these were investigated at relatively low surfactant concentration. A self-consistent analysis in accordance with the work of Schurtenberger and Cavaco[18] was performed and the persistence length was obtained by extrapolation to zero concentration. It was found that the type of counter ion (penetrating or nonpenetrating into the head group region) had a large influence on both flexibility and micellar length. The 2,6-dichlorobenzoate ion penetrates and reduces the net head group charge and thus lowers the persistence length and enhances the growth. A growth exponent was determined for pure CTA26ClBz in 2 M NaCl. It was found to be 1.1, much higher than predicted by mean-field theory. However, the value is in good agreement with those found in several other systems.[13,17,21,47,56,57]

A very large range of concentrations was considered in a study of half-ionized TDAO in a 0.1 M NaCl heavy water solution.[21] Also in this case, the high salt concentration screens the electrostatic interactions. The data were analyzed with the RPA (Equation (6.31)). The data were very well fitted by this expression even at the highest concentration, which corresponds to 65 times C^* (Figure 6.16). Polydispersity was not included in the model. At the highest concentration, the data are heavily suppressed at low q due to the interaction effects, and the data

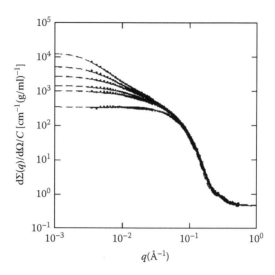

FIGURE 6.16 SANS data for different concentration of half-ionized TDAO in 0.1 M NaCl heavy water solution.[21] The concentrations are from top to bottom 2.88, 8.51, 14.7, 23.7, 29.8, and 62 mg/mL, respectively. The curves are the model fits using the RPA Equation (6.31). Reproduced from Ref. 21 with permission of the American Chemical Society.

resemble those of more globular particles. In order to perform a more extensive comparison with theory and simulations, the correlation length was also derived from the SANS data. It was divided by the radius of gyration of the single chains calculated using the growth law to allow a direct comparison with theory and simulation (Figure 6.17). There is excellent agreement between the experimental data for TDAO and the theory and simulation results. This suggests that the analysis performed on this system is indeed self-consistent.

The RPA equation was also used by Stradner et al.[57] for describing concentration effects in SANS data from aqueous solutions of the alkylpolyglucoside $C_{8/10}G_{1.5}$ (Glucopon 215CSUP). In this system the micellar length is controlled by addition of varying amounts of hexanol. The expression could also in this case reproduce the measured data very well, and this even for quite short chains with contour length $L \approx b$, where the stiff cylinder expressions also should be applicable.

Wormlike micelles of SDS induced by addition of various concentrations of salt (NaBr) were investigated by Arleth et al.[58] by SANS. Concentrations of 0.08 to 8.6% vol SDS in NaBr aqueous solutions at salinities from 0.6 to 1.0 M were studied. The data were analyzed with the RPA/PRISM expression with a rod scattering function for the direct correlation function. Polydispersity was included in both single-chain form factor and in the structure factor, by using the same polydisperse function for $P_{KP}(q)$ in both expressions. The growth exponent was in agreement with the mean-field value.

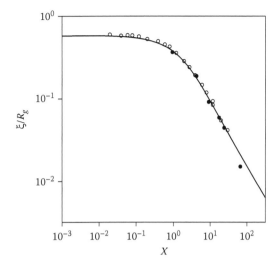

FIGURE 6.17 Results for half-ionized TDAO in 0.1 M NaCl heavy water solution.[21] Comparison of correlation length ξ divided by the radius of gyration R_g (solid circles) versus reduced concentration X with Monte Carlo simulations (open circles) for chains with $L/b = 30$[20] and the renormalization group theory (solid line).[3] Reproduced from Ref. 21 with permission of the American Chemical Society.

For micelles formed by fully ionized TDAO micelles in 0.1 M NaCl heavy water solutions,[56] the local electrostatic interactions are so strong that the flexibility of the micelles is suppressed and they can thus be described as stiff rods. Combined SANS and SLS data from this system were analyzed using Equation (6.33) with the direct correlation function (Equation (6.32)). The forward scattering was derived from a self-consistent analysis of the $q = 0$ data using the cylinder expression of Equation (6.25).

6.4 IONIC SYSTEMS: MICELLES AS EQUILIBRIUM POLYELECTROLYTES

Ionic micelles constitute an interesting model system for polyelectrolytes. The charge of the micelles, the ionic strength, and the concentration can easily be varied in such systems. Moreover, as the scattering cross section is quite large for SANS and light scattering, it is comparatively easier to obtain good scattering data on such systems even at low concentration, that is, in the dilute regime, when compared to conventional organic polymer-based systems. However, the length scales in micellar systems are quite different from those of the conventional polyelectrolytes (except for the Debye–Hückel screening length), and models therefore have to be more specific and incorporate some molecular details. Consequently, this has the unfortunate effect that the results obtained will be less general. In the following, Monte Carlo simulations for models of ionic micelles are reviewed with emphasis on the scattering functions. Both single-chain and many-chain systems are considered.

6.4.1 MONTE CARLO SIMULATIONS OF CHARGED SEMIFLEXIBLE WLM

A quite simple way to mimic the polar groups of ionic surfactant molecules in solution is to add a number Z of elementary charges on each sphere of the chain and treat the electrostatic interactions in a mean-field manner, whereas the solvent is considered as a continuum dielectricum with permittivity ε_{sol}. This approach has the valuable computational advantage of reducing the long-ranged Coulomb potential energy to the short-ranged Debye–Hückel (D-H) expression given by

$$U = \begin{cases} \dfrac{(Ze)^2}{4\pi\varepsilon_0\varepsilon_{sol}} \dfrac{e^{2d/\lambda_D}}{(1+d/\lambda_D)^2} \dfrac{e^{-r/\lambda_D}}{r} & \text{if } r > 2d \\ \\ \infty & \text{otherwise,} \end{cases} \tag{6.34}$$

where e is the elementary charge, ε_0 the permittivity of vacuum, r the center-to-center distance of two spheres, and

$$\lambda_D = \left(\frac{\varepsilon_0\varepsilon_r k_B T}{2e^2 I(C,C_{salt},Z)} \right)^{1/2} \tag{6.35}$$

is the Debye screening length which fixes the range of the interaction. In the above expression $I(C,C_{salt},Z)$ is the ionic strength (in mol/L) of the solution generated by the counter-ions, and clouds of salt ions (co-ions) when present. The exponential decay in Equation 6.34 allows one to introduce a cutoff for the energy calculation which further speeds up the simulations. Note that the effects of counter-ions and co-ions only enter implicitly through λ_D; therefore, their only effect is to screen the electrostatic interactions between the charged beads. The validity of the D-H approximation has been seriously questioned. Since it is obtained by linearizing the Poisson–Boltzmann equation, it is not expected to describe phenomena like counter-ion condensation, which are highly nonlinear. In general this approximation is expected to lose accuracy as the magnitude of the electrostatic interactions increases. There are also some more subtle assumptions regarding the symmetry of the electric field in the vicinity of a macro-ion which cannot be *a priori* justified and thereby may pose some concerns too. As a result, it is clear that the D-H potential should be employed very carefully. Nevertheless, in the range of validity of the approximations, it is hard to find a more accurate alternative that can be employed with comparable efficiency in computer simulations. On the contrary, a complete treatment of the electrostatics

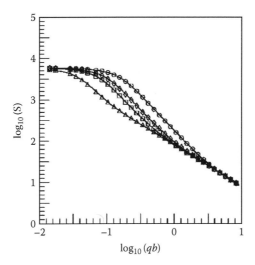

FIGURE 6.18 Scattering functions for a chain of length $L/b = 240$ and for four different ionic strengths.[22] The full lines are fits performed with the wormlike chain function with excluded volume.[15] Triangles $I = 10^{-4}$ M, squares $I = 10^{-3}$ M, diamonds $I = 2.5 \times 10^{-3}$ M, and circles $I = 5 \times 10^{-2}$ M. Reproduced from Ref. 22 with permission of the American Physical Society.

should consider explicitly the counter-ions and co-ions in the system, interacting with the full Coulomb potential. The price of such an approach is quite high as the energy calculation becomes very time consuming, even if the Ewald summation technique is employed; consequently, only very few short chains can in practice be simulated in this way. But in most cases, they are not representative of the macroscopic system considered, and therefore the results are of little interest.

The size and the conformation of WLM in salt solutions have been systematically studied by MC simulations using the model introduced above.[22] As expected, the analysis of the radius of gyration and the end-to-end distance show that charged micelles are more elongated than uncharged ones due to the electrostatic repulsion of the charges on the chain. The single-chain scattering functions have also been sampled during the Monte Carlo simulations (Figure 6.18). A remarkable result found is that the single micelle form factor can be fitted by the parameterized form factor of uncharged, semiflexible, self-avoiding chains.[15] This demonstrates that an effective persistence length and an effective excluded-volume strength (binary cluster integral) are sufficient to account for electrostatics, and validate the basic assumption of the Skolnick, Fixman, and Odijk (OSF) theory[67,68,69] that treats the interactions between neighboring charges as a local effect that increases the persistence length, whereas those between charges located far apart along the chain are incorporated in the excluded volume. A more quantitative analysis of the results shows, however, important discrepancies from the OSF theory. Specifically, the theory predicts the Kuhn length to be

$$b_{tot} = b + \frac{\lambda_B}{2}\left(\frac{\lambda_D}{d}\right)^2 \tag{6.36}$$

where $\lambda_B = e^2/(4_{0r}k_BT)$ is the Bjerrum length and d the distance between charges. Monte Carlo data show that Equation (6.36), generally, underestimates the Kuhn length. In the limit of low salt concentration, that is, in the rigid rod regime, however, the simulation data tend to agree with the predictions. This is likely to happen because OSF theory neglects the entropic contribution to the free energy.

Simulations have also been done on many chain systems in order to study the concentration effects.[23] At high concentration, the screening of electrostatic interactions is very pronounced and cannot be considered to be a small effect like the screening of excluded-volume interactions of neutral polymers. The simulations have revealed that the screening sets in at about the overlap concentration C^* and that the size of the chains approaches that of neutral chains at a volume fraction of about 0.1.

The scattering functions of ionic micelles at finite concentration look similar to those of polyelectrolyte solutions (see Figure 6.19), with a pronounced peak

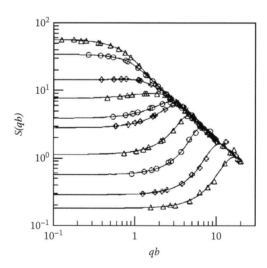

FIGURE 6.19 The full system scattering function $S(q)$ as a function of concentration as obtained by MC simulations.[23] The number of charges per Kuhn length is 25, and the ionic strength is 0.001 M. The volume fractions are from above to below 0.0001, 0.001, 0.002, 0.003, 0.005, 0.008, 0.01, 0.02, 0.04, 0.1, and 0.2. The symbols are sampled data, and the full lines are fits using the sampled single-chain scattering functions (Figure 6.8) and the direct correlation function given by Equation (6.37). Reproduced from Ref. 23 with permission of the American Chemical Society.

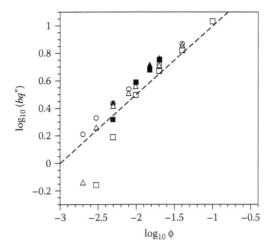

FIGURE 6.20 Position of the correlation peak of $S(q)$ as a function of volume fraction ϕ from MC simulations.[23] The ionic strength is 0.001 M. Data shown for different number of charges per Kuhn length: 25 (open square), 50 (open triangle), and 75 (open circle). Experimental data from $C_{16}EO_6$ doped with two different amounts of the ionic surfactant $C_{16}SO_3Na$[3]: 3% (filled square) and 6% of ionic surfactant (filled triangle). The line has a slope of 1/2. Reproduced from Ref. 23 with permission of the American Chemical Society.

at sufficiently high concentration and low ionic strength. This peak is likely to be produced by the onset of a short-range, liquidlike order of locally rigid structures, which interact through mutual repulsions. Many scattering experiments on polyelectrolyte (but not exclusively) have found that the position of the peak q^* scales as $C^{1/2}$ in agreement with theoretical predictions. The same result has been found for charged WLM by both Monte Carlo simulations[23] (see Figure 6.20) and SANS experiments[3] (see next section).

Again, this is a sound indication that the model is able to reproduce universal structural properties of the system. Another important feature evident from Figure 6.19 is the decreasing of the forward scattering intensity $S(0)$ on increasing concentration as a result of the interchain interactions. This effect is more pronounced when the electrostatic interactions are stronger, that is, at lower ionic strength and at higher charge density. Physically, this corresponds to the increase of the osmotic compressibility with concentration.

Monte Carlo simulations also allowed the direct correlation function to be determined and analyzed (Figure 6.21). It follows a slightly more complex form than that of uncharged chain:

$$c(q) = \frac{\sin(qR_c)}{qR_c}\exp(-\sigma^2 q^2) \tag{6.37}$$

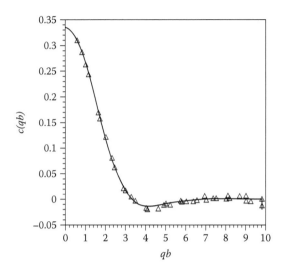

FIGURE 6.21 The direct correlation $c(q)$ (symbols) calculated from the simulation data.[23] The volume fraction of the chains is 1%, the ionic strength is 0.001 M, and the number of charges per Kuhn length is 25. The full line is a fit of Equation 6.37 to the data. Reproduced from Ref. 23 with permission of the American Chemical Society.

The parameters obey the empirical relation $R_c = 2\log\sigma + 2$ and depend on concentration, charge density, and salt concentration. Equations (6.27) and (6.28), with $c(q)$ given by Equation (6.38), can be fitted very well (by fitting one parameter, e.g., R_c) to the MC data (see full line in Figure 6.21), when using the sampled single-chain form factor. This result suggests that the above procedure can, in principle, be employed to achieve a complete parameterization of the scattering function in the same way done for the uncharged micelle.[62] It is, however, not trivial to make such a parameterization since there is a pronounced screening of electrostatic interaction effects when the concentration is increased. This is demonstrated in Figure 6.22, which shows a large influence on the low q part of the function and a smaller influence at higher q.

The forward scattering follows a universal behavior similar to that of neutral chains, provided that the volume fraction ϕ is properly rescaled. Specifically,

$$S(0)^{-1} = 1 + \frac{1}{8}\left(9X - 2 + \frac{2\ln(1+X)}{X}\right)\exp\left\{0.8\left[\frac{1}{X} + \left(1 - \frac{1}{X^2}\right)\ln(1+X)\right]\right\} \quad (6.38)$$

where $X \approx 42.1\phi_r$, with

$$\phi_r = \left(\frac{\lambda_D + R}{R}\right)\left(\frac{R_g}{R_{g,u}}\right)^3\phi \quad (6.39)$$

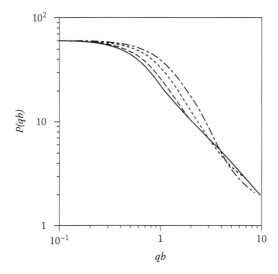

FIGURE 6.22 Single-chain scattering function in a charged system as a function of concentration.[23] The ionic strength is 0.001 M and the number of charges per Kuhn length is 50. The volume fractions are 0.0001 (full line), 0.01 (broken line), 0.04 (dotted line), and 0.1 (broken dotted line). Reproduced from Ref. 23 with permission of the American Chemical Society.

In this expression R_g and $R_{g,u}$ are the radius of gyration of the charged and uncharged chains at infinite dilution, respectively. The rescaled volume fraction (Equation (6.39)) accounts for the swelling of the chain through the ratio $R_g/R_{g,u}$, and for the increased excluded volume through the contribution of the Debye length which adds up to the radius of the hard spheres, that is, to the intrinsic EV. Note that the relation between R_g and $R_{g,u}$ for a given contour length can be estimated using the OSF theory in all cases where this theory is applicable. Figure 6.23 shows the data for the forward scattering as a function of the rescaled volume fraction collapsing on a single curve.

6.4.2 COMPARISON WITH EXPERIMENTS

At low concentration the interchain interference effects are small and one can, as for the scattering functions from simulation, attempt to fit the single-chain scattering function for neutral chains with excluded-volume effects. Sommer et al.[3] have performed a systematic study of $C_{16}EO_6$ micelles doped with the $NaC_{16}SO_3$ anionic surfactant in $NaCl/D_2O$ solutions. Doping levels of 3, 6, and 9% were investigated and the salt concentration was varied from 0.001 M to 0.1 M. A polydisperse single-chain form factor with a circular cross-section structure was fitted to the SANS data.

Figure 6.24 shows fits to low concentration data for different ionic strengths. The fits are very good and provide an apparent contour length and an apparent

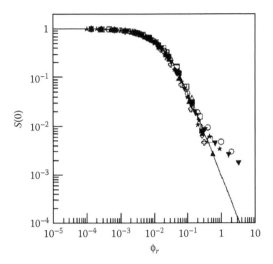

FIGURE 6.23 Plot of $S(0)$ versus rescaled volume fraction ϕ_r defined by Equation (6.39) with $\delta = 1$.[23] The full line is expression 6.38. In the semidilute regime, the scaling is $S(0) \propto \phi_r^{-1.8}$ Reproduced from Ref. 23 with permission of the American Chemical Society.

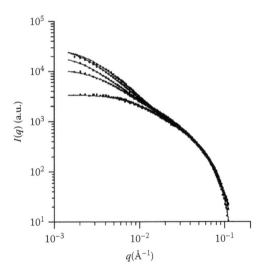

FIGURE 6.24 SANS data for $C_{16}EO_6$ doped with 6% $C_{16}SO_3Na$ ionic surfactant in NaCl/D$_2$O solutions with a surfactant concentration of 0.6 mg/mL. From top to bottom the ionic strength is 0.1, 0.01, 0.005, 0.0025, and 0.001 M. The full curves are model fits using a polydisperse single-chain form factor and a cross-section scattering function for a homogeneous circular cross section.[3] Reproduced from Ref. 3 with permission of the American Physical Society.

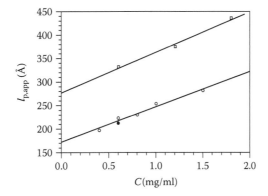

FIGURE 6.25 Apparent persistence length $l_{p,app}$ versus concentration for uncharged $C_{16}EO_6$ micelles [open circle: $T = 26°C$; filled circle $T = 35°C$] and $C_{16}EO_6$ micelles charged with 6% $C_{16}SO_3Na$ at $T = 26°C$ (open square).[47] In the limit of C going to zero, the linear fits give in $l_p = 170$ Å for the uncharged micelles and $l_p = 280$ Å for charged micelles. Reproduced from Ref. 47 with permission of the American Chemical Society.

persistence length together with a cross-section radius. When fitting a series of data at low concentration, one finds that there is a small but significant and systematic dependence of the persistence length on concentration. The results for $C_{16}EO_6$ micelles in D_2O and for $C_{16}EO_6$ doped with 6% $NaC_{16}SO_3$ anionic surfactant in 0.01 M $NaCl/D_2O$ are shown in Figure 6.25. Comparisons with data generated by Monte Carlo simulations have shown that this is an artifact caused by intermicellar interference effects that are present even at very low concentration.[20] In order to obtain the true value for the persistence length the values therefore have to be extrapolated to zero concentration. For the present data, linear fits give the true values of $l_p = 170$ Å for the uncharged micelles and $l_p = 280$ Å for charged micelles. There is thus a significant influence of electrostatics even at a low doping level. The results as a function of charge and ionic strength were in qualitative agreement with the predictions of the SFO polyelectrolyte theory,[67,68,69] which justified the assumption that the total persistence length is the sum of an intrinsic contribution and an electrostatic contribution. The influence of electrostatics was determined to be systematically larger than predicted by the SFO theory.

The fact that the experimentally determined micelle form factor can be fitted by the parameterized form factor of uncharged, semiflexible, self-avoiding chains is again remarkable.[15] It confirms that the persistence length and the effective excluded-volume strength due to electrostatics are influenced in the same way. Therefore it is enough to have a single parameter (an effective binary cluster integral) for describing the influence of the electrostatic interactions.

To our knowledge, there are very few studies of ionic micelles in broad ranges of surfactant concentration and salt concentration. The exception is the SANS and light scattering study by Sommer et al.[3] (see Figure 6.24) which has just been discussed in connection with applications of single-chain form factors. Static light scattering was used for investigating the forward scattering as a function of concentration. The behavior is qualitatively the same as for neutral systems. There is an increase in the forward scattering with increasing concentration at low concentration below the overlap concentration C^*, whereas for higher concentration the forward scattering decreases again due to structure factor effects. In general, the electrostatics lead to a reduction of the micellar growth. At the time of the experiments, the theoretical expression (6.38) for the structure factor effects was not yet available and therefore the data were not analyzed any further.

SANS was applied for obtaining the full q dependence of the scattering intensity.[3] An example as a function of concentration is shown in Figure 6.4. At high concentration (and low ionic strength) a pronounced structure factor peak is observed. The scattering vector position q^* of the peak shifts to higher values of q at higher surfactant concentration, as the peak becomes more pronounced. This peak is absent at higher salt concentrations due to the screening of the electrostatic interactions by the salt, and the system exhibits classical polymer behavior. The peak location q^* is plotted in Figure 6.20 as a function of the surfactant concentration for different doping levels. q^* follows a power law of the form $q^* \propto C^{1/2}$, that is, one observes the same behavior as reported for classical polyelectrolytes[70] and salt-free ionic wormlike micelles.[32] Figure 6.20 also contains the peak position from the MC simulations[23] and excellent agreement is found.

6.5 CONCLUSIONS AND OUTLOOK

In this chapter we have reviewed the concepts and theories used in the analysis and interpretation of scattering experiments from giant polymer-like micelles. During the last decades there has been a large degree of synergy between the research fields of polymer science and of giant wormlike micelles. It was realized at an early stage that giant wormlike micelles have many structural features in common with traditional covalently bonded organic polymers. Scattering experiments covering a broad range of length scales provided crucial information in this direction. An important difference is that organic polymers under usual conditions have a fixed and often narrow size distribution, whereas wormlike micelles have a concentration-dependent, and often very broad, size distribution. The use of expressions from renormalization group theory of polymers[19,72] made it possible to obtain the growth law of giant micelles from scattering experiments for a broad range of concentrations, even above the overlap concentration, where the micelles are strongly entangled.[17,18] The scattering vector dependence of the form factors measured at low concentration revealed the presence of excluded-volume effects and semiflexibility. However, theoretical scattering functions for analyzing such data were not available and therefore such data (like data for organic polymers) could not be described in detail. This motivated a Monte Carlo study of excluded-volume semiflexible chains for which the parameters were chosen to describe typical wormlike micelles.[15,16] The scattering form factors provided an important tool for obtaining structural information on the micelles, and furthermore it turned out that the form factors could also be applied for interpretation of scattering from organic polymers. The Monte Carlo study also gave information on the chain expansion due to excluded-volume effects and the expansion's scaling dependence on chain length. This information could be used for checking previous theoretical predictions. Additional simulation studies using the same simple chain model were performed with the aim of studying concentration effects.[20] The derived scaling relations confirmed in general the renormalization group theory calculations,[72] and also provided information on the influence of local stiffness and the breakdown of flexible polymer theories at high concentration. The parameterization of the results were applied both in the interpretation of scattering data from wormlike micelles and organic polymers.[21,57,58,62] For the former, some empirical extensions to the polydisperse case were made.

The above description of the development within the field demonstrates the fruitful interplay between the research field of giant wormlike micelles and of conventional polymers. The generality of theory and simulations is due to the simplicity of the models, which only include short-range interactions of nonspecific character and furthermore does not include any direct dependence on the solvent properties.

A systematic study of charged wormlike micelles was initiated as a model study of polyelectrolyte chains.[3] The advantage of this system compared to organic polyelectrolyte polymers is that the micelles have hydrophilic surfaces and therefore the frequently observed clustering in solutions of the organic polyelectrolyte chains can be avoided. Moreover, their charge density can easily be tuned, and their much larger scattering contrast allows for measurements even in the dilute region, where single-chain properties such as the persistence length can be obtained. However, a drawback of wormlike micelles as models for equilibrium polyelectrolytes arises from the fact that their aggregation state depends not only on concentration, but also on charge density and ionic strength of the solution. In addition, the range of charge densities that can be obtained is limited, and there always has to be some added salt in the solvent so that the salt-free case cannot be studied.

The charge density and the solvent salt concentration in the system are characterized by real physical length scales, and these have to be introduced in a model for the system. As a consequence, the radius and the persistence length of the chains also have to have the correct magnitudes relative to the other physical length scales. This makes the model quite specific and the results obtained with it much less general. However, some general properties can be derived and, with proper rescaling of the length scales, they should also to some extent be applicable to other systems and to organic polyelectrolyte systems. The relatively large cross-section diameter of the micelles makes it reasonable to use a coarse-grained model and to attempt to use screened Coulomb potentials for describing the electrostatic interactions.[22,23] The comparison between the simulation results obtained with the model and experimental results shows close agreement, which supports the validity of the model.

The present chapter is a review of the present status of the field of scattering from giant polymer-like micelles covering both theory and applications. Our hope is that it is useful for scientists working in the field and that it also gives guidance to identifying the open questions that remain to be addressed.

REFERENCES

1. Safran, S. A. In *Structure and Dynamics of Strongly Interacting Colloids and Supramolecular Aggregates in Solution,* Chen, S. H., Huang, J.S., Tartaglia, P., Eds., Kluwer Academic Publishers, Dordrecht, 1992, Vol. 369, p. 237.
2. Porte, G., Marignan, J., Bassereau, P., May, R. *J. Phys. (France)* 1988, *49*, 511.
3. Sommer, C., Pedersen, J. S., Egelhaaf, S. U., Cannavacciuolo, L., Kohlbrecher, J., Schurtenberger, P. *Langmuir* 2002, *18,* 2495.
4. Cates, M. E., Candau, S. J. *J. Phys.: Condens. Matter* 1990, *2*, 6869.
5. Schurtenberger, P., Cavaco, C. *J. Phys. Chem.* 1994, *98*, 5481.
6. Cates, M. E. *Macromolecules* 1987, *20*, 2289.
7. Magid, L. In *Dynamic Light Scattering: The Method and Some Applications,* Brown, W., Ed., Clarendon Press, Oxford, 1993, p. 554.
8. Schurtenberger, P. In *Light Scattering: Principles and Development,* Brown, W., Ed., Clarendon Press, Oxford, 1996, p. 293.
9. *Structure and Flow in Surfactant Solutions*, Herb, C.A., Prud'homme, R.K., Eds., American Chemical Society, Washington, DC, 1994.
10. Odijk, F. *Curr. Opinion Colloid Interface Sci.* 1996, *1*, 337.
11. Lequeux, F. *Curr. Opinion Colloid Interface Sci.* 1996, *1*, 341.
12. Cates, M. E. *J. Phys.: Condens. Matter* 1996, *8*, 9167.
13. Jerke, G., Pedersen, J. S., Egelhaaf, S. U., Schurtenberger, P. *Phys. Rev. E* 1997, *56*, 5772.
14. Cates, M. E. *J. Phys. (France)* 1988, *49*, 1593.
15. Pedersen, J. S., Schurtenberger, P. *Macromolecules* 1996, *29*, 7602.
16. Pedersen, J. S., Laso, M., Schurtenberger, P. *Phys. Rev. E* 1996, *54*, R5917.
17. Schurtenberger, P., Cavaco, C. *J. Phys. II (France)* 1993, *3*, 1279.
18. Schurtenberger, P., Cavaco, C. *J. Phys. II (France)* 1994, *4*, 305.
19. Ohta, T., Oono, Y. *Phys. Lett.* 1982, 89A, 460.
20. Pedersen, J. S., Schurtenberger, P. *Europhys. Lett.* 1999, 45, 666.
21. Garamus, V. M., Pedersen, J.S., Kawasaki, H., Maeda, H. *Langmuir* 2000, *16,* 6431.
22. Cannavacciuolo, L., Sommer, C., Pedersen, J. S., Schurtenberger, P. *Phys. Rev. E* 2000, *62*, 5409.
23. Cannavacciuolo, L., Pedersen, J. S., Schurtenberger, P. *Langmuir* 2002, *18*, 2922.
24. See, for example, Hamley, I. W. *Introduction to Soft Matter,* Wiley, New York, 2000.
25. Langevin, D. *Annu. Rev. Phys. Chem.* 1992, *43*, 341
26. Olsson, U., Wennerström, H. *Adv. Colloid Interface Sci.* 1994, *49*, 113.
27. Evans, D. F., Mitchell, D. J., Ninham, B. W. *J. Phys. Chem.* 1986, *90*, 2817.

28. Israelachvili, J. *Intermolecular and Surface Forces,* Academic Press, London, 1992.
29. Israelachvili, J.N., Mitchell, D. J., Ninham, B.W. *J. Chem. Soc. Farad. Trans. II* 1976, *72,* 1525.
30. Lindman, B., Olsson, U., Wennerström, H., Stilbs, P. *Langmuir* 1993, *9,* 625.
31. Safran, S. A., Pincus, P. A., Cates, M. E., MacKintosh, F. C. *J. Phys. (Paris)* 1990, *51,* 503.
32. Bellour, M., Knaebel, A., Munch, J. P., Candau, S. J. *Eur. Phys. J. E* 2000, *3,* 111.
33. Binder, K., Ed. *Monte Carlo and Molecular Dynamics Simulations in Polymer Science,* Oxford University Press, Oxford 1995.
34. Stellman, S. D., Gans, P. J. *Macromolecules* 1972, *5,* 516.
35. Stellman, S. D., Froimowitz, M., Gans, P. J. *J. Comput. Phys.* 1971, *7,* 178.
36. Cannavacciuolo, L., Pedersen, J. S. *J. Chem. Phys.* 2002, *117,* 8973.
37. Kratky, O., Porod, G. *Rec. Trav. Chim. Pays-Bas* 1949, *68,* 1106.
38. Porod, G. *Acta Phys. Austriaca* 1948, *2,* 255.
39. Pedersen, J. S., Schurtenberger, P. *J. Appl. Cryst.* 1996, 29, 646.
40. Harada, S., Nakajima, T., Komatsu, T., Nakagawa, T. *J. Solution Chem.* 1978, *7,* 463.
41. Vass, S., Török, T., Jákli, G., Berecz, E. *J. Phys. Chem. B* 1989, *93,* 6559.
42. Maccarini, M., Briganti, G. *J. Phys. Chem. B* 2000, *104,* 11451.
43. Glatter, O. *J. Appl. Crystallogr.* 1977, *10,* 415.
44. Glatter, O. *J. Appl. Crystallogr.* 1981, *14,* 10.
45. Schurtenberger, P., Jerke, G., Cavaco, C., Pedersen, J. S. *Langmuir* 1996, *12,* 2433.
46. Pedersen, J. S., Schurtenberger, P. *J. Appl. Cryst.* 1996, *29,* 646.
47. Jerke, G., Pedersen, J. S., Egelhaaf, S. U., Schurtenberger, P. *Langmuir* 1998, *14,* 6013.
48. von Berlepsch, H., Mittelbach, R., Hoinkis, E., Schnablegger, H. *Langmuir* 1997, *13,* 6032.
49. Mittelbach, R., Glatter, O. *J. Appl. Crystallogr.* 1998, *31,* 600.
50. Iampietro, D. J., Brasher, L. L., Kaler, E. W., Stradner, A., Glatter O. *J. Phys. Chem. B* 1998, *102,* 3105.
51. Huber, K., Stockmayer, W. H. *Macromolecules* 1987, *20,* 1400.
52. Cotter, M. A., Martire D. E. *J. Phys. Chem.* 1970, 52, 1902.
53. Cotter, M. A., Martire D. E. *J. Phys. Chem.* 1970, 52, 1909.
54. Wiltzius, P., Haller, H. R., Cannell, D. S. *Phys. Rev. Lett.* 1983, *51,* 1183.
55. Brown, W., Nicolai, T. *Colloid Polym. Sci.* 1990, *268,* 977.
56. Garamus, V. M., Pedersen, J. S., Maeda, H., Schurtenberger, P. *Langmuir* 2003, *19,* 3656.
57. Stradner, A., Glatter, O., Schurtenberger, P. *Langmuir* 2000, *16,* 5354.
58. Arleth, L., Bergström, M., Pedersen, J. S. *Langmuir* 2002, *18,* 5343.
59. Arleth, L., Bauer, R., Øgendal, L. H., Egelhaaf, S. U., Schurtenberger, P., Pedersen, J. S. *Langmuir* 2003, *19,* 4096.
60. Zimm, B. H. *J. Chem. Phys.* 1948, *16,* 1093.
61. Schweizer, K. S., Curro, J. G. *Adv. Polym. Sci.* 1994, *116,* 319.
62. Pedersen J. S., Schurtenberger P. *J. Polym. Sci., Polym. Phys. Ed.* 2004, *42,* 3081.
63. Utiyama, H., Tsunashima, Y., Kurate, M. *J. Chem. Phys.* 1971, *55,* 3133.
64. Pedersen, J. S. (unpublished).
65. Magid, L. J., Li, Z., Butler, P. D. *Langmuir* 2000, *16,* 10028.

66. Magid, L. J., Li, Z., Butler, P. D. *J. Phys. Chem. B* 2000, *104*, 6717.
67. Odijk, T. *J. Polym. Sci.*, *Polym. Phys. Ed.* 1977, *15*, 477.
68. Skolnick, J., Fixmann, M. *Macromolecules* 1977, *10*, 944.
69. Fixman, M., Skolnick, J. *Macromolecules* 1978, *11*, 863.
70. Morfin, I., Reed, W. F., Rinaudo, M., Borsali, R. *J. Phys. II* 1994, *4*, 1001.
71. Schurtenberger, P., Cavaco, C., Tiberg, F., Regev, O. *Langmuir* 1996, *12*, 2894.
72. Schäfer L. *Macromolecules* 1984, *17*, 1357.

7 Phase Behavior of Systems with Wormlike Micelles

Florian Nettesheim and Eric W. Kaler

CONTENTS

7.1 INTRODUCTION

Soluble surfactants self-assemble in solution once their critical micellization concentration is reached. Above this concentration the shape of the micelles is governed by the geometrical constraints of the surfactant molecules. The concept of a packing parameter, introduced by Israelachvili, can be used as a qualitative tool to predict aggregate shape.[1] The packing parameter is defined as the ratio of the volume of a surfactant molecule V to the product of the length l of the hydrophobic moiety and headgroup area a.

$$p = \frac{V}{al} \tag{7.1}$$

The values for the basic building blocks of a wormlike micellar network, such as hemispherical endcaps, cylindrical sections, and threefold junctions, are shown in Figure 7.1 together with the corresponding values for the packing parameter.

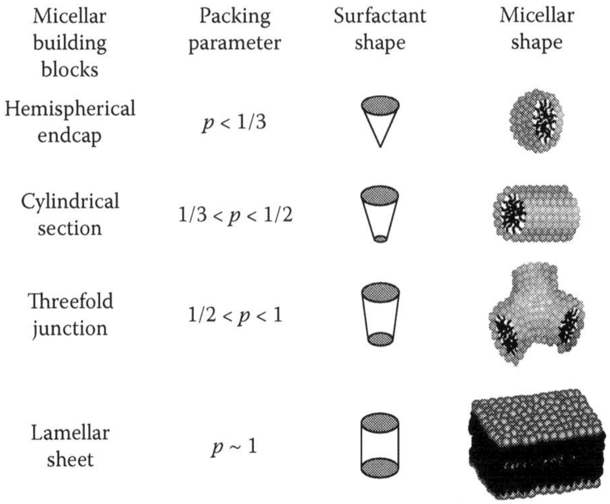

Micellar building blocks	Packing parameter	Surfactant shape	Micellar shape
Hemispherical endcap	$p < 1/3$		
Cylindrical section	$1/3 < p < 1/2$		
Threefold junction	$1/2 < p < 1$		
Lamellar sheet	$p \sim 1$		

FIGURE 7.1 Values of the packing parameter, surfactant shapes, and micellar building blocks. The branched micellar structure is intermediate between the cylindrical and the lamellar aggregate shapes.

A molecular-based discussion of packing is given in detail in Chapter 2. An equivalent language based on the concept of the state of curvature of the surfactant aggregate can also provide a good description of micellar structures. The two are linked by the relation[2,3]

$$p = 1 - \langle H \rangle l + \frac{\langle K \rangle l^2}{3} \qquad (7.2)$$

where $\langle H \rangle$ and $\langle K \rangle$ denote the bending and the Gaussian modulus of the surfactant film, respectively.

As surfactant molecules aggregate above the critical micelle concentration (cmc), they generally initially form spherical micelles but, as their aggregation number grows, they extend in two dimensions and form disklike micelles or, much more commonly, they grow in one dimension into cylinders. In the latter case, the micelles may grow to microns in length and be described as semiflexible chains that above an overlap concentration form entangled networks. In analogy to the behavior of polymer chains, many concepts, such as rubber elasticity theory, have been adapted from polymer physics to describe the rheology of wormlike micellar solutions, as discussed in Chapters 8 and 9.

This chapter explores the phenomenological connections between the growth and interactions of these giant micelles in solution and the macroscopic thermodynamic phase behavior that their solutions display. The linkage between microscopic structure and macroscopic phase separation can at least initially be thought of as arising from either *inter*micellar interactions or *intra*micellar rearrangements, although in practice both may change simultaneously. In the first case, the growth of micelles leads to an increase in the net van der Waals interaction between the micelles. Should this attractive interaction become strong enough, then the initially homogenous micellar solution can separate into a micelle-rich and a micelle-lean phase, in analogy to a gas-liquid phase separation.

In the second case, changes in the conformations or the interactions between the surfactant molecules themselves as a function of concentration or a field variable such as temperature can drive a phase separation. Examples of this kind of phase separation include the cloud-point phenomenon of ethoxylated surfactants (and ethoxylated polymers), wherein the hydrophilic group takes on less polar configurations as the temperature increases and eventually becomes insoluble, or the salting out of ionic surfactants from solution by the addition of electrolyte. A subtler example is a change in the net curvature of the aggregates that favors the formation of saddle-shaped junctions that initially branch micelles and eventually connect micellar segments into a micellar network. The statistical mechanics of that network can lead to phase separation upon dilution. Finally, flow fields can affect the stability of solutions of giant micelles, and the features of shear-induced phase separation (SIPS) and how it is linked to the equilibrium phase behavior will be discussed below briefly.

The phenomena above can all occur in dilute solutions. As the surfactant concentration is increased, excluded volume effects begin to play an important role and orientational order is induced in the surfactant solution. Consequently, nematic and hexagonal phases are formed, as shown in the seminal schematic phase diagram of Candau (Figure 7.2).[4] In this chapter we will focus on the central part of this phase diagram, where wormlike micelles are found and form either entangled or branched networks that may ultimately phase-separate as a function of concentration or a field variable.

7.2 RELATION OF A SINGLE VISCOSITY MAXIMUM TO PHASE BEHAVIOR

7.2.1 Molecular Packing

In the dilute regime the micelles grow into threads, mildly affecting the solution viscosity. These micelles can essentially be described in analogy to dilute polymer solutions. The transition into the semidilute regime is accompanied by a drastic change in properties such as an increase in zero shear viscosity η_0 and elasticity. In the semidilute regime, that is, above the overlap concentration C^*, micelles form an entangled network. In this regime, in addition to reptation, the self-assembled nature of this network introduces a second relaxation path that reflects the fact the micelles can break and reform (see Chapter 10). If this breakage time is much shorter than the reptation time, the relaxation can be described by a single time constant τ_r.[5–7] This leads to the Maxwellian behavior observed for a large number of entangled wormlike micellar solutions.

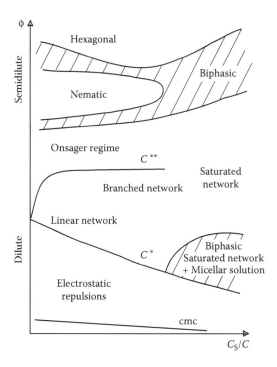

FIGURE 7.2 Schematic surfactant phase diagram redrawn from Ref. 4 with permission from the American Chemical Society.

Herb et al.[8] explored the correlation between solution rheology and phase behavior and examined mixed anionic/nonionic, anionic/cationic, and cationic/hydrotrope mixtures that formed viscoelastic solutions with a single terminal relaxation time, that is, Maxwellian fluids. As is the case for many such mixtures, the authors found a maximum in the solution viscosity as a function of a composition variable, in this case either the ratio of the concentrations of the two surfactants at a fixed total concentration (dashed arrow b in Figure 7.3a) or as a function of total surfactant concentration at a fixed concentration ratio (dashed arrow a in Figure 7.3a). They found the same result for the relaxation time as a function of composition, and the loci of viscosity maxima or relaxation times can be plotted on the ternary phase diagram (Figure 7.3). A key observation is that the maximum in viscosity is related to protrusion of a two-phase region, wherein the micellar L_1 phase and the hexagonal phase coexist, into the L_1 phase. The iso-viscosity and iso-relaxation time contours illustrate the connection between phase behavior and rheology.

Herb et al. relate this observation to the values of packing parameter, where the values of the chain length l and the hydrophobic volume V are gotten from the correlations of Tanford[9] and the area per head group a is determined from surface tension experiments.

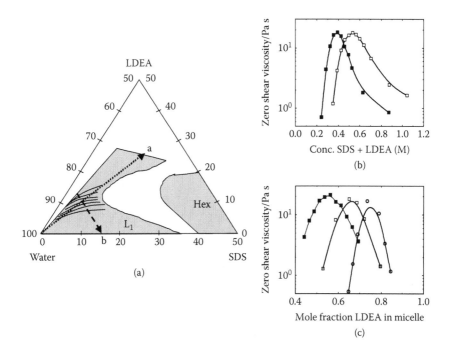

FIGURE 7.3 (a) Ternary phase diagram for the system lauryl diethanol amide (LDEA) / sodium dodecyl sulfate (SDS) / H_2O at 25°C. The micellar phase (L_1) and the hexagonal phase (Hex) are separated by a two-phase region that protrudes into the L_1 phase. The lines in the L_1 region depict contours of constant viscosity in Pa s. (b) Zero shear viscosity as a function of total surfactant concentration at 25% (solid squares) and 33% (open squares) SDS. (c) Zero shear viscosity as a function of mole fraction LDEA in the micelle at total surfactant concentration of 10 wt% (solid squares), 15 wt% (open squares) and 20 wt% (open circles). Data and phase diagram reproduced from Ref. 8 with permission from the American Chemical Society.

Figure 7.4 shows the contours of constant packing parameter, which is assumed to mean contours of constant average surfactant geometry. These lines closely mimic the contours of constant viscosity and relaxation time. The correlation between the packing parameter and the viscosity can be interpreted in terms of changes in the packing parameter. As the packing parameter increases from 1/3 to 1/2, the energy cost of an endcap is increased over that of forming a cylindrical section of a micelle. This will reduce the number of endcaps, which causes the micelles to elongate and thus increase the solution viscosity. The observation of a maximum in viscosity along a particular composition path is a question of how one moves in the phase diagram relative to the lines of constant average surfactant geometry and hence likely similar micellar structure. In particular, paths in the phase diagram where either the ratio of surfactants (dashed arrow b in Figure 7.3a) or total surfactant concentration (dashed arrow a in Figure 7.3a) is varied often cross lines of constant packing parameter. Consequently, the maximum in viscosity is correlated to the maximum in packing

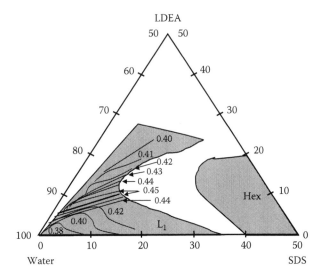

FIGURE 7.4 Same phase diagram as in Figure 7.3, but now the lines in the L_1 region indicate contours of constant surfactant packing parameter. Data and phase diagram reproduced from Ref. 8 with permission from the American Chemical Society.

value, or the structure with the least curvature. Examples are displayed in Figure 7.3, where either the total surfactant concentration (Figure 7.3b) or the mole fraction of lauryl diethanol amide (LDEA) in the micelle (Figure 7.3c) is varied. In both cases maxima in zero shear viscosity are observed. It is interesting to note that the hexagonal phase protrudes most deeply into the two-phase region along the extrapolated line of the highest packing number value.

While this explanation of the correlation of viscosity to surfactant packing is quite plausible for the case of varying surfactant ratio at constant concentration (dashed arrow b), there are other explanations for the variation at a fixed ratio as a function of total surfactant concentration (dashed arrow a). In this case, adding more surfactant also increases the total ionic strength of the solution and can cause the surfactant film to assume a lower curvature, or equivalently a higher value of the packing parameter. As the value of the packing parameter increases beyond 1/2 (Figure 7.1), branches may form and lead to a reduction in viscosity (see Section 7.2.2). In many cases, especially for cationic surfactants with hydrotropic salts, there are two maxima in viscosity and relaxation time as a function of concentration, and these are discussed in Section 7.3.

7.2.2 Cationic/Anionic Mixtures

The surface charge of a micelle has a strong influence on its shape and its micellar phase behavior. The influence of binding or nonbinding salts such as sodium salicylate (NaSal) or sodium chloride (NaCl), respectively, on the rheological

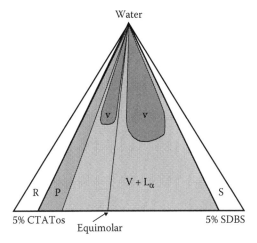

FIGURE 7.5 The water-rich corner of the CTATos-SDBS phase diagram. R indicates the presence of rodlike or wormlike micelles, P a precipitate, V regions where vesicles form, S small globular micelles, and L_α the lamellar phase. Reproduced from Ref. 10 with permission from the American Chemical Society.

properties and phase behavior of cationic surfactant systems clearly show the effect of charge density. However, it is useful to control the charge of the micelle itself more directly, which can be accomplished by using mixtures of cationic and anionic surfactants. The cetyltrimethylammonium tosylate (CTATos) and sodium dodecylbenzene sulfonate (SDBS) system is a well-studied candidate.[10, 11] A general phase diagram is shown in Figure 7.5.

This surfactant pair in water shows many of the same rheological signatures as the cationic wormlike micellar systems containing a hydrotrope, as is discussed below. Additionally, intrinsic micellar properties, such as the persistence length, can be directly controlled by the ratio of both surfactant concentrations. With respect to CTATos, SDBS is more effective in driving micelle growth than either sodium tosylate (NaTos, a binding salt) or NaCl (a nonbinding salt) because of its ability to screen the surface charge of the micelle more effectively and thereby reduce the effective size of the surfactant headgroup (this is a stronger example of the attractive interactions that exist between sodium dodecyl sulfate (SDS) and LDEA as studied by Herb et al.[3]). The resulting higher packing parameter would favor the formation of longer micelles and, in this case, threefold junctions over the formation of hemispherical endcaps. Indeed, the steady decrease in η_0 and τ_r upon addition of SDBS (Figure 7.6d) is consistent with the transition to a branched network immediately after addition of SDBS to the already quite viscous solution of CTATos. The addition of either a screening salt (NaCl) or a binding salt (NaTos) affects the transition from an entangled to a branched network in the same way by more or less screening the positive charge of the micelle and therefore rendering it more flexible. In this respect NaTos is more effective in

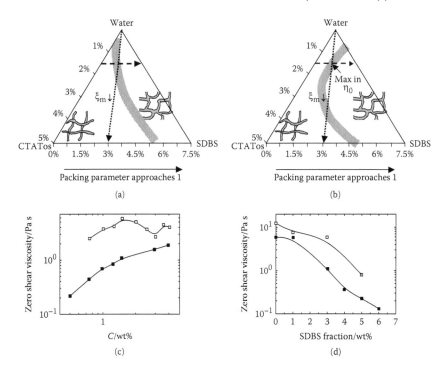

FIGURE 7.6 (a) An expanded section of Figure 7.5 at SDBS fractions below 7.5% with 0.25% NaTos. The dotted arrow indicates the dilution line at constant CTATos/SDBS weight ratio (97/3). The dashed arrow indicates the variation of surfactant composition at constant overall surfactant concentration (1.5 wt%). The hatched curve indicates the zone of maximum viscosity. (b) Same as (a) with 0.25% NaCl instead of NaTos. (c) Viscosity as a function of surfactant concentration along the dotted arrow. (d) Viscosity as a function of SDBS fraction along the dashed arrow. Solid symbols in (c) and (d) refer to 0.25% NaTos and open symbols to 0.25% NaCl. Data reproduced from Ref. 11 with permission from the American Chemical Society.

driving the formation of branches than NaCl, as it associates more closely with the micelle and thus screens the charge more effectively.

A more detailed look at the phase diagram enables other observations. Figure 7.6 shows a schematic of the transition between the two regimes. Along the line of constant surfactant composition, a maximum occurs in the zero shear viscosity only in the presence of NaCl, and it is relatively weak. In the presence of NaTos, the viscosity increase is continuous over the range of surfactant concentrations studied. By increasing the surfactant concentration the number density of entanglements increases, and accordingly the plateau modulus G_0 increases (see Chapters 8 and 9).

Variation of the surfactant composition at fixed concentration yields a decrease in the zero shear viscosity for both NaTos and NaCl salts. Here, the density of the network does not change (and thus G_0 is constant), so the change in viscosity can only be due to a change in the relaxation time of the system. This can be realized by introducing an alternative relaxation mechanism that relaxes stress faster than can either reptation or the scission-recombination mechanism. Such a relaxation can be introduced by threefold junctions in the network. While for an entangled network the viscosity increases with surfactant concentration due to an increase in the mean micellar length $\bar{L}$, the viscosity decreases as branches successively replace entanglements. The distance between crosslinks $\bar{L}_c$, which is then a decreasing function of surfactant concentration, replaces the contour length $\bar{L}$ as the appropriate micellar length scale. For a saturated network the reptation model breaks down, which suggests that the system becomes very fluid.[12]

Additionally, a crosslink or branch point can slide along the contour of a micelle without energy penalty.[12] The presence of such branches was first suggested by Porte et al.,[13] and rheology measurements[14] as well as cryogenic electron micrographs[15] provide more evidence for their presence.

The shaded line in the phase diagrams in Figure 7.6a and b indicate the transition between entangled and branched networks. Note that the maximum in zero shear viscosity in the case of NaCl can only be realized if the transition line between entangled and branched network crosses the investigation line twice, as indicated in Figure 7.6b. Otherwise, there would be a continuous increase in zero shear viscosity as is observed for the case of NaTos (Figure 7.6a). The change in slope in the viscosity curve (Figure 7.6c) may be associated with the transition from a branched to an entangled structure.

7.2.3 NONIONIC/IONIC MIXTURES

Mixing nonionic surfactants such as alkyl polyglucosides with ionic surfactants allows variation of the charge density of the micelles over a wider range than is possible in the mixed anionic/cationic systems, which form a precipitate when the net charge is near zero (or near equal molar concentrations as shown in Figure 7.5). The nonionic/ionic mixtures therefore allow for a more detailed study of the structural changes from an entangled to a branched network of micelles. Again, lines at constant as well as varying surfactant composition of the surfactants $C_{12}\beta G_1$, a nonionic monoglucoside surfactant with a 12 carbon aliphatic tail, and sodium dodecyl sulfate (SDS) were investigated.[16] The paths are shown in Figure 7.7 along with zero shear viscosities as a function of a compositional parameter.

The phase diagram of $C_{12}\beta G_1$/SDS/water displays a lamellar phase close to the binary $C_{12}\beta G_1$/water edge of the ternary diagram. As required, two-phase regions flank the lamellar phase. The lines in Figure 7.7a were constructed from viscosity measurements (Figure 7.7b and Figure 7.7c) along dilution lines (dotted) and the line of constant surfactant concentration (dashed). From such a diagram one can easily see why both experiments, whether along the dilution line or along the line of varying composition, lead to a maximum in viscosity. Along the dilution lines one continually increases the density of the network (congruent with an increase in the plateau of the storage modulus, not shown).

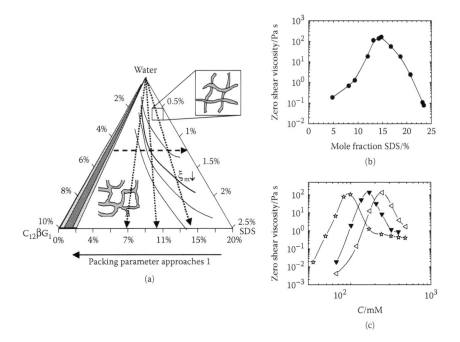

FIGURE 7.7 (a) The water rich corner of the $C_{12}\beta G_1$/SDS/H$_2$O phase diagram. The hatched areas in the $C_{12}\beta G_1$ corner of the diagram are the miscibility gap (light gray) and the two-phase region between the lamellar phase (gray area) and the isotropic L$_1$ region. The arrows indicate the paths of investigation. Three lines at constant surfactant composition (dotted) and one at varying composition (dashed) but constant overall surfactant concentration were investigated. The curved lines indicate contours of constant viscosity (or relaxation times), with the bold one being the contour of maximum viscosity. (b) Zero shear viscosity as a function of SDS mole fraction, that is, along the dashed arrow. (c) Zero shear viscosity as a function of total surfactant concentration at three different mole fractions of SDS along the dotted arrows (stars: 9% SDS; solid triangles: 14% SDS; and open triangles: 19% SDS). Data reproduced with permission from Ref. 16.

At the point where the viscosity passes its maximum, new entanglements are replaced by branch points, which are free to slide and therefore decrease the zero shear viscosity.

The observations made in the two systems presented here (CTA-Tos/SDBS/water in Section 7.2.2 and $C_{12}\beta G_1$/SDS/water in this section) can be organized in analogy to the discussion of Herb et al.[8] The presence of the lamellar phase in the $C_{12}\beta G_1$/SDS/water systems makes the discussion in terms of surfactant geometry clearer as it reflects the presence of surfactant packing number near unity. (Note that there are also bilayer phases present in the CTA-Tos/SDBS/water system as well, both in the form of vesicles at more equimolar concentrations and as lamellar phases at higher concentrations than shown in Figure 7.6.) As one moves toward the lamellar phase, the packing parameter

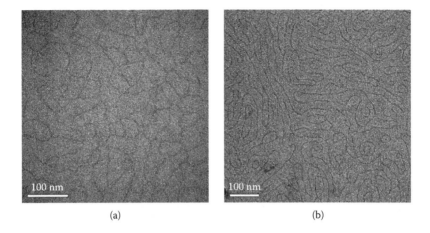

(a)   (b)

FIGURE 7.8 Cryo-TEM images of two $C_{12}\beta G_1$/SDS samples with increasing SDS mole fraction at 5 mol% total surfactant concentration. The micrographs were taken by Travis Hodgdon.

increases toward unity, which first leads to an increase in micellar length and thus to an increase in viscosity. At a certain packing parameter the formation of branched micelles is facilitated by the lower formation energy of junctions with respect to that of endcaps. In both Figure 7.6 and Figure 7.7 the succession of structures is such that the branched structure is found at lower charge density— which in the CTATos/SDBS case corresponds to high SDBS content or a high concentration of screening salt. With the glucoside, the branched network is favored in the regions of low SDS concentration.

The presentation of the phase diagrams above and the arguments made in favor of a transition from an entangled to a branched network remain mostly conjectural. Cryo-TEM can provide direct visualization of these situations, however, and Figure 7.8 shows images of two samples along the line of varying surfactant composition, that is, along the dashed arrow in Figure 7.7. The progression from a sample that consists essentially of one highly branched micelle and no endcaps (Figure 7.8a) to a sample with only very few branches, and conversely a high number of entanglements and endcaps (Figure 7.8b) can be observed clearly. The increasing charge density and thus the stronger coulomb repulsion leads to an increase in the curvature of the surfactant film (or equivalently a decrease in the packing parameter), which in turn favors the entangled over the branched network.

This transition to a micellar network seems to be common to a variety of systems ranging from cationic surfactant/binding salt such as the cetylpyridinium chloride (CPyCl) with NaSal[17] and erucyl-bis(hydroxyethyl)methylammonium chloride (EHAC) with NaSal,[18, 19] as described in Section 7.3, mixed cationic/anionic (CTATos/SDBS with or without screening or hydrotropic salt) to nonionic wormlike networks with inclusion of charge ($C_{12}\beta G_1$/SDS and LDEA/SDS) and mixtures of nonionic surfactants such as the sucrose alkanoates with $C_{16}EO_4$, as discussed next. The role of charge is merely to influence the spontaneous curvature of the surfactant aggregate, which thus leads to the observed transition.

Similar packing arguments should apply to binary nonionic surfactant systems as well, where the packing is solely controlled by the hydration of the headgroup and not by charge. These systems therefore show pronounced temperature dependence and no significant dependence of salt concentration. It is thus interesting to compare the phase behavior of nonionic surfactants to the ionic surfactant systems, as will be done in the following section.

7.2.4 NONIONIC SURFACTANTS AND MIXTURES

Wormlike micelles frequently form in solutions of nonionic surfactants with a long aliphatic tail, as for example the polyglycol monoalkylethers, denoted C_mEO_n, where m indicates the number of carbons in the hydrophobic tail and n the number of ethylene oxide groups in the hydrophilic head group. There are interesting similarities to the ionic systems presented above in that the nonionic solutions interestingly show a nonmonotonous evolution of viscosity with temperature.[20]

The hydration of the ethylene oxide groups decreases as the temperature increases. Therefore, the size of the headgroup of C_mEO_n surfactants decreases with increasing temperature (the packing parameter increases). As a consequence the lamellar phase for C_mEO_n surfactants is found in the center of the phase diagram (packing parameter ~1). In some cases the lamellar phase intersects the micellar region and divides it into two pieces, normally denoted as L_1 and L_2, or the lamellar phase is an island in the L_1 phase envelope. The more highly curved hexagonal phase is found at lower temperatures and lower surfactant concentrations. Representative phase diagrams are shown in Figure 7.9 for the surfactants $C_{12}EO_5$ and $C_{12}EO_8$.

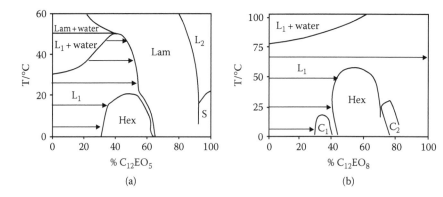

FIGURE 7.9 Phase diagrams of $C_{12}EO_5$ (a) and $C_{12}EO_8$ (b) in H_2O. The arrows indicate paths of investigation using pulsed gradient spin echo NMR. Reproduced from Ref. 21 with permission from the American Chemical Society.

Using pulsed field gradient NMR, Nilsson et al.[21] studied the change in the self-diffusion coefficient D_c as a function of surfactant concentration in the L_1 phase of $C_{12}EO_5$ and $C_{12}EO_8$ at constant temperature (see arrows in Figure 7.9). Interestingly, the self-diffusion coefficient for $C_{12}EO_5$ displays a minimum as a function of concentration, and this minimum shifts to lower concentrations with increasing temperature, that is, as the cloud point T_c is approached. The authors argue that the change in D_c can be explained by an initial increase followed by a slow decrease in the apparent size of the micellar aggregates. The existence of branches at higher concentrations may explain the increase in D_c (and conversely the decrease in apparent size) as well. It is also plausible that the minimum in D_c would shift to lower concentrations as T_c is approached. The explanation of the results of Nilsson, however, may be more complex. Zana and Weill have shown[22] that the aggregation number N of C_mEO_n micelles increases with temperature, first slowly and then fairly rapidly starting at about 35°C below T_c. At the same time extensive exchange occurs between micelles via intermicellar collisions and temporary merging of the micelles. The lifetime of such a connection increases as T_c is approached. As the C_mEO_n concentration increases, one observes a decrease of the temperature at which the rapid increase of N occurs. This study highlights the importance of time scales. While fluorescence probing accesses times in the submicrosecond range and therefore can see individual micelle collisions, other techniques such as rheology, cryo-TEM, and light scattering see either an average or a frozen state.

A more compelling argument can be made if one compares the phase behavior of the two systems $C_{12}EO_5$ and $C_{12}EO_8$ (see Figure 7.9) and the concentration dependence of the respective self-diffusion coefficients. In the case of $C_{12}EO_5$, where a minimum in D_c is observed for all investigated temperatures, there is a lamellar phase present at higher surfactant concentrations, whereas it is absent in the case of $C_{12}EO_8$. The self-diffusion coefficient of $C_{12}EO_8$ increases monotonically for the lower temperatures and is almost independent of concentration close to the cloud point. Based on packing arguments, the curvature of the surfactant film in the $C_{12}EO_5$ system approaches 1, at least at the higher temperatures studied, as the phase boundary of the lamellar phase is approached. For $C_{12}EO_8$, the changes in curvature are less pronounced and lead to linear growth of the micelles (and a monotonic increase in D_c). The absence of the lamellar phase in the phase diagram indicates that the packing parameter, which for $C_{12}EO_8$ is smaller than for $C_{12}EO_5$, also never reaches values close to unity.

Similar observations supporting a transition to a branched network were made by Ambrosone et al.,[23] who measured the mean squared displacement of soybean lecithin in reverse micelles using pulsed field gradient NMR. Using a detailed analysis developed by Anderson and Wennerström[24] they show that the diffusional time dependence of a multiconnected network differs from that of an entangled network. For a multiconnected or branched network a linear dependence of the mean squared displacement on the observation time is found at times, t, longer than the time it takes for a molecule to diffuse in a micelle from one branch to the next. This dependence results from a complete randomization of the molecule's trajectory, such that the diffusion can be treated as Gaussian. For shorter observation times, the diffusion is curvilinear and has a $t^{1/2}$ dependence. The transition between the two regimes in a branched network depends on concentration.

For entangled networks, however, the diffusion is curvilinear for all observation times, since the surfactant molecules are confined to one micelle, if one disregards the relatively slow exchange of molecules between micelles. Ambrosone et al.[23] studied the lecithin/water/isooctane system along oil dilution lines at water to lecithin molar ratios of $W_0 = 2$ and 3. For $W_0 = 3$ they found that the amphiphile self-diffusion measurements were independent of the time scale of observation (scaling with $t^{1/2}$) and essentially independent of the micellar volume fraction. These results would be expected if an entangled network is present. In the case of $W_0 = 2$ they find a transition from curvilinear ($\propto t^{1/2}$) to Gaussian diffusion ($\propto t$), indicating the presence of a branched network.

Kato et al.[25] make a compelling argument for the transition to a branched network with increasing temperature by analysis of the self-diffusion coefficient of $C_{16}EO_7$ measured by pulsed gradient spin echo NMR (see phase diagram in Figure 7.10). From an Arrhenius relation, the activation barrier for an exchange of a $C_{16}EO_7$ molecule between two micellar segments can be determined, and is found to vary monotonically as a function of both concentration and temperature. Kato et al. attribute the decrease in the activation barrier with temperature to the continuous transition from an entangled to a branched micellar network. The iso-activation lines in Figure 7.10 are nearly perpendicular to the phase boundary of the hexagonal phase, while they are almost parallel to the boundary to the lamellar phase. This agrees very well with the increase in packing parameter with temperature toward the lamellar phase ($p = 1$), and thus the change from morphology

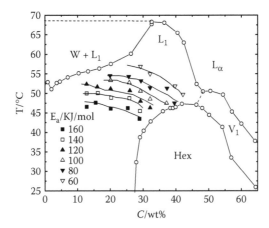

FIGURE 7.10 Phase diagram of $C_{16}EO_7$ in D_2O with iso-activation energy lines in the high temperature region of the L_1 phase. Reproduced from Ref. 25 with permission from the American Chemical Society.

with high curvature to one with lower curvature, that is, the formation of branches. Conversely, the iso-activation contours only show a weak dependence on concentration, which can be attributed to the shorter distance between micelles.

Aside from this effect, no morphological transition is expected, at least at the lower temperatures, since the packing parameter does not change upon approaching the hexagonal phase. The probable reason for the lower activation energy for diffusion in the multiconnected case is the possibility of network points sliding along the contour of a micelle. This scenario can be described by the progression of curvature and the relative energy costs of forming a hemispherical endcap to the cylindrical body of a micelle and finally a threefold junction point. If one extrapolated this progression of curvature one would eventually find locally flat bilayers, and indeed the lamellar phase is found in the vicinity of the L_1 phase. In a more detailed study, Kato et al.[26] associated the structural transition in the binary nonionic wormlike micellar system with an increase of the lifetime of the transient threefold junctions.

The relation of the rheological properties to phase behavior can now be discussed in more detail. In the preceding paragraphs, the existence of branches in nonionic micellar networks has been established, and it is clear they are found in the vicinity of the lamellar phase, as is the case for the ionic micellar networks. With a similar progression of structures, except now as a function of temperature and not as a function of salinity, the viscosity displays a maximum, whose position depends on surfactant concentration (Figure 7.11).[20] In general the viscosity drops significantly when the cloud point temperature is approached. Constantin et al.[20] additionally conducted high frequency dynamic rheology experiments on aqueous $C_{12}EO_6$ binary mixtures and were able to show that the data are best described

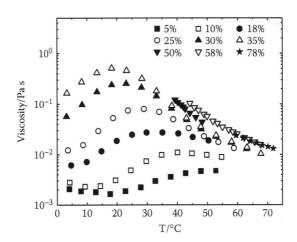

FIGURE 7.11 Viscosity of $C_{12}EO_6$ solutions as a function of temperature at various concentrations. Reproduced from Ref. 20 with permission from the American Chemical Society.

using bimodal relaxation spectra, which agrees with the idea that entanglements and branches coexist in the region of decreasing viscosity.

The same trends in micellar morphology are observed in other surfactant systems as well, such as sucrose alkanoate/C_mEO_n[27, 28] or gemini/C_mEO_n surfactant mixtures.[29] The same relation between surface charge, packing parameter, and micellar morphology and thus viscosity apply there, too, as the data by Maestro et al. show.[27, 30] Although not explicitly shown in their phase diagrams, the authors refer to the nearby presence of a micellar-lamellar coexistence region where the fraction of nonionic surfactant $C_{16}EO_n$ (with n = 2–4) is increased in solutions with sucrose alkanoate (sucrose dodecanoate and sucrose hexadecanoate) solutions. The addition of $C_{16}EO_n$, much like in the case of adding SDBS to CTATos, reduces the effective headgroup size of the surfactant and therefore increases the packing parameter, with the consequential formation of less curved aggregates. A hexagonal phase exists at higher sucrose alkanoate concentrations, corresponding to regions of higher curvature.

In the sucrose dodecanoate/$C_{16}EO_n$/water systems the viscoelastic region is located close to the phase boundary to the lamellar and hexagonal phases, where it extends furthest close to the wormlike micellar to lamellar transition. In the case of the sucrose hexadecanoate/$C_{16}EO_n$/water systems, however, the region of higher viscosity is located closer to the binary sucrose hexadecanoate/water axis of the ternary diagram. A more detailed examination shows that the viscosity in both systems passes through a maximum as a function of $C_{16}EO_n$ content, but the maximum is at much lower nonionic surfactant concentrations for the sucrose hexadecanoate systems. Less $C_{16}EO_n$ is needed to reach curvatures that favor a branched over an entangled morphology since the packing parameter of sucrose hexadecanoate is already closer to 1 than is that of sucrose dodecanoate, owing to its longer hydrophobic tail.

7.3 RELATION OF MULTIPLE VISCOSITY MAXIMA TO PHASE BEHAVIOR

In many situations the addition of an additive such as a hydrotrope to ionic surfactant solutions produces not one but two maxima in viscosity, as shown for the cetylpyridinium chloride/sodium salicylate (CPyCl/NaSal) system well studied by Hoffmann et al. (Figure 7.12).[17] The first maximum, as discussed in the previous section, can be explained by the increase in packing parameter and thus by the initial growth of linear micelles, which at a certain value of the packing parameter greater than 1/2 begin to branch. The presence or absence of the two maxima seems to depend on the strength of binding of the additive and it also appears only in ionic mixtures.

Some of the equilibrium structures in the semidilute regime, in particular the transition from an entangled to a branched network as a function of salt concentration, have been inferred from rheological observations, but direct and reliable evidence (such as cryo-TEM; see Chapter 5) is still scarce. Moreover, this hypothesis is not the only possible explanation for the observed nonmonotonic

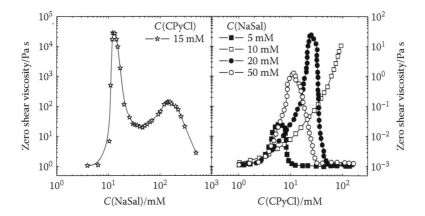

FIGURE 7.12 Zero shear viscosity as a function of NaSal concentration (left) and of total surfactant concentration (right) for the CPyCl/NaSal system. Reproduced from Ref. 17 with permission from the American Chemical Society.

behavior of relaxation times and zero shear viscosities as a function of salt concentration. The second maximum, however, must reflect additional rearrangements of the micellar morphologies. Because the second maximum occurs when the molar ratio of hydrotrope to surfactant is greater than one (the minimum in viscosity is found at about equimolar conditions), it is usually attributed to the strong binding of the hydrotrope to the micelle. Thus, it reflects the reverse progression of structure from branched to ultimately individual unentangled micelles at very high salt concentration. In this progression the net charge of the micelle reverses from positive at low salt concentrations, where the first maximum in viscosity is observed, to negative at higher salt concentrations, and passes through zero near the position of the viscosity minimum.[31] Thereafter, increasing the salt leads to an increase in the repulsion between headgroups, and thus a decrease in the packing parameter. The effect of increasing salt concentration in this regime of already high ionic strength is expected to be less dramatic than it is around the first viscosity maximum, and indeed the second maximum in viscosity is normally weaker.

The propensity of a wormlike micellar system to display phase separation at equimolar conditions of cationic surfactant and additive such as a binding salt depends on the strength of the counterion binding to the micelle. Hoffmann et al. studied the use of a homologous series of para-alkyl benzoates as binding salts.[32] The viscosity displays two maxima in the case of 4-ethyl benzoate, yet there is no phase separation at intermediate salt concentrations. Interestingly a miscibility gap opens up when 4-pentyl benzoate is used instead, and the viscosity maxima and the minimum between them become more pronounced.[32]

As also pointed out by Rehage and Hoffmann,[32] the presence of these maxima, or more accurately the minima between them, is related to the phase behavior of these mixtures. One of the better studied mixtures in which two viscosity maxima are found is that of erucyl-bis(hydroxyethyl)methylammonium chloride (EHAC) with a number of different salts including NaTos and NaSal.[18, 19, 33] For the case of NaCl (Figures 7.13a and b) only a single maximum is seen, and there is ultimately phase separation at high salt concentrations. The authors also studied the rheology at fixed surfactant and varying hydrotrope concentrations at compositions where a two-phase region is approached and actually crossed (see Figures 7.13c–f). The line at 100 mM EHAC in Figure 7.13c represents the case where no two-phase region is crossed, yet there is a minimum in viscosity (Figure 7.13d) that reflects the near presence of a two-phase region. The two-phase region can be seen at 45°C as illustrated in Figure 7.14. The line at 40 mM shows a viscosity curve interrupted by a two-phase region (Figure 7.13d). The changes in packing parameter described above likely occur on both sides of the two-phase lobe inferred by the viscosities (see Figures 7.13c and d). Similar observations are made using different hydrotropes, such as NaSal (Figures 7.13e and f). In these systems the two-phase region also protrudes into the phase triangle, as is the case for the NaTos system. Although the two-phase region is much larger in the case of NaSal the observations are qualitatively the same.

The contours of constant packing parameter in the phase triangles can be replaced by iso-surfaces in a phase prism (made of compositions with temperature as the vertical axis), which can show how a two-phase region at higher temperatures influences the packing parameter variation as a function of composition (that is, salt concentration). The maximum in the packing parameter and thus a minimum in viscosity are expected at the composition where the two-phase region comes closest to the isothermal plane of investigation. The order of events can be summarized as follows: The viscosity initially increases due to micellar growth as the packing parameter approaches 1/2. When branching sets in, the viscosity decreases and if network saturation is reached, phase separation occurs. However, saturation and phase separation need not occur as can be seen for several systems such as the CPyCl/NaSal system studied by Rehage et al.,[17] and the 100 mM EHAC system with varying NaTos concentration presented here (Figure 7.14). On the other side of the two-phase region, the geometry evolves in the reverse fashion, that is, a second maximum is found, although less pronounced, and the packing parameter decreases again. One can assume that in the regions between the maximum in viscosity and the two-phase region the samples contain a branched network. In the notation of a packing parameter, one has to assume that it increases from about 1/2 towards 1 as the two-phase region is approached and the reverse happens on the other side.

An interesting aspect of these systems is the phenomenon of shear-induced phase separation (SIPS), which can be observed for samples that are closest to the two-phase region, that is, for branched networks.[34] This phenomenon is particularly pronounced for 40 mM EHAC and 300 mM NaSal, which is on the increasing branch of the viscosity curve past the two-phase region. This sample shows turbidity at fairly moderate shear rates (see Figure 7.15), and it is interesting

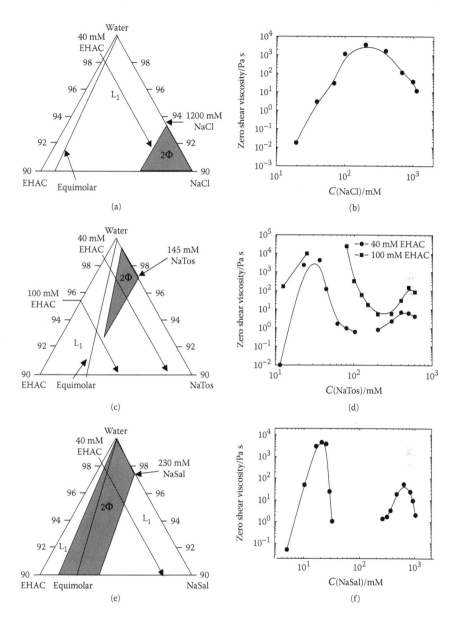

FIGURE 7.13 Ternary phase diagrams of the water-rich corners of EHAC/NaCl/H$_2$O (a), EHAC/NaTos/H$_2$O (c), and EHAC/NaSal/H$_2$O (e) and the corresponding zero shear viscosities in (b), (d), and (f). Arrows in the ternary phase diagrams indicate the path of investigation in rheological experiments. Reproduced from Ref. 19 with permission from the American Chemical Society.

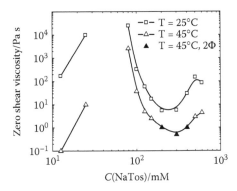

to compare to a sample of 40 mM EHAC and 1000 mM NaSal that has the same
zero shear viscosity but does not show SIPS.[35]

Under the influence of shear the nearby saturation point of the network may
be shifted to higher salt concentrations such that phase separation occurs in the
40 mM EHAC/300 mM NaSal system. At higher salinity the accessible shear
rates are not sufficient to shift the saturation point far enough to induce phase
separation. The shift of the saturation point is equivalent to a shift of the spinodal
line, which has been proposed as the mechanism for shear induced phase separation
in polymer solutions,[36–40] in colloidal systems,[41,42] for the isotropic to nematic

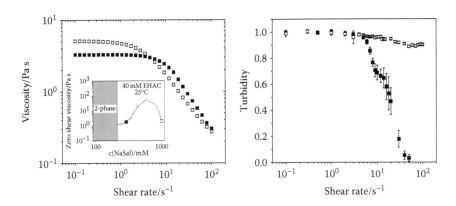

FIGURE 7.15 Shear induced phase separation in a 40 mM EHAC and 300 mM NaSal
sample (filled symbols). A 40 mM EHAC/1000 mM NaSal sample with a very similar
flow curve (right) is shown for comparison (open symbols). The left panel displays the
turbidity as a function of shear rate. The inset indicates the location of the two samples
with respect to their phase diagram. Data are courtesy of M. Liberatore.[35]

transition in wormlike micellar solutions,[43,44] and for the sponge to lamellar transition.[45] A similar explanation, that is that the pretransitional local order is enhanced by shear, is often invoked to describe the shear-induced isotropic to nematic transition in wormlike micellar solutions.[44,46]

7.4 THEORETICAL CONSIDERATIONS

The observations presented in the previous section can be put in a more general theoretical framework. As mentioned before, theory predicts a transition from linear entangled networks to branched networks driven by a subtle interplay of the energies and entropies of formation of endcaps, cylindrical micelles, and branch points. Schematic phase diagrams can be calculated using this approach, and Safran and Tlusty et al. have used variational approaches to predict the succession of local symmetries in micellar solutions from locally spherical to cylindrical to network to ultimately lamellar.[47–50] The phase diagram in Figure 7.16a qualitatively describes the succession of local symmetries found in wormlike micellar systems. As shown in the cryo-TEM image in Figure 7.16b, micellar systems are indeed forming branched networks under certain conditions. They have to be regarded as dynamic in nature (see sketch in Figure 7.16b) and any cryo-TEM image is thus a snapshot of the situation during the vitrification process.

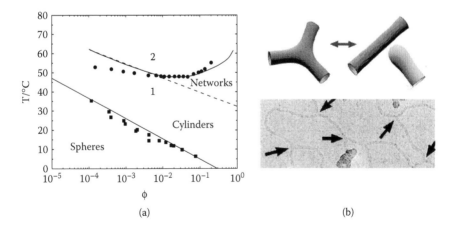

FIGURE 7.16 (a) Succession of aggregate shapes in the $C_{12}EO_6/H_2O$ system as a function of surfactant volume fraction reproduced from Ref. 50 with permission from the American Chemical Society. The upper line is the spinodal of the phase separation between a network and excess solvent in the $C_{12}EO_6$/water system; the dashed line predicts the transition from cylinders to a network; and the lower line marks the sphere to cylinder transition. Symbols are data for the respective transitions. (b) Schematic of a dynamic crosslink (top) and a cryo-TEM image of a wormlike micellar sample containing branches (bottom) reproduced from Ref. 48 with permission from the American Physical Society.

Approaching the two-phase region in a semidilute wormlike micellar solution thus leads to micellar branching. The theory of Safran and co-workers shows that entropic effects lead to an attractive force between branch points. Owing to this entropic attraction, there is a correlation between the location of the branch points that produces an average correlation length that is manifest in the structure factor S(q). These attractive forces will also eventually lead to phase separation when the network is saturated with junctions. The two-phase region is then neighbored by a lamellar phase as captured by the theoretical phase diagram. Drye and Cates[7] have devised a Flory–Huggins-type description of networks that also predicts phase separation caused by network saturation.

The approaches to explain the relaxation behavior presented by Cates et al. are based on a simple scission-recombination and end interchange model. The main caveat of this approach is that it cannot describe the maximum in viscosity or respectively in the network relaxation time τ_r. Two explanations have been offered: The first, by Porte et al.[14] suggests that the activation energy for a branch point sliding along the contour of the micelle is very low, and that this mechanism could lead to a decrease in τ_r. Shikata et al.,[51] on the other hand, proposed a mechanism in which an entanglement could relax stress by the two chains passing each other like ghosts, with the intermediate being a transient junction (see also Chapter 8, Section 8.2.2). To date, there is no conclusive experiment or model calculation supporting either hypothesis. Micellar junctions nonetheless clearly play an important role in setting both the rheology and phase behavior of wormlike micellar solutions.

7.5 CONCLUSIONS

The phase behavior of wormlike micellar solutions is characterized by subtle transitions from spherical to rodlike micelles, which can eventually entangle and branch. The growth is governed by a change in surfactant geometry with either changing surfactant or salt concentration. The role of either salt or temperature is to change the surfactant packing parameter by changing the headgroup size of the surfactant and therefore driving changes in micellar morphology. These changes in micellar morphology have consequences for the thermodynamic stability of the micellar phases, and neighboring phases, such as the hexagonal or lamellar phases found in the vicinity of the L_1 phase, give important insight in the evolution of the packing parameter. It is possible to correlate regions within the L_1 phase with a particular morphology. Two examples are linear entangled wormlike micellar solutions found close by the hexagonal phase, or branched networks that are located closer to the lamellar phase.

Three distinct scenarios can be distinguished:

I. Linear growth of the micelles leads to an entangled network and as a result the zero shear viscosity and the plateau modulus increase significantly.

II. Linear growth is followed by either a decrease in micellar contour length or a transition to a branched network. The zero shear viscosity after the growth regime would decreases in both cases, while the plateau modulus is a function of the network density and therefore remains constant in the case of branches replacing entanglements. In the case of decreasing micellar length it would decrease.

III. Linear growth is followed by branching and in the presence of a strong binding salt, such as NaSal, phase separation occurs at about equimolar conditions. A reentrant behavior is observed and the reverse evolution of micellar morphologies is observed with further increases in salt concentration.

Type I phase behavior displays a monotonically increasing viscosity, as is for instance observed for nonionic surfactants as a function of surfactant concentration or for ionic surfactants with an uncharged cosurfactant.[32]

Type II phase behavior displays a maximum in zero shear viscosity and past the maximum may, but not necessarily will, display phase separation.

Type III phase behavior displays two maxima in zero shear viscosity. Branching is widely believed to be the origin of the intermittent decrease in η_0. Phase separation is not a requirement for the presence of the two maxima, but in all cases a two-phase region is nearby.

The factors influencing phase behavior are first the surfactant geometry, that is, its packing parameter, or equivalently the curvature of the surfactant film. Secondly, charge density influences the packing parameter by tuning the repulsion between the surfactant headgroups. Finally, the strength of binding of the counterion leads to more effective screening of the charge of the surfactant headgroup, and this can ultimately lead to charge inversion. The second and third factors together lead to the reentrant phase behavior observed for the ionic surfactants with strong binding counterions (either hydrotropic salt or oppositely charged surfactant).

ACKNOWLEDGMENTS

This work was supported by grants from Unilever and from the National Science Foundation (CTS – 0625047). The cryo-micrographs in Figure 7.8 were generously provided by Travis Hodgdon.

REFERENCES

1. Israelachvili, J.N., Mitchell, D.J. *Biochim. Biophys. Acta* 1975, *389*, 13.
2. Hyde, S., Anderson, S., Larsson, K., Blum, Z., Landh, T., Lidin, S., Ninham, B.W. *The Language of Shape, The Role of Curvature in Condensed Matter: Physics, Chemistry and Biology.* 1997, Elsevier: Amsterdam.
3. Testard, F., Zemb, T. *J. Colloid Interface Sci.*, 1999, *219*, 11.
4. Lequeux, F., Candau, S.J. In *Structure and Flow in Surfactant Solutions*, Prud'homme, R.K., Herb, C.A., Eds., 1994, American Chemical Society: Washington DC, pp. 51–62.

5. Cates, M.E. *J. Phys. Chem.* 1990, *94*, 371.
6. Granek, R., Cates, M.E. *J. Chem. Phys.* 1992, *96*, 4758.
7. Drye, T.J., Cates, M.E. *J. Chem. Phys.* 1992, *96*, 1367.
8. Herb, C.A., Chen, L.B., Sun, W.M. In *Structure and Flow in Surfactant Solutions*, Prud'homme, R.K., Herb, C.A., Eds., 1994, American Chemical Society: Washington DC, pp. 153–166.
9. Tanford, C. *The Hydrophobic Effect: Formation of Micelles and Biological Membranes.* 1980, John Wiley & Sons: New York, pp. 51–53.
10. Kaler, E.W., Herrington, K.L., Murthy, A.K., Zasadzinski, J.A.N. *J. Phys. Chem.* 1992, *96*, 6698.
11. Schubert, B.A., Kaler, E.W., Wagner, N.J. *Langmuir* 2003, *19*, 4079.
12. Candau, S.J., Oda, R. *Colloids Surf. A* 2001, *183*, 5.
13. Porte, G., Gomati, R., Elhaitamy, O., Appell, J., Marignan, J. *J. Phys. Chem.* 1986, *90*, 5746.
14. Appell, J., Porte, G., Khatory, A., Kern, F., Candau, S.J. *J. Phys. II* 1992, *2*,1045.
15. Danino, D., Talmon, Y., Levy, H., Beinert, G., Zana, R. *Science* 1995, *269*, 1420.
16. Schubert, B.A. *Microstructure and Rheology of Charged, Wormlike Micelles*, In *Department of Chemical Engineering.* 2003, University of Delaware: Newark.
17. Rehage, H., Hoffmann, H. *J. Phys. Chem.* 1988, *92*, 4712.
18. Raghavan, S.R., Kaler, E.W. *Langmuir* 2001, *17*, 300.
19. Raghavan, S.R., Edlund, H., Kaler, E.W. *Langmuir* 2002, *18*, 1056.
20. Constantin, D., Freyssingeas, E., Palierne, J.F., Oswald, P. *Langmuir* 2003, *19*, 2554.
21. Nilsson, P.G., Wennerstrom, H., Lindman, B. *J. Phys. Chem.* 1983, *87*, 1377.
22. Zana, R., Weill, C. *J. Phys. Lett.* 1985, *46*, L953.
23. Ambrosone, L., Angelico, R., Ceglie, A., Olsson, U., Palazzo, G. *Langmuir* 2001, *17*, 6822.
24. Anderson, D.M., Wennerstrom, H. *J. Phys. Chem.* 1990, *94*, 8683.
25. Kato, T., Taguchi, N., Terao, T., Seimiya, T. *Langmuir* 1995, *11*, 4661.
26. Kato, T., Nozu, D. *J. Mol. Liq.* 2001, *90*, 167.
27. Maestro, A., Acharya, D.P., Furukawa, H., Gutierrez, J.M., Lopez-Quintela, M.A., Ishitobi, M., Kunieda, H. *J. Phys. Chem. B* 2004, *108*, 14009.
28. Rodriguez-Abreu, C., Aramaki, K., Tanaka, Y., Lopez-Quintela, M.A., Ishitobi, M., Kunieda, H. *J. Colloid Interface Sci.* 2005, *291*, 560.
29. Acharya, D.P., Kunieda, H., Shiba, Y., Aratani, K. *J. Phys. Chem. B* 2004, *108*, 1790.
30. Rodriguez, C., Acharya, D.P., Maestro, A., Hattori, K., Aramaki, K., Kunieda, H. *J. Chem. Eng. Jap.* 2004, *37*, 622.
31. Olsson, U., Soderman, O., Guering, P. *J. Phys. Chem.* 1986, *90*, 5223.
32. Rehage, H., Hoffmann, H. *Mol. Phys.* 1991, *74*, 933.
33. Raghavan, S.R., Fritz, G., Kaler, E.W. *Langmuir* 2002, *18*, 3797.
34. Schubert, B.A., Wagner, N.J., Kaler, E.W., Raghavan, S.R. *Langmuir* 2004, *20*, 3564.
35. Liberatore, M.W., Nettesheim, F., Porcar, L., Kaler, E.W., Wagner, N.J. in preparation, 2006.
36. Onuki, A. *J. Phys. Soc. Jap.* 1990, *59*, 3427.
37. Onuki, A., Yamamoto, R., Taniguchi, T. *J. Phys. II* 1997, *7*, 295.
38. Onuki, A. *J. Phys.-Condensed Matter* 1997, *9*, 6119.
39. Criadosancho, M., Jou, D., Casasvazquez, J. *J. Non-Equil. Thermod.* 1993, *18*, 103.

40. Criadosancho, M., Casasvazquez, J., Jou, D. *Polymer* 1995, *36*, 4107.
41. Dhont, J.K.G. *Phys. Rev. Lett.* 1996, *76*, 4269.
42. Dhont, J.K.G., Lettinga, M.P., Dogic, Z., Lenstra, T.A.J., Wang, H., Rathgeber, S., Carletto, P., Willner, L., Frielinghaus, H., Lindner, P. *Faraday Disc.* 2003, *123*, 157.
43. Lenstra, T.A.J., Dogic, Z., Dhont, J.K.G. *J. Chem. Phys.* 2001, *114*, 10151.
44. Berret, J.F., Roux, D.C., Lindner, P. *Eur. Phys. J. B* 1998, *5*, 67.
45. Cates, M.E., Milner, S.T. *Phys. Rev. Lett.* 1989, *62*, 1856.
46. Berret, J.F., Roux, D.C., Porte G. *J. Phys. II* 1994, *4*, 1261.
47. Tlusty, T., Safran, S.A. *J. Phys.-Condensed Matter* 2000, *12*, A253.
48. Tlusty, T., Safran, S.A., Strey, R. *Phys. Rev. Lett.* 2000, *84*, 1244.
49. Tlusty, T., Safran, S.A., Menes, R., Strey, R. *Phys. Rev. Lett.* 1997, *78*, 2616.
50. Zilman, A., Safran, S.A., Sottmann, T., Strey, R. *Langmuir* 2004, *20*, 2199.
51. Shikata, T., Hirata, H., Kotaka, T. *Langmuir* 1988, *4*, 354.

8 Linear Rheology of Aqueous Solutions of Wormlike Micelles

Martin In

CONTENTS

8.1   INTRODUCTION

Aqueous solutions of surfactants often present pronounced viscoelastic properties because amphiphilic molecules in water form micelles of different shapes and possibly ordered.[1-4] They belong to the family of complex fluids that have attracted the attention of physical chemists for decades. The rheological properties of surfactant solutions have long been studied because they are of great importance in almost each step that leads from the reactor to the shelf at home. They are crucial for many engineering processes and when trying to satisfy consumer needs or to influence consumer preferences in home care or personal care products (see Chapters 17 and 18). From the practical point of view, it is often required to dispose of concentrated solutions with rather low viscosity that are easy to handle in the formulation process and at the same time dilute solutions with sufficient viscosity during application. However, ordered mesophases formed at high concentration are often very viscous. On the other hand more dilute solutions of surfactants in the micellar phase generally have low viscosity and Newtonian behavior, unless they form giant micelles.[5-11]

It is now well established that surfactant micellar phase become viscoelastic when long cylindrical wormlike micelles form and entangle just as polymers do.[12-15] In several instances, surfactant solutions are actually used for the same purposes as polymeric liquids: as fracturing fluids for oil drilling (Chapter 15); for drag reduction (Chapter 16); or to enhance the viscosity of personal or home care products (Chapters 17 and 18). The viscosity of these solutions can be increased by several orders of magnitude by only small variations of composition parameters such as concentration, ionic strength, and additives.

Recognition of the analogy between wormlike micelles and polymers shed much light in the understanding of the rheological behavior. Models originally developed to describe the rheological properties of polymers were adapted to wormlike micelle solutions.[16,17] An important peculiarity of wormlike micelles is the extreme narrowing of the relaxation spectrum. Very often indeed, the relaxation modulus of wormlike micelle solutions is monoexponential, which means that those systems are mechanically equivalent to the Maxwell model consisting of a spring and a dashpot in series. The transient nature of the wormlike micelles and the role of reversible scission were early recognized as the origin of this peculiar behavior. Micelles are continuously breaking apart and merging and this provides an efficient way to relax stress. Despite two decades of intense research pioneered by Candau, Hoffmann and Shikata, a quantitative assessment of the dominant process in the mechanical relaxation of wormlike micelles remains to be done. It is important to guide the design of effective surfactant structures and formulations to meet particular requirements in high performance formulations but also in order to use rheological measurements as a reliable tool for micellar growth characterization.

Viscoelasticity of surfactant solutions is related to the growth of cylindrical wormlike micelles. Therefore all factors governing the packing parameter are to be considered to account for the viscoelastic behavior. Wormlike micelles are obtained with all kinds of surfactants. Nevertheless, cationic surfactants-based systems have been the most studied for basic research and are found in large-scale applications, more recently entering the composition of fracturing fluids (see Chapter 15). Anionic surfactants such as laurylethersulfate and zwitterionic surfactants were used the most to enhance the viscosity of personal care products. In most cases the addition of salts, with strongly binding counterions or cosurfactants of opposite charge is required to get viscoelastic solutions. Some exceptions are provided for instance by cationic gemini surfactants whose rheological properties have recently been reviewed.[18]

The chapter is organized as follows. In Section 8.2 experimental and theoretical background is briefly recalled. Experiments to characterize the viscoelastic behavior of wormlike micelle solutions are described. Current schemes to account for the peculiar relaxation dynamics of wormlike micelles are also reviewed. Section 8.3 summarizes the rheological performances of different surfactant systems. It will be pointed out that elasticity is a universal feature shared by almost all surfactant systems (except those with branched micelles), while relaxation dynamics is very system specific. In Section 8.4 interpretations of the rheological properties in terms of micelle growth, kinetics of recombination and branching are discussed.

8.2 BACKGROUND

8.2.1 EXPERIMENTAL RHEOLOGY

Rheology is the science of deformation and flow of matter. It aims at establishing and understanding the mechanical response function of materials.[19,20] The knowledge of the response function is important for processing while its understanding in terms of molecular transport properties can provide guidance for the development of new molecular structures with improved performances. While fluid mechanics studies simple fluids in complex flow, rheology looks at complex fluids under simple flow. Experimentally, the simplest strain is the simple shear. The sample is held between two surfaces separated by a small gap. One surface is mobile and the other stationary (Figure 8.1a). Shear strain γ is defined as the ratio of the displacement over the gap between the two surfaces that hold the sample. Figure 8.1 shows simple shear experiments in the frequency domain (Figure 8.1b) and in the time domain (Figure 8.1c and Figure 8.1d). When a sinusoidal strain is applied at an angular frequency ω (rad/s), the resulting stress (Pa) oscillates at the same frequency, but with a phase shift δ. The stress can be decomposed in an in-phase component, the storage modulus G', and an out-of-phase component, the loss modulus G''. The loss tangent is $\tan\delta = G''/G'$. In a relaxation

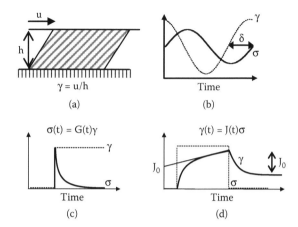

FIGURE 8.1 Simple shear experiments. (a) Simple shear strain γ; (b) Dynamic experiment under sinusoidal strain yields $G^*(\omega)$; (c) Relaxation experiment after a step strain yields $G(t)$; (d) Creep experiment followed by a creep recovery experiment yields $J(t)$.

experiment, the relative stress $G = \sigma/\gamma$ induced by an instantaneously set constant strain γ is measured over time. $G(t)$ is called the relaxation modulus. Alternatively, when a step stress σ is applied to the sample (creep experiment), the resulting strain normalized by the stress is the compliance $J(t)$ (Pa^{-1}). For a fluid, the steady state flow is reached when the rate of strain $d\gamma/dt$ becomes constant (equal to σ/η). If one stops applying the stress (creep recovery experiment) the elastic part of the deformation is recovered. It is called the recoverable compliance J_0.

All these experiments contain the same information, which can be summarized by two rheological parameters: the first one characterizes the elasticity ($G_0 = G(0)$ or $1/J_0$) and the second one is the characteristic time of the relaxation dynamics, τ. The zero shear viscosity η_0 (in Pa·s) is well approximated by the product $G_0\tau$. Exact relations exist between $G(t)$, $J(t)$, and $G^*(\omega) = G' + iG''$.[20] For example, $G_0 = k/J_0$, where k depends on the width of the relaxation spectrum of the system ($k = 1$ for Maxwellian systems). To characterize materials, experiments have to be performed in the linear regime, which corresponds to small amplitude strain, and where the response is proportional to the stimulus. The normalized response functions, $G(t)$, $J(t)$, or $G^*(\omega)$ do not depend on the amplitude of the imposed stimulus.

The relaxation modulus of entangled polymeric systems involves phenomena over a very broad time scale and this makes log-log representation more suitable. One generally distinguishes three regions in the relaxation spectrum of polymeric systems: the first one, the transition region at short time scale, does not exist in entangled wormlike micelle systems. Then follows an intermediate region, where the relaxation modulus decreases very slowly and looks constant in log-log representation. This so-called plateau region where polymeric systems exhibit an

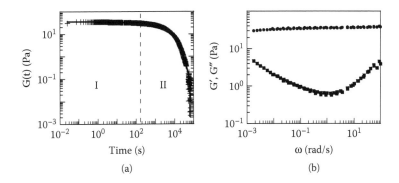

FIGURE 8.2 (a) Relaxation modulus of a 5% w/w 16-3-16, 2Br solution at 45°C. I: Plateau region; II: Terminal region; (b) Loss modulus G''(■) and storage modulus G'(●) versus angular frequency for the same sample in the frequency domain.

elastic behavior and temporarily store energy is typical of entangled polymers or wormlike micelle solutions (Figure 8.2). At this time scale, each wormlike micelle feels topological restrictions imposed by the surrounding wormlike micelles, and the systems behave like a rubber. This region is characterized by the corresponding value of the relaxation modulus, namely the plateau modulus G_0. The plateau modulus is not an equilibrium modulus and polymeric liquids store energy only temporarily. Strained entanglements can eventually be resolved by diffusive motion of the chains along their own contour (reptation). That happens in the so-called terminal region, which is characterized by the terminal relaxation time τ. Figure 8.2a illustrates the above explanations with data obtained for a wormlike micelle solution of the gemini surfactant 16-3-16, 2Br. Data obtained on the same sample but in the frequency domain are presented in Figure 8.2b. The storage modulus G' is almost constant over the whole accessible frequency range and the loss modulus G'' goes through a minimum. For this sample, the terminal region is not accessible in the frequency domain. It is necessary to integrate the relaxation modulus $G(t)$ over the entire time scale to get the zero shear viscosity η_0.[19,20]

More often, data obtained on wormlike micelles look like the ones in Figure 8.3. Both plateau region and terminal region are accessible. Several features in this rheogram are typical for Maxwellian behavior: first the G' plot crosses the G'' plot when the latter is maximum. Second, the maximum value of G'' is half that of the plateau modulus (taken as the G' value at the angular frequency ω_{min} where G'' is minimum and noted G_N thereafter). η_0/G_N (which is the number average relaxation time $\langle\tau\rangle_n$)[19,20] and $1/\omega_{G'=G''}$ are equal within 5%, and confirm that the relaxation spectrum is very narrow and that the system is Maxwellian. Figure 8.3b presents a Cole–Cole plot (G'' versus G') of the same data.[12–22]

Figure 8.4 presents data for a solution of another gemini surfactant, the 12-3-12, 2Br. The terminal region spans almost the entire accessible range of frequencies. At low frequency, G' and G'' vary respectively like ω^2 and ω^1. The

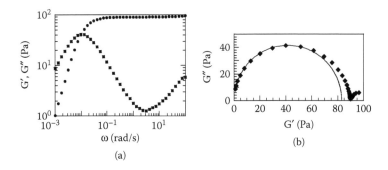

FIGURE 8.3 (a) Loss modulus G″(■) and storage modulus G′ (●) vs. angular frequency for a 6% w/w 12-3-12-3-12, 3Br solution at 25°C; (b) Cole–Cole representation of the same data.

frequency at which the plots intercept is still observable and its inverse corresponds to an average relaxation time. The complex viscosity can be accurately extrapolated to zero frequency to yield the zero shear viscosity η_0 (Figure 8.4b). The elasticity of the system is described either by the product $G_0 = \eta_0 \omega_{G'=G''}$ or by the inverse of recoverable compliance J_0, obtained by extrapolation of J' (the real part of $J^*(\omega) = 1/G^*(\omega)$) to zero frequency (Figure 8.4b). J_0 is often the only accessible parameter to characterize the elasticity of low concentration samples. $J_0\eta_0$ is the weight average relaxation time $<\tau>_w$[19,20] and is slightly larger than $1/\omega_{G'=G''}$. This indicates that the system is not strictly Maxwellian. Fits of similar data to the Maxwell model are often reported in the literature. However, it must be realized that this procedure gives more weight to the high frequency data that are often biased by inertia effects, particularly for low concentration samples.

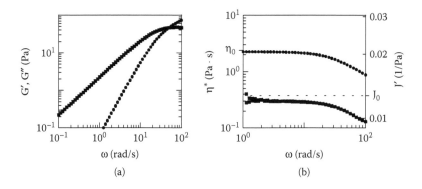

FIGURE 8.4 (a) Loss modulus G″(■) and storage modulus G′ (●) vs. angular frequency for a 9% w/w 12-3-12, 2Br solution at 25°C; (b) Complex viscosity (●) and real part of the complex compliance (■) versus angular frequency from the same experiment.

Low frequency data can be more reliable than high frequency ones, despite the fact that they are obtained closer to the limits of the rheometer sensitivity and phase shift resolution. Extrapolating Cole–Cole plots to determine the elastic modulus is correct for highly elastic samples, but overestimates the elasticity of low concentration samples, leading to wrong scaling exponents.

Viscoelastic properties of wormlike micelles have also been obtained from dynamic light scattering studies (DLS).[23-26] DLS measures the lifetime of the fluctuations of concentration, which are coupled with the viscoelasticity of the medium. The time autocorrelation function of the scattered field is found bimodal. The slower relaxation process can be accounted for by the theoretical models describing the coupling of concentration fluctuations to viscoelasticity. It has been shown for different systems that the dependence of the slow viscoelastic mode on temperature, ionic strength, and concentration is the same as that of the terminal time of the stress relaxation measured by a conventional rheometer.[23-26]

Diffusing wave spectroscopy (DWS), another light scattering-based method, has been applied to characterize the viscoelastic properties of wormlike micelle solutions up to 10^5 rad/s.[27,28] DWS consists in following the mean square displacement of a probe particle and analyzing it through a generalized Stokes–Einstein relation, which involves the complex modulus G*. With a multidetector such as a CCD camera, signal accumulation time can be shortened because time averaging can be compensated by multispeckle averaging. This is of great practical use for slow systems and broadens the accessible range of frequency toward lower frequency.[27] In another method based on laser interferometry, lateral displacements of embedded particles under thermal fluctuations are followed by optical microscopy in differential interference contrast.[29] This method has been applied to wormlike micelles and allows measurements over six decades in frequency up to about 100 kHz. The method was validated by comparison with macrorheological experiments carried out on a custom-built piezorheometer.[29]

A complete understanding of the macroscopic viscoelastic behavior of wormlike micelles requires other tools of investigation, and rheological studies have often been associated with other methods. Several of them (small angle scattering, transmission electron microscopy, temperature-jump, etc.) are described in this volume. Other ones include birefringence measurements,[30] dielectric relaxation,[31] or NMR.[32]

8.2.2 MODELS FOR THE RELAXATION OF LIVING POLYMERS

The striking features of the viscoelastic properties of surfactant wormlike micelles solutions are the following: First, the plateau region of their relaxation spectrum is very similar to that of entangled polymers. Second, the terminal region is characterized by a very narrow relaxation spectrum despite the high polydispersity of the micelle size. The transient nature of the wormlike micelles and the role of reversible scission were early invoked to explain this peculiar behavior.[33]

Shikata et al.[17] argued that if a diffusion process prevailed in the relaxation mechanism, the terminal relaxation time would be strongly dependent on the

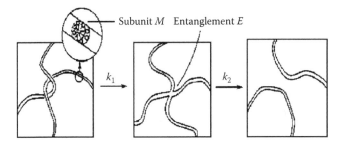

Subunit M Entanglement E

k_1 k_2

FIGURE 8.5 Chemorheological process of disentanglement. (Adapted with permission from Ref. 74. © 2001 by Elsevier.)

wormlike micelle length and on concentration, contrary to their results. They qualitatively explained their results by the quasi-network model schematically depicted in Figure 8.5. The elasticity in the plateau region would result from the formation of entanglements described as a second-order reaction between two strands. The order of the reaction would explain the concentration dependence of the plateau modulus ($G_0 \propto C^2$). The relaxation time of the terminal region would be determined by the kinetics of resolution of the entanglement that takes place in a second step. For the second step, two possible loci for the scission of the micelles have been considered: anywhere along the cylinder or preferentially at the entanglement junctions. The authors were led to conclude that the scission occurs at the junctions because no endcaps were observed in their TEM pictures.[17] Micelles would thus be able to pass through each other like ghosts by temporarily merging and then breaking apart. The scission step is a zero-order reaction, which rate could be somehow increased by an excess of counterions. This process corresponds to bond interchange in Cates model (Chapter 4). It did not receive due attention but recent simulations support it (see Chapter 3 and below). However, without knowledge of the rate constant of each step it is difficult to formulate the viscoelastic properties more specifically and this model remains essentially qualitative.

At the same time, Cates showed that the relaxation modulus can be monoexponential while reptation still has a significant contribution to the relaxation process.[16] The terminal relaxation time τ could result from the interplay between the dynamics of polymer diffusion (with characteristic time τ_{rep}) and the kinetics of scission/recombination (with characteristic time τ_b).[16,22,34,35] Different schemes of recombination were considered (see Chapter 4). The ratio $\zeta = \tau_b/\tau_{rep}$ determines the broadness of the relaxation spectrum. When the micelle lifetime is long ($\zeta \gg 1$) the limiting step for relaxation is the reptation of the wormlike micelle over its whole length as in classical polymers. The smearing with an exponential distribution of size yields a stretched exponential relaxation modulus. When the micelle lifetime is short ($\zeta \ll 1$) the relaxation time is shorter and the spectrum very narrow. Considering reversible scission as the sole chemical path for the length evolution, it is conveniently assumed that a wormlike micelle can break with equal probability per unit time and per unit length, at any point along its length.[16,34] The frequency of scission is then proportional to the average micelle length $<L>$, and given by $1/\tau_b \propto <L>$. It is also assumed that no correlation exists between a scission and the recombination of the resulting ends. The terminal relaxation process is then characterized by a single time $\tau = (\tau_b \tau_{rep})^{1/2}$.

Corrections to monoexponential behavior have been computed for the intermediate case where $\zeta \approx 1$ and allow estimation of τ_b from viscoelastic data.[34,35] Further corrections to the Maxwell behavior have been introduced to describe relaxation at shorter time scale, when the diffusive process is no longer reptation but Rouse motion.[35-37] The crossover to this time scale is reflected in the frequency domain by a minimum in the $G''(\omega)$ plot as seen in Figure 8.3. The inverse of the loss tangent at this frequency, G_N/G''_{min} is related to the ratio $<L>/l_e$ (see Section 8.4.1.3), where l_e is the entanglement length.

Scaling laws for concentration (C) dependence of the rheological properties have been proposed in the frame of the reversible scission-reptation model[38,39] and many experimental studies were later carried out to test them.

Extension of this model to weakly branched wormlike micelles[40] has been proposed to quantify the conjecture that branching would decrease the viscosity.[41] Recall that in the reptation model, the dynamics of relaxation of the stress is considered to be equivalent to the kinetics of annihilation of defects when they meet chain ends after diffusion onto the chain. In classical polymers, a branching point is an insurmountable barrier for the diffusion of defects. In wormlike micelles, owing to the fact that junctions can slide, the main consequence of branching is essentially to increase the number density of ends. Therefore, when the micelles are branched, defects have to diffuse smaller distances to reach an end and the system would relax as if the micelles were shorter.[41]

The coupled scission-reptation model involves several assumptions that are important to keep in mind. First, the chain scission probability is supposed to be uniform along the contour of wormlike micelles. Second, the predominant mode of stress relaxation is reptation, and constraint release by tube renewal is neglected. Third, the chain scission and recombination events are uncorrelated, meaning that two newly created ends have equal probability to recombine with one another (self-recombination) as with other ones around.

The first assumption leads to the conclusion that the breaking frequency is proportional to $<L>$. The dependence of the prefactor on concentration may have a strong influence on the concentration dependence of the viscosity and remains to be specified. The second assumption is still questionable: why should the reptation phenomenon, which accounts for diffusion of polymers at large scale while preserving their topological integrity, be relevant for wormlike micelles whose topology is transient? The third assumption has been shown to fail[42] especially in situations where wormlike micelle ends are very rare, as for example, at high concentration. In that case, self-recombination of the two newly created ends is more likely at short time. This seems to be confirmed by simulation.[43]

Recent particle based simulations[44,45] suggest that under stress, wormlike micelles brought to contact at an entanglement point, fuse rather than break (see Chapter 3). Bond interchange would be the dominant process of recombination and this is more consistent with the fact that chain ends involved in the other processes are extremely rare. The four-armed branch point resulting from a fusion event could split into two threefold junctions[44] (see Figure 3.3, Chapter 3). It could also break, and the relative frequency of scission and fusion at an entanglement point depends on the set of forces introduced to simulate the Newtonian mechanics of the interacting particles and on the boundary conditions of the simulation.[45] The interesting point here is that the activated state in the process of stress relaxation is not the formation of two ends but the formation of junctions, which correspond to smaller curvature than ends or cylinders. This reconciles the puzzling fact that triggering an increase of the micelle length often leads to lower viscosity due to faster relaxation, because the triggering tricks consist essentially in reducing the preferred curvature of the micelles.

8.3 RHEOLOGICAL PERFORMANCES OF WORMLIKE MICELLE SOLUTIONS

The viscoelastic behavior of micellar solutions shows a considerable chemical specificity with respect to the surfactant structure, counterion nature, and presence of additives. In dilute solutions the micelle morphology is determined by the packing parameter, which results from the surfactant molecular structure and from the interaction between surfactants.

In most cases addition of salt is necessary to obtain giant micelles. The specificity of a simple salt to induce the growth of ionic surfactant micelles is related to the tendency of the added counterion to bind to the micelle surface. Organic counterions often strongly bind and can be considered as cosurfactants since they partially penetrate into the micelle core. Addition of oppositely charged surfactant ions (catanionic systems) also reduce the preferred curvature most often leading to vesicles but also to wormlike micelles.

Two situations have to be distinguished when salts with binding organic counterions or oppositely charged surfactant ions are involved. They can be simply added to a solution of ionic surfactant. In that case, the composition of the system is specified by the concentrations of surfactant C and of added salt C_S. In the other cases, the binding counterion or the oppositely charged surfactant substitutes the initial counterion and no excess salt is present (salt-free systems).

Beyond a certain concentration C^*, the zero shear viscosity η_0 increases rapidly with C, sometimes by several orders of magnitude. In the literature, the C-dependence of η_0 is often fitted by a power law: $\eta_0 \propto C^\alpha$. In many cases η_0 goes through a smooth maximum at a concentration C^{**}. The rheological performances of a system can be characterized by the relative viscosity η_R (viscosity of the sample divided by the solvent viscosity at the same temperature) at C^{**} and by the minimum value of the loss tangent $\tan\delta_{min}$. As new applications concern high temperature conditions, the activation energy E_a related to the temperature

TABLE 8.1
Viscoelastic Characteristics of Various Surfactant Solutions

Systems	C^*	C^{**}	$\eta_{rel\ Max}$	$\tan\delta_{min}$	Ref.
CTAB, 30°C	12%	20%	3.5×10^3	—	51
CTAB, 0.25 M KBr, 30°C	< 3%	—	$> 10^5$	—	46
CTASal, 20°C	0.13%	—	$> 10^6$	—	80
CPyCl+NaSal 1:1,[a] 25°C	3 mM	9 mM	10^4	0.025	11
CTAC$_7$SO$_3$, 20°C	0.24%	0.79%	10^5	0.080	81
CTAB+CTAHNC 0.95:1,[a] 55°C	4%	15%	1.5×10^3	—	82
CTATos, 30°C	0.6%	2–3%	2×10^5	0.05	73
CTAC+NaTos 1:0.8,[a] 30°C	1mM	400 mM	3.5×10^5	—	76
CTA3,5DClBz, 25°C	0.1%	20 mM	10^6	0.024	70
EHAC+NaSal 1:0.5,[a] 60°C	—	5–6 mM	10^7	0.016	88
EHAC (IPAF), 0.4M KBr, 25°C	0.05%	—	10^6	0.18	90
FOSTEA C$_8$F$_{17}$SO$_3$N(C$_2$H$_5$)$_4$	10 mM	85 mM	6.5×10^4	—	1, 116
12-2-12, 2Br, 25°C	1.6%	7%	2×10^5	0.05	39
12-3-12, 2Br, 25°C	4%	10%	10^4	—	99
12-3-12-3-12, 3Br, 25°C	1%	5%	10^7	0.014	99
ChEO$_{10}$+C$_{12}$EO$_3$ 6:4,[a] 25°C	—	60 mM	2×10^6	0.08	132
ChEO$_{15}$+NMEA-12 4:6,[a] 25°C	—	60 mM	10^5	0.1	133

[a] Molar ratio.

dependence of the viscosity $\eta_0 = \eta\exp(-E_a/k_BT)$ is also a good indicator of the rheological performances. Some of these indicators are listed in Table 8.1.

8.3.1 WORMLIKE MICELLES OF CATIONIC SURFACTANTS

Cationic surfactants have been most used for basic research on the rheological properties of wormlike micelles. Cetyltrimethylammonium bromide (CTAB) and cetylpyridinium chloride (CPyCl) have been associated with various additives and salts with organic counterions. Later on, as higher performances were sought, longer chain surfactants received much attention. Also studies of new model systems such as gemini surfactants provided useful insight into the rheological properties of wormlike micelle solutions.

8.3.1.1 Cationic Surfactants with Simple Salts

The rheological properties of CTAB have been studied in a wide range of C and C_S, and at several temperatures.[15,21,46,47] The systems were found to be Maxwellian and this suggested that scission/recombination reactions contribute to the relaxation process. The elastic modulus G_0 (or G_N) scales like $C^{9/4}$ as expected for entangled linear polymers. The concentration dependence of the zero shear viscosity could also be expressed as a power law: $\eta_0 \propto C^\alpha$, with α decreasing when the ionic

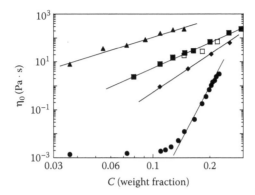

FIGURE 8.6 Zero shear viscosity η_0 of CTAB solutions vs. surfactant concentration C, at different KBr concentrations. ($\bullet$) no KBr;[51] ($\blacklozenge$) 0.1 M;[15] ($\blacksquare$) 0.25 M;[46] ($\square$) 0.25 M;[47] ($\blacktriangle$) 1.5 M.[21]

strength increases (Figure 8.6). In the absence of added salt, $C^* = 12$ wt% is quite high but then the concentration dependence of η_0 is very strong ($\eta_0 \propto C^{12}$).[51] Two explanations were given to the high exponent obtained at low ionic strength:[15] (i) the reptation would be the main mechanism of relaxation but the corresponding time would be determined solely by the average length and would not be affected by the distribution of length, and (ii) the higher exponent results from the simultaneous increase in volume fraction and ionic strength. At intermediate salinity (0.25 M KBr)[46, 47] $\alpha \approx 3.5$, as expected by Cates model,[38] but depends on temperature.[15] At high salt content (1.5M KBr), $\eta_0 \propto C^{2.5}$. This weaker concentration dependence has been interpreted as due to branching.[21] A similar decrease of α with salt concentration has been observed with CPyClO$_3$ in the presence of NaClO$_3$.[48]

The relative viscosity η_R of CTAB solution in brine can reach 10^6 with relaxation times of the order of seconds. The temperature dependence of the relaxation time obeys Arrhenius law: $\tau = \tau_0 \exp(-E_\tau/T)$. E_τ decreases from $18k_BT$ to $6k_BT$ as the salt concentration increases.

The breaking time τ_b was estimated[46,47] (see Section 8.4.2) and found to increase with concentration, while T-jump experiments[49] yielded an opposite result ($\tau_b \propto C^{-1/2}$). This was interpreted as due to the fact that at high concentration the mesh size of the entanglement network becomes shorter than the persistence length. This should have led to phase separation,[50] and concentrations are actually very close to the transition to nematic phase in CTAB.[51,52]

The counterion tendency to bind onto the micelle surface and to promote micelle growth goes in the sequence ClO$_3^-$ > NO$_3^-$ > Br$^-$ > Cl$^-$ > F$^-$.[53,54] Even though NO$_3^-$ [54,55] and ClO$_3^-$ [41] are better triggers than Br$^-$ for micelle growth, they do not always lead to an increase in viscosity. Moreover, when salt is added to cationic surfactant solutions (KBr to CTAB,[48] NaClO$_3$ to CPyCl,[56] NaNO$_3$ to CPyCl,[55] NaClO$_3$ to CPyClO$_3$[24,41]) at fixed C, the viscosity goes through a pronounced maximum with increasing C_S (Figure 8.7). The decrease of viscosity

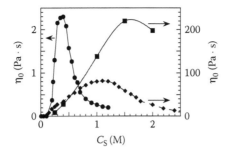

FIGURE 8.7 Zero shear viscosity η_0 versus added salt concentration C_S at fixed surfactant concentration. (●) CPyCl (0.3 M)/NaClO$_3$, at 32°C;[56] (◆) CTAB (0.3 M)/NaNO$_3$, at 30°C;[55] (■) CTAB (0.35 M)/KBr, at 35°C.[48]

upon addition of NaNO$_3$ to CPyCl solutions was attributed to a decrease in micelle size because the plateau modulus also decreased.[55]

The solutions of CPyClO$_3$ micelles in 0.1 and 1 M NaClO$_3$ brine have been studied by light scattering,[24] rheology,[41,56] and fluorescence recovery after bleaching.[57] This system is the first one for which branching was proposed to explain the observed low viscosity. At 20% w/w, η_R reaches 10^4 at 35°C in 0.1 M NaClO$_3$ (with $\eta \propto C^2$) but does not exceed 5×10^2 in 1 M NaClO$_3$ (with $\eta \propto C^1$). In order to reconcile the structure revealed by scattering techniques and the fluidity of the system, a multiconnected (rather than an entanglement) network structure has been proposed. Upon shearing, the induced stress could relax almost instantaneously by the sliding of the connections through the viscous flow of surfactants, as suggested in the sponge phase.[58] The mechanism proposed by Shikata[17] was also invoked[57] and was actually retained in the most recent study of this system.[24]

8.3.1.2 Cationic Surfactants with Organic Counterions

Many organic anions, especially aromatic ones, have been used to induce micellar growth of CTA or CPy micelles. Salicylate (Sal) and alkylbenzoate are among the most efficient ones and are often used in pharmaceutical or personal care products. Hydrophobic counterions influence the micelle microstructure in a way that strongly depends on their fine structure. For example, methyl *para*benzoate is much more efficient at increasing the viscosity of CPyCl solutions than the *meta* isomer, while the *ortho* isomer has almost no effect.[6] Hydroxybenzoate ions trigger the sphere-to-rod transition with an efficiency that varies in the order: 2-hydroxy > 3-hydroxy > 4-hydroxy benzoate.[59–61] The reverse sequence is found for chlorine-substituted benzoates: 4-chlorobenzoate induces viscoelasticity while 2-chlorobenzoate does not.[62]

Sodium Salicylate. NaSal has been added to CTAB[12,17] as well as CPyCl[13] solutions. In both instances, the relaxation time varies from tens of milliseconds to hundreds of seconds and η_R reaches 10^6 for some compositions. Viscoelastic

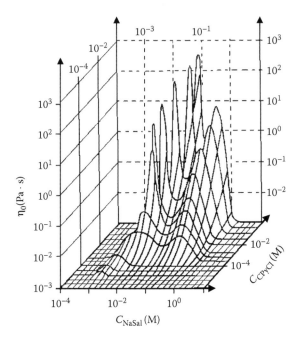

FIGURE 8.8 Zero shear viscosity η_0 of aqueous CPyCl solution vs. concentration C_{CPyCl} and vs. sodium salicylate concentration C_{NaSal} at 32°C. (Adapted with permission from Ref. 13. © 1988 by American Chemical Society.)

properties of CTAB/NaSal solutions have been reported in a wide range of concentration and of molar ratio [NaSal]/[CTAB] = C_S/C.[12,17] At constant C, the relaxation time varied with C_S in a complicated way, with two maxima and one minimum. The same complexity was observed in the rheological behavior of CPyCl/NaSal solutions (see Figure 8.8).[13] The zero shear viscosity η_0 goes through a maximum when increasing C at constant C_S. It goes successively through a maximum, a minimum, and a second maximum and then goes down to values close to that for water, when increasing C_S at constant C.[13] The solutions become Maxwellian just at the first maximum. The shear modulus is then independent of the ratio C_S/C and solely dependent on C.[13] This crossover to Maxwellian behavior was interpreted as due to a crossover to a regime where reversible scission dominates the relaxation process.[12,17,13]

The studies of the CTAB/NaSal system can be summarized as follows.[12,17] At low C_S, the ratio C_S/C is the relevant parameter to account for the rheological properties, because it describes the composition of the micelle that determines their growth. At $C_S/C = 1$ micelles are saturated with salicylate ions. Beyond, the additional salicylate ions remain free in solution and mediate the first step of the relaxation process described as a ghost-like crossing of strands of the network (bond interchange), as shown in Figure 8.5.[12,17] The rheological characteristics of the system

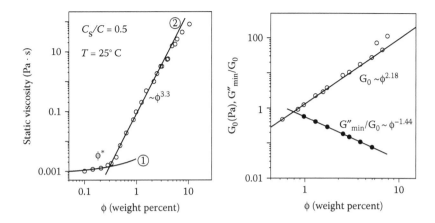

FIGURE 8.9 Rheological properties of aqueous solutions of CPyCl/NaSal at ratio $C_S/C=0.5$, in water + 0.5M NaCl at 25°C. Weight fraction dependence of (left) the zero shear viscosity η_0 and (right) the elastic modulus G_0 and the minimum in loss tangent G''_{min}/G_0. (Reproduced with permission from Ref. 64. © 1993 by American Chemical Society.)

become independent of C_S/C: the elasticity depends only on C, while the relaxation time depends only on the free salicylate ion concentration that is, $C_S - C$.

Simple relations between C_S and C at the maxima and at the minimum in viscosity were pointed out in CPyCl/NaSal solutions.[13] The decrease of viscosity at high C_S/C ratio was attributed to the shortening of the micelles resulting from an inversion of the sign of the micelle charge, suggested by NMR experiments.[63] In the presence of an additional simple salt such as NaCl, the concentration dependences of the rheological characteristics of CPyCl/NaSal solutions at $C_S/C = 0.5$,[64] are very close to those expected theoretically[38] (see Figure 8.9).

Additional studies have been performed at constant C_S/C ratio and increasing C. The viscosity of CTAC/NaSal solutions at $C_S/C = 0.6$, varied nonmonotonically with C, reaching a maximum at about $C = 0.2M$ and then decreasing rapidly.[65] The same nonmonotonic behavior was observed in the presence of NaCl.[66] It was noticed that a certain concentration must be reached for the system to become Maxwellian. Moreover increasing temperature broadened the relaxation spectrum in the terminal region.[66] Understanding the effect of the salt content requires the knowledge of the dependence of the rate constants of reversible scission on the salt content. The results of this study again suggest that they increase with ionic strength.[66]

Replacing NaSal by HSal (salicylic acid) resulted in a biexponential relaxation process at high HSal/CTA concentration ratio.[67–69] The short time corresponds to that observed with NaSal and depends only on the free HSal. The longer time (by a factor 10 to 100) is more strongly dependent on the free HSal but also depends on C. NMR measurements suggest that HSal molecules reside at two different locations once incorporated into the micelles.[67] Up to a ratio of

0.5, the location is the same as NaSal. At higher ratio, HSal is less ionized and may enter deeper in the micelle. At the same ratio of Sal/CTA, additions of NaBr considerably decrease the relaxation time of HSal-based systems but have almost no effect on NaSal-based systems.[68] This was interpreted as due to the difference in micelle charge, neutral with NaSal and positive with HSal. This charge hinders the fusion of micelles needed to relax the mechanical stress. It would also explain the existence of a slow mode of relaxation. Ion specific electrodes showed that 2Br⁻ ions are released from CTAB micelles upon adding one HSal to the solution, while 1Br⁻ ion is released by one added NaSal.[68] The same conclusions were drawn in a study of CTAC/HSal system in salted water (0.1 M), which combined rheological measurements with cryo-TEM.[69]

Chlorobenzoates (ClBz). Highly viscoelastic wormlike micelle solutions have been obtained with CTA mono- and dichlorobenzoates.[70] The three organic anions having chlorine substituents in *para* and/or *meta* with respect to the carboxylate group (3,5-diClBz, 3,4-ClBz, and 4,5-diClBz) give rise to fully entangled worm-like micelles below 5 mM. The plateau modulus G_0 scales as $C^{9/4}$ and the value of $\tan\delta_{min}$ is as low as 0.024. η_0 goes through a smooth maximum, where η_R reaches 10^6 and then decreases down to 2×10^3. This nonmonotonic behavior is a common feature of salt-free systems of wormlike micelles. For CTA3,5-diClBz, $C^* \leq 2$ mM (0.1%). Relative viscosities of CTA2,6-diClBz and CTA2-ClBz solutions do not exceed 100.[70] The rheological behavior of CTAC in the presence of sodium 3,4-diClBz, 4,4-dimethylbenzoate and 3,5-dichlorosalicylate has also been reported.[71]

Tosylate (Tos). The linear viscoelastic properties of CTATos solutions were determined from $C = 1\%$ up to $C = 27\%$.[72,73] Beyond, the system crosses over to a hexagonal phase. At 30°C, above $C^* = 0.6\%$, the viscosity increases with C and becomes constant ($\eta_R = 2 \times 10^5$) at 3%. The relaxation time goes through a maximum at a concentration that shifts from 2 to 4%, when the temperature increases from 25°C to 40°C. The influence of free counterions on the terminal relaxation time has also been studied in systems containing NaTos.[74–76] Adding NaBr, NaTos,[77] or NaCl[78] to CTATos solutions decreases the relaxation time. The plateau modulus is also influenced by the salt concentration.[77,78]

8.3.1.3 Catanionic Systems

Mixtures of oppositely charged surfactants are well known to show synergistic enhancement of rheological properties.[79] Catanionic systems are prone to form vesicles but wormlike micelles are obtained by either increasing the temperature or finely tuning the composition (for example by adding spherical micelle-forming surfactants). The linear viscoelastic behavior of micellar solutions of CTAX surfactants, where the counterion X is a surfactant anion such as 3-hydroxy-2-naphthalenecarboxylate (HNC)[80] or alkylsulfonate ($C_nSO_3^-$),[81] has been reported. The mixtures CTAHNC/CTAB[81,82] and CTATos/sodium dodecylbenzenesulfonate (SDBS)[83,84] have also been characterized.

From the rheological point of view, catanionic mixtures behave like salt-free systems of fluorinated (Section 8.2.2) or gemini (Section 8.3.1.5) surfactants: the relaxation time varies nonmonotonically with concentration going through a maximum, which is quite shallow for CTAHNC at high temperature, but rather pronounced for $CTAC_nSO_3$.[81] The strong increase in viscosity at low concentration has been associated to the rapid growth regime theoretically expected for charged wormlike micelles.[85] It is believed to result from a proper combination of endcap energy E_c and charge density of the micelle surface[81] (see Section 8.4.1.1 for a discussion on this interpretation). The decrease of η_0 beyond C^{**} is attributed to either branching or ionic strength dependence of the reversible scission kinetics, since addition of salt speeds up the relaxation process.[81] It was also noticed that the values of C^{**}/C^* fall between 3 and 6 for the systems investigated.[81]

Adding NaBr increases the viscosity at low C but decreases it at high C,[81] a result already reported in simpler systems (see Figure 8.7). The viscosity of CTAOH solutions increases by up to six orders of magnitude upon addition of 2-hydroxy-1-naphthoic acid (2,1-HNC) as the molar ratio R =[2,1-HNC]/[CTAOH] is raised from 0.5 to 0.6.[86] It remains constant up to R = 0.70 and then decreases as the system approaches the phase separation composition (R = 0.77).

Dynamic viscoelastic experiments have been performed on CTATos/SDBS[83] and CTATos/CTAHNC[87] mixtures in order to understand the effect of the counterion hydrophobicity on the aggregate structure. In the CTATos/CTAHNC system at C_{total} = 0.1 M, the zero shear viscosity, η_0, and the stress relaxation time, τ, varied nonmonotonically with the HNC mole fraction x. The maximum of η_0 corresponds to a 5-order of magnitude increase and is reached at $x = 0.3$. η_0 then decreases by 2 orders of magnitude. The elastic modulus remains constant up to $x = 0.5$ and then increases rapidly from 50 to 200 Pa. This was interpreted as an increase of the persistence length of the wormlike micelles. The temperature dependence of the viscosity follows Arrhenius law, with an activation energy decreasing from $80k_BT$ to $10k_BT$ as x increases from 0 to 0.7.[87]

In the study of the CTATos/SDBS system,[83] at constant C_{total} = 1.5% and increasing SDBS molar fraction R = 100[SDBS]/([CTATos] + [SDBS]), η_0 and τ again go through a maximum at R = 4, while G_N is constant and $\tan\delta_{min}$ goes through a minimum at R = 5 (see Figure 8.10). This was interpreted as due to branching. When C_{total} is increased at constant R = 3, η_0 and τ vary nonmonotonically with a maximum at C_{total} = 2%. The authors retain the hypothesis of a shortening of the micelles as the most probable explanation for this observation. However, G_0 varies as expected for fully entangled linear polymers.

8.3.1.4 Cationic Surfactants for Higher Performance

With the new application of viscoelastic surfactant solutions as fracturing fluids (Chapter 15), the monounsaturated long-chain (C_{22}) cationic surfactant erucyl-bis(hydroxyethyl)methylammonium chloride (EHAC) has recently received considerable attention.[88–93]

As with other cationic surfactants, organic counterions are more efficient in triggering high viscosity. At 60°C, simple salts increase η_R to 4×10^4,[92] while

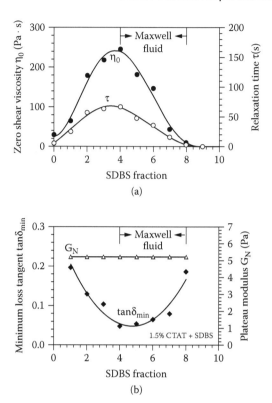

FIGURE 8.10   Rheological characteristics of the CTATos/SDBS system at a fixed CTA-Tos concentration. (a) Zero shear viscosity and relaxation time versus SDBS weight fraction (percent); (b) Plateau modulus G_N and minimum loss tangent versus SDBS weight fraction (percent) at 25°C. (Adapted with permission from Ref. 83. © 2000 by American Chemical Society.)

NaSal raises it to 10^7. The zero shear viscosity varies nonmonotonically with concentration at a fixed ratio $C_S/C = 0.5$ and an optimal composition consisted of 2.7% EHAC plus 0.5% NaSal.[88]

The viscosity goes through a maximum upon increasing the concentration of added NaCl[88,89] or KCl[90] at a fixed EHAC concentration. Further increase in simple salt concentration leads to phase separation.[89] Upon addition of counterions such as Sal,[88] Tos,[89] or HNC,[93] η_0 shows a complex behavior with two maxima flanking a minimum,[89] as reported for CPyCl/NaSal and CTAB/NaSal systems (Section 8.3.1.2). Moreover, a two-phase region is observed around the minimum in viscosity. (A miscibility gap for a certain range of C_S/C ratio was also reported for the CPyCl/alkylbenzoate system.[6]) Cryo-TEM imaging suggests that the maximum in viscosity is due to branching.[90] However branching was observed between stretched micelles and could be transient structures resulting from the splitting of four-arms junctions after fusion at entanglement points (see Figure 3.3, Chapter 3). Branching has also been proposed to explain phase separation in EHAC/organic counterion systems upon heating.[89,93] This behavior is reminiscent of the one of nonionic surfactants and could be interpreted as an increasing of branching density.[94]

In EHAC/NaSal[88] and EHAC/NaHNC,[93] unusual trends were also found in the influence of temperature. Instead of decreasing monotonically over the entire temperature range, the viscosity shows an initial enhancement at low temperature followed by a subsequent decrease at higher temperature. This initial increase is correlated with a growth of the contour length evidenced by SANS measurements and is explained by the increased solubility of NaHNC with temperature. At low temperature, NaHNC is poorly soluble and most of it forms mixed micelles with EHAC. As temperature increases, some NaHNC is released in the solution and the mixed micelles recover the optimum composition ratio for growth.[88]

The endcap energy of the EHAC/NaSal system at $C_S/C = 0.5$ is estimated at $65k_BT$.[88] The flow activation energy is about $80k_BT$ for EHAC/NaSal ($C_S = 0.06$M and $C_S/C = 0.5$).[88] It decreases to about to $36k_BT$ when the KCl concentration is increased from 0.3M to 1.6M and remains constant beyond.[90]

For practical reason EHAC is supplied mixed with isopropyl alcohol (IPA). The influence of the presence of IPA has been considered.[90] Also, adding $C_{18}EO_{18}$ to KCl solutions of EHAC shortens the micelles.[95] η_0 decreases by 4 orders of magnitude at the molar ratio $[C_{18}EO_{18}]/[EHAC] = 1$. A rheological study of an analog of EHAC with a shorter C_{18} chain with added NaSal has been reported[96] as well as a comparison of EHAC with erucyltrimethylammonim chloride.[88]

8.3.1.5 Model Cationic Surfactants: Gemini and Surfactant Oligomers

Gemini surfactants consist of two amphiphilic moieties tethered at the level of the head group.[97] The length of the tethering alkyl chain is variable. In quaternary ammonium gemini surfactants with a polymethylene spacer $(CH_2)_s$, referred to as m-s-m (m = alkyl chain carbon number), the average optimal surface area imposed by short tethers (s ≤ 3) happens to be just appropriate for wormlike micelle formation. Cationic gemini surfactants and their higher oligomeric analogs with short polymethylene spacers form wormlike micelles and display very strong viscoelasticity in the absence of any additives.[18]

The concentration dependence of η_0 of solutions of quaternary ammonium dimers, trimer, and tetramer[39,98,99] is shown in Figure 8.11. C^* decreases upon decreasing spacer length and increasing alkyl chain length or degree of oligomerization at constant s and m.[18,39,99] For the asymmetrical surfactants m-2-m', C^* increases with the difference m − m'.[100,101] Micellar growth is completely inhibited when m < m'/2.[100] The viscosity of long-chain gemini surfactants (up to C_{22}) has been reported.[102]

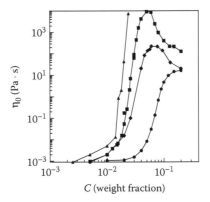

FIGURE 8.11 Weight fraction dependence of the zero shear viscosity of aqueous solutions of the surfactants 12-3-12, 2Br (●), 12-2-12, 2Br (◆), 12-3-12-3-12, 3Br (■), and 12-3-12-4-12-3-12, 4Br (▲) at 25 °C. (Reproduced with permission from Ref. 18. © 2004 by Marcel Dekker.)

The plateau modulus of 12-2-12, 2Br solutions was found to scale as C^3.[39] The exponent is higher than expected for classical polymers, and the difference has been interpreted in terms of electrostatic orientational correlations that reduce the elastic modulus and that are stronger at low concentration.[39] Measurements of the elastic properties of 12-3-12, 2Br and 12-3-12-3-12, 3Br solutions in a wider range of concentration revealed a more complex C-dependence (see Figure 8.12).[98] At low C, the experimental data lie below the universal curve and show a stronger dependence on C. The irregular increase of the elastic properties of the 12-3-12-3-12, 3Br solution has been attributed to the presence of branches that have been evidenced by cryo-TEM.[103,104]

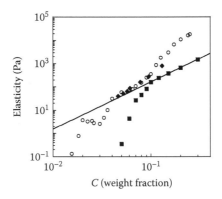

FIGURE 8.12 Weight fraction dependence of the elasticity of aqueous solutions of the surfactants 12-3-12, 2Br ($1/J_0$; ■);[98] 12-3-12-3-12, 3Br ($1/J_0$; ○)[98] and 12-2-12, 2Br (G_0; ◆)[39] at 25 °C. (Reproduced with permission from Ref. 18. © 2004 by Marcel Dekker.)

For the low concentration data lying below the scaling law, it has been proposed that a significant fraction of the wormlike micelles are still not entangled.[98] The concentration C_e where the elasticity data merge the universal curve would mark the onset of the fully entangled state. For cationic gemini systems, it coincides with the concentration C^{**} where the viscosity is a maximum (Figure 8.11 and Figure 8.12) and where the relaxation becomes Maxwellian.[18,98] Similar correlations were reported in interpreting results for the CTAB/NaSal[12] and CPyCl/NaSal systems.[13]

The addition of hexanol to aqueous solutions of 12-2-12, 2Br leads to the formation of vesicles.[105] This has the interesting consequence of making the zero shear viscosity of the system more robust against temperature increase. The transition from vesicles to wormlike micelles compensates for the activation of the flow and the shortening of the micelles that are both thermally induced.

The rheology of mixed wormlike micelles of 12-2-12, 2Br and CTAB in 0.15 M KBr has been studied as a function of the mole fraction x of 12-2-12, 2Br.[106] C^* goes through a minimum at $x = 0.5$. The terminal relaxation time goes through a shallow maximum at $x = 0.4$, at $C = 5\%$. The elastic shear modulus increases monotonically, from less than 20 Pa up to 100 Pa, as x increases from 0.25 to 0.75. The latter result has been interpreted as due to a decrease of the mixed micelle diameter upon increasing x, that is, content of the surfactant with the shorter alkyl chain. Indeed, for the same total surfactant weight fraction, the total contour length of wormlike micelles is larger for a shorter chain surfactant.[106] However, the elastic modulus seems to decrease below the value expected for pure CTAB system, and an alternative explanation could be that the shortening of the micelles lowers the fraction of entangled micelles.

8.3.2 WORMLIKE MICELLES OF ANIONIC SURFACTANTS

As with cationic surfactants, in most cases additives are necessary to get significant viscoelastic properties. They consist of multivalent cations (Ca^{2+} [107] or Al^{3+} [108,109]), organic cations such as p-toluidine hydrochloride (pTHC)[110], or nonionic[111] or cationic[112] cosurfactants.

Fatty alcohol ether sulfate (FAES)[26,113,114] yield viscoelastic solutions in the presence of salt and are currently used to enhance the viscosity of personal care products.

The long-chain soap sodium oleate, NaOA, has been studied in the presence of triethylammonium chloride (Et_3NHCl), or KCl,[115] and in catanionic systems.[112] The relative viscosity η_R of a 50 mM NaOA solution increases upon addition of KCl and levels off at 3×10^4 for 400 mM KCl. The same qualitative behavior is obtained by adding Et_3NHCl but η_R does not exceed 8×10^2. Et_4NBr and Bu_4NBr do not enhance the viscosity although they do bind to the micelle.

Wormlike micelles and bilayer structures are formed in mixtures of NaOA and cationic alkyltrimethylammonium bromide surfactants (C_mTAB).[112] The synergistic growth of wormlike micelles is maximized for m = 8. At $C = 3\%$, η_R reaches 2×10^6 for a weight ratio NaOA/C_8TAB = 70/30. The peak is shifted to lower fraction of C_8TAB when the total surfactant concentration increases or in salted water. $NaOA/C_6TAB$ mixtures display a much lower viscosity, while increasing m to 10 or 12 leads to phase separation over a broad range of composition.[112]

The influence of inorganic salts with different cations, $AlCl_3$, $MgCl_2$, $CaCl_2$, and NaCl, on the viscosity of sodium dodecyltrioxyethylenesulfate (SDES) micellar solutions was investigated.[109] The ability of cations to enhance the viscosity followed the order of $Al^{3+} > Mg^{2+} > Ca^{2+} > Na^+$. The relative viscosity of a 0.2 M SDES solution can reach $10^4 - 3 \times 10^4$ at 25°C. An optimum ratio C_S/C is observed as in cationic surfactant-based systems. The study concludes by pointing out the importance of the trioxyethylene spacer between the sulfate group and the alkyl chain, since the same cations have shown much less efficiency in enhancing the viscosity of SDS, SDBS, and other alkylsulfonates.

Maxwellian viscoelastic behavior was still reported for wormlike micelles formed in SDS solutions in the presence of $Al(NO_3)_3$.[108] SDS micelles also grow in the presence of pTHC.[110] Despite the very strong concentration dependence of $\eta_0 \sim C^9$, the viscoelastic properties are not pronounced and η_R does not exceed several hundreds.

Fluorinated surfactants such as tetramethylammonium and tetraethylammonium perfluorooctylsulfonate (FOSTEA)[1] are examples of additive-free viscoelastic systems of anionic wormlike micelles. FOSTEA shows an increase of η_0 by almost 5 orders of magnitude between $C^* = 0.1$ M and $C^{**} = 0.85$ M and then η_0 falls down to 3Pa·s at $C = 500$ mM.[1,116] The nonmonotonic behavior is also pronounced for the relaxation time with an increase by 3 orders of magnitude (from 40 ms to 4 s and down to 2 ms). The plateau modulus follows the scaling expected for entangled polymer solution $G_N \sim C^{9/4}$. The viscoelastic properties of this system depend strongly on the counterion[117] and are modified in the presence of an excess of salt. Free TEA counterions would play a similar catalytic role on the recombination of FOS wormlike micelles as Sal for wormlike micelles of CPy.[118]

The rheological behavior of solutions of the partially fluorinated anionic surfactant sodium 1-oxo-1[4-(tridecafluorohexyl)phenyl]-2-hexanesulfate shows complex concentration and temperature dependences.[119,120] This has been elucidated as due to the presence of small disklike micelles in equilibrium with wormlike micelles.[120]

8.3.3 WORMLIKE MICELLES OF ZWITTERIONIC SURFACTANTS

The rheological and micellar properties of viscoelastic solutions of alkyldimethylamineoxide (C_mDMAO) were studied as a function of concentration and in the presence of an ionic surfactant (SDS and $C_{14}TAB$).[121,122] The η_0 vs. C plot displays no maximum, but a clear break in slope is observed in a log-log representation. This has been interpreted as a switch of the dominant mechanism of the relaxation process, from diffusion-controlled to kinetics-controlled. η_R can reach 10^6 for $C_{16}DMAO$ at $C = 1$ M and $C_{18}DMAO$ at $C = 0.1$ M at 25°C.[121] The modulus shows the scaling behavior expected for entangled polymers, while the structural relaxation time τ is independent of concentration. The plateau modulus is little affected when the micelles are doped with SDS and $C_{14}TAB$, while η_0 and τ first increase and then decrease with increasing charge density. The viscoelastic properties are more affected by SDS than by $C_{14}TAB$. Adhesive contacts between the micelles are supposed to explain the strong increases of viscosity and the high values of τ (hundreds of seconds).[122] Protonation of $C_{14}DMAO$ has a marked effect on the viscoelastic behavior.[123] The zero shear viscosity, the elastic modulus, and the relaxation time all reached a maximum for an ionization degree of 0.5.

SDS was shown to enhance the viscoelastic properties of N-alkyl-N,N-dimethylbetaine solutions[124] and of tetradecyldimethylammonium propanesulfonate solution,[125] with maximum efficiency at equimolar composition.

The viscoelasticity of aqueous zwitterionic dimer acid betaine solutions has been studied by stress relaxation experiments as a function of surfactant and salt concentrations.[126] This system never shows the monoexponential relaxation behavior typical for kinetically controlled relaxation mechanisms, even in the presence of salt or at high temperature. The relaxation modulus is accurately fitted to a stretched exponential with an exponent close to 0.7. The scaling of the relaxation modulus suggests that the system consists of entangled wormlike micelles, but other shapes cannot be ruled out. The viscosity increases as $C^{5.3}$ in the whole C-range and η_R reaches 3×10^7 for a 15% solution at 20°C. It was concluded that this system relaxes essentially via reptation.[126]

8.3.4 WORMLIKE MICELLES OF NONIONIC SURFACTANTS

Formation of wormlike micelles from nonionic surfactant requires high concentration and η_R seldom exceeds 10^3. This is true for polyoxyethylene monoalkylether C_mEO_n [127,128] as well as for sugar surfactant solutions.[129–131] Nonionic surfactants are however versatile additives to tune the rheological properties of various surfactant systems and to modify they temperature dependence.

8.3.4.1 Mixtures of Nonionic Surfactants

In the presence of C_mEO_n surfactants, polyoxyethylene cholesteryl ethers ($ChEO_n$, n = 10 and 15) form Maxwellian viscoelastic solutions.[132,133] Adding $C_{12}EO_n$ (n = 1-4) at a fixed $ChEO_{10}$ concentration induces a steep increase of η_0.[132] The relative viscosity reaches more than 2×10^6, the relaxation time tens of minutes, and $G''_{min}/G_0 = 0.08$ at $C_{total} = 0.06$ M and $C_{12}EO_3$ mole fraction 0.4. Stiff gels are obtained from $ChEO_{10}/C_{12}EO_3$ mixtures. The sharp increase of η_0 is obtained for a mixing fraction of $C_{12}EO_n$ that increases in the order: $C_{12}EO_1 \approx C_{12}EO_2 < C_{12}EO_3 < C_{12}EO_4$.[132] Besides, this increase is shifted to relatively higher $C_{12}EO_3$ mole fraction on going from $ChEO_{10}$ to $ChEO_{15}$.[132]

Alkanoyl-N-methylethanolamide (NMEA) also enhances the viscosity of $ChEO_{15}$.[133] Relative viscosity above 10^5, relaxation time of about 5 s, and $G''_{min}/G_0 = 0.10$ have been measured at 0.06 M $ChEO_{15}$, for a ratio of NMEA-12 of 0.6. In $ChEO_n/C_{12}EO_n$ and $ChEO_n/NMEA$ mixtures η_0 goes through a maximum because a lamellar phase nucleates upon further addition of cosurfactant.

Spherical micelles present in dilute solution of sucrose dodecanoate and hexadecanoate surfactants turn into giant micelles in the presence of $C_{12}EO_n$ with n = 1-3.[134] This is reflected in the dependence of η_0 on the weight ratio R = $C_{12}EO_n$/sucrose. The viscosity increases up to an optimum ratio where η_R reaches 4×10^5 for the sucrose hexadecanoate/$C_{12}EO_1$ system. The optimum ratio decreases and the viscosity at the maximum increases as n decreases. The elastic modulus increases rapidly up to the optimum ratio and more slowly beyond. The results have been interpreted as due to the onset of the entanglement regime below the optimum ratio followed by increasing influence of the branching. The branching hypothesis is supported by the phase diagram of the ternary mixture which revealed the nucleation of a lamellar phase upon enrichment in $C_{12}EO_n$.[134]

Addition of C_mEO_n surfactants, with (m, n) = (12, 3), (12, 4), and (16, 4) to aqueous solutions of an anionic gemini-type surfactant with no spacer group, disodium 2,3-didodecyl-1,2,3,4-butanetetracarboxylate increases the viscosity by several orders of magnitude.[135] Relative viscosities as high as 10^7 are reached with $C_{12}EO_3$ and $C_{16}EO_4$.

8.3.4.2 Nonionic Surfactants as Neutral Cosurfactants in Other Systems

Addition of dodecanoyl-N-methylethanolamide (NMEA-12) and hexadecanoyl-N-methylethanolamide (NMEA-16) to DTAB or CTAB induces micellar growth.[136] The relative viscosity reaches values above 10^5 for CTAB/NMEA-12, above 10^4 for DTAB/NMEA-16, and about 300 for DTAB/NMEA-12 mixtures, with a relaxation time around 5 s and $G''_{min}/G_N = 0.1$-0.05. At a fixed concentration of cationic surfactant, the viscosity goes through a maximum as the mole fraction of neutral cosurfactant increases. For the CTAB/NMEA-12 mixture, the NMEA-12 mole fraction x at the maximum increases from 0.3 to 0.5 as the CTAB concentration decreases from 0.2 to 0.01 M. However, the value of the viscosity at the maximum is constant. The phase diagrams of these ternary systems show that alkanolamide addition induces the formation of a lamellar phase and branching was proposed as a possibility to explain the decrease in viscosity at values of x beyond the maximum. A qualitative model was presented to describe the micelle length dependence on x.[136]

Wormlike micelles also form in CTAB/$C_{12}EO_n$ mixtures.[137,138] The amount of $C_{12}EO_n$ necessary to significantly enhance the viscosity increases with n. However adding too much of $C_{12}EO_n$ leads to a lamellar phase. In the presence of salt, η_0 shows a maximum at a certain mixing fraction of $C_{12}EO_3$ which depends on salt concentration.[138]

Viscosity measurements have been used to study the effect of short chain (C_3-C_8) alcohols and amines on micelle growth in a 0.2 M CPyCl solution, in the presence and absence of 0.1 M KCl and at several temperatures.[139] Micelle growth occurs only for additive chain longer than butyl. For a given chain length, alcohols are more efficient than amines, but the efficiency of these additives to enhance viscosity is limited.

8.3.5 WORMLIKE MICELLES IN THE PRESENCE OF POLYMERS

The need to address the viscoelastic properties of wormlike micelles at high temperature led to the development of formulations containing both living and dead polymers. Two situations must be distinguished. In the first one the dead polymers bridge the wormlike micelles through few hydrophobic stickers they carry along their backbone and that anchor into the micelles. This leads to a transient network of crosslinked wormlike micelles with enhanced elasticity. In the second situation, the polymer adsorbs onto or anchors into the wormlike micelles all along its contour. This enhances the cohesion of the wormlike micelles and leads to reinforced wormlike micelles with slower dynamics.

8.3.5.1 Crosslinked Wormlike Micelles

Viscosity enhancement by bridging wormlike micelles was first reported with hydrophobically grafted polyacrylamide.[140] Synergistic effects were also observed in the rheology of mixtures of EHAC with hydrophobically modified hydrox-ypropyl guar, HMHPG,[141] or with hydrophobically modified polyacrylamide, HMPAM.[92,142] At 25°C zero shear viscosity values of 0.06Pa·s and 2Pa·s were measured for 0.35% solution of HMHPG and EHAC, respectively, whereas their mixture at the same total concentration had a viscosity of almost 100Pa·s. The relaxation time and the plateau modulus vary with $x = C_{EHAC}/(C_{EHAC} + C_{HM-HPG})$ with a bell-shaped curve with a maximum at $x = 0.2$, irrespective of temperature (25 to 80°C). The system is described as an interpenetrating polymer network with adhesive contacts at the sticker locations. Lesser or suppressed synergy is observed at high temperature with HMHPG,[141] while the HMPAM/EHAC[142] system remains highly viscoelastic at elevated temperature. The latter system demonstrates a 10^4-fold increase in viscosity as compared to pure component solutions, the effect being more pronounced for polymers with less blocky distribution of hydrophobic units.[142]

Other crosslinked wormlike micelles have been obtained with cationic gemini surfactants and hydrophobically modified telechelic polymers (HMTP) consisting of polyethylene glycol (M = 35,000), extended by diisocyanates and endcapped with hexadecyl chains.[143] The addition of DTAB, 12-3-12, 2Br or 12-3-12-3-12, 3Br to a 5% solution of HMTP enhances its viscosity (Figure 8.13). The increase is more pronounced for wormlike micelles of 12-3-12, 2Br and 12-3-12-3-12, 3Br than for spherical micelles of SDS or DTAB. The shape of the curves reveals the sphere-to-rod transition of gemini surfactant micelles. The rheological behavior of such transient networks is determined by the residence time of the hydrophobic stickers in the micelles and the connectivity of the surfactant/polymer mixed network. The latter strongly depends on the averaged functionality of the micelles (Figure 8.13).[144,145] Upon addition of surfactant, mixed micelles form and the distance between them is sufficiently reduced for the polyethylene glycol chain to bridge two micelles, giving rise to the initial increase in viscosity. Upon further addition of surfactant, the mixed micelles contain less and less polymer stickers. When the micelles remain spherical, the average micelle functionality decreases. This explains the bell-shaped curves

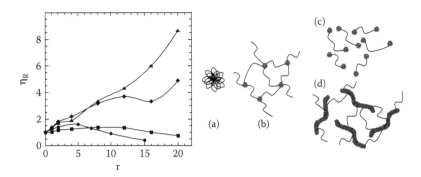

FIGURE 8.13 Left part: relative viscosity of SDS (●), DTAB (■), 12-3-12,2Br (◆), and 12-3-12-3-12, 3Br (▲), in the presence of 5% w/w hydrophobically modified telechelic polymers (HMTP) at 25°C (η_R = zero shear viscosity of the mixture normalized by the zero shear viscosity of the 5% HMTP solution; r = number of surfactant hydrophobic chains/number of HMTP hydrophobic chains). Right part: Evolution of the functionality f of mixed surfactant/HMTP micelles with increasing surfactant concentration. (a) flower micelles of HMTP ($f = 0$); (b) crosslinked mixed spherical micelles ($f > 2$); (c) isolated pairs of spherical micelles ($f = 1$); (d) crosslinked wormlike micelles ($f > 2$).

obtained for SDS and DTAB. The transformation into wormlike micelles upon surfactant addition keeps the number of micelles small and thereby preserves their functionality at a high value.[146] That is why, in Figure 8.13, the viscosity keeps increasing upon addition of wormlike micelle-forming surfactants. However, this increase should not be misinterpreted: the viscosity of the mixture HMTP/12-3-12-3-12, 3Br is lower than the viscosity of the pure surfactant solution. The gain in rheological performance results from a compromise between increasing connectivity and shortening of micelles possibly induced by the anchoring of the polymers.

8.3.5.2 Reinforced Wormlike Micelles

When the added polymer carries anchoring groups all along its backbone, it can act as a splint for the wormlike micelles and improve their rheological properties as noted for several wormlike micelle systems.

The viscoelastic properties of aqueous solutions of cetyltrimethylammonium bromide (CTAB) in the presence of sodium tosylate (NaTos) have been enhanced by addition of poly(p-vinylbenzoic acid) (PVB).[147] Also, addition of partially sulfonated polystyrene (P(St/NaSS)) to CTAB solution leads to strongly viscoelastic systems.[148,149] With PVB, the rheological behavior is pH dependent and the relaxation time and elastic modulus are higher than in pure CTAB/NaTos only at pH lower than the pK_a of the polymer. The enhancement of viscoelastic properties results from a fine tuning of the interaction between surfactant and polymer to avoid possible phase separation (frequent in mixture of surfactants and oppositely charged polymers) and shortening of the micelle. With PVB the

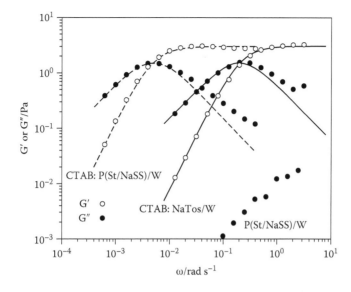

FIGURE 8.14 Storage modulus G' (○) and loss modulus G'' (●) vs. angular frequency for CTAB/NaTos (1:1) wormlike micelles in the presence and in the absence of partially sulfonated polystyrene. (Reproduced with permission from Ref. 148. © 2003 by American Chemical Society.)

charge density of the polymer was tuned through pH control. With P(St/NaSS), the degree of sulfonation allows a fine tuning of the charge density of the polymer. In both cases, incorporation of polymer into wormlike micelles hinders the fusion/scission reactions and delays the relaxation process (Figure 8.14). In PVB systems, an increase in the rigidity of the hybrid micelles as compared to pure CTAB micelles was proposed to explain the increase in elasticity. PVB might actually both stitch as well as crosslink the wormlike micelles. But P(St/NaSS) does not enhance the elasticity of the system (Figure 8.14).

8.4 CHARACTERIZATION OF WORMLIKE MICELLES THROUGH RHEOLOGY

Rheology can be a powerful tool for physicochemical characterization. Below are discussed procedures that have been proposed to determine the endcap energy, the branching density, and the kinetics of reversible scission of wormlike micelles systems.

8.4.1 MICELLAR GROWTH AND ENDCAP ENERGY

Wormlike micelles are obtained when the preferred curvature of surfactants in the prevailing physicochemical conditions is cylindrical. Hemispherical endcaps correspond to an excess of curvature which is associated with an excess free energy, often called the endcap energy E_c. A compromise between this term and the mixing entropy of the system leads to the following growth law: $<L> = C^{1/2}\exp(E_c/2k_BT)$, where $<L>$ is the average length of the wormlike micelles. E_c might depend on C especially in the case of unscreened charged micelles (see Chapter 4).[85] Various procedures to obtain E_c based on rheological measurements have been proposed.

8.4.1.1   Concentration Dependence of the Zero Shear Viscosity

The C-dependence of η_0 often presents several concentration regimes delineated by C^* and C^{**}. C^* has often been considered to be the crossover concentration to the semidilute regime (see Figure 8.11). For classical polymers C^* decreases with increasing degree of polymerization. In wormlike micelles, since $<L>$ increases with C to an extent determined by E_c, the temperature dependence of C^* permits an estimate of E_c. In EHAC/KCl (0.4 M) for instance, it leads to $E_c = (23 \pm 6)k_BT$.[90]

For unscreened charged surfactant, a relation between C^* and E_{c0} (E_c at zero concentration) and involving the temperature has been tentatively applied to the gemini surfactant 12-2-12, 2Br.[39] C^* is certainly a good qualitative indicator of the micelle tendency to grow. Whether it corresponds to a crossover to the semidilute regime or to the rapid growth regime for charged surfactants is still unclear. It is now well established by light scattering[150] and cryo-TEM that very long wormlike micelles exist at concentrations below C^* as determined by rheology. Moreover, when the crossover to the semidilute regime is determined by several techniques, the values obtained from rheology are always the highest ones.[57,128]

A quantitative interpretation of the concentration dependence of viscosity in terms of E_c has been proposed for gemini surfactants[39] and for CTAHNC/CTAB ($E_c = 43k_BT$).[80] All the proposed relations, however, assumed $G_N \sim C^{2.2}$, that is, fully entangled wormlike micelles. This hypothesis was shown not to hold,[98] and another relation between η_0 and $<L>$ was used instead to take into account this fact and the polyelectrolyte character of charged micelles. This led to $E_c = 40k_BT$ for 12-3-12, 2Br and $E_c = 80k_BT$ for the trimer 12-3-12-3-12, 3Br.

In both approaches just described, the wormlike micelles are considered either fully entangled or completely independent of each other. The reality is certainly in between and the strong increase of viscosity with concentration does not reflect just the tendency to grow (i.e., E_c), but also a smooth crossover to the fully entangled regime. It is sometime argued that the micelle growth is too fast to permit the observation of an intermediate region of viscoelastic behavior.[80] Nevertheless, evidence for this intermediate regime is numerous and has to be recalled. The first is given in Figure 8.6 and concerns CTAB micelles in deionized water. In such physicochemical conditions, CTAB micelles are certainly not expected to show a great tendency to grow. This explains the high value $C^* = 12\%$.

On the other hand, the C-dependence of η_0 is very strong: $\eta_0 \sim C^{12}$.[51] At $C >$ 12%, the micelles are rather close to each other and thus do not need to grow much in order to get entangled. This naturally translates into a high rate of increase of the viscosity.

Moreover, in the range of concentration where η_0 increases rapidly, the plot for the elastic modulus (or the inverse of the recoverable compliance) is often below the universal curve predicted for fully entangled micelles (Figure 8.12).[98,151] This observation is reminiscent of the behavior of cationic surfactant solutions in the presence of organic anions: the elastic modulus becomes independent of the organic anion concentration just when the viscosity reaches the first maximum.[6,12,13] It suggests that all micelles are not entangled. The strong increase in viscosity might therefore also reflect the incipient entanglement behavior. This point is further discussed in Section 8.4.1.2. It calls for caution when interpreting the rate of increase of viscosity in term of endcap energy.

8.4.1.2 Incipient Entanglement Regime

Figure 8.12 and Figure 8.15 show that the plot of the elastic modulus merges the expected universal scaling law curve only at a certain concentration C_e. This has been also observed with other systems and indicates that the micelles are weakly entangled.[6,12,13,151] Other observations confirm this interpretation: rheological behavior gets closer to Maxwellian as the temperature decreases.[66] The plateau modulus is sometime observed to depend on temperature,[27] which translates in the fact that the temperature dependence of the viscosity does not follow Arrhenius law[39] or leads to very high values of E_a.[89] Figure 8.15a shows that as

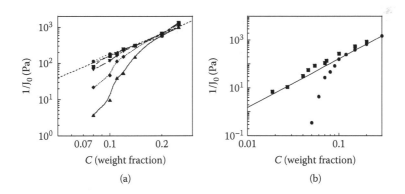

FIGURE 8.15 Incipient entanglement behavior. (a) Weight fraction dependence of the elastic properties of 12-3-12, 2Br solutions at 15°C (●), 25°C (■), 35°C (▼), 40°C (♦) and 45°C (▲). The broken straight line corresponds to the universal scaling law $1/J_0 \sim C^{9/4}$. (b) Comparison of the elastic properties of 12-3-12, 2Br solutions in water (●) and in 0.1 M KBr (■), at 25°C. (Reproduced with permission from Ref. 18. © 2004 by Marcel Dekker.)

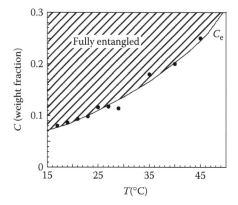

FIGURE 8.16 State diagram of 12-3-12, 2Br wormlike micelles in the concentration-temperature plane. (Reproduced with permission from Ref. 18. © 2004 by Marcel Dekker.)

temperature decreases, the elasticity of low concentration samples get closer to the scaling law. In some reports the elastic modulus is observed to increase with salt concentration.[18,123,152] Figure 8.15b shows that this increase of elasticity levels off once the data merge the universal curve.

All these observations suggest that wormlike micelle systems show a cross-over from weakly entangled regime to fully entangled regime. The concentration C_e where the elasticity data merge the universal curve marks the onset of the fully entangled state, and depends on temperature as shown in Figure 8.16.[18] The endcap energy can be obtained from elasticity data by looking at the temperature dependence of C_e. For classical polymers, it has been established that the entanglement concentration scales as $C_e = <L>^{-4/5}$.[153] Inserting this relation into the growth law for equilibrium polymers leads to $C_e = a \exp(-2E_c/7k_BT)$. This relation accounts well for the boundary in the state diagram of 12-3-12, 2Br in Figure 8.16, with $E_c = 44k_BT$.[18]

Another way of estimating E_c from the incipient entanglement behavior is to model the concentration dependence of the elastic modulus over the entire C-range.[151] At each concentration, only the fraction p_{eff} of wormlike micelles that are longer than the mesh size of the network are supposed to be elastically active. The fraction p_{eff} is given by the average length of the micelles, which depends on C through the endcap energy E_c, and by the form of the size distribution, which is known to be exponential. The elastic modulus is then expressed as $G_N = A(p_{eff}C)^{9/4}$. The two fitting parameters of the procedure are the prefactor A and the endcap energy. This procedure was used to estimate the influence of the amphiphilic copolymers Pluronics on the growth tendency of CPyCl/NaSal micelles in salted water.[151] In the absence of copolymer it yielded $E_c = 32k_BT$ and E_c decreased linearly with increasing fraction of copolymer.

8.4.1.3 Minimum Loss Tangent

In the frequency domain, the plateau region where G' is almost constant is also characterized by a minimum in G''. At frequencies higher than the frequency corresponding to the minimum loss tangent, $\tan\delta_{min} = G''_{min}/G_0$, the sample is probed at a length scale that corresponds to Rouse modes of strands between entanglements. At lower frequency, reptation proceeds. Therefore $\tan\delta_{min}$ is related to the ratio $l_e/<L>$, where l_e is the entanglement length. For conventional polymers, $\tan\delta_{min}$ decreases with increasing molecular weight and shifts toward low frequency.[20] In the case of wormlike micelles it has been shown[22,36] that $G''_{min}/G_0 = (l_e/<L>)^{5/4}$. This relation allows a semiquantitative estimate of the micelle length since l_e can be estimated from the plateau modulus or can be obtained from the correlation length determined by scattering techniques.[22,36] This method has been applied to estimate the length of CTAB, CPyCl/NaSal, and CTATos[73] micelles.[22,36] A surprisingly weak C-dependence of the length of CTAB micelles obtained with this approach has been pointed out.[36] In this system $\tan\delta_{min}$ goes through a minimum as KBr concentration increases, an observation that has been explained by micellar branching.[21] In several systems (gemini surfactants,[18,39] CTATos/SDBS[83]) $\tan\delta_{min}$ goes through a minimum as C increases (see Figure 8.10). This type of result has been interpreted as due to either branching or shortening of the micelles. However, shortening is very unlikely when concentration rises. The kinetics of scission-recombination becomes probably too fast for the minimum loss tangent to be related to the total length of the micelles.[18]

The endcap energy has been obtained from the temperature dependence of $\tan\delta_{min}$ for EHAC/NaSal ($C_S/C = 0.5$) solutions ($E_c = 65k_BT$)[88] and EHAC/KCl (0.4 M) solutions ($E_c = (31\pm4)k_BT$).[90] This approach applied to gemini surfactant led to the surprising conclusion that E_c decreases as C increases.[39]

The above equation is correct as long as $\tan\delta_{min}$ is small enough to ensure that the system is strongly entangled and that the breaking time τ_b is much larger than the entanglement time τ_e (Rouse time associated with l_e). Another relation has been established when τ_b is close to τ_e, but it involves too many unknown parameters to be practically useful.[36]

8.4.1.4 The Activation Energy of Flow

The temperature dependence of the terminal relaxation time follows Arrhenius law with activation energy E_a (see Section 8.3.1.1). This activation energy is the same as the one obtained from an Arrhenius plot of the viscosity vs. temperature, provided the system remains in the fully entangled regime when T increases. A procedure has been proposed to combine rheological data with T-jump results in order to estimate E_c.[46,47]

In Cates model of coupled reversible scission-reptation, the terminal relaxation time is given by $\tau = (\tau_b\tau_{rep})^{1/2}$, when $\tau_b<<\tau_{rep}$ (Section 8.2.2). The temperature dependence of all the three characteristic times involved in this equation follows Arrhenius law, and the three activation energies are related by $E_a = 1/2(E_b + E_{rep})$.

Considering the growth law for wormlike micelles and the length dependence of the reptation time, the relation $E_a = 1/2(E_b+3/2E_c)$ is readily obtained. E_b can be obtained from T-jump experiments at various temperatures,[66] since τ_b is proportional the relaxation time after a T-jump.[154]

In CTAB/NaSal system with added salt, $E_a = 67k_BT$ with a slight tendency to decrease with increasing salt.[66] According to the above equation, this yields $E_c = 51k_BT$. However, this value was considered unrealistic because it would result in unrealistically long micelles.[66] The values $E_a = 34k_BT$ and $E_c = 25k_BT$ were reported for the EHAC/KCl system.[90] In CTAHNC, $E_c = 48k_BT$.[82] Note that this approach assumes that reversible scission is the main mechanism of relaxation in T-jump as well as in mechanical relaxation experiments.

8.4.2 KINETICS OF REVERSIBLE SCISSION

Tuner and Cates presented a quantitative method for estimating the time τ_b in wormlike micelle systems by analyzing their viscoelastic response in the linear regime.[34] The method relies upon the departure from the Maxwellian behavior (see Figure 8.3b) observed at high frequency when τ_b and τ_{rep} are of the same order of magnitude. A portion of the Cole–Cole plot is theoretically expected to be a straight line described by the equation: $G'' = B-G'$. The determination of τ_b proceeds as follows:[34] the left part of the Cole–Cole plot obtained after rescaling of G' and G" by G_N is then fitted to a semicircle constrained so that its center lies on the abscissa axis. A relation between the diameter D of the fitted semicircle and $\zeta = \tau_b/\tau$ has been established by numerical calculation and τ_b can be determined from D.

For the CTAB/KBr system,[21,47] this method yielded τ_b values ranging from 40 ms at 0.25 M KBr to 300–700 ms for 1.5 M KBr at 35°C and showed the presence of a maximum as the salt concentration was increased. In CTAC$_7$SO$_3$, τ_b decreases from 44 s to 0.1 s as the weight fraction increases from 1% to 3.4%.[80] In the CTATos system, τ_b ranges from 0.1 s to 1 s.[73] For the gemini surfactant 12-2-12, 2Br at 25°C τ_b decreases from 3.3 to 0.2 s as C increases from 4% to 10%.[39]

More results are given in Chapter 10.

8.4.3 BRANCHING

Branching of wormlike micelles received considerable attention from the theoretical[40,155–157] as well as from the experimental point of view.[39,90,98,158–160] First proposed as a plausible explanation of the surprisingly low viscosity of some wormlike micelles solutions,[41] branching has been invoked to explain any unusual concentration or ionic strength dependence of the rheological properties. However, no theoretical model accounts quantitatively for the decrease of the relaxation time as concentration or salinity increases.

Transition from an entangled network to a multiconnected network[128,161,162] has often been considered close to the liquid-liquid phase separation in nonionic surfactant solution at high temperature. However, apart from cryo-TEM very little independent evidence for branching has been presented. Interpretation of TEM micrographs offers the possibility to distinguish between equilibrium branching points and shear induced branching points. The former ones are isolated and connect waggling wormlike micelles,[158] while the latter come often by pair and connect stretched wormlike micelles.[90]

As branching was also proposed as an intermediate state in the transition from wormlike micelles to bilayers,[163] the vicinity of a lamellar phase in the phase diagram might be a good indication that the system is prone to branching.

Since the elastic modulus is the only universal property of wormlike micelles, enhancement of elasticity has been proposed as a reliable indicator of branching, just as for conventional polymers. Elasticity enhancement was indeed observed in the trimeric surfactant 12-3-12-3-12, 3Br,[98] which was known from cryo-TEM to form branched micelles.[103] When looking at Figure 8.12, elasticity enhancement is also clear for 12-2-12, 2Br solutions, where branching point had been suggested from simulation[164] and confirmed by cryo-TEM.[165] It is important to stress that an enhancement of elasticity can be interpreted as due to branching only when the elastic modulus is higher than what would be observed for fully entangled wormlike micelles. Any increase in elasticity that would still lie below the scaling law $G_N \sim C^{9/4}$ could also and more simply be interpreted as due to growth.

The enhancement of elasticity as compared to universal scaling law can be analyzed in terms of branching density. Assuming additivity for the entanglement and branching contribution to elasticity, one can write: $G_N = (v_b + v_e)k_B T$. The density of entanglement v_e being given by the universal scaling law, the density of branching v_b can be determined.[18,98]

The variation of elastic modulus upon addition of NaCl to CTATos system[78] might also reflect branching. The modulus is first constant, then increases and eventually decreases rapidly at high salt content. This behavior was interpreted as due to an increase in the micelle persistence length and a modification of micelle interactions resulting from the release of tosylate ions. An alternative interpretation by the present author is the following: since the plateau modulus first does not depend on salinity it can be concluded that the micelles are fully entangled even in the absence of salt. Adding salt triggers branching. This first leads to an increase of the elastic modulus. But at higher salinity, connections become so numerous that the system becomes a multiconnected network, without entanglements. This translates macroscopically by an abrupt fall of the elastic modulus.

8.5 CONCLUSION

Aqueous solutions of entangled wormlike micelles can present strong viscoelastic Maxwellian properties. As for entangled polymer solutions, their relaxation modulus shows a plateau region and a terminal region. The terminal relaxation region has the distinctive characteristic to be monoexponential.

The magnitude of the plateau modulus is rather insensitive to the chemical nature of the surfactant, and the concentration scaling of the plateau modulus is a very universal feature of the viscoelasticity of wormlike micelle solutions. This is expected since it depends on neither the (not completely understood) relaxation mechanism nor the micelle length. It measures essentially the number density of entanglement points in the solution and characterizes the wormlike micelles at short time. At short time living polymers behave like dead ones. It is therefore more reliable to characterize micelle growth or branching from the elastic properties rather than from the relaxation dynamics.

When one seeks performances, relaxation dynamics makes the difference and shows strong specificity to surfactant molecular structure, composition of the solution, and physicochemical conditions. Maxima of viscosity as a function of surfactant or salt concentration are profuse in the literature. Sometimes, instead of a maximum, a marked change of the rate of rise in viscosity is reported. Explanations of these experimental facts remain to be unambiguously established. Two explanations are often proposed: shortening or branching. Micelles shortening is very unlikely, since it would violate Le Chatelier principle. An increase of concentration or salinity should always result in longer micelles unless charge inversion occurs as a result of a variation in micelle composition. Branching is a more satisfying explanation because it corresponds to lower curvature and, in some sense, to a more condensed state of the surfactants. However, it is often invoked to explain any unexpected result without any independent evidence for it.

Back to interpretations proposed earlier, maxima in viscosity vs. concentration might reflect a crossover from a diffusion-controlled to a recombination-controlled regime of relaxation. Incipient entanglement behavior would explain the rapid increase of viscosity at low concentration that levels off or even reverses once the system is fully entangled.

Recent simulations confirm early ideas that wormlike micelles would relax faster than polymers not because they break, but rather because they fuse. This reconciles the puzzling fact that increasing the micelle length often translates in a decrease in viscosity. Triggering micelle growth might as well trigger fusion. Branching would still have a great importance, but care must be taken to distinguish equilibrium versus induced transient branching. In any case, the role of branching remains to be understood in a theoretical frame that would give a reinforced role to reptation.

Despite these difficulties for quantitatively accounting for the experimental data, our qualitative understanding is good enough to guide the formulation of higher performance wormlike micelle-based systems. They involve longer chain surfactants or mixed systems containing amphiphilic copolymers that bridge or reinforced wormlike micelles.

Model systems, where growth and branching could be controlled by composition parameters, remain to be developed. This would help to reach a larger consensus on the relative contribution of reversible scission and diffusion in the relaxation process. Linear rheology would then become the most valuable approach to characterize giant micelles.

REFERENCES

1. Hoffmann, H. *Adv. Colloid Interface Sci.* 1990, *32*, 123.
2. Hoffmann, H. *Adv. Mater.* 1994, *6*, 116.
3. Hoffmann, H., Thuning, C., Schmiedel, P., Munkert, U., Ulbricht, W. *Tenside Surfact. Det.* 1994, *31*, 389.
4. Hoffmann, H., Ulbricht, W. *Curr. Opin. Colloid Interface Sci.*, 1996, *1*, 726.
5. Cates, M. E., Candau, S. J. *J. Phys. C: Cond. Matt.* 1990, *2*, 6869.
6. Rehage, H., Hoffmann, H. *Mol. Phys.* 1991, *74*, 933.
7. Hoffmann, H., Rehage, H., Rauscher, A. In *Structure and Dynamics of Strongly Interacting Colloids and Supramolecular Aggregates in Solution*, Chen S. H., Ed., Kluwer Academic Publishers, Dordrecht, The Netherlands, 1992, p. 493.
8. Lequeux, F., *Curr. Opin. Colloid Interface Sci.* 1996, *1*, 341.
9. Magid, L. J., *J. Phys. Chem. B* 1998, *102*, 4064.
10. Walker, L. M., *Curr. Opin. Colloid Interface Sci.* 2001, *6*, 451.
11. *"Structure and Flow in Surfactant Solutions,"* Herb, C.A, Prud'homme, R. K., Eds., ACS Symposium Series 578, ACS, Washington, 1994.
12. Shikata, T., Hirata, H., T. Kotaka, *Langmuir* 1987, *3*, 1081.
13. Rehage, H., Hoffmann H. *J. Phys. Chem.* 1988. *92*, 4712.
14. Candau, S. J., Hirsch, E., Zana, R. *J. Phys. France* 1984, *45*, 1263.
15. Candau, S. J., Hirsch, E., Zana, R., Adam, M. *J. Colloid Interface Sci.* 1988, *122*, 430.
16. Cates, M. E. *Macromolecules* 1987, *20*, 2289.
17. Shikata, T., Hirata, H., Kotaka, T. *Langmuir* 1988, *4*, 354.
18. In, M. In *Gemini Surfactants. Synthesis, Interfacial and Solution-Phase Behavior, and Applications,* Zana, R., Xia, J., Eds., Marcel Dekker Inc., New York, 2004, Chap. 8.
19. Macosko, C. W. *Rheology. Principles, Measurements, and Applications*, VCH, New York, 1994.
20. Ferry, J. D. *Viscoelastic Properties of Polymers*, 3rd ed., John Wiley & Sons, New York, 1980.
21. Khatory, A., Lequeux, F., Kern, F., Candau, S. J. *Langmuir* 1993, *9*, 1456.
22. Granek, R, Cates, M. *J. Chem. Phys.* 1992, *96*, 4758.
23. Buhler, E., Munch, J. P., Candau, S. J. *J. Phys. II* 1995, *5*, 765.
24. Buhler, E., Munch, J. P., Candau, S. J. *Europhys. Lett.* 1996, *34*, 251.
25. Nemoto, N., Kuwahara, M., Yao, M. L., Osaki, K. *Langmuir* 1995, *11*, 30.
26. Clancy, S. F., Fuller, J. G., Scheidt, T., Paradies, H. H. *Z. Phys. Chem. Int. J. Res. Phys. Chem. Chem. Phys.* 2001, *215*, 905.
27. Cardinaux, F., Cipelletti, L., Scheffold, F., Schurtenberger, P. *Europhys. Lett.* 2002, *57*, 738.
28. Hassan, P. A., Bhattacharya, K., Kulshreshtha, S. K, Raghavan, S. R. *J. Phys. Chem. B* 2005, *109*, 8744.
29. Buchanan, M., Atakhorrami, M., Palierne, J. F., MacKintosh, F. C., Schmidt, C. F. *Phys. Rev. E* 2005, *72*, 11504.
30. Kim, W. J., Yang, S. M. *J. Colloid Interface Sci.* 2000, *232*, 225 and references therein.
31. Shikata, T., Imai, S. *Langmuir* 2000, *16*, 4840 and references therein.
32. Angelico, R., Burgemeister, D., Ceglie, A., Olsson, U., Palazzo, G., Schmidt, C. *J. Phys. Chem. B* 2003, *107*, 10325 and references therein.

33. Löbl, M., Thurn, H., Hoffmann, H. *Ber. Bunsenges. Phys. Chem.* 1984, *88*, 1102.
34. Turner, M. S., Cates, M. *Langmuir* 1991, *7*, 1590.
35. Turner, M. S., Cates, M. E. *J. Phys. France* 1992, *2*, 503.
36. Granek, R. *Langmuir* 1994, *10*, 1627.
37. Granek, R. *Macromolecules* 1995, *28*, 5370.
38. Cates, M. E. *J. Phys. France* 1988, *49*,1593.
39. Kern, F., Lequeux, F., Zana, R., Candau, S. J. *Langmuir* 1994, *10*, 1714.
40. Lequeux, F. *Europhys. Lett.* 1992, *19*, 675.
41. Appell, J., Porte, G. *J. Phys. Lett. (Paris)* 1983, 44, L-689; Appel, J., Porte, G., Kathory, A., Kern, F., Candau, S. J. *J. Phys. France,* 1992, *2*, 1045.
42. O'Shaughnessy, B., Yu, J. *Phys. Rev. Lett.* 1995, *74*, 4329.
43. Padding, J. T., Boek, E. S. *Europhys. Lett.* 2004, *66*, 756.
44. Briels, W. J., Mulder, P., den Otter, W. K. *J. Phys. C: Cond. Matt.* 2004, *16*, S3965.
45. Yamamoto, S., Hyodo, S. *J. Chem. Phys.* 2005, *122*, 204907.
46. Candau, S. J., Hirsch, E., Zana, R., Delsanti, M. *Langmuir* 1989, *5*, 1225.
47. Kern, F., Lemarechal, P., Candau, S. J., Cates, M. E. *Langmuir* 1992, *8*, 437.
48. Candau, S. J., Khatory, A., Lequeux, F., Kern, F. *J. Phys. France IV* 1993, *3*, 197
49. Candau, S. J., Merikhi, F., Waton, G., Lemarechal, P. *J. Phys. Paris* 1990, *51*, 977.
50. Shikata, T., Pearson, D. S. *Langmuir* 1994, *10*, 4027.
51. Cappelaere, E., Cressely, R., Decruppe, J. P. *Colloids Surf. A* 1995, *104*, 353.
52. Hertel, G., Hoffmann, H. *Progr. Colloid Polym. Sci.* 1988, *76*, 123.
53. Thalody, B., Warr, G. G. *J. Colloid Interface Sci.* 1995, *175*, 297 and references therein.
54. Porte, G. Appell, J. In *Surfactants in Solution,* Mittal, K. L., Lindman, B., Eds., Plenum Press, New York, 1984, Vol. 2, p. 805.
55. Cappelaere, E., Cressely, R. *Colloid Polym. Sci.* 1998, *276*, 1050.
56. Cappelaere, E., Cressely, R. *Rheol. Acta* 2000, *39*, 346.
57. Khatory, A., Kern, F., Lequeux, F., Appell, J., Porte, G., Morie, N., Ott, A., Urbach, W. *Langmuir* 1993, *9*, 933.
58. Snabre, P., Porte, G. *Europhys. Lett.* 1990, *13*, 641.
59. Larsen, J. W., Magid, L. J., Patyton, V. *Tetrahedron Lett.* 1973, *29*, 2663.
60. Ulmius, J., Wennerstrom, H., Johansson, L. B.-A., Lindblom, G., Gravsholt, S. *J. Phys. Chem.* 1979, *83*, 2232.
61. Lin, Z., Cai, J. J., Scriven, L. E, Davis, H. T. *J. Phys. Chem.* 1994, *98*, 5984.
62. Gravsholt, S. *J. Colloid Interface Sci.* 1976, *57*, 575.
63. Olsson, U., Söderman, O., Guering, P. *J. Phys. Chem.* 1986, *90*, 5223.
64. Berret, J. F., Appell, J., Porte, G. *Langmuir* 1993, *9*, 2851.
65. Ali, A. A., Makhloufi, R. *Phys. Rev. E* 1997, *56*, 4474.
66. Kern, F., Zana, R., Candau, S. J. *Langmuir* 1991, *7*, 1344.
67. Shikata, T., Hirata, H., Kotaka, T. *Langmuir* 1989, *5*, 398.
68. Shikata, T., Hirata, H., Kotaka, T. *J. Phys. Chem.* 1990, *94*, 3702.
69. Clausen, T. M., Vinson, P. K., Minter, J. R., Davis, H. T., Talmon, Y., Miller, W. G. *J. Phys. Chem.* 1992, *96*, 474.
70. Carver, M., Smith, T. L., Gee, J. C., Delichere, A., Caponetti, E., Magid, L. J. *Langmuir* 1996, *12*, 691.
71. Lin, Z. Q., Zakin, J. L., Zheng, Y., Davis, H. T., Scriven, L. E., Talmon, Y. *J. Rheol.* 2001, *45*, 963.
72. Soltero, J. F. A., Puig, J. E., Manero, O., Schulz, P. C. *Langmuir* 1995, *11*, 3337.

73. Soltero, J. F. A., Puig, J. E., Manero, O. *Langmuir* 1996, *12*, 2654.
74. Imai, S., Shikata, T. *J. Colloid Interface Sci.* 2001, *244*, 399.
75. Shikata, T., Shiokawa, M., Imai, S. *J. Colloid Interface Sci.* 2003, 259, 367.
76. Ali, A. A., Makhloufi, R. *Colloid. Polym. Sci.* 1999, *277*, 270.
77. Shikata, T., Shiokawa, M., Imai, S. *J. Colloid Interface Sci.* 2003, *259*, 367.
78. Bandyopadhyay, R., Sood, A. K. *Langmuir* 2003, *19*, 3121.
79. Barker, C. A., Saul, D., Tiddy, G. J. T., Wheeler, E. W. *J. Chem. Soc. Faraday I* 1974, *70*, 154.
80. Narayanan, J., Manohar, C., Kern, F., Lequeux, F., Candau, S. J. *Langmuir* 1997, *13*, 5235.
81. Oda, R., Narayanan, J., Hassan, P. A., Manohar, C., Salkar, R. A., Kern, F., Candau, S. J. *Langmuir* 1998, *14*, 4364.
82. Hassan, P. A., Valaulikar, B. S., Manohar, C., Kern, F., Bourdieu, L., Candau, S. J. *Langmuir* 1996, *12*, 4350.
83. Koehler, R. D., Raghavan, S. R., Kaler, E. W. *J. Phys. Chem. B* 2000, *104*, 11035.
84. Schubert, B. A., Kaler, E. W., Wagner, N. J. *Langmuir* 2003, *19*, 4079.
85. MacKintosh, F. C., Safran, S. A., Pincus, P. A. *Europhys. Lett.* 1990, *12*, 697.
86. Abdel-Rahem, R., Gradzielski, M., Hoffmann, H. *J. Colloid Interface Sci.* 2005, *288*, 570.
87. Hassan, P. A., Candau, S. J., Kern, F., Manohar, C. *Langmuir* 1998, *14*, 6025.
88. Raghavan, S. R., Kaler, E. W., *Langmuir* 2001, 17, 300.
89. Raghavan, S. R., Edlund, H., Kaler, E. W. *Langmuir* 2002, *18*, 1056.
90. Croce, V., Cosgrove, T., Maitland, G., Hughes, T., Karlsson, G. *Langmuir* 2003, *19*, 8536.
91. Couillet, I., Hughes, T., Maitland, G., Candau, F., Candau, S. J. *Langmuir* 2004, *20*, 9541.
92. Shashkina, J. A., Philippova, O. E., Zaroslov, Y. D., Khokhlov, A. R., Pryakhina, T. A., Blagodatskikh, I. V. *Langmuir* 2005, *21*, 1524.
93. Kalur, G. C., Frounfelker, B. D., Cipriano, B. H., Norman, A. I., Raghavan, S. R. *Langmuir* 2005, *21*, 10998.
94. Drye, T. J., Cates, M. E. *J. Chem. Phys.* 1992, *96*, 1367.
95. Croce, V., Cosgrove, T., Dreiss, C. A., Maitland, G., Hughes, T., Karlsson, G. *Langmuir* 2004, *20*, 7984.
96. Lin, Z. Q., Lu, B., Zakin, J. L., Talmon, Y., Zheng, Y., Davis, H. T., Scriven, L. E. *J. Colloid Interface Sci.* 2001, *239*, 543.
97. In, M. In *Reactions and Synthesis in Surfactant Systems*, Texter, J., Ed., Marcel Dekker, New York, 2001, p. 59.
98. In, M., Warr, G. G., Zana, R. *Phys. Rev. Lett.* 1999, *83*, 2278.
99. In, M., Bec, V., Aguerre-Chariol, O., Zana, R. *Langmuir* 2000, *16*, 141.
100. Oda, R., Huc, I., Candau, S. J. *Chem. Commun.* 1997, 2105.
101. Oda, R., Huc, I., Homo, J.-C., Heinrich, B., Schmutz, M., Candau, S. *Langmuir* 1999, *15*, 2384.
102. Han, L. J., Chen, H., Luo, P. Y. *Surf. Sci.* 2004, *564*, 141.
103. Danino, D., Talmon, Y., Levy, H., Beinert, G., Zana, R., *Science* 1995, *269*, 1420.
104. In, M., Bec, V., Aguerre-Chariol, O., Zana, R. *Langmuir* 2000, *16*, 141.
105. Oda, R., Bourdieu, L. and Schmutz, M. *J. Phys. Chem. B* 1997, *101*, 5913.
106. Oelschlaeger, C., Buhler, E., Waton, G., Candau, S. J. *Eur. Phys. J. E* 2003, *11*, 7.
107. Mu, J. H., Li, G. Z. *Chem. Phys. Lett.* 2001, *345*, 100.

108. Angelescu, D., Khan, A., Caldararu, H. *Langmuir* 2003, *19*, 9155.
109. Mu, J. H., Li, G. Z., Jia, X. L., Wang, H. X., Zhang, G. Y. *J. Phys. Chem.* B 2002, *106*, 11685.
110. Hassan, P. A., Raghavan, S. R., Kaler, E. W. *Langmuir* 2002, *18*, 2543.
111. Rodriguez, C., Acharya, D. P., Hattori, K., Sakai, T., Kunieda, H. *Langmuir* 2003, *19*, 8692.
112. Raghavan, S. R., Fritz, G., Kaler, E. W. *Langmuir* 2002, *18*, 3797.
113. Balzer, D., Varwig, S., Weihrauch, M. *Colloids Surf. A* 1995, *99*, 233.
114. Janeschitz-Kriegl, H., Papenhuijzen, J. M. P. *Rheol. Acta* 1971, *10*, 461.
115. Kalur, G. C., Raghavan, S. R. *J. Phys. Chem.* B 2005, *109*, 8599.
116. Angel, M., Hoffmann, H., Kramer, U., Thurn, H. *Ber. Bunseges. Phys. Chem.* 1989, *93*, 184.
117. Hoffmann, H., Platz, G., Rehage, H., Reizlein, K., Ulbricht, W. *Makromol. Chem.* 1981, *182*, 451.
118. Watanabe, H., Sato, T., Osaki, K., Matsumoto, M., Bossev, D. P., McNamee, C. E., Nakahara, M. *Rheol. Acta* 2000, *39*, 110.
119. Tobita, K., Hideki, S., Yukishige, K., Norio, Y., Keiji, K., Nobuyuki, M., Masahiko, A. *Langmuir* 1997, *13*, 4753.
120. Danino, D., Weihs, D., Zana, R., Oradd, G., Lindblom, G., Abe, M., Talmon, Y. *J. Colloid Interface Sci.* 2003, *259*, 382.
121. Hashimoto, K., Imae, T. *Langmuir* 1991, *7*, 1734.
122. Hoffmann, H., Rauscher, A., Gradzielski, M., Schulz, S. F. *Langmuir* 1992, *8*, 2140.
123. Maeda, H., Yamamoto, A., Souda, M., Kawasaki, H., Hossain, K. S., Nemoto, N., Almgren, M. *J. Phys. Chem.* B 2001, *105*, 5411.
124. Iwasaki, T., Ogawa, M., Esumi, K., Meguro, K. *Langmuir* 1991, *7*, 30.
125. Lopez-Diaz, D., Garcia-Mateos, I., Velazquez, M. M. *Colloids Surf. A* 2005, *270*, 153.
126. Fischer, P., Rehage, H., Gruning, B. *J. Phys. Chem.* B 2002, *106*, 11041.
127. Imae, T., Sasaki, M., Ikeda, S. *J. Colloid Interface Sci.* 1989, *127*, 511.
128. Kato, T., Terao, T. and Seimiya, T., *Langmuir* 1994, *10*, 4468.
129. Schulte, J., Enders, S., Quitzsch, K. *Colloid. Polym. Sci.* 1999, *277*, 827.
130. Ericsson, C. A., Soderman, O., Ulvenlund, S. *Colloid. Polym. Sci.* 2005, *283*, 1313.
131. Choplin, L., Sadtler, V., Marchal, P., Sfayhi, D., Ghoul, M., Engasser, J. M. *J. Colloid Interface Sci.* 2006, *294*, 187.
132. Acharya, D. P., Kunieda, H. *J. Phys. Chem.* B 2003, *107*, 10168.
133. Acharya, D. P., Hossain, M. K., Jin, F., Sakai, T., Kunieda, H. *Phys. Chem. Chem. Phys.* 2004, *6*, 1627.
134. Maestro, A., Acharya, D. P., Furukawa, H., Gutierrez, J. M., Lopez-Quintela, M. A., Ishitobi, M., Kunieda, H. *J. Phys. Chem.* B 2004, *108*, 14009.
135. Acharya, D. P., Kunieda, H., Shiba, Y., Aratani, K. *J. Phys. Chem.* B 2004, *108*, 1790.
136. Acharya, D. P., Hattori, K., Sakai, T., Kunieda, H. *Langmuir* 2003, *19*, 9173.
137. Rodriguez, C., Acharya, D. P., Maestro, A., Hattori, K., Aramaki, K., Kunieda, H. *J. Chem. Eng. Jap.* 2004, *37*, 622.
138. Kunieda, H., Rodriguez, C., Tanaka, Y., Kabir, M. H., Ishitobi, M. *Colloids Surf. B* 2004, *38*, 127.
139. Kabir-ud-Din, Kumar, S., Kirti, Goyal, P. S. *Langmuir* 1996, *12*, 1490.
140. Peiffer, D. G. *Polymer* 1990, *31*, 2353.

141. Couillet, I., Hughes, T., Maitland, G., Candau, F. *Macromolecules* 2005, *38*, 5271.
142. Shashkina, Y. A., Philippova, O. E., Smirnov, V. A., Blagodatskikh, I. V., Churo-chkina, N. A., Khokhlov, A. R. *Polym. Sci. Ser. A* 2005, *47*, 1210.
143. Michaud, F., In., M., unpublished results.
144. Tanaka, F., Edwards, S.F. *J. Non-Newtonian Fluid Mech.* 1992, *43*, 247.
145. Annable, T., Buscall, R., Ettelaie, R., Shepherd, P., Whittlestone, D. *Langmuir* 1994, *10*, 1060.
146. Panmai, S., Prud'homme, R. K., Peiffer, D. G. *Colloids Surf. A* 1999, *147*, 3.
147. Nakamura, K., Yamanaka, K., Shikata, T. *Langmuir* 2003, *19*, 8654.
148. Nakamura, K., Shikata, T. *Macromolecules* 2003, *36*, 9698.
149. Nakamura, K., Shikata, T. *Macromolecules* 2004, *37*, 8381.
150. Oelschlaeger, C., Waton, G., Candau, S. J., Cates, M. E. *Langmuir* 2002, *18*, 7265.
151. Massiera, G., Ramos, L., Ligoure, C., *Europhys. Lett.* 2002, *57*, 127.
152. Schmidt V, Lequeux F. *J. Phys. II France* 1995, *5*, 193.
153. Kavassalis, T. A., Noolandi, J. *Macromolecules* 1989, *22*, 2709.
154. Turner, M. S., Cates, M. E. *J. Phys. France* 1990, *51*, 307.
155. Drye, T. J., Cates, M. E. *J. Chem. Phys.* 1992, *96*, 1367.
156. Elleuch, K., Lequeux, F., Pfeuty, P. *J. Phys. I France* 1995, *5*, 465.
157. May, S., Bohbot, Y., Ben-Shaul, A. *J. Phys. Chem. B* 1997, *101*, 8648.
158. Bernheim-Groswasser, A., Wachtel, E., Talmon, Y. *Langmuir* 2000, 16, 4131.
159. Lin, Z. *Langmuir* 1995, *12*, 1729.
160. Constantin, D., Freyssingeas, E., Palierne, J. F., Oswald, P. *Langmuir* 2003, *19*, 2554.
161. D'Arrigo, G., Briganti, G. *Phys. Rev. E* 1998, *58*, 713.
162. Constantin, D., Freyssingeas, E., Palierne, J. F., Oswald, P. *Langmuir* 2003, *19*, 2554.
163. Porte, G., Gomati, R., El Haitamy, O., Appell, J., Marignan, J. *J. Phys. Chem.* 1986, *90*, 5746.
164. Karaborni, S, Esselink, K., Hilbers, P. A. J., Smit, B., Karthäuser, J., van Os, N. M., Zana, R. *Nature* 1994, *266*, 254.
165. Bernheim-Groswasser, A., Zana, R., Talmon, Y. *J. Phys. Chem. B* 2000, *104*, 4005.

9 Nonlinear Rheology of Giant Micelles

Jorge E. Puig, Fernando Bautista, J. Felix Armando Soltero, and Octavio Manero

CONTENTS

9.1 INTRODUCTION

The rheology and dynamics of giant micellar solutions have attracted special attention because subtle changes in surfactant, cosurfactant, temperature, added electrolyte, and counterion can produce enormous changes in dimensions, flexibility, and interactions.[1] Since giant micelles exhibit shear thickening in the dilute regime, strong viscoelasticity, high viscosity at low shear rates and shear thinning at higher concentrations, they have been commercially used as drag-reducing agents, in home and care products, and in oil field applications (see Chapters 15 to 18 in this volume).

The complex rheological behavior of micellar solutions is a consequence of micellar growth, micellar interactions, and their dynamic nature (see Chapter 10).

To comprehend this behavior it is convenient to distinguish three concentration regimes: dilute, semidilute, and concentrated. The dilute regime ends at the so-called *entanglement or overlapping concentration C**, and the semidilute concentration spans from *C** to concentrations where the entangled network has mesh sizes larger than the persistence length, typically around 10 wt% for common surfactant systems.[2] Above *C**, the long and flexible wormlike micelles can entangle, similarly to polymer solutions. However, and in contrast to polymers in solution, wormlike micelles break and reform continuously. Two relaxation processes control the viscoelastic response of these solutions: a kinetic one due to micellar breaking and reformation, and a diffusional one associated with the reptation of the micelles. The ratio of the characteristic times of these processes, τ_{Break}/τ_{Rep} (τ_{Break}, τ_{Rep} = breaking and reptation times, respectively), determines the rheological response of these solutions (see Chapter 4). When the kinetic process controls the relaxation, that is, $\tau_{Break}/\tau_{Rep} \ll 1$, the linear viscoelastic response at low and intermediate frequencies is Maxwellian with a single structural relaxation time, $\tau_R = (\tau_{Break}\tau_{Rep})^{1/2}$, but when reptation is the dominant mechanism, deviations from the Maxwell model are observed and there is a distribution of relaxation times (see Chapter 4). When the shear rate is high enough to perturb the equilibrium structure, nonlinear viscoelastic response is observed. At low concentrations well below *C**, the response is Newtonian at low shear rates, but at a critical value, pronounced shear thickening caused by the induction of structure is observed.[3,4] Above *C**, the nonlinear response includes flow phenomena such as shear thinning and shear-banding flow.[2]

This chapter is focused on recent developments on the nonlinear rheology of giant micelles. First the general rheological patterns of these systems (Section 9.2) and of mixed surfactant solutions (Section 9.3) are presented. In Section 9.4 the relevant aspects of the shear-thickening transition and the possible morphology and mechanism of this transition are discussed. Last, in Section 9.5, the shear-banding transitions in semidilute and concentrated systems are examined.

9.2   GENERAL RHEOLOGICAL BEHAVIOR

Even though the rheological behavior of micellar solutions is varied and complex, it is possible to distinguish three basic patterns of the zero-shear viscosity (η_0) as a function of concentration, where this variable can be the total surfactant concentration *C*, salt or additive concentration C_S at a fixed *C*, and so forth. Figure 9.1 (inset) illustrates these patterns.

These patterns share common features up to the first maximum in viscosity, mainly: low viscosity similar or slightly larger than that of the solvent, a sharp increase in η_0 at the overlapping concentration *C** and a viscosity maximum. Above *C** and up to the concentration where the viscosity maximum is detected, η_0 increases with *C* according to a power law of the form $\eta_0 \sim (C/C^*)^a$, where the exponent *a* takes values from 1 to 10.[5–11] Hoffmann and Ulbricht reported values of 8.5 ± 0.5 for a variety of ionic surfactant systems and concluded that this power law exponent must be controlled by electrostatic interactions.[12] For

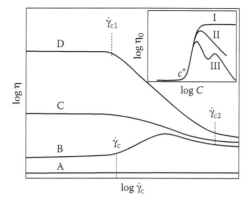

FIGURE 9.1 Plots of shear viscosity versus shear rate. Inset: Typical patterns of the zero-shear viscosity versus concentration curves for micellar solutions. The different patterns are explained in the text.

higher concentrations, several trends are observed, depending on the particular micellar solution (inset in Figure 9.1). In Type-I behavior, the viscosity saturates after the first maximum reaching a plateau until phase separation occurs. Type-II behavior, which is the most common, is characterized by a decrease in viscosity after the maximum. Type-III behavior is more complex, where two maxima at lower and higher concentrations and a minimum at intermediate concentrations are observed.

Micellar solutions of anionic surfactants with added electrolytes exhibit Type-I behavior.[10,13] Dilute solutions of sodium oleate (NaOA) exhibit a viscosity plateau at the maximum viscosity with increasing concentration of KCl or triethylammonium chloride (Et$_3$NHCl), but the plateau viscosity is two decades larger with KCl; moreover, a cloud point, which is rarely observed in ionic surfactant solutions, is detected only in the presence of Et$_3$NHCl.[14] The authors suggested that electrostatic screening leads to micellar growth and to a viscosity increase, and because the head group area reaches a minimum at high salt concentration, further salt addition has no influence on electrostatic head group interactions and so, the plateau develops. However, Type-II behavior is reported for a similar system (KOA/KCl).[15] Type-II behavior has been reported at fixed sodium dodecylsulfate (SDS)[10] or sodium dodecyl trioxyethylene sulfate[16,17] concentration and varying concentration of AlCl$_3$, mixtures of anionic and cationic surfactants,[7,18,19] mixtures of anionic surfactant and hydrotropic salt,[20] mixtures of nonionic surfactants,[21–24] mixtures of nonionic and ionic surfactants,[25–27] mixtures of a gemini anionic surfactant with nonionic surfactants[28] and in mixtures of zwitterionic with anionic or cationic surfactants.[29] Type-III behavior has been reported in solutions of cationic surfactants with inorganic or organic salts.[12,30–33] The solutions of cationic surfactants with strong binding counterions[34] and erucyl-bis-(hydroxyethyl)methyl ammonium chloride alone or mixed with an associative polymer[35] exhibit a continuous increase in the viscosity with surfactant concentration.

The increase in η_0 with concentration for $C > C^*$ has been interpreted as the result of micellar growth and entanglements that yield a transient network formation and high viscoelasticity. However, the proposed mechanism of micellar growth depends on the nature of the species involved. In ionic surfactant solutions, micellar growth is promoted by electrostatic screening that reduces the repulsion between the charged polar heads upon addition of electrolytes and by reduction of the micellar surface charge density by adding salts with strongly binding counterions.[10,12,16,17,20,35-38] In mixtures of anionic and cationic surfactants, both electrostatic screening and reduction of surface charge density promote micellar growth and the rise in viscosity.[7,18,19] Micellar growth can also be induced by addition of nonionic surfactant to a dilute solution of ionic surfactant by reducing electrostatic repulsion by the insertion of the nonionic molecules at the micellar surface.[25-27]

The decrease in viscosity after the maximum has been attributed to the occurrence of branched micelles, which causes the formation of sliding points that allow faster relaxation.[10,16,17-19,21,22,35-41] A reduction in the micellar length due to breaking has also been suggested to explain the viscosity decrease.[16,17,40] Kaler and co-workers have explained the viscosity fall by an increase in micellar flexibility and the polyion nature of wormlike micelles, in catanionic systems.[7] Kunieda and co-workers have proposed structural changes in the vicinity of phase transitions for mixed nonionic surfactant systems.[24]

The increase in viscosity after the minimum and the second maximum in Type-III behavior are not completely understood. For solutions of cetylpyridinium chloride (CPyCl) containing increasing concentration of sodium salicylate (NaSal), the micelles are positively and highly charged at the first maximum, neutral at the minimum, and negatively charged at the second maximum, suggesting that electrostatic interactions are important.[12] This behavior, on the other hand, can be understood in terms of the dependence of the structural relaxation time (τ_R) and the plateau elastic modulus (G_0) with concentration inasmuch as η_0 results from a combined structure and dynamic behavior. In particular, for systems exhibiting Type-III behavior with increasing electrolyte concentration, it is found that τ_R decreases after the first maximum, goes though a minimum and then increases again with increasing electrolyte concentration, whereas G_0 is fairly insensitive to this parameter.[12] However, as pointed out before, no details on the morphological (or topological) structure in this minimum-and-second-maximum concentration range has been elucidated so far.

The dependence of η_0 on temperature, T, is more universal and better understood. Above C^*, the average contour length follows an Arrhenius-type behavior with temperature and with the square root of surfactant concentration, at least for electrostatically screened and noncharged systems (see Chapter 4). According to the mean-field theory, the average micellar length ($\bar{L}$) diminishes with increasing temperature according to an Arrhenius law (see Chapter 4). Hence, the viscosity, which is a function of $\bar{L}$, also follows an Arrhenius-type behavior.[7,42-44]

From the slope of plots of log (η_0) against T^{-1}, the activation energy can be obtained and values ranging from 20 to 70 k_BT have been reported.[5,38,41,42,45,46] Exceptions of the Arrhenius dependence of the viscosity have been reported in solutions of nonionic polymer surfactants[47] and of nonionic glucoside surfactants[48] because the micelles grow with increasing T.

Shear flow induces alignment and structural changes in giant micelles solutions; hence, the nonlinear rheology is quite complex. Nevertheless, the behavior of most systems can be reduced to the patterns shown in Figure 9.1. At low concentrations near the critical micelle concentration, cmc, Newtonian behavior as a function of shear rate is always detected (plot A). The shear viscosity η follows Einstein equation with increasing hydrated volume fraction of the spherical micelles.[49] Above the second cmc, cmc_2, and up to the neighborhood of C^*, where locally rodlike micelles are present, shear thickening is observed above a critical shear rate $\dot{\gamma}_c$ and at higher shear rates, shear thinning is detected (plot B).[3,4,50] The shear-thickening transition is discussed in detail in Section 9.4. For $C > C^*$, a transient network of entangled giant micelles forms. In this situation, the fluid is Newtonian at low shear rates and shear thins at higher shear rates (plot C). For short-chain surfactants, however, the Newtonian viscosity increases with concentration but no shear thinning is detected within the examined range of shear rates.[26] When the relaxation is controlled by the kinetics of reformation of the micelles (the so-called *fast-breaking regime*), the slope of the shear-thinning region becomes closer or equal to 1, indicating that a stress plateau develops between two critical shear rates, $\dot{\gamma}_{c1}$ and $\dot{\gamma}_{c2}$ (plot D).[2] This flow discontinuity, which is accompanied by long transients and oscillations in the viscosity, is referred to as shear banding and is discussed in Section 9.5.

9.3 MIXED SURFACTANT SYSTEMS

The addition of lipophilic nonionic surfactant to hydrophilic nonionic surfactant solutions induces micellar growth and the formation of viscoelastic solutions. Kunieda and co-workers have investigated the effects of the size of the surfactant head group of polyoxyethylene monoalkyl ethers (C_mEO_n) and of the length of the lipophilic chain of alkanoyl-N-methylethanolamide (NMEA-m) on the micellar growth of polyethylene cholesteryl ethers ($ChEO_x$) and sucrose alkanoates.[21–24] These systems exhibit Type-II behavior. For low values of the mole fraction of the nonionic surfactant (X), the solution behaves as a low-viscosity Newtonian fluid: η_0 rises slowly with X up to a critical value where η_0 increases rapidly with X. Above this critical or overlapping concentration, the solutions exhibit strong viscoelasticity and shear-thinning behavior due to micellar entanglement and the formation of a transient network. The critical mole fraction in these mixed surfactant systems shifts to higher values as the size of the EO_n group increases or as the length of the lipophilic chain decreases.[21–24] These results can be attributed to a decrease in the effective area per surfactant molecule or to an increase in chain length of the lipophilic core, which reduces the curvature and induces micellar growth. For the $ChEO_x$-NMEA and the sucrose-$C_{12}EO_n$ systems, η_0 goes through a maximum and then diminishes, which was interpreted as a result of micellar branching.[24]

Although there are some reports of significant micellar growth induced by the addition of nonionic amphiphiles (medium- and long-chain alcohols and amines) to dilute ionic surfactant solutions,[51] the formation of highly viscoelastic dilute micellar solutions was only recently documented. For instance, the addition of small amounts of NMEA-m or $C_{12}EO_n$ nonionic surfactants to dilute anionic (SDS),[25] cationic (DTAB and CTAB),[26,27] or gemini cationic surfactants[28] solutions yields highly viscoelastic wormlike micellar solutions. With no added nonionic species, the ionic micellar solutions exhibit Newtonian behavior and low viscosities due to the low surfactant concentrations. Upon addition of NMEA-m (m = 12 or 16 hydrocarbon tail) or $C_{12}EO_n$ (n = 1, 2, 3, or 4), Type-II behavior is observed. The overlapping concentration decreases as the EO chain length in the $C_{12}EO_n$ diminishes or as the chain length in the NMEA surfactants increases, that is, as the lipophilicity of the nonionic surfactants increases.[25–28]

Kaler and co-workers investigated recently the phase and rheological behavior of mixtures of anionic and cationic surfactants.[7,18,19] These systems are of special interest because they allow a fine-tuning of the electrostatic interactions by varying the ratio of the oppositely charged surfactants. Mixtures of sodium oleate (NaOA) and alkyltrimethylammonium bromide surfactants (C_mTAB) exhibit a strong synergistic rheological behavior and a Type-II behavior with concentration.[18] With the C_8-homologue, a millionfold increase in viscosity relative to the single-component solutions is observed when the $NaOA/C_8TAB$ weight ratio is 70/30; the maximum in viscosity increases with increasing total surfactant concentration and shifts to lower $NaOA/C_8TAB$ ratios. Moreover, this system does not precipitate at the equal molar ratio of these surfactants. By reducing the tail to C_6, on the other hand, the viscosity maximum is only 50 times the viscosity of water. With the C_{10}- and C_{12}-homologues, there is phase separation at intermediate-weight ratios and only the raising branch of the viscosity at low-weight ratios and the decreasing branch at high-weight ratios are observed. Hence, the asymmetry of the tails of the surfactant pair plays a crucial role in the phase and rheological behavior of these surfactant mixtures. Mixtures of cetyltrimethylammonium tosylate (CTAT) and sodium dodecylbenzyl sulfonate (SDBS) were investigated as a function of total surfactant concentration and CTAT/SDBS varying ratio without electrolyte or by addition of NaCl or sodium tosylate.[7,19] The addition of SDBS to CTAT solutions induces a Type-II behavior with a viscosity maximum at a CTAT/SDBS weight ratio of 97/3, which is one order of magnitude larger than the viscosity of the pure CTAT solution. Upon increasing surfactant concentration, keeping the CTAT/SDBS = 97/3, Type-II behavior is also observed; the relaxation time goes through a maximum and Maxwell behavior is achieved after the maximum. Small-angle neutron scattering (SANS) studies of the quiescent solutions reveal that the micelles become more flexible at higher C and higher CTAT/SDBS ratios. By a combination of rheology, rheo-optics and SANS, Schubert et al.[19] determined the variations of the relevant length scales (contour length, entanglement length, mesh size, persistence length, and cross-sectional radius) with surfactant concentration and composition as well as with electrolyte concentration, and found that these parameters systematically affect the self-assembled microstructure in a way that a transition in rheological behavior from nonionic micelles to polyelectrolytes was spanned.

9.4 THE SHEAR-THICKENING TRANSITION

9.4.1 EXPERIMENTAL OBSERVATIONS

Since its discovery in the 1980s by Hoffmann and co-workers,[52,53] the shear-thickening transition (STT) in surfactant solutions has received a great deal of attention because of its implications in drag reduction in turbulent flow and in the fundamental understanding of the role of instabilities and the coupling between flow and concentration fluctuations.[54,55] This transition is characterized by an increase of viscosity by a factor up to 20–50 when the solution is sheared above a critical shear rate $\dot{\gamma}_c$ or a critical shear stress σ_c due to the formation of shear-induced structures (SIS). σ_c and $\dot{\gamma}_c$ have a one-to-one correspondence since the same pair of values are obtained whether measurements are made by strain- or stress-controlled conditions.[3] The critical shear rate increases with T and diminishes with increasing C.[50,53,56–59] For dilute CTAT solutions[57] it has been reported that $\dot{\gamma}_c \sim C^\alpha \exp(-E_a/k_B T)$, with an activation energy $E_a =$ of 123 ± 3 $k_B T$. Similar dependences of $\dot{\gamma}_c$ on T and C have been documented for other surfactant systems.[56,58,60] $\dot{\gamma}_c$ also varies upon addition of salts, but no clear tendency was found: it decreases with some salts (NaBr, NaNO$_3$, and NaBr) but increases with others (NaCl and NaF).[61,62] When the solution is sheared above $\dot{\gamma}_c$ or σ_c, shear thickening is not instantaneous but an induction time, t_{ind}, ranging from seconds to several minutes is necessary for the inception of the viscosity growth.[3] After the induction time, the steady viscosity is reached after another long time, often referred to as the saturation time, t_{Sat}.[3] Both t_{ind} and t_{Sat} depend on the shear and thermal histories of the sample[63,64] and they become longer as the shear rate is closer to $\dot{\gamma}_c$. A power law, $t_{ind} \sim \dot{\gamma}^{-m}$, has been reported by several authors.[3,65–68] The effect of the flow field on the STT of dilute CTAT solutions has also been investigated.[69] No shear thickening occurred under strong elongation flow fields, which emphasizes the importance of shear in the formation of the SIS. Moreover, a synergistic viscosity enhancement was noticed when the solution was forced to flow through a porous medium where the flow field contains both shear and extensional components.

Most of the research on the STT has been performed using cationic surfactants in the presence of electrolytes or salts containing strongly binding counterions such as salicylate or tosylate.[52,53,56] Dilute solutions of gemini[59,70] and fluorocarbon surfactants[64] also exhibit pronounced shear thickening in the absence of electrolytes. In these systems, large aggregates, which are considered precursors of the SIS, have been discovered at concentrations near the cmc. The influence of the micellar surface charge density on the STT has been examined by mixing anionic and cationic surfactants at different ratios in dilute solutions.[7] Shear-thickening behavior has been observed in uncharged micelles of lecithin in n-decane swollen with water (organogels), in which slow kinetics is observed; in addition, the solutions did not revert to the original state after cessation of flow, even after several hours.[8] This was attributed to a substantial growth of wormlike micelles that affected their alignment. Truong and Walker were able to control $\dot{\gamma}_c$ and the apparent steady state viscosity by adding nonionic polymers (hydroxylpropyl cellulose and polyethylene oxide) to dilute CTAT solutions.[62] The shear-thickening behavior of dilute solutions of CTAT has been examined exhaustively in recent years.[57,63,67–69,71]

Besides strain-controlled and stress-controlled rheometry, the shear-thickening transition and the *SIS* morphology has been investigated by flow birefringence,[52,58,63,67] rheo-optics,[67] flow electrical conductivity,[58] light scattering,[66,70,72,73] SANS at rest and under shear,[50,57,59,74] cryo-transmission electron microscopy,[58,75] and flow visualization by particle image velocimetry (PIV).[68,71,76]

Four regimes have been identified in steady shear (Figure 9.2, inset).[3,68,71] Regime I is Newtonian (no *SIS* formation) and occurs for shear stresses or shear rates smaller than σ_c or $\dot{\gamma}_c$. Regime II is detected only under stress-controlled conditions as a re-entrant region over intermediate stresses ($\sigma_c < \sigma < \sigma_f$) where large fluctuations with long time scales in the shear viscosity are detected; these fluctuations are associated with the inhomogeneous formation of flow-induced

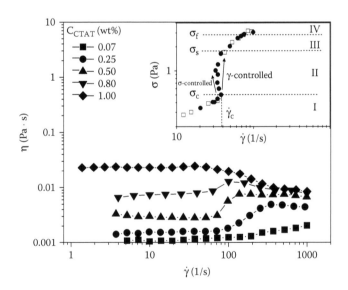

FIGURE 9.2 Experimental and predicted shear stress versus shear rate measured in cone and plate geometry at 25°C for various CTAT concentrations (wt%): 0.07 (■), 0.25 (●), 0.5 (▲), 0.8 (▼), 1.0 (◆). Inset: Steady stress rate versus shear-rate curves measured in stress-controlled or strain-controlled rheometer. (Adapted from Ref. 77.)

structures.[3] Under strain-controlled conditions, there is a discontinuity or jump from σ_c to σ_f (inset in Figure 9.2).[3,73] However, others have found by strain-controlled rheometry a continuous increase in the viscosity with increasing shear rate followed by shear thinning at high $\dot{\gamma}$; as surfactant concentration is increased, the shear-thickening behavior vanishes and only shear-thinning behavior is observed (Figure 9.2).[57,61,63,67,68,71] In fact, the shape of the experimental curve of σ versus $\dot{\gamma}$ depends on the geometry of the sheared cell and whether strain-controlled or stress-controlled rheometry is used.[3,4,65,67,68] In regime III, the shear rate increases monotonically with shear stress (Figure 9.2) and the shear-induced viscous phase forms by homogeneous nucleation.[3,4] In this regime, the time scale and the intensity of the viscosity fluctuations decrease and regions of low and high velocity coexist.[68] The viscosity in Regime III increases with the thickness of the gap in Couette flow,[4,60] but it decreases with the cone angle in cone-and-plate geometry.[77,78] This effect has been explained in terms of (1) the size increase of micellar structures when the gap is increased,[79] (2) the slipping of the *SIS* along the walls of the flow cells through a thin lubricating fluid layer[4] and (3) stratified shear flows.[78] Regime IV, occurring for $\sigma > \sigma_f$, is characterized by the mechanical breakdown of the *SIS* and shear thinning.[4]

Macías et al.[68] investigated the shear thickening of dilute solutions of CTAT (0.25 wt%) by rheometry and PIV. In parallel-plate flow, the four regions of flow behavior reported by Hu and co-workers[3] were identified. In pipe pressure flow, PIV reveals a superposition of two parabolic regions of the velocity profile located near the center and close to the pipe walls, and a transition region where again strong fluctuations in the velocity profile are observed. In the parallel-plate fixture, as the stress at the rim of the plates corresponds to that within region II of flow behavior (at the onset for shear thickening) and for times longer than the induction time, the angular component of the velocity deviates from linearity, indicating a 30% increase in the apparent viscosity. When the fluid is subjected to stress within region III, corresponding to the largest increase in viscosity with shear rate, the tangential velocity of the fluid in the radial position deviates strongly from linearity and fluctuates spatially. The velocity fluctuations can be associated to changes in time of the viscosity. The velocity oscillates between values displayed by the low-viscosity fluid and those corresponding to the high-viscosity fluid. For stresses at the rim of the plate corresponding to region IV of flow behavior, the fluid is subjected to stresses covering the four regions due to the linear variation of the stress with radial distance. PIV data collected in less than a second exhibit strong fluctuations and solidlike displacements. The recorded fluctuations in the tangential velocity are much larger than those of the viscosity.

Initially, for small radial distances the tangential velocity increases linearly. As the radial distance increases, the velocity profile becomes nonlinear and shows a maximum where regions of high velocity coexist with regions of low tangential velocity.

The *STT* occurs at concentrations where locally cylindrical micelles exist as deduced by SANS experiments of quiescent solutions,[50] that is, from the cmc$_2$ up to the neighborhood of C^*. However, an independent *in situ* measurement of the micelle length is necessary to confirm this conclusion.[1] Recent results, on the other hand, have discovered large aggregates below C^* in the quiescent state, which have been suggested as the precursors of the high viscous phase.[59,64] Strong electrostatic repulsion competing against endcap energy also appears to be a requirement for the occurrence of the shear-thickening phenomenon. Troung and Walker showed by SANS that when the electrostatic interactions are adequately screened by the addition of an electrolyte, the *STT* disappears.[74] However, the formation of *SIS* in nonionic surfactant systems indicates that electrostatic interactions may not be the only mechanism that leads to the *SIS* formation.[80]

9.4.2 SIS MORPHOLOGY

One of the most controversial issues in shear thickening is about the morphology of the *SIS*. Using SANS at rest and under shear, Berret et al. found that for semidilute CTAT solutions, the structure factor is not affected by the flow, whereas for dilute shear-thickening solution there is a systematic shift of the well-defined maximum at the wave vector q_{MAX} in the shear-thickening regime (5 s^{-1} < $\dot{\gamma}$ < 50 s^{-1}, see Figure 9.3) indicating that the short rodlike micelles at rest become elongated and strongly aligned in the shear-thickening regime.[81] Similarly, Oda et al. detected in gemini surfactant solutions a shift in the SANS correlation peak in the sheared solutions and an increase in the scattered intensity at small q, but these modifications were only observed below the overlap concentration.[58]

Oda and co-workers proposed a different morphology for dilute solutions of a gemini cationic surfactant from time-resolved SANS studies after cessation of flow.[59] These authors found that the local anisotropy relaxes in about 100 s and

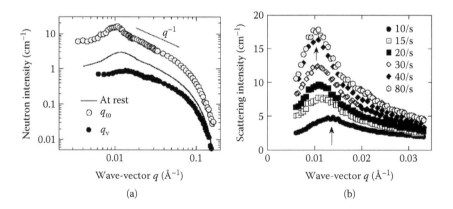

(a)

(b)

FIGURE 9.3 (a) Scattering intensities obtained at $\dot{\gamma}$ = 188 s^{-1} along the velocity (closed symbols) and vorticity (open symbols) directions. The continuous curve is the scattering in the absence of flow. SANS results on a CTAT/D$_2$O solution at 0.26 wt% under shear. (b) Intensity scattered in the vorticity direction at shear rates ranging from 8 to 80 s^{-1}. (Reproduced from Ref. 81 with permission of EDP Science.)

found evidence of a shear-induced viscous phase that is stable for a few hours. These results were interpreted in terms of a viscous phase viewed as a network of anisotropic aggregates connected by links that are much more labile than the aggregates themselves. A possible mechanism to explain the formation of these aggregates and their stability is a counterion-mediated attraction between rodlike micelles.[82] To further investigate the structure of the gel-like phase reported by Oda and co-workers,[59] the same gemini surfactant system was studied using small-angle light scattering (SALS) under shear in a strain-controlled rheometer.[70] The relevant results of this SALS study were: (1) a streak pattern of scattered light in the direction perpendicular to the flow appears when $\dot{\gamma} > \dot{\gamma}_c$; (2) this pattern begins to show up after an induction time (t_{ind}) with the simultaneous rise in viscosity; (3) t_{ind} diminishes with increasing shear rate and decreases or even vanishes when the solutions were previously sheared above $\dot{\gamma}_c$; (4) the steady state values of the average streak intensity and of the apparent shear viscosity follow the same behavior with increasing $\dot{\gamma}$, going through a maximum and then diminishing at higher shear rates (Figure 9.4); and (5) the increase in $\dot{\gamma}$ does not modify the shape of the streak but simply its intensity level. The qualitative comparison between the experimental two-dimensional light scattering intensities and those predicted by a model of light scattering from a set of rods in a shear flow indicated that the experimental data cannot be interpreted in terms of large anisotropic objects aligned by the flow, but they are consistent with an aligned heterogeneous gel-like layer in the gap. Weber and Schosseler suggested that the gel-like phase might form and grow at shear rates below the critical value followed by a subsequent structuration and alignment at $\dot{\gamma}_c$ or, alternatively, that the gel-like phase might form at $\dot{\gamma}_c$ (homogeneous nucleation).[70]

Prötzl and Springer suggested the existence of a gel-like phase below $\dot{\gamma}_c$ with a structure that is not substantially different from that of the original solution before shear.[66] Weber et al. reported evidence by various scattering techniques of

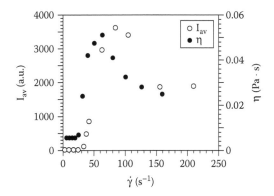

FIGURE 9.4 Comparison of the evolution of the average streak intensity and of the apparent shear viscosity as a function of shear rate ($C = 18.3$ mM, $T = 25°C$). (Reproduced from Ref. 70 with permission of the American Chemical Society.)

the coexistence of small, intermediate, and very large micellar sizes in the concentration range between the cmc and C^*.[83] The presence of large aggregates has also been reported in dilute shear-thickening fluorocarbon surfactant solutions.[64] Analysis of static and dynamical scattering data on these micellar systems led to conclude that the large aggregates are ringlike micelles, which coexist with smaller ringlike or short wormy micelles.[64] The presence of these large aggregates could trigger the gel-like phase growth at $\dot{\gamma} \geq \dot{\gamma}_c$. However, the microscopic nature of this gel-like phase and the mechanism of formation remain still unclear. Barentin and Liu suggested that the *SIS* consist of interconnected bundles or aggregates of interconnected bundles, which form as a result of counterion-assisted intermicellar attractions.[82]

9.4.3 KINETICS OF *SIS*

Another important aspect of *SIS* is its kinetics of formation after inception of flow and of disappearance after cessation of flow. Current results on these topics demonstrate that these phenomena are complex and intriguing. The onset of flow birefringence[67] and the appearance of a bright streak pattern[70] coincide with the increase of viscosity. However, the direct visualization of the birefringent *SIS* occurs a short time after the viscosity begins to increase; this time lapse diminishes as $\dot{\gamma}$ increases,[3] which could be due to the sensitivity of this method. The induction times detected by flow birefringence[67] and SALS[70] coincide with those detected by rheometry. In steady state, the apparent viscosity and the scattered light-streak intensity shift in a similar way and remain correlated.[70] Moreover, the proportion of the shear-induced phase derived from SANS measurements under shear increases monotonically with shear rate and fills the whole gap at the shear rate where shear thinning is detected.[50] These results suggest that the *SIS* grows progressively with increasing shear rate. However, steady state intensity profiles indicate that the gap is filled with the birefringent phase after the induction time and that they increase in intensity.[67] The kinetics of *SIS* formation in dilute CTAT solutions was followed by measuring the flow birefringence Δn and the extinction angle χ as a function of time and spatial coordinates in a 1-mm gap of a Couette cell with a spatial resolution of 15 μm.[67] Typical results are shown in Figure 9.5. The values, A, B, C, D, and E, correspond to different positions within the gap (inset of Figure 9.5a). The first reliable measurement can be made 4 s after t_{ind}. Already at this time, $\chi \sim 25°$ in the whole gap indicating a noticeable alignment and it decreases rapidly to its stationary state (ca. 0.7°) (Figure 9.5a). Moreover, χ does not depend on shear rate or the position in the gap. The spatial evolution of Δn (Figure 9.5b) shows that the kinetics of growth is more rapid close to the inner (rotating) cylinder (point E) than close to the outer cylinder (point A). Nevertheless, the saturation of the flow birefringence occurs simultaneously through the gap.

From these results, Berret and co-workers concluded that the kinetics of growth of *SIS* occurs in three stages: (1) induction with the incipient formation

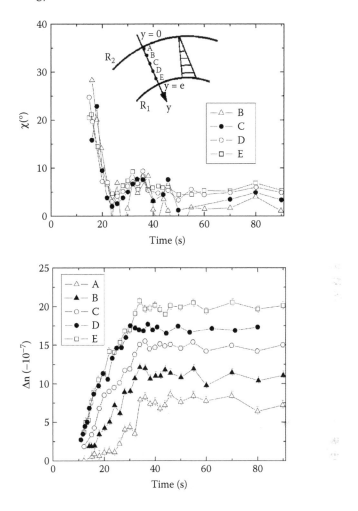

FIGURE 9.5 Time dependence of (top) the extinction angle χ and (bottom) the flow birefringence Δn at different points in the gap of the Couette cell for a 0.16 wt% CTAT solution sheared at 100 s^{-1}; at this shear rate, $t_{Ind} = 11 \pm 1$ s and $t_{Sat} = 35 \pm 1$ s. Inset: Points A to E correspond to the coordinates y: (A) 0.04 mm, (B) 0.2 mm, (C) 0.4 mm, (D) 0.6 mm and (E) 0.8 mm, respectively. (Reproduced from Ref. 67 with permission of the American Chemical Society.)

of a birefringent phase close to the rotating cylinder, which requires a latency time for the *SIS* to appear; (2) growth occurring in time lapse of the order of $t_{sat} - t_{ind}$, where the *SIS* develops until the gap is filled with the birefringent phase; and (3) saturation or steady state (Figure 9.6).[67] The fact that the birefringent phase occupies the whole gap during the whole kinetic process suggests that the *SIS* nucleation occurs in a homogeneous fashion, as suggested by Hu et al. in

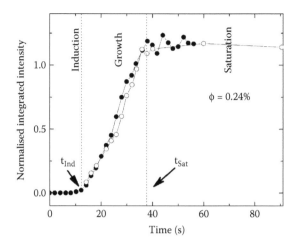

FIGURE 9.6 Time evolution of the normalized integrated intensity in step shear-rate experiments at 60 s⁻¹. The experiments were performed on two different 0.24 wt% CTAT stock solutions (full and empty symbols). (Reproduced from Ref. 67 with permission of the American Chemical Society.)

regime III.[3] However, in stress-controlled rheometry, a regime II where the nucleation is apparently nonhomogeneous has been reported.[3] Hence, it will be enlightening to examine the kinetics of *SIS* formation under stress-controlled conditions. The fact that both viscosity and flow birefringence exhibit short time and short scale fluctuations around the average profiles suggests that the flow in the shear-thickening regime is nonhomogeneous and shear banded.[67] Macías et al. demonstrated by PIV that the flow is inhomogeneous in the shear-thickening region of CTAT solutions.[68,71] The detection of turbid (high-viscosity) and clear (low-viscosity) shear bands was reported for semidilute (30-80 mM/mL) solutions of CPyCl/NaSal in equimolar ratio that exhibit shear thickening.[72,73] A unique behavior of these bands is that they alternate in a very regular fashion, which leads to periodic oscillations in the shear rate under stress-controlled conditions. However, the kinetics of formation of these bands is still a challenge.

The kinetics of disappearance of the *SIS* has been studied in gemini cationic surfactant solution by time-resolved SANS after cessation of flow.[59] These authors found that the flow birefringence decays in about 10 s, the local anisotropy measured by SANS relaxes in ca. 100 s, and some shear-induced features of the SANS intensity are stable for more than 1000 s. From these results, a loosely connected network of long-lived aggregates with typical sizes of 1 μm was proposed, where counterion mediated attraction between rodlike micelles could explain the formation of these aggregates and their long lifetime. The existence of long-lived structures can explain the slow relaxation processes following a shearing at $\dot{\gamma} > \dot{\gamma}_c$ or a thermal pretreating of the solutions (Figure 9.7).[63,64] Cates and Candau argue that the perturbation induces a change in the balance between

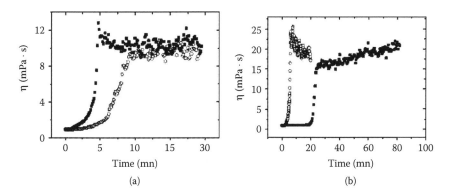

FIGURE 9.7 (a) Effect of shear cycling for two solutions of perfluoro octylbutane trime-thylammonium bromide at 25°C: $C = 0.2 \times 10^{-2}$ g cm^{-3} ($C/C_e \sim 0.15$, C_e = concentration at which the viscosity starts increasing rapidly), $\dot{\gamma} = 30$ s^{-1}, at equilibrium (O) and restarted after 5 minutes (■). (b) Effect of thermal cycling on a sample sheared at 25°C: $C = 0.6 \times 10^{-2}$ g cm^{-3}; $\dot{\gamma} = 30$ s^{-1}. (O) at equilibrium; (■) preheated at 50°C. (Reproduced from Ref. 64 with permission of the American Chemical Society.)

the large aggregates (ringlike micelles) and the small rodlike micelles and that the slow relaxation is controlled by reversible scission kinetics.[84]

9.4.4 ORIGIN OF SIS

The disagreement in the morphology of the *SIS* has led to divergent views about the origin and mechanism of the shear-thickening phenomenon, even though most shear-thickening systems share common features: (1) the starting structure is that of rod or short wormlike micelles; (2) electrostatic interactions play an important role; (3) shear thickening occurs above a critical shear rate, which is independent of thermal and strain histories, after a latency or induction time; (4) the time to reach steady state is hundredsfold larger than the relaxation time of the system; and (5) there are large fluctuations in the shear viscosity, birefringence, and scattered light intensity at the steady state point. The two current leading points of view to explain the origin of the shear-thickening behavior observed in dilute surfactant solutions are the *shear-induced transition* (*SIT*) and the *shear-induced phase* (*SIP*). In the *SIT*, two structures within the same phase region, rodlike micelles and aligned wormy micelles, coexist in the gap of the cell above $\dot{\gamma}_c$.[50] As $\dot{\gamma}$ is increased above $\dot{\gamma}_c$, the aligned region grows from the high-shear region and progressively fills the gap as shear rate is increased.[50] In the *SIT*, the induction and saturation times can be interpreted as the times related to the alignment of the small micelles, aggregation, and growth of the aligned wormlike micelles.[67] The aggregation is controlled by hydrodynamics and it can begin at $\dot{\gamma} < \dot{\gamma}_c$; however, the viscosity increase does not occur until a critical aggregation size has been reached.[59,66] This scenario is consistent with the continuous increase of the apparent viscosity and thickness of the birefringent phase as the shear rate is increased in strain-controlled experiments. Also, SANS measurements under shear strongly supports the formation of this aligned wormlike micellar solution, similar to that detected in the semidilute regime.[62,63] Moreover, the extinction angle of the birefringent phase measured by rheo-optics techniques is close to 0° indicating a strongly aligned phase.[67] However, this picture does not explain successfully the relaxation times on the order of hours after cessation of shear flow of the long-lived aggregates formed in the shear-thickening process.[59] On the other hand, the *SIP* suggests a shear-induced phase separation into a gel phase and another consisting of small rodlike micelles.[3,4] This scenario is reminiscent to the shear-induced phase mechanism suggested to explain shear-banding phenomena; however, in the *SIP*, the shear stress is multivalued,[3] whereas in the shear-banding transition the shear rate is multivalued.[2] In this view, the induction and saturation times correspond to the nucleation and development times of the highly viscous gel phase. The existence of a structured gel phase has been deduced from the interpretation of the time-resolved SANS studies under cessation of shear flow and SALS measurements.[64,70] However, the microscopic structure of this viscous phase remains still unknown.

Evidently, additional theoretical and experimental work is required. The use of several scattering techniques in start-up, steady, and cessation of flow is necessary to further understand the origin and mechanism of formation and destruction of the *SIS* as well as its nature. Clearly, even though there are many common features in the shear-thickening phenomena, subtle differences such as electrostatics, flexibility, counterion nature, and so forth, play important roles that introduce differences in the origin of the *SIS* and the mechanism of formation. In this context, the recent work of Fischer and co-workers is relevant.[72,73]

9.4.5 MODELING

Microscopic theories and models to analyze shear thickening in dilute surfactant solutions are scarce. Preliminary attempts to describe this phenomenon include theories that incorporate nonequilibrium shear-induced phase transitions and hydrodynamic instabilities.[85,86] Macroscopic models attribute the *STT* to shear-induced gelation into a network of long micelles. Early models have considered a growth mechanism involving the collinear fusion of micelles aligned in the flow.[87–89] However, the predicted critical shear rate is not within the range of experiments because interactions among micelles were not considered. Nevertheless, progress has been achieved in the understanding of short-range attractive electrostatic interactions between rodlike particles bearing the same net charge.[90] Barentin and Liu have attributed the shear-induced gelation to counterion-mediated intermicellar interactions by estimating the interaction potential between wormlike micelles.[82] In this situation, the predicted $\dot{\gamma}_c$ falls within the range of observed values.

Cates and Candau have proposed the existence of large ringlike micelles in the quiescent solutions as the precursors of the gel-like phase.[84] The Goveas and Pine model considers the growth of a shear-induced gel phase in which there is no flow.[91] This model predicts a re-entrant region in the stress-shear rate flow curve only under stress-controlled conditions and a discontinuity under strain-controlled conditions.

A simple model consisting of the upper-convected Maxwell constitutive equation coupled to a kinetic equation that accounts for the structural modifications induced by flow has been proposed to explain the *STT* and the velocity variations observed in the pipe and parallel-plates flows.[92] This model predicts a continuous increase in apparent viscosity with increasing shear rate as reported by several researchers,[57,61,63,67,68] but it cannot predict the re-entrant region in regime II detected by Hu et al.[3] In parallel-plate flow, the model can account for the average steady values of the viscosity in the structured thickened state.[68] In Poiseuille flow, the model predicts in the shear-thickening regime, a transition region between two parabolas where a change in curvature in the velocity profile is observed. The shear rate in the vicinity of the wall is smaller than that resulting from extrapolation of the parabola positioned at the pipe center indicating that a more viscous liquid layer flows close to the walls, corresponding to the shear-thickening regime of region III (inset in Figure 9.2) of high viscosity. This determines the region of the pipe where shear thickening manifests itself. In the region of transition corresponding to that in regime II (inset in Figure 9.2), fluctuations in the measured point velocities are detected, whereas the predictions depict the change in curvature in the transition region. In the vicinity of the wall, there is an apparent viscosity increase and the predictions of the model match the change in curvature of the experiments. Other interesting predictions of the model are the flow curves for various values of the length/diameter ratio of the pipe.[68] For short pipe lengths, the fluid does not shear thicken because the residence time is smaller than the induction time and its behavior is Newtonian. As the residence time in the pipe increases, a gradual manifestation of shear thickening takes place. These predictions illustrate the transient development of shear-thickening phase, in which an induction time, related to the residence time of the fluid in the pipe, is a necessary condition for shear thickening to develop.

9.5 SHEAR-BANDING FLOW

Shear-banding flow is the most intriguing and important manifestation of the nonlinear rheology of wormlike micelles. This flow phenomenon has been observed in many wormlike micellar systems in both the semidilute and concentrated regimes.[2] Shear banding appears as a discontinuity in the shear stress-shear rate flow curve, in which a stress plateau (σ_p) develops between two critical shear rates, $\dot{\gamma}_{c1}$ and $\dot{\gamma}_{c2}$; below and above these shear rates, the flow is homogeneous and usually Newtonian (Figure 9.8), although apparent shear thickening in narrow range of shear rates for $\dot{\gamma} > \dot{\gamma}_{c2}$ has been recently reported on the CTAB/NaSal solutions in the semidilute regime.[93] Moreover, under certain conditions, a meta-stable branch is detected instead of the stress plateau up to a stress value known as *stress jump* where the system becomes unstable and the shear rate begins to increase rapidly.[94,95] Long transients and oscillations accompany this flow

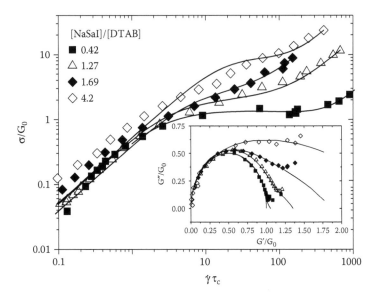

FIGURE 9.8 Reduced shear stress versus reduced shear rate of micellar solutions with different values of the [NaSal]/[DTAB] molar ratio. Inset: Cole–Cole plots for samples shown in figure. (Reproduced from Ref. 36 with permission of the American Chemical Society.)

phenomenon.[34,39,96–103] Shear banding is not unique to wormlike micelles. This transition has been recognized in many complex fluids such as lamellar phases,[104] associative polymer solutions,[105] and bicontinuous microemulsions.[106] Hence, this flow phenomenon appears to be quite general for complex fluids. Shear-banding flow in wormlike micelles has been identified by flow birefringence,[98] NMR velocimetry,[97,107] small-angle light, neutron, and x-ray (SAXR) scattering techniques,[108–110] PIV, and particle tracking velocimetry (PTV).[110]

Even though there are scores of papers on shear-banding flow,[2] there are still various controversies and open questions. One is related to the physical origin of the stress plateau associated with shear banding. Other unanswered issues are: (1) the link between the model viscoelastic behavior in the linear regime often observed and the occurrence of a stress plateau in the nonlinear regime; (2) the selection of the stress plateau; and (3) the kinetics of shear-banding formation. These issues will be addressed below.

9.5.1 Mechanisms of Shear Banding

Several authors have interpreted the isothermal shear-banding flow as a nonequilibrium isotropic-nematic phase transition,[6,39,85,86,102,111–113] but others argue that it is a mechanical instability.[114–120] Concentration fluctuation and layering have also been proposed to explain this phenomenon.[121] Wall slippage has been invoked to explain the discontinuity in the flow curve in polymer solutions.[122] However, the gradual increase of the high-shear band with increasing shear rate, determined by flow birefringence, NMR velocimetry, and PTV in both semidilute and concentrated regimes, rules out this explanation.[110,123,124] Interestingly, in both regimes, flow birefringence measurements overestimate the width of the high-shear band and so it appears that this technique can only be used as a qualitative tool.

The most commonly invoked mechanism for shear banding is the shear-induced phase transition. The arguments that support this mechanism are: (1) a shear-induced structural or nonequilibrium phase transition is typically observed; (2) the stress plateau is unique and robust in the sense that it is independent of shear history; (3) the low transients and oscillations of the stress in start flows resemble the kinetics of nucleation and growth of a second phase, similar to phenomena reported in equilibrium first-order phase transitions; (4) the generalized *flow-phase diagram* (Figure 9.8), obtained by applying a temperature–concentration superposition, strongly resembles the phase diagram of an equilibrium system undergoing phase separation.[111] This mechanism appears conceivable for concentrated systems near the isotropic–nematic phase transition. However, it is unlikely to occur in the semidilute regime.

Shear banding in semidilute micellar solutions is usually attributed to a constitutive instability of purely mechanical origin along the negative slope of the flow curve.[2] This mechanical instability can be enhanced by concentration fluctuations governed by osmotic free energy.[113] In this context, shear banding has been suggested to be a class of shear-induced demixing of the Cahn–Hilliard type. In general, shear banding has been associated with a double-valuedness in the constitutive behavior, and both approaches, mechanical instability versus nonequilibrium phase transition, are not totally incompatible since there is the possibility that the high-shear rate fluid phase nucleates much before the criterion of mechanical instability is reached.[114] Recently, Hu and Lips demonstrated conclusively that shear banding is provoked by a mechanical instability triggered by chain disentanglement in CPyCl/NaSal solutions with added NaCl in the semidilute regime.[110] However, no similar measurements have been performed in semidilute solutions with no added salt where the electrostatic interactions are important and in concentrated systems to generalize this mechanism.

9.5.2 Link between Linear and Nonlinear Rheology

The critical shear rate for the onset of shear-banding flow is of the order of the rate of disentanglement time, τ_{Rep}^{-1}. However, this characteristic time depends strongly on molecular weight ($\tau_{Rep} \sim M_W^3$) and so, molecular weight polydispersity can strongly damp the onset of shear banding. At $\dot{\gamma} \gg \dot{\gamma}_{c1}$, other dynamical processes, such as Rouse modes and local motion of the micelles, cause a stress upturn and the higher shear branch emerges. The Rouse characteristic time, τ_{Rouse}, is also sensitive to polydispersity as $\tau_{Rouse} \sim M_W^2$. Hence, polydispersity can mask the shear-banding transition. Stress relaxation in wormlike micelles occurs from the interplay of the reptation and breaking processes. In the fast-breaking regime

($\tau_{Break} \ll \tau_{Rep}$) applicable to a large class of micellar solutions, the stress relaxation is single exponential with a characteristic relaxation time given by $\tau_R = (\tau_{Rep}\tau_{Break})^{1/2}$.[88] This implies that in the fast-breaking regime τ_R is independent of micelle length polydispersity because the chain breaking and recombination average the micelle length distribution. Hence, it is expected that systems in the fast-breaking regime should exhibit a stress or quasi-stress (not completely flat) plateau. On the other extreme, in the slow-breaking regime, the relaxation is characterized by multiexponential relaxation, and single Maxwellian behavior is no longer observed and the stress plateau should be expected to fade. Figure 9.8 depicts steady shear data for DTAB/NaSal solutions as a function of R = [NaSal]/[DTAB] molar ratio; Cole–Cole plots for the same systems are reported in the inset.[36] The system departs from the fast-breaking regime, as ascertained by the early departures from the osculating semicircle, as R is increased. The stress plateau, which is well defined at low R-values, tends to fade as R increases and eventually it vanishes at large R-values, in agreement with the arguments given here. Hence, it appears that model linear viscoelastic regime (fast breaking) and shear-banding flow are closely linked in wormlike micellar systems. This behavior has been observed in semidilute and concentrated systems.[2] However, there are anomalous systems in which there is no correspondence between the occurrence of fast breaking and stress plateau. Berret examined these anomalous behaviors in a recent review on the shear-banding transition.[2]

9.5.3 SELECTION OF THE STRESS PLATEAU

Along the unstable range where $d\sigma/d\dot{\gamma} < 0$, various scenarios are possible: the flow may never become steady with time, or it may become inhomogeneous, or both. The system generates a coexistence between layers supporting different shear rates, and the crucial question is the corresponding shear stress that they must share. To set the stress plateau, the chemical potential is considered a function of concentration and flow. In the framework of the extended irreversible thermodynamics (EIT), the generalized Gibbs free energy of a fluid system of volume V, subjected to isothermal flow, is given by:[125]

$$dG = V\tau_R \underline{\sigma} : \underline{\Gamma} \qquad (9.1)$$

where τ_R is the disentanglement relaxation time, and $\underline{\sigma}$ and $\underline{\Gamma}$ are the stress and the rate of strain tensors, respectively. Under steady shear flow and neglecting the contributions of normal stresses, this equation can be integrated to give:

$$\Delta G = V\tau_R \int \frac{\dot{\gamma}}{\varphi} d\dot{\gamma} \qquad (9.2)$$

Here φ is the steady state fluidity. For the Bautista et al. model, the steady state fluidity is the solution of $\varphi^2 - \varphi_0\varphi - k_0\lambda(\varphi_\infty - \varphi)\dot{\gamma}^2(1 + \vartheta\dot{\gamma}) = 0$, where φ_0 and φ_∞

are the zero- and very high shear rates, respectively, λ is a structure relaxation time, k_0 is a kinetic constant for structure breakdown, ϑ can be interpreted as a *shear-banding intensity* parameter.[126] This model predicts a sigmoid nonmonotonic curve of stress versus shear rate with three stable solutions. The extended Gibbs free energy versus shear-rate curve exhibits one or two minima, depending on whether the shear rate lies outside or inside the multivalued region. The criterion for bands coexistence is the equality of the extended Gibbs free energy of the bands. The predicted stress plateau coincides remarkably well with experimental data.[36,112] Olmsted and Lu have derived the phase diagrams for complex fluids undertaking phase separation in shear-banding and shear-thickening transitions by coupling flow and concentration gradients as the driving mechanisms for phase separation and established a criterion to set the stress plateau.[85,127]

9.5.4 KINETICS OF SHEAR BANDING

The kinetics of shear banding in the plateau region has distinctive features. The characteristic time for the evolution of the stress toward the steady state value is much longer than the reptation time and it is accompanied with oscillations.[34,39,96-103] These long transients and oscillations are characteristic of the nucleation and growth of the high-shear band.[103] In stress relaxation at applied shear rates within the plateau region, fast decay, a transition, and slow decay are detected.[36,103] The fast decay region is sensitive to the applied shear rate, whereas the slow decay region is governed by the reptation time.[34,36] The long time tail, observed in some systems, can be fitted to a stretched exponential governed by a characteristic time of the nucleation and growth processes, and hence this time is longer than the reptation time.[6] The fast decay scales with the high shear-rate band, whereas the slow decay scales with the low shear-rate band. These findings indicate that models derived from a criterion of complete mechanical instability disagree with experimental data.

Hu and Lips reported a detailed investigation on the kinetics of band formation in a wormlike micellar solution in the semidilute regime (5.9 wt% CPyCl and 1.4 wt% NaSal) in steady and transient shear flow by a combination of PTV, SALS, microscopic visualization, and flow birefringence.[110] From these measurements, performed in both controlled-strain and controlled-stress rheometers with Couette geometries, time-resolved local velocity and shear-rate profiles (Figure 9.9), constitutive curve (constructed from local shear rates) (Figure 9.10), as well as direct visualization of positions across the gap of the bands and structures were obtained. They propose a two-stages chain disentanglement mechanism for shear banding: (1) shear tilting, in which the local shear rate changes smoothly across the gap and tilts toward high shear rates (rotating inner cylinder) and lower shear rates (fixed outer cylinder); this tilt is driven by the coupling between disentanglement and shear rate; and (2) shear banding, which begins when the local shear rate reaches the value of the shear rate for the onset of shear banding. Then, a low shear-rate band develops and expands across the gap due to the continuous disentanglement of the solution in the inner gap until steady

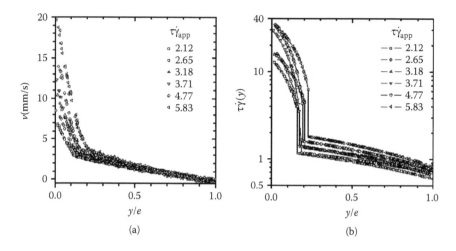

FIGURE 9.9 (a) Snapshots of velocity profiles across the flow cell gap in the steady state measured at 23 ± 0.1°C; (b) local shear rates $\tau\dot\gamma(y)$ extracted from the velocity profiles. (Reproduced from Ref. 110 with permission of The Society of Rheology.)

state is reached. The constitutive curves (Figure 9.10) show that there is not a unique relationship between stress and shear rate in the shear band coexisting regime, suggesting a constitutive instability. Moreover, they found from transient experiments monitored by PTV that the effective lifetime of the bands is comparable to that of the chain disentanglement.

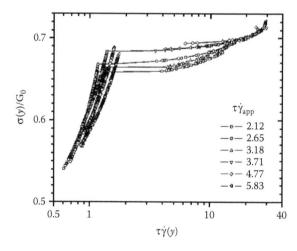

FIGURE 9.10 Constitutive curve constructed using local shear rates measured at 23 ± 0.1°C from PTV and local shear stresses. (Reproduced from Ref. 110 with permission of The Society of Rheology.)

9.5.5 MODELING

Models to explain shear banding have been proposed[49,114] that are based on a nonmonotonic constitutive relation for σ as a function of $\dot{\gamma}$ and account for the existence of the stress plateau. The maximum in σ is explained to occur due to the saturation of the stress as the deformation increases that prevents the total stress to increase indefinitely with higher $\dot{\gamma}$. The death rate increases abruptly, so tube segments are renewed faster and this causes the stress to fall. The decrease of σ with $\dot{\gamma}$ means that the flow is unstable in this region. Steady simple shear in the unstable region can only be supported if the system forms two or more shear bands. These are layers of high and low shear-rate material of equal shear stress that coexist at fractions arranged to match the imposed macroscopic shear rate. The first normal stress difference (N_1) shows a linear increase with shear rate that is consistent with the shear-banding mechanism. N_1 increases in proportion to the volume fraction of high shear material required to maintain the imposed average shear rate. The details of these models are discussed and described in Chapter 4 of this volume and therefore, they are not given further attention here.

In a simple shear device, such as a Couette apparatus, the flow has the same shear rate everywhere, but this is not possible if the applied $\dot{\gamma}$ lies in the region of decreasing σ. In simple shear, σ is constant and, so, two shear rates may coexist if they correspond to the same stress. This implies that the flow takes a banded form containing regions of two shear rates with two volume fractions x_A and $x_B = 1 - x_A$ such that the average shear rate is given by a lever rule of the form, $\dot{\gamma}_{appl} = x_A \dot{\gamma}_A + x_B \dot{\gamma}_B$.

If a banded flow is present, σ is not determined by $\dot{\gamma}_{appl}$, but it can lie between the top jumping and bottom jumping stresses. For increasing $\dot{\gamma}_{appl}$, the stress follows an upper trajectory, while for decreasing $\dot{\gamma}_{appl}$ the stress follows a lower trajectory, describing a hysteretic behavior. Experiments show a stress that initially increases with $\dot{\gamma}_{appl}$ and then goes smoothly into a plateau. The stress of the homogeneous phase immediately prior to band formation is the same as those of the low shear band immediately after.

Experimental work on CPyCl/NaSal (100 mM/60 mM) shows the existence of a metastable branch with a stress larger than the plateau stress.[94,95] The time scale for a relaxation of a metastable state onto the true steady state flow is far longer than the Maxwell time of the fluid, which is consistent with shear banding. There is a small window of shear rates for which this relaxation time is effectively infinite.

Greco and Ball analyzed the Johnson and Segalman constitutive equation in Couette geometry and obtained two bands in steady state.[115] Three characteristic steady state stresses in the system develop: the stresses at the inner and outer cylinders, and the stress at the interface as a function of velocity of the inner cylinder. Experimental results by Berret[39] show that the Johnson and Segalman model plus a Newtonian stress (JSN model) gives better agreement with his experiments than the reptation-reaction model of Cates. Yuan has also examined this model to study the shear banding in the context of a constitutive instability.[128] A linear stability analysis shows that the shear rate in the region where the stress decreases, past the maximum, splits into bands of low and high shear rate. This was called mechanical phase separation. In analogy to the van der Waals theory, a phase diagram for the mechanical phase separation is suggested, where spinodal and coexistence lines can be drawn. The former one is calculated from the maximum and minimum of the flow curve and the latter needs a criterion to select the plateau stress. The equal-area criterion is used to select the plateau stress.

The kinetics of formation of inhomogeneous flows in the stress plateau range has distinctive features. The characteristic time for the sigmoid evolution of the stress toward a steady-state plateau was found to be much longer than the reptation time. According to Porte et al. such kinetics are characterized by a nucleation and growth process, usually associated with first order phase transitions.[86] In some systems, the nucleation and growth processes lead to a kind of spinodal instability. Hence, the robustness of the plateau and slow sigmoidal onset kinetics are distinctive features of a first-order phase transition. Additional experimental verification is the stretched exponential relaxation of the stress, which involves the characteristic time for nucleation and growth processes.[86]

Shear-enhanced concentration fluctuations around the critical point (located at the onset of the plateau in the shear stress) can be explained with the models developed by Helfand and Fredrickson,[129] Onuki,[130] Doi and Onuki,[131] and Milner,[132,133] based on the two-fluid model of Brochard and de Gennes.[134] This model was originally developed to explain polymer migration under nonhomogeneous flows.[135] The mechanism of migration rests on the assumption that the polymer molecules interact with each other strongly enough that the entire mass of polymer can be regarded as a temporary gel that can support elastic stresses (one fluid), which is distinct from the solvent medium (the other fluid). This approach describes well the dynamic light scattering in semidilute theta solutions of polymers[136] and, in addition, it predicts the time scale of the relaxation of the structure factor in micellar systems.[137]

Flow can induce not only migration of polymer molecules, but also turbidity that may lead to demixing. The isotropic component of this scattering is the signal of flow-induced phase separation, as observed by Wu et al.[138] Phenomena seen in scattering measurements at low shear rates can be explained by the theory of Helfand and Fredrickson.[129] This model predicts the anisotropic enhancement of concentration fluctuations that result in turbidity when the correlation length of the fluctuations becomes comparable to the wavelength of light. Heterogeneities in the micelle concentration are produced because of the long-wave concentration fluctuations. The model constitutive equation is that of a second-order fluid, akin to Rouse dynamics of nonentangled chains, and it can describe qualitatively the behavior of semidilute wormlike micellar solutions. An inherent assumption in the model is that the time scale of the stress relaxation is much smaller than the time scale of the relaxation of concentration fluctuations, the so-called adiabatic approximation. These fluctuations decay with diffusion coefficient D, as in a simple two-fluid mixture.

Milner's model deals with a constitutive equation written in terms of the strain tensor with a relaxation time for the stress, which leads to two coupled

Langevin equations describing concentration and stress fluctuations.[132,133] Within a two-fluid approach, the time scale of generation and decay of concentration fluctuations is predicted to become shorter than the stress relaxation time. Hence, for wave vectors larger than k^*, the theory should predict a structure factor $S(k, \dot{\gamma})$ that decreases with increasing $k > k^*$ down to $S(k,0)$. The model predicts a peak in the structure factor with $k \approx k^*$ along the 45° orientation angle, with a peak width of the order of k^*. The presence of this peak in the structure factor is in striking agreement with light scattering experiments in polymer solutions.[138]

Time-dependent small-angle light scattering data of CTAB/NaSal exhibit butterfly patterns similar to those observed in polymer solutions, attributed to concentration fluctuations that are coupled to the mechanical stress.[138,139] The presence of bright streaks is indicative of long elongated structures aligned in the flow direction.[140] These patterns appear in the region where the stress is still increasing with shear rate. This is in apparent agreement with the statement that nucleation takes place before the mechanical instability appears.

The relaxation rate as a function of the wave vector agrees well with predictions of the Brochard and de Gennes model of the transient gel, that is, with the lower of the two characteristic decay rates observed in dynamic light scattering.[134] For small k, all the weight is in the slowly decaying signal, while at large k, the fast and slow decay rates have comparable weight when the plateau and osmotic moduli are comparable. If the former is much larger than the latter, then the slow decay signal is also dominant at large wave vectors. The following approximate forms give the slow (Ω_s) and fast (Ω_f) modes:

$$\Omega_s = \left[\frac{1}{Dk^2} + \frac{D_g}{D}\lambda_\tau\right]^{-1}, \quad \Omega_f = \frac{1}{\lambda_\tau} + D_g k^2 \qquad (9.3a, b)$$

where D_g is the diffusion coefficient of the transient gel. These two modes of the relaxation of the structure factor have also been observed in micellar systems.[141] In fact, owing to the similarity of wormlike micelles to polymers, the Brochard–de Gennes model, originally developed for polymer systems, can be used to predict data in wormlike micellar systems. Kadoma et al. applied SALS to a semidilute solution of wormlike micelles.[141] The relaxation of $S(k, t)$ after flow cessation was studied at several wave numbers. Data can be represented according to the following expression (wherein the slow mode dominates the fast mode), which agrees with Equation 9.3a:

$$S(k,t) = A\exp(-t/\tau_1) \qquad (9.4)$$

$$\tau_1(k) = \tau_s + \frac{1}{D_c k^2} \qquad (9.5)$$

Schmitt et al. were the first to predict the enhancement of flow instabilities by concentration coupling.[121] This model assumes that the chemical potential (μ) in the diffusion equation is a function of both shear rate and concentration. Moreover, the shear stress depends on both shear rate and concentration according to:

$$\sigma(C,\dot{\gamma}) = \sigma(C_0,\dot{\gamma}_0) + \eta_d \delta\dot{\gamma} + (\partial\tau/\partial C)_{C=C_0}\,\delta C \qquad (9.6)$$

where $\eta_d \equiv (\partial\sigma/\partial\dot{\gamma})_{\dot{\gamma}=\dot{\gamma}_0}$.
A coupling term arises in the dispersion equation given by:

$$F = \rho^{-1}(\partial\mu/\partial\dot{\gamma})_{\dot{\gamma}=\dot{\gamma}_0}\,(\partial\sigma/\partial C)_{C=C_0} \qquad (9.7)$$

The first derivative on the right-hand side of Equation 9.7 accounts for migration of the dispersed phase. This coupling term represents the feedback of the different concentrations and viscosities present in the layered solution during shear flow. The sign of F determines type of instability: if it is positive, the sheared solution is less viscous than the initial one, and if it is negative, the sheared solution is more viscous. In fact, a classification scheme is suggested. This stability analysis is based on a Newtonian expression, and it does not consider a relaxation time in their scheme.[121] This is the so-called *adiabatic approximation*.[132]

For wormlike micellar systems, Fielding and Olmsted proposed a model that is basically the Milner model with added Newtonian terms plus the Johnson and Segalman equation with a diffusion term for the stress.[142] This model is described in detail in Chapter 4.

The prediction of the rheological behavior of wormlike micelles by constitutive equations is still a challenging issue. Elsewhere we have proposed a simple model that couples the convected Maxwell constitutive equation with a time-dependent equation for the structure changes, which is able to predict the basic rheological behavior exhibited by the micellar systems in simple shear and Poiseuille flows.[112,126,143] The evolution equation for structural changes was conceived to account for the kinetic process of scission and reformation of the micelles under flow. The model was shown to give excellent fit to experimental observations in both steady shear and small amplitude oscillatory flows for CTAT and CTAB/NaSal wormlike micellar solutions.[36,112]

Extended irreversible thermodynamics (EIT) can provide an approach to derive hydrodynamic equations that govern the flow-induced concentration changes produced by inhomogeneous stresses in a complex fluid.[144] One of the most relevant effects arising from these inhomogeneous flows is manifested in flow-induced concentration fluctuations. In addition, it has been demonstrated that the generalized extended Gibbs free energy variation with shear rate provides a criterion to predict the stress plateau under shear-banding flow.[112] The criterion of bands coexistence is the equality of the extended Gibbs free energy of the bands when normal stresses can be neglected. The predicted stress plateau coincides remarkably well with experimental data.

In complex fluids that present flow-induced changes in their internal structure, the stress constitutive equation is also coupled to an evolution equation of a scalar representing the flow-induced modifications on the internal structure of the fluid. By applying the usual procedure of EIT to lowest order in the nonconserved variables and neglecting diffusion, the following equations for the structure parameter ς^0 and stress tensor $\underline{\sigma}$ are obtained:

$$\frac{d\varsigma^0}{dt} = \frac{1}{\tau_0}(1-\varsigma^0) + k_0(\mu-\varsigma^0)\underline{\sigma}{:}\underline{D} \tag{9.8}$$

$$\varsigma^0\underline{\sigma} + \tau_2\overset{\triangledown}{\underline{\sigma}} = 2\eta_0\underline{D} \tag{9.9}$$

In these equations, $\overset{\triangledown}{\underline{\sigma}}$ is the upper-convected derivative of the stress tensor, $\underline{D}$ is the rate of deformation tensor, τ_0 and τ_2 are the structure and stress relaxation times, respectively, and η_0 is the zero-shear-rate viscosity. μ and k_0 are phenomenological coefficients. The parameters of the model can be estimated from independent rheological experiments.[112]

The Bautista et al. model has been used to predict the rheological behavior of complex fluids in which the internal structure is modified by the external flow and in which diffusion effects are negligible.[36,112] In this model, the structure parameter is made proportional to a measure of the structure under flow. The fluidity (inverse of the viscosity) is here chosen to represent a given state of structure modification. The structure variable is given in terms of the ratio of the fluidity to that of a reference state, that is, $\varsigma^0 = \varphi/\varphi_0$, where φ_0 is the zero-strain-rate fluidity (reference state). Substitution of this definition into Equation 9.8 and Equation 9.9 and making $\mu \equiv \varphi_\infty/\varphi_0$ gives

$$\frac{d\varphi}{dt} = \frac{1}{\tau_0}(\varphi_0 - \varphi) + k(\varphi_\infty - \varphi)\underline{\sigma}{:}\underline{D} \tag{9.10}$$

$$\underline{\sigma} + \tau_2\frac{\varphi_0}{\varphi}\overset{\triangledown}{\underline{\sigma}} = \frac{2}{\varphi}\underline{D} \tag{9.11}$$

In the Bautista et al. model, the parameter k in Equation 9.10 and the relaxation time in Equation 9.11 are variables. Consistent with the thermodynamic scheme presented, the phenomenological coefficients are either constants or functions of the scalar invariants of the tensor variables. In this case, k is a linear function of the second invariants of the rate of deformation tensor and that of the stress tensor according to $k = k_0(1 + \vartheta II_D + \upsilon II_\tau)$. Hence, the expansion is restricted up to first order. However, although the thermodynamic scheme may provide nonlinear terms, new phenomenological coefficients must be determined from experimental data. The simplest approach is then the first approximation from which the coefficients can be determined straightforwardly from experiments. Accordingly, Equation 9.10 and Equation 9.11 become

$$\frac{d\varphi}{dt} = \frac{1}{\tau_0}(\varphi_0 - \varphi) + k(II_D, II_\tau)(\varphi_\infty - \varphi)\underline{\underline{\sigma}}:\underline{\underline{D}} \tag{9.12}$$

$$\underline{\underline{\sigma}} + \frac{1}{G_0\varphi}\overset{\triangledown}{\underline{\underline{\sigma}}} = \frac{2}{\varphi}\underline{\underline{D}} \tag{9.13}$$

Equation 9.12 and Equation 9.13 are the constitutive relations of the Bautista et al. model.

One of the relevant and particular characteristics of the shear-banded state is its intrinsic unsteadiness or transient behavior under flow. The time to achieve a true steady state is sometimes of the order of hours or a wealth of Maxwell relaxation times, so it is not contradictory to regard it as a true metastable and time evolving flow regime. A relevant prediction of the Bautista et al. model is the possibility to identify a singularity in the time to achieve steady state with increasing pressure gradient under Poiseuille flow. In fact, as shown elsewhere, the residence time in the pipe to achieve a developed shear-banded flow varies with imposed pressure gradient.[112] When the applied pressure gradient is below $\dot{\gamma}_{c1}$, steady state is achieved very rapidly within fractions of the Maxwell relaxation time of the sample. However, once the pressure gradient reaches values corresponding to the critical shear stress at which shear banding begins to develop, the time required to arrive to the steady state conditions increases rapidly, reaching hundreds of relaxation times. In fact, at the critical pressure gradient no steady state is ever reached. Such long transients are also observed at the onset for shear banding in Poiseuille[145] and in cone and plate flow.[34] Under simple shear flow, the same type of unsteadiness is observed and predicted by the model.[112] Predictions of the stress growth upon inception of shear flow for shear rates equal or larger than $\dot{\gamma}_{c1}$ show that the time required to reach steady state is several tens of the main relaxation time of the sample. As a matter of fact, these long transients appear simultaneously to the oscillations in the stress in which the period becomes larger with increasing shear rate. The oscillations and long transients disappear at very high shear rates, as the second critical shear rate $\dot{\gamma}_{c2}$ is approached. Similar slow transients have been observed elsewhere.[31,95]

In stress relaxation following cessation of simple shear flow, agreement exists between model predictions and experimental data. Figure 9.11 shows that the stress relaxation curve is single exponential and can be reproduced with the Maxwell model using the relaxation time obtained from oscillatory measurements

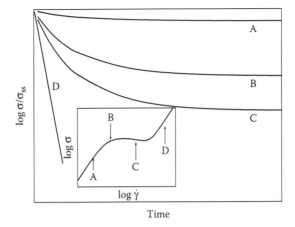

FIGURE 9.11 Stress relaxation after cessation of flow for different applied shear rates as indicated in the inset.

(curve A) when the applied shear rate is within the low shear Newtonian region of the flow curve (point A in the inset). At high shear rates (point D in the inset) above the multivalued region, the relaxation is fast and single exponential also (curve D). However, when the applied shear rate is within the multivalued region (points B and C in the inset), the stress exhibits two main relaxation mechanisms, one fast and another slow (curves B and C). Notice that the slopes of the fast and slow relaxation mechanisms in the shear-banding region are similar to those of the very high and very low shear rate Newtonian regions, respectively. This suggests that two structures coexist within the shear-banding region. Elsewhere, a relaxation curve with two main relaxation mechanisms was also found.[36,103]

The Bautista et al. model has been recently applied to extensional flow.[146] In uniaxial extension, a thickening region at a nondimensional strain rate of order one followed by a thinning region at high strain rates have been predicted. At the inception of extensional flow, the model predicts overshoots larger than those seen in shear flows. Recent experiments in flow through porous media qualitatively confirm some of the predictions of the model.[146]

9.6 CONCLUSIONS

In this chapter we have reviewed the nonlinear rheology of giant micelles including the shear-thickening and the shear-banding transition, which are the characteristic nonlinear viscoelastic features of giant micellar solutions. The variation of the zero-shear viscosity as a function of concentration and temperature are reported. It was found that the viscosity dependence on concentration could be reduced to three patterns. Because of synergistic effects on phase and rheological behavior, mixed surfactant systems have received considerable attention in the last 10 years; hence, the rheology of mixed surfactant solutions was thoroughly examined and discussed. An important conclusion is that the adequate mixture of two surfactants, whether anionic-nonionic, nonionic-nonionic, or anionic-cationic, can lead to enormous viscosity rise and strong viscoelastic effects.

The nonlinear rheology of giant micellar systems includes complex systems that exhibit shear-thickening and shear-banding flow. The shear-thickening transition involves the formation of shear-induced structures whose morphology has been investigated by a number of experimental techniques. The origin and kinetics of *SIS* formation have been given special attention and the few theoretical explanations of the phenomenon have been briefly described. The mechanisms suggested for shear-banding flow involve the link between linear and nonlinear rheological responses and the analysis of the kinetics of shear-banding formation. The origin of this phenomenon with regard to the first-order phase transition and a pure mechanical stability has also been discussed, mentioning the available experimental evidence from several experimental techniques, notably light scattering under flow. The comparison of these data with current theories and models has also been given special attention. Our contribution to the understanding of the rheology of these complex fluids has also been included.

REFERENCES

1. Walker, L. M. *Curr. Opin. Colloid Interface Sci.* 2001, *6*, 451.
2. Berret, J.-F. In *Molecular Gels. Materials with Self-Assembled Fibrillar Networks*, Weiss, R. G., Terech, P., Eds., Springer, Secaucus, 2006.
3. Hu, Y. T., Boltenhagen, P., Pine, D. J. *J. Rheol.* 1998, *42*, 1185.
4. Hu, Y. T., Boltenhagen, P., Matthys, E., Pine, D. J. *J. Rheol.* 1998, *42*, 1209.
5. Kern, F., Zana, R., Candau, S. J. *Langmuir* 1991, *7*, 1344.
6. Berret, J.-F., Roux D. C., Porte G. *J. Phys. II France* 1994, *4*, 1261.
7. Koehler, R. D., Raghavan, S. R., Kaler, E. W. *J. Phys. Chem. B* 2000, *104*, 11035.
8. Shchipunov, Y. A., Hoffmann, H. *Rheol. Acta* 2000, *39*, 542.
9. Candau, S. R., Oda, R. *Colloids Surf. A* 2001, *5*, 183.
10. Angelescu, D., Khan, A., Caldararu, H. *Langmuir* 2003, *19*, 9155.
11. Ericsson, C. A., Söderman, O., Ulvenlund, S. *Colloid Polym. Sci.* 2005, *283*, 1313.
12. Hoffmann H., Ulbricht, W. In *Structure-Performance Relationships in Surfactants*, Kesumi, K., Ueno, M., Eds., Surfactant Sci. Series Vol. 70, Marcel Dekker, New York 1997, Chap. 7, 285.
13. Mu, J.-H., Li, G.-Z. *Colloid Polym. Sci.* 2001, *279*, 872.
14. Kalur, G. C., Raghavan, S. R. *J. Phys. Chem. B* 2005, *109*, 8599.
15. Flood, C., Dreiss, L. A., Croce, V., Cosgrove, T. *Langmuir* 2005, *21*, 7646.
16. Mu, J.-H., Li, G.-Z., Wang, Z.-W. *Rheol. Acta* 2002, *41*, 493.
17. Mu, J.-H., Li, G.-Z., Jia, X.-L., Wang, H.-X., Zhang, G. -Y. *J. Phys. Chem. B* 2002, *106*, 11685.
18. Raghavan, S. R., Fritz, H., Kaler, E. W. *Langmuir* 2002, *18*, 3797.
19. Schubert, B. A., Kaler, E. W., Wagner, N. J. *Langmuir* 2003, *19*, 4079.
20. Hassan, P. A., Raghavan, S. R., Kaler, E. W. *Langmuir* 2002, *18*, 2543.
21. Acharya, D. P., Kunieda, H. *J. Phys. Chem. B* 2003, *107*, 10168.

22. Acharya, D. P., Hossain, M. K., Feng, J., Sakai, T., Kunieda, H. *Phys. Chem. Chem. Phys.* 2004, *6*, 1627,

23. Maestro, A., Acharya, D. P., Furukawa, H., Gutiérrez, J. M., López-Quintela, M. A., Ishitobi, M., Kunieda, H. *J. Phys. Chem. B* 2004, *108*, 14009.

24. Rodríguez-Abreu, C., Aramaki, K., Tanaka, Y., López-Quintela, M. A., Ishitobi, M., Kunieda, H. *J. Colloid Interface Sci.* 2005, *291*, 560.

25. Rodriguez, C. Acharya, D. P., Hattori, K., Sakai, T., Kunieda, H. *Langmuir* 2003, *19*, 8692.

26. Rodriguez, C., Acharya, D. P., Maestro, A., Hattori, K., Aramaki, K., Kunieda, H. *J. Chem. Eng. Jpn.* 2004, *37*, 622.

27. Acharya, D. P., Hattori, K., Sakai, T., Kunieda, H. *Langmuir* 2003, *19*, 9173.

28. Acharya, D. P., Kunieda, H. Shiba, Y., Aratani, K. *J. Phys. Chem. B* 2004, *108*, 1790.

29. Hoffmann, H., Rauscher, A, Gradzielski, M., Schulz, S. F. *Langmuir* 1992, *8*, 2140.

30. Shikata, T., Hirata, H., Kotaka, T. *Langmuir* 1987, *3*, 1081.

31. Rehage, H., Hoffmann, H. *Mol. Phys.* 1991, *74*, 933.

32. Hartmann, V., Cresseley, R. *Colloid Polym. Sci.* 1998, *276*, 169; *Rheol. Acta* 1998, *37*, 115.

33. Abdel-Rahem, R., Gradzielski, M., Hoffmann, H. *J. Colloid Interface Sci.* 2005, *288*, 570.

34. Soltero, J. F. A., Bautista, F., Puig, J. E., Manero, O. *Langmuir* 1999, *15*, 1604.

35. Shashkina, J. A., Philippova, O. E., Zaroslov, Y. D., Khokhlov, A. R., Pryakhina, T. A., Blagodatakikh, I. V. *Langmuir* 2005, *21*, 1524.

36. Escalante, J. I., Macias, E. R., Bautista, F. , Pérez-López, J. H., Soltero, J. F. A., Puig, J. E., Manero, O. *Langmuir* 2003, *19*, 6620.

37. Bandyopadhyay, R., Sood , A. K. *Langmuir* 2003, *19*, 3121.

38. Couillet, I., Hughes, T., Maitland, G. *Macromolecules* 2005, *38*, 5271.

39. Berret, J.-F. *Langmuir* 1997, *13*, 2227.

40. In, M., Warr, G. G., Zana, R. *Phys Rev. Lett.* 1999, *83*, 2278.

41. Raghavan S. R., Kaler, E. W. *Langmuir* 2001, *17*, 300.

42. Kern, F., Lequeux, F., Zana, R., Candau, S. J. *Langmuir* 1994, *10*, 1714.

43. Magid, L. J. *J. Phys. Chem. B* 1998, *102*, 4064.

44. Oda, R., Narayanan, J., Hassan, P. A., Manohar, C., Salkar, R. A., Kern, F., Candau, S. J. *Langmuir* 1998, *14*, 4364.

45. Oelschlaeger, C. L., Waton, G., Candau, S. J. *Langmuir* 2003, *19*, 10495.

46. Soltero, J. F. A., Puig, J. E., Manero, O. *Langmuir* 1996, *12*, 2654.

47. Guo, L., Colby, R. H., Lin, M. Y., Dado, G. P. *J. Rheol.* 2001, *45*, 1223.

48. Ericsson, C. A., Söderman, O., Ulvenlund, S. *Colloid Polym. Sci.* 2005, *283*, 1313.

49. Larson, R. G. *The Structure and Rheology of Complex Fluids*, Oxford University Press, Oxford, 1999.

50. Berret, J.-F., Gámez-Corrales, R., Oberdisse, J., Walker, L. M., Lidner, P. *Europhys. Lett.* 1998, *41*, 677.

51. Kim, W.-J., Yang, S.-M., Kim, M. *J. Colloid Interface Sci.* 1997, *194*, 108.

52. Rehage, H., Hoffmann, H. *Rheol. Acta* 1982, *21*, 561.

53. Wunderlich, I., Hoffmann, H., Rehage, H. *Rheol. Acta* 1987, *26*, 532.

54. Lin, Z. Q., Zakin, J. L., Zheng, Y., Davis, H. T., Scriven, L. E., Talmon, Y. *J. Rheol.* 2001, *45*, 963.

55. Myska, J., Lin, Z. Q., Stepanek, P., Zakin, J. L. *J. Non-Newtonian Fluid Mech.* 2001, *97*, 251.

56. Hartmann, V., Cresseley, R. *J. Phys. II Fr.* 1997, *7*, 1087.
57. Gámez-Corrales, R., Berret, J.-F., Walker, L. M., Oberdisse, J. *Langmuir* 1999, *15*, 6755.
58. Oda, R., Panizza, P., Schmutz, M., Lequeux, F. *Langmuir* 1997, *13*, 6407.
59. Oda, R., Weber, V., Lindner, P., Pine, D. J., Mendes, E., Schosseler, F. *Langmuir* 2000, *16*, 4859.
60. Wunderlich, A. M., Brunn, P. O., *Colloid Polym. Sci.* 1989, *267*, 627.
61. Hartmann, V., Cresseley, R. *Europhys. Lett.* 1997, *40*, 691.
62. Truong, M. T., Walker, L. M. *Langmuir* 2000, *16*, 7998.
63. Berret, J.-F., Gámez-Corrales R., Lerouge, S., Decruppe, J.-P. *Eur. Phys. J. E* 2000, *2*, 343.
64. Oelschlaeger, D., Waton, G., Buhler, E., Candau, S. J., Cates, M. E. *Langmuir* 2002, *18*, 3076.
65. Boltenhagen, P., Hu, Y. T., Matthys, E. F., Pine, D. J. *Europhys. Lett.* 1997, *38*, 389.
66. Pröztl, B., Springer, J. *J. Colloid Interface Sci.* 1997, *190*, 237.
67. Berret, J.-F., Lerouge, S., Decruppe, J.-P. *Langmuir* 2002, *18*, 7279.
68. Macías, E. R., Bautista, F., Soltero, J. F. A., Puig, J. E., Attané, P., Manero, O. *J. Rheol.* 2003, *47*, 643.
69. Müller, A. J., Torres, M. F., Sáez, A. E. *Langmuir* 2004, *20*, 3838.
70. Weber, V., Schosseler, F. *Langmuir* 2002, *18*, 9705.
71. Macías, E. R., González, A., Manero, O., González-Nuñez, R., Soltero, J. F. A., Attané, P. *J. Non-Newtonian Fluid Mech.* 2001, *101*, 149.
72. Fischer, P., Wheeler, E. K., Fuller, G. G. *Rheol. Acta* 2002, *41*, 35.
73. Herle, V., Fischer, P., Windhab, E. J. *Langmuir* 2005, *21*, 9051.
74. Truong, M. T., Walker, L. M. *Langmuir* 2002, *18*, 2024.
75. Zheng, Y., Lin, Z., Zakin, J. L., Talmon, Y., Davis, H. T., Scriven, L. E. *J. Phys. Chem. B.* 2000, *104*, 5263.
76. Marín-Santibañez, B. M., Pérez-González, J., de Vargas, L., Rodríguez-González, F., Huelsz, G. *Langmuir* 2006, *22*, 4015.
77. Hu, Y. T., Matthys, E. F. *Rheol. Acta* 1995, *34*, 450.
78. Lee, J.-Y., Magda, J. J., Hu, H., Larson, R. G. *J. Rheol.* 2002, *46*, 195.
79. Brunn, P. O., Wunderlich, A. M. *Colloid Polym. Sci.* 1989, *267*, 627.
80. Hu, Y. T., Matthys, E. *J. Rheol.* 1997, *41*, 151.
81. Berret, J.-F., Gámez-Corrales, R., Séréro, Y., Molino, F., Lindner, P. *Europhys. Lett.* 2001, *54*, 605.
82. Barentin, C., Liu, A. J. *Europhys. Lett.* 2001, *55*, 432.
83. Weber, V., Narayanan, T., Mendes, E., Schosseler, F. *Langmuir* 2003, *19*, 992.
84. Cates, M. E., Candau, S. J. *Europhys. Lett.* 2001, *55*, 887.
85. Olmsted, P. D., Lu, C.-Y., D. *Phys. Rev. E* 1999, *60*, 4397.
86. Porte, G., Berret, J.-F., Harden, J. L. *J. Phys. II Fr.* 1997, *7*, 459.
87. Wang, S. Q. *Macromolecules* 1991, *24*, 3004.
88. Cates, M. E. *J. Phys. Chem.* 1990, *94*, 371.
89. Cates, M. E., Turner, M. S. *Europhys. Lett.* 1990, *11*, 681.
90. Ha, B.-Y., Liu, A. J. *Phys. Rev. Lett.* 1999, *46*, 624.
91. Goveas, J. L., Pine, D. J. *Europhys. Lett.* 1999, *48*, 339.
92. Bautista F., de Santos, J. M., Soltero, J. F. A., Puig, J. E., Manero, O. *J. Non-Newtonian Fluid Mech.* 1999, *80*, 93.
93. Azzouzi, H., Decruppe, J. P., Lerouge, S., Greffier, S. *Eur. Phys. J. E* 2005, *17*, 507.
94. Grand, C. Arrault, J., Cates, M. E. *J. Phys. II Fr.* 1997, *6*, 551.

95. Britton M. M., Mair, R. W., Lambert, R. K., Callaghan, P. T. *J. Rheol.* 1999, *43*, 897.
96. Cates, M. E., McLeish, T. C. B., Marucci, G. *Europhys. Lett.* 1993, *21*, 451.
97. Callaghan, P. T., Cates, M. E., Rofe, C. F., Smoulders, J. B. F. A. *J. Phys. II* 1996, *6*, 375.
98. Makhloufi, R., Decruppe, J.-P., Aït-Ali, A., Cresseley, R. *Europhys. Lett.* 1995, *32*, 253.
99. Grand, C., Arrault, J., Cates, M. E. *J. Phys. II* 1996, *7*, 1071.
100. Fischer, P., Rehage, H. *Rheol. Acta* 1997, *36*, 13.
101. Aït-Ali, A., Makhloufi R. *J. Rheol.* 1997, *41*, 307.
102. Decruppe, J.-P., Cappelare, E., Cresseley, R. *J. Phys. II* 1997, *7*, 257.
103. Berret, J.-F., Porte, G. *Phys. Rev. E.* 1999, *60*, 4268.
104. Escalante, J. I., Hoffmann, H. *Rheol. Acta* 2000, *39*, 209.
105. Berret, J.-F., Séréro, Y. *Phys. Rev. Lett.* 2001, *87*, 0483031.
106. Krishnan, K., Chapman B., Bates, F. S., Lodge, T. P., Almadal, K., Burghardt, W. *J. Rheol.* 2002, *46*, 529.
107. Mair, R., Callaghan, P. T. *Europhys. Lett.* 1996, *36*, 719.
108. Berret, J.-F., Roux, D. C., Lidner P. *Eur. Phys. J. B* 1998, *5*, 67.
109. Decruppe, J.-P., Lerouge, S., Berret, J.-F. *Phys. Rev. E* 2001, *63*, 0225011.
110. Hu, Y. T., Lips, A. *J. Rheol.* 2005, *49*, 1001.
111. Berret, J.-F., Porte, G., Decruppe, J. P. *Phys. Rev. E* 1997, *55*, 1668.
112. Bautista, F., Soltero, J. F. A., Manero, O., Puig, J. E. *J Phys. Chem. B* 2002, *106*, 13018.
113. Fielding, S. M., Olmsted, P. D. *Eur. Phys. J. E* 2003, *11*, 65.
114. Spenley, N. A., Yuan, X. F., Cates, M. E. *J. Phys. II Fr.* 1996, *7*, 1071.
115. Greco, F., Ball, R. C. *J. Non-Newtonian Fluid Mech.* 1997, *69*, 195.
116. Español, P., Yuan, X. F., Ball, R. C. *J. Non-Newtonian Fluid Mech.* 1996, *65*, 93.
117. Boger, D. V. *Annu. Rev. Fluid Mech.* 1987, *19*, 157.
118. McKinley, G. H., Raiford, W. P., Brown, R. A., Armstrong, R. C. *J. Fluid Mech.* 1991, *223*, 411.
119. Larson, R. G. *Rheol. Acta* 1992, *31*, 213.
120. Byars, J. A., Oztekin, A., Brown, R. A., McKinley, G. H. *J. Fluid Mech.* 1994, *271*, 173.
121. Schmitt, V., Marques, C. M., Lequeux, F. *Phys. Rev. E.* 1995, *52*, 4009.
122. Vinagradov, G. V. *Rheol. Acta* 1973, *12*, 273.
123. Cappelare, E., Berret, J.-F., Decruppe, J.-P., Cresseley, R., Lidner, P. *Phys. Rev. E* 1997, *56*, 1869.
124. Fischer, E., Callaghan, P. T. *Phys. Rev. E.* 2001, *64*, 0115011.
125. Jou, D., Casas-Vázquez, J., Criado-Sancho, M. *Thermodynamics of Fluids Under Flow*, Springer-Verlag, Berlin, 2000.
126. Bautista, F., Soltero, J. F. A., Pérez-López, J. H., Puig, J. E., Manero, O. *J. Non-Newtonian Fluid Mech.* 2000, *94*, 57.
127. Olmsted, P. D., Lu, C.-Y. D. *Faraday Discuss.* 1999, *112*, 183.
128. Yuan, X. F. *Europhys. Lett.* 1999, *46*, 542.
129. Helfand, E., Fredrickson, G. H. *Phys. Rev. Lett.* 1989, *62*, 2468.
130. Onuki, A. *J. Phys. Soc. Jpn.* 1990, *59*, 3423.
131. Doi, M., Onuki, A. *J. Phys. II,* 1992, *2*, 1631.
132. Milner, S. T. *Phys. Rev. Lett.* 1991, *66*, 1477.
133. Milner, S. T. *Phys. Rev. E* 1993, *48*, 3674.

134. Brochard, F., de Gennes, P. G. *Macromolecules* 1977, *10*, 1157.
135. Larson, R. G. *Rheol. Acta* 1992, *31*, 497.
136. Adam, M., Delsanti, M. *Macromolecules* 1985, *18*, 1760.
137. Kadoma, I. A. *Phys. Rev. Lett.* 1996, *76*, 4432.
138. Wu, X. I., Pine, D. J., Dixon, P. K. *Phys. Rev. Lett.* 1991, *66*, 2408.
139. Dixon, P. K., Pine, D. J., Wu, X. I. *Phys. Rev. Lett.* 1992, *68*, 2239.
140. Wheeler, E. K., Izu, P., Fuller, G. G. *Rheol. Acta* 1996, *35*, 139.
141. Kadoma, I. A., van Egmond, J. W. *Phys. Rev. Lett.* 1992, *76*, 4432.
142. Fielding, S. M., Olmsted, P. D. *Phys. Rev. Lett.* 2003, *90*, 224501.
143. Manero, O., Bautista, F., Soltero, J. F. A., Puig, J. E. *J. Non-Newtonian Fluid Mech.* 2002, *106*, 1.
144. Manero, O., Rodríguez, R. F. *J. Non-Equilib. Thermodyn.* 1999, *24*, 177.
145. Hernández-Acosta, S., González-Álvarez, A., Manero, O., Méndez-Sánchez, A. F., Pérez-González, J., de Vargas, L. *J. Non-Newtonian Fluid Mech.* 1999, *85*, 229.
146. Boek, E. S., Padding, J. T., Anderson, V. J., Tardy, P. M. J., Crawshaw, J., Pearson, J. R. A. *J. Non-Newtonian Fluid Mech.* 2005, *126*, 39.

10 Relaxation in Wormlike Micelle Solutions

Gilles Waton and Raoul Zana

CONTENTS

10.1 INTRODUCTION

In aqueous solution surfactants self-associate into micelles at concentration above the critical micelle concentration (cmc). Micelles are spherical or spheroidal at concentration close to the cmc but their shape can be modified by acting upon

323

parameters such as temperature, ionic strength, surfactant concentration, or nature of the counterion for ionic surfactants. Besides, additives, such as alcohols or aromatic compounds, can modify the micelle shape. In most instances these changes induce a growth of the micelles and a change of shape from spheroidal to elongated, often resulting in the formation of *wormlike* micelles that can be microns long (*giant* micelles). Evidence for the coexistence of spherical and wormlike micelles has been obtained through indirect methods[1,2] and also directly visualized by transmission electron microscopy at cryogenic temperature (cryo-TEM).[3,4]

Spherical and wormlike micelles are not frozen objects. They are in dynamic equilibrium with free (nonmicellized) surfactant. Surfactant is constantly exchanged between micelles and intermicellar solution via the so-called *exchange process*. Also micelles, whether spherical or elongated, constantly *form and break down*. Owing to the existence of these two processes, two steps generally characterize the relaxation in micelle solutions submitted to a fast perturbation, with a rapid relaxation associated with the exchange process and a much slower relaxation associated with the micelle formation/breakdown.

Extensive investigation of the relaxation in micelle solutions has provided very important information on the kinetics of the exchange process and on the steps involved in micelle formation and breakdown. These studies have been recently summarized in a review that mostly dealt with spherical micelles and touched only briefly upon wormlike micelles.[5]

This chapter is devoted to the relaxation in aqueous solutions of wormlike micelles formed by surfactants. Information about the dynamics of micelles formed by amphiphilic block copolymers can be found in Chapter 14, Section 14.3.3 of this volume and in Ref. 6. Here, it is simply recalled that in aqueous solution only block copolymers with a weakly hydrophobic block, such as poly(propylene oxide), and of low molecular weight give rise to labile micelles, for which the formation/breakdown can be conveniently investigated by chemical relaxation methods.

The chapter is organized as follows. Section 10.2 briefly recalls the main conclusions of theoretical and experimental studies of the relaxation in spherical micelle solutions. Section 10.3 presents the theoretical aspects of the relaxation in solutions of wormlike micelles. Section 10.4 reviews the experimental studies of these systems. The "Conclusions" section emphasizes the importance of the kinetics of wormlike micelles for explaining the properties of these systems.

10.2   RELAXATION IN SPHERICAL MICELLE SOLUTIONS

10.2.1   Historical Aspects

The first study that showed unambiguously that micelle solutions of ionic surfactants are characterized by *two* well-separated relaxation processes dates back to 1975.[7] Several authors[8–10] independently assigned the fast process to the exchange equilibrium represented by reaction 10.1, where A_1 represents a free surfactant, A_s and A_{s-1} refer to two micelles of aggregation numbers s and $s - 1$, k_s^- is the surfactant exit rate constant from micelle A_s, k_s^+ and is the surfactant entry rate constant in micelle A_{s-1}.

$$A_{s-1} + A_1 \underset{k_s^-}{\overset{k_s^+}{\rightleftarrows}} A_s \qquad (10.1)$$

The slow relaxation was attributed to the micelle formation/breakdown that can be schematically represented by reaction 10.2:

$$sA_1 \rightleftarrows \text{Intermediate species} \rightleftarrows A_s \qquad (10.2)$$

Nakagawa[9] was the first to relate the two relaxation processes to shifts in the micelle size distribution curve induced by a perturbation as shown in Figure 10.1, where $[A_s]$ is the concentration (or number density) of micelles A_s. The initial distribution curve (a) is close to Gaussian. The trough (region 2) that separates micelles proper (region 3) from oligomers (region 1) is responsible for the two-step relaxation response of micelle solutions to a perturbation. Indeed a fast perturbation first induces a rapid variation of the aggregation number of the micelles proper that leaves the number of micelles *constant*. The resulting

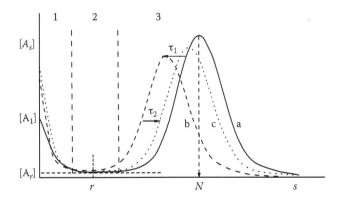

FIGURE 10.1 Distribution curve of the micelle aggregation number and its shifts after a very rapid step perturbation. (a) Distribution curve before perturbation with a maximum at $s = N$ and a minimum at $s = r$ (see text). (b) Distribution curve for the state of pseudo-equilibrium of the system reached upon equilibration of the micelles proper at constant number of micellar species. (c) Distribution curve for the state of final equilibrium after equilibration between micelles and oligomers. Regions 1 and 3 correspond to the oligomers and the micelles proper and region 2 to the species around the minimum of the size distribution curve.

distribution curve (b) is not the equilibrium distribution as micelles and oligo-mers still have to equilibrate through a relaxation process that brings about a variation of the number of micelles. This process is slow because it involves many steps. Aniansson and Wall[10] derived the first analytical expressions of the relaxation times characterizing the two processes based on this model.

Extensive chemical relaxation data for solutions of ionic surfactants were first interpreted on the basis of this theory in 1976.[11] The conclusions reported in this study still constitute the basis for the present understanding of the dynamics of micelles. Aniansson and Wall theory was later refined and extended to ionic surfactant micelles taking into account the presence of the counterions and of added electrolyte.[12,13] It was also extended to include *scission/fusion* (or *fragmentation/coagulation*) reactions 10.3 by which a micelle A_s can break reversibly into two daughter micelles A_{s-k} and A_k, with the rate constants a_k^s and b_k^s.[14]

$$A_s \underset{b_k^s}{\overset{a_k^s}{\rightleftarrows}} A_{s-k} + A_k \qquad (10.3)$$

The existence of reactions 10.3 has been demonstrated experimentally (see Ref. 5, pp. 99–101). The contribution of these reactions to the micelle kinetics becomes important for systems where collisions between micelles can occur.

10.2.2 ANIANSSON AND WALL THEORY OF MICELLE KINETICS

This section recalls only the essential aspects and assumptions of this theory. Details can be found in Refs. 5, 10, and 11.

Aniansson and Wall[10] consider that the various micellar species present in solution are in dynamic equilibrium through a series of stepwise reactions 10.1. The expression of the relaxation time τ_1 for the exchange process was obtained assuming the following:

(i)   The amplitude of the perturbation is small.
(ii) No exchange of surfactant between oligomers and micelles proper (regions 1 and 3 in Figure 10.1) occurs during the rapid relaxation.
(iii) The shape of the distribution curve around the maximum is Gaussian, that is

$$[A_s] = [A_s]^0 \exp[-(N - s)^2/2\sigma^2] \qquad (10.4)$$

σ^2 is the variance of the distribution that characterizes the polydisper-sity of micelles. N is close to the average micelle aggregation number, an experimentally accessible quantity. Equation 10.4 is consistent with the results of theoretical calculations of the size distribution function for dilute solutions where micelles are *spherical* or close to spherical.[15]

(iv) Counterions are not included (case of nonionic surfactants).

(v) The rate constants k_s^+ and k_s^- are independent of s and equal to k^+ and k^- in the range of micelles proper. This assumption reduces the spectrum of relaxation times associated with the fast process to a single relaxation time, given by:

$$1/\tau_1 = (k^-/\sigma^2) + (k^-/N)a \tag{10.5}$$

The reduced surfactant concentration, a, is given by:

$$a = (C - \overline{[A_1]})/\overline{[A_1]} \tag{10.6}$$

In Equation 10.6, C is the total surfactant concentration and $\overline{[A_1]}$ the equilibrium concentration of free surfactant which can be taken as the cmc (nonionic surfactants) or calculated if the micelle ionization degree is known (ionic surfactants). Thus $1/\tau_1$ should increase linearly with C provided that all other quantities in Equation 10.5 are independent of C. From such plots one can obtain k^- and σ^2, provided that N is known. Note that $T_R = N/k^-$ is the *average residence time of a given surfactant in micelle A_N*, whereas $1/k^-$ is the residence time of *any* surfactant in micelle A_N. Moreover, since $\overline{[A_1]} \cong$ cmc, k^+ can be obtained from

$$k^+ \cong k^-/\text{cmc} \tag{10.7}$$

The expression of the relaxation time τ_2 for the micelle formation/breakdown was obtained using assumptions (i)–(v) together with the approximation that this process occurs via a series of stepwise reactions 10.1. Reactions 10.3 of fragmentation/coagulation are excluded. The calculations yielded the expression:

$$1/\tau_2 \cong N^2\{R \, \overline{[A_1]}[1 + (\sigma^2/N)a]\}^{-1} \tag{10.8}$$

with

$$R = \sum_{s_1+1}^{s_2} (k_s^- \overline{[A_s]})^{-1} \tag{10.9}$$

The summation extends to all species in range 2 separating oligomers and micelles proper (see Figure 10.1). The quantity R is analogous to a resistance to the transfer of monomers between oligomers and micelles proper. Information on the species in the range separating oligomers and micelles proper can be obtained from the value of R.

Aniansson[16] further derived the expression of the surfactant residence time T_R in a micelle of aggregation number N:

$$T_R = N/k^- = (l_b^2/D)\exp(\varepsilon/k_B T) \tag{10.10}$$

D is the diffusion coefficient of the free surfactant, l_b is a molecular distance (0.1–0.2 nm), and ε is the free energy of exit of one surfactant from the micelle. The model used suggested that ε should vary with the surfactant alkyl chain length (characterized by its carbon number m) exactly as the free energy of transfer ΔG_{tr}^0 of the surfactant alkyl chain from the micelle core to water, that is, linearly with m, and with the same increment per methylene group.

Another important contribution of Aniansson[17] to micelle dynamics was the derivation of the relationship between the micelle lifetime T_M and τ_2 in the case of dilute micelle solutions:

$$T_M = N\tau_2 a/[1 + (\sigma^2/N)a] \tag{10.11}$$

As stated by Aniansson,[17] "except close to the cmc $a/[1 + (\sigma^2/N)a]$ is of the order of one so that generally the order of magnitude of the micelle lifetime is determined by the product $N\tau_2$." Since N is often close to 100, the micelle lifetime is therefore much longer than τ_2. At this stage it is noteworthy that Equation 10.5, Equation 10.8, and Equation 10.11 strictly apply to dilute micelle solutions (thus spheroidal micelles) of nonionic surfactants.

10.2.3 MICELLE FORMATION/BREAKDOWN VIA SIMULTANEOUS STEPWISE REACTIONS AND SCISSION/FUSION REACTIONS

The inclusion of fragmentation/coagulation reactions 10.3 in the theory of the kinetics of micelle formation/breakdown process was carried out by Kahlweit et al.[14,18,19] The contribution of reactions 10.3 is significant in systems where the intermicellar interactions are attractive or weakly repulsive. This is the case of solutions of nonionic surfactants and of ionic surfactants at high surfactant concentration and/or ionic strength or in the presence of divalent counterions, and of water-in-oil microemulsions.

The treatment of Kahlweit et al.[14,18,19] makes use of the fact the contributions of reactions 10.1 and 10.3 to the relaxation of the system add in the same manner as parallel resistances:

$$1/\tau_2 = 1/\tau_{21} + 1/\tau_{22} \tag{10.12}$$

In Equation 10.12, τ_{21} corresponds to micelle formation/breakdown via reactions (1) and is given by Equation 10.8 or its modified form for ionic surfactants, whereas τ_{22} corresponds to reactions 10.3 and is given by:

$$1/\tau_{22} = \beta Na/[1 + (\sigma^2/N)a] \tag{10.13}$$

In Equation 10.13 β is a measure of the mean scission rate constant for reactions 10.3. Its expression is obtained using DLVO theory. The authors further showed that at low salt concentration Equation 10.12 takes the form:

$$1/\tau_2 = Q_1([X]/\overline{[A_1]})^{-q_1} + Q_2([X]/[X]_0)^{q_2} \qquad (10.14)$$

Q_1 and Q_2 are two constants; $[X]_0$ is the concentration of free counterions at the onset of coagulation; and q_1 and q_2 are two positive numbers. As a result, the change in $1/\tau_2$ with the concentration of surfactant or added salt is expected to be V-shaped and to level off as $[X]$ tends toward $[X]_0$. Several studies reported this type of variation.[5]

10.2.4 MAIN CONCLUSIONS DRAWN FROM STUDIES OF RELAXATION IN SPHERICAL MICELLE SOLUTIONS

Extensive studies of the fast relaxation showed that the rate constant k^+ for the entry (association) of a surfactant into the micelles is controlled by diffusion for surfactants with an alkyl chain containing up to 14 carbon atoms. For some surfactants this limit extends to 18 carbon atoms. The exit rate constant k^- depends strongly on the surfactant alkyl chain length and decreases by a factor close to 3 per additional methylene group. This factor corresponds to the well-known free energy increment of about 1.1 $k_B T$ per methylene group for the transfer of an alkyl chain from the micelle core to water and is consistent with the prediction made on the basis of Equation 10.10.

The measurement of the slow relaxation time indicated that the species A_r at the minimum of the size distribution curve (see Figure 10.1) are present at very low level (10^{-10}–10^{-14} M) and that the values of r fall in the range 7–10.

More details can be found in Ref. 5.

10.3 THEORETICAL ASPECTS OF THE RELAXATION IN WORMLIKE MICELLE SOLUTIONS

10.3.1 INTRODUCTION

Giant micelles are expected to exchange surfactant with the bulk phase and also to form and break down via a series of exchange reactions 10.1 and/or fusion/scission reactions 10.3 as discussed above (Figure 10.2, scheme a). However, the possible contribution of three other processes, specific to giant micelles, has been discussed in the literature. The first process, *end evaporation kinetics*,[20,21] is analogous to the surfactant exchange but would only occur at the end-caps of elongated micelles. Recall that the end-caps of elongated micelles have the shape of truncated spheres with a diameter that is slightly larger than the cylindrical part (see Chapters 1 and 2).[3,4,15,22-24] The treatment of end evaporation kinetics was carried out on the assumption that surfactants cannot enter/exit an elongated

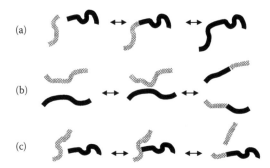

FIGURE 10.2 Schematic representation of the processes of scission/fusion (a), end interchange (b), and bond interchange (c) occurring in wormlike micelle solutions.

micelle into/from the cylindrical part or do so at a much slower rate. However, such an assumption is not supported by the fact that the free energy of transfer of a surfactant from water to a spherical micelle is only marginally different from water to an elongated micelle of the same surfactant, by 0.05 to 0.3 k_BT.[24–26] This value is to be compared to the free energy change of about -15 k_BT associated to the transfer of a surfactant with a dodecyl chain from water to a micelle.[11] The other two processes concern the formation/breakdown of giant micelles. They have been discussed mainly in studies dealing with the rheological behavior of these micelles (see Chapters 4, 8, and 9). The so-called *bond interchange reaction* can "occur when two micelles come into contact and react at some point along their arc length, chosen at random. A transient structure resembling a four-armed star polymer is formed briefly, then decays to give two new micelles" (Figure 10.2, scheme b).[27,28] The *end interchange reaction* "occurs when the end of one micelle bites into a second micelle at a random position along its length. A transient structure resembling a three-arm star polymer is formed briefly and decays to give two new micelles" (Figure 10.2, scheme c).[28] Note that such processes do not change the number of micelles in the solution.

All approaches to the relaxation of solutions of giant micelles upon perturbation consider that exchange reactions 10.1 and scission/fusion reactions 10.3 control the variation of the number density or concentration [A_s] of micelles A_s. The entry/exit of one surfactant in/from a micelle is easier than the micelle scission/fusion because it involves minor changes of the micelles. It can therefore be safely assumed that exchange reactions 10.1 are much faster than scission/fusion reactions 10.3.

The kinetics of scission/fusion reactions has been dealt with by Kahlweit et al.[14,18,19] and Cates et al.[29,30] using two very different approaches. Kahlweit et al. do not specifically deal with wormlike micelles but introduce the possibility that small aggregates can detach/associate from/to micelles proper that are still assumed to be spherical or close to spherical. Cates et al. consider explicitly wormlike micelles with an exponential distribution of length. It is clear that the equilibrium state of the system has an important bearing on its relaxation behavior upon perturbation. In this respect the two preceding approaches are not completely satisfactory because it is now well established from the theoretical as well as the experimental viewpoints that spherical micelles can coexist with elongated ones, that is, the distribution of micelle length or aggregation number can be bimodal.[3,4,22,23] This fact gives a distinct advantage to a recent approach of the relaxation in wormlike micelle solutions by a method which considers the relaxation (time dependence) of the different moments of the micelle size distribution.[31] Indeed this distribution can be described in terms of either the number density $[A_s]$ of aggregate A_s that is a function of s, or the moments of the distribution. In the moment description, the moment of order 0 is the total number Z of aggregates and this moment is not affected by the exchange reactions. On the contrary, the moments of higher order are affected by the surfactant exchange. It is shown below that this approach permits one to obtain the equations reported by Kahlweit et al.[14,18,19] or Cates et al.,[29,30] when using the same set of assumptions.

Irrespective of the approach used, the relaxation in micelle solutions can be split in two parts corresponding to two time scales. The first time scale is taken as slower than the kinetics of exchange between micelles and intermicellar solution, and as faster than the scission/fusion process. In this time scale, after a perturbation, the number of micelles $Z = (C - [A_1])/N$ is not at its equilibrium value but the other moments, driven by the monomer exchange are at their equilibrium values. In the second and slower time scale the change of Z with time can be observed. In this time scale the exchange process imposes relations between $[A_1]$, $[A_s]$, and Z.

10.3.2 SURFACTANT EXCHANGE PROCESS

The perturbation results in time dependent values of $[A_1]$, $[A_s]$ and Z given by Equations 10.15, with $[\overline{A_1}]$, $[\overline{A_s}]$, and $\overline{Z}$ being the equilibrium values.

$$[A_1(t)] = [\overline{A_1}] + \Delta[A_1(t)]; \quad [A_s(t)] = [\overline{A_s}] + \Delta[A_s(t)] \quad \text{and} \quad Z(t) = \overline{Z} + \Delta Z(t) \quad (10.15)$$

In the first time range the fast exchange process imposes the relation

$$\frac{[A_1][A_{s-1}]}{[A_s]} = K_s \quad (10.16)$$

where K_s is the equilibrium constant of reaction 10.1. When a weak perturbation is imposed on Z, Equation 10.15, Equation 10.16, and the mass conservation equation result in the relationships:

$$\frac{\Delta[A_1(t)]}{[A_1]} = -\frac{1}{((N_W + 1/a) - N)} \frac{\Delta Z(t)}{\overline{Z}} \tag{10.17}$$

$$\frac{\Delta[A_s(t)]}{[A_s]} = ((N_W + 1/a) - s) \frac{\Delta[A_1(t)]}{[A_1]} \tag{10.18}$$

In Equation 10.17 and Equation 10.18 N is the average aggregation number and N_w is the weight average aggregation number.

10.3.3 TIME DEPENDENCE OF THE NUMBER OF MICELLES Z

The variation of Z with time is given by:

$$\frac{dZ}{dt} = P_S - P_F \tag{10.19}$$

where P_F is the number of micelles that disappear per unit time due to fusion reactions and P_C the number of micelles created per unit time by scission reactions.

From reaction 10.3 the number of micelles A_{s-k} that fuse per unit time with micelles A_k to give micelles A_s is given by

$$P_F(s,k) = a_k^s [A_{s-k}][A_k] \tag{10.20}$$

Likewise the number of micelles A_s that break into two micelles A_{s-k} and A_k is given by

$$P_S(s,k) = b_k^s [A_s] \tag{10.21}$$

At equilibrium, reaction 10.3 leads to

$$a_k^s \overline{[A_{s-k}][A_k]} = b_k^s \overline{[A_s]} \tag{10.22}$$

For a weak perturbation, the first order expansion of the concentrations and of Z according to Equations 10.15, the use of Equation 10.17 and Equation 10.18, and the summing of $P_F(s,k)$ and $P_S(s,k)$ over all values of s and k lead to

$$\frac{d\Delta Z(t)}{dt} = -\frac{N_W + 1/a}{((N_W + 1/a) - N)} \sum_{s,k} \overline{[A_s]} b_k^s \frac{\Delta Z}{\overline{Z}} \tag{10.23}$$

Inserting in Equation 10.23 the average rate constants of fusion, α, and of scission, β, and the quantity θ, defined by Equations 10.24 yields the simple relaxation Equation 10.25

$$\alpha = \sum_{s,k} \frac{\overline{[A_{s-k}][A_k]}}{\overline{Z}\overline{Z}} a_k^s \qquad \beta = \frac{\sum_{s,k} \overline{[A_s]}b_k^s}{\overline{Z}} = \overline{Z}\sum_{s,k} \frac{\overline{[A_{s-k}][A_k]}}{\overline{Z}\overline{Z}} a_k^s \qquad \theta = \frac{N_W + 1/a}{N} \tag{10.24}$$

$$\frac{d\Delta Z(t)}{dt} = -\frac{\theta}{\theta-1}\beta\Delta Z(t) \tag{10.25}$$

Therefore, after a perturbation the value of Z goes exponentially to its equilibrium value with the reciprocal relaxation time

$$1/\tau_2 = [\theta/(\theta - 1)]\beta \tag{10.26}$$

This model shows that the slow relaxation process is mono-exponential because it is associated to a single moment of the size distribution. The relaxation of Z controls the slow relaxation process. The other moments have one or many fast relaxation times which correspond to the kinetics of the exchange process and a slow relaxation time equal (in first approximation) to the relaxation time of Z because at long times they all follow the variation of Z.

If the surfactant concentration is much larger than the cmc ($1/a << N_W - N$), as is usually the case for systems of wormlike micelles, $1/\tau_2$ is given by

$$\frac{1}{\tau_2} = \alpha \frac{d(C - [A_1])}{dN} = \beta \frac{d \log(C - [A_1])}{d \log N} \tag{10.27}$$

Since $N \propto C^\nu$ with ν around 0.5,[29] Equation 10.27 predicts an *increase* of $1/\tau_2$ with concentration. Recall that the Aniansson and Wall theory of micelle formation/breakdown via stepwise reactions 10.1 predicts a *decrease* of $1/\tau_2$ with increasing C.[10]

10.3.4 Asymptotic Behaviors for Giant Micelles: Cates Equation[29,30]

For very long (giant) micelles, the distribution of lengths obeys an exponential law and $\theta = 2$. The rate constant of scission b_k^s is independent of the position on the giant micelle where the scission occurs, and the probability of scission of a micelle is proportional to its length. The average rate constant of scission writes:

$$\beta = \frac{\sum_s s \overline{[A_s]}k_b}{\overline{Z}} = Nk_b \tag{10.28}$$

where k_b is the scission rate constant per unit length. In this situation the expression of the relaxation time is the same as that reported by Cates et al.:[29,30]

$$\tau_2 = 1/2Nk_b \qquad (10.29)$$

In this model, the kinetics of surfactant exchange has been supposed to be very fast, but this assumption may be wrong for giant micelles. Indeed, Marques et al.[20] have shown that relaxation time for N_W via the end evaporation process is proportional to N^2 (N_W is equivalent to the second order moment of the micelle size distribution when the number of surfactant exchanged between bulk and micelles is low). The relaxation time associated to the scission/fusion process is inversely proportional to N and for a critical length the two relaxation times are equal. However, the end evaporation process is not the only exchange process occurring in the system. The surfactant exchange between cylindrical part and bulk phase can have a large effect for giant micelles because the probability that the exchange occurs in the cylindrical part grows as N, whereas the probability of end evaporation is constant.

10.3.5 SHORT ELONGATED MICELLES: KAHLWEIT ET AL. EQUATION[14,18,19]

When the micelles are short and the distribution of micelle length is narrow, N_W is close to N and their difference is given by:

$$N_W - N = \sigma^2/N \qquad (10.30)$$

where σ^2 is the variance of the size distribution (Section 10.2.2). With these notations Equation 10.26 becomes:

$$1/\tau_2 = \beta N_W/(\ 1/a + \sigma^2/N) \approx \beta Na\ /(1 + a\sigma^2/N) \qquad (10.31)$$

Equation 10.31 is identical to Equation 10.13 first derived by Kahlweit et al.[14,18,19]

10.3.6 TRANSITION FROM SPHERICAL TO ELONGATED MICELLES

For spherical micelles with a narrow distribution width, the aggregation number increases very slowly with concentration; to a first approximation, $N \propto C^v$, with $0 \leq v \ll 1$, and

$$1/\tau_2 \propto \alpha C^{1-v}/v \cong \alpha C/v \qquad (10.32)$$

Using the detailed balance, Equation 10.32 yields

$$1/\tau_2 \propto \beta/v \qquad (10.33)$$

Kahlweit et al. assume that β does not depend on concentration.[14,18,19] The detailed balance $\alpha Z = \beta$ implies that the average probability of fusion decreases when the number of micelles increases with C. In turn, this assumption implies that only the micelles in the vicinity of the maximum of the size distribution can break and that only the smallest aggregates, whose concentration varies very little with the surfactant concentration, can merge with the micelles. This model implicitly assumes that the micelles having a size larger than that at the maximum of the size distribution do not contribute to the relaxation process. This assumption is not valid when the transition between spherical and elongated micelles is observed.

In Cates et al. derivation of τ_2 (Equation 10.29) the scission rate constant b_k^s is assumed to be proportional to the length of the cylindrical part of micelle A_s and a constant value is used for a_k^s. If the same assumptions are retained when micelles tend to become spherical, $1/\tau_2$ tends to vary as the concentration of spherical micelles.

10.3.7 NUMERICAL SIMULATIONS

The recursive relation between $[A_s]$ and $[A_{s-1}]$ yields:

$$[A_s] = \prod_{s=1}^{s} K_{s-i}[A_1] = F(s)[A_1]^s \text{ with } F(s) = \prod_{s=1}^{s} K_{s-1} \qquad (10.34)$$

The value of the equilibrium constant $F(s)$ is directly related to the energy of formation of aggregate A_s. Assuming a distribution for $F(s)$ and values for the rate constants of micelle scission and fusion, the detailed balance and the mass conservation equation permit the calculation of the variation of the static and dynamic parameters with the concentration.

In numerical simulations of micelles, in order that the calculations be not too lengthy, the routines often use a lower (or higher) cutoff value of the aggregation number below (or above) which the aggregate concentration is negligible. For the calculation of the relaxation time, new limits have been added to the numerical routines that correspond to values above (or below) which $P_F(s,k)$ and $P_S(s,k)$ become negligible (see Equation 10.20 and Equation 10.21).[31]

Reference 31 reports the calculation of micelle length and relaxation time for three simple energy distributions. However, the impact of a potential barrier to fusion/scission reactions was not fully considered. Such a barrier can result from repulsive interactions between aggregates in the case of ionic micelles and/or, more generally, from the existence of an intermediate state in the scission/fusion process that has a free energy higher than the initial and final states. Some insight into the physical nature of this intermediate state can be found in the work of Eriksson and Ljunggren[22] (see also Chapter 2, Section 2.5). These authors considered the fluctuations of diameter (constrictions) of the micelle cylindrical part, an aspect that has been little discussed in the literature. The surfactants at such constrictions are locally assembled with a negative curvature and thus have a free energy higher than in end-caps which have a positive curvature. Although most of the constrictions are of rather small amplitude owing to the fixed length of the surfactant alkyl chain, the largest ones (corresponding to the state of highest energy) may be large enough for the micelle to break in two daughter micelles.

For short elongated micelles it is anticipated that scissions mostly occur close to the micelle end-caps, and indeed, the end-caps of wormlike micelles have a diameter larger than the cylindrical body.[3,4,15,22,24] This favors scission at the junction between end-cap and cylindrical body if a constriction occurs just there or reaches there during its lifetime, and explains the experimentally observed or inferred coexistence of spherical and elongated micelles at not too high surfactant concentration.[1-4] Note that a recent study of the dynamics of wormlike micelles in aqueous solutions of a diblock copolymer did show that the micelles break down at the micelle end-caps, resulting in the release of spherical micelles.[32] Nevertheless, as elongated micelles grow at the expense of spherical micelles[3] it is likely that for extremely long (giant) micelles scission occurs anywhere in the cylindrical part, as postulated in theoretical treatments.[18,19,29,31]

Figure 10.3 represents schematically the two situations discussed above. The top drawing corresponds to a potential barrier to scission independent of position on the micelle cylindrical part with no scission possible *in* the end-cap. The scission rate constant is then constant and scission occurs with an equal probability anywhere in the micelle cylindrical part. In the bottom drawing scission is favored in a short micelle section connecting an end-cap to the cylindrical part (position dependent potential barrier). These two situations have been numerically simulated and the results are summarized below. Note that assuming a constant potential barrier simply affects the value of $1/\tau_2$ but not the qualitative aspect of its variation with concentration.

The size distribution sets the ratio between the scission and fusion rate constants:

$$b_k^s = a_k^s \frac{F(k)F(s-k)}{F(s)} \qquad (10.35)$$

The values of the rate constants are proportional to a factor that depends on the energy barrier to fusion/scission reactions.

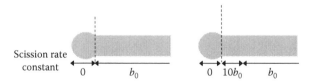

Scission rate constant

FIGURE 10.3 Schematic representation of the variation of the scission rate constant along a wormlike micelle in the situation where the potential barrier to scission is independent of the position on the micelle (left) and where it is easier at the connection between end-cap and cylindrical part (right).

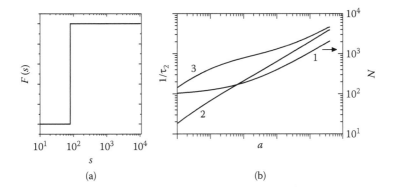

FIGURE 10.4 Numerical simulations assuming that the surfactant free energy is the same in the maximum spherical micelle and in end-caps. (a) Normalized variation of the equilibrium constant $F(s)$, assuming a surfactant for which $N_S = 80$. (b) Relative variations with the reduced surfactant concentration a of the average micelle aggregation number N (right ordinate, plot 1) and of the reciprocal relaxation time (left ordinate) assuming a potential barrier independent of position of scission (plot 2) and assuming a position-dependent potential barrier, here 2.3 $k_B T$ higher in the cylindrical part than at end-caps (plot 3). The same value of the scission rate constant has been used in the both cases.

Consider first the situation where the micelle size distribution corresponds to a free energy of scission and a potential barrier that are independent of the position where scission occurs on wormlike micelles. $F(s)$ is then constant for N values larger than the number N_S of surfactants making up the maximum spherical micelle (Figure 10.4a). In this case, the size distribution is exponential at any surfactant concentration. The simulations show that the exponent in the scaling law relating $1/\tau_2$ to concentration quickly tends to 0.5 and it is below 0.6 for N around $2N_S$ (Figure 10.4b, plot 2).

Consider now the situation where the potential barrier for scission at an end-cap is different from that in the cylindrical part. If scission at end-caps is easier so is fusion between spherical micelles and end-caps of wormlike micelles (detailed balance), and the average rate constants for fusion and scission decrease as N increases. This possibility is illustrated by plot 3 in Figure 10.4b that shows the variation of $1/\tau_2$ with the reduced concentration a in the case of a potential barrier that is taken as 2.3 $k_B T$ lower for a scission close to an end-cap than in the cylindrical body. In this case, the average rate constant for fusion at low N values is about 10 times larger than for large N values. For aggregation numbers up to $10N_S$, the observed exponent is about 0.36. For giant micelles the contribution of scission at end-caps becomes negligible and the variation of the relaxation time with concentration is not affected (power law exponent 0.5).

If the surfactants making up short micelles have a free energy lower than those in end-caps of wormlike micelles the $F(s)$ distribution has the shape represented in Figure 10.5a. In this case if the fusion rate constant is independent

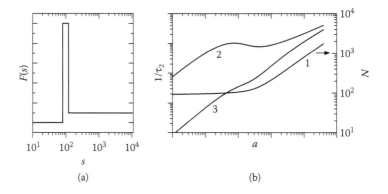

(a)   (b)

FIGURE 10.5 Numerical simulations assuming that the surfactant free energy is lower in the maximum spherical micelle than in end-caps. (a) Normalized variation of the equilibrium constant $F(s)$, assuming a surfactant for which $N_S = 80$. (b) Relative variations with the reduced surfactant concentration a of the average micelle aggregation number N (right ordinate, plot 1) and of the reciprocal relaxation time (left ordinate) assuming a potential barrier independent of position of scission (plot 2) and assuming a position-dependent potential barrier, here 2.3 $k_B T$ lower in the cylindrical part than at end-cap (plot 3). The same value of the scission rate constant has been used in both cases.

of size and the potential barrier independent of position, the scission rate constant is higher close to or at end-caps in order to satisfy the detailed balance. The sphere-to-cylinder transition of micelle shape becomes steeper and results in a nonmonotonous variation of the reciprocal relaxation time (Figure 10.5b, plot 2). On the contrary, if the scission rate constant is independent of size, the potential barrier is higher at the end-caps and for the fusion of short micelles the variation of the reciprocal relaxation time is very different from that in the preceding situation. It is monotonous with an apparent power law exponent of 0.8 up to a concentration corresponding to an aggregation number around $20N_S$ (Figure 10.5b, plot 3).

10.3.8 Other Theoretical Treatments

Recently Rusanov et al.[33] published a series of papers dealing with the slow relaxation process in systems with coexisting spherical and cylindrical micelles. The authors considered that the equilibrium between spherical and elongated micelles occurs *only* via exchange reactions 10.1. This model predicts the existence of two slow relaxation processes for surfactant solutions at concentrations well above that for which the first cylindrical micelles occur in the solution. To our knowledge this behavior has never been observed. Besides, micelle formation/breakdown via scission/fusion is not accounted for in this model, even though its existence has been experimentally demonstrated (Ref. 5, pp. 99–101).

Ilgenfritz et al.[34] derived the expression of τ_2 as given by Cates et al. on the basis of the so-called *random micellar aggregation model* which assumes that micelles form and break down via all possible reactions 10.3 with s and k going from 1 to infinity. This model predicts that N and τ_2 scale as $C^{0.5}$.

10.4 REVIEW OF THE EXPERIMENTAL RESULTS

10.4.1 Methods Used in the Investigations

The studies that reported on the relaxation in solutions of wormlike micelles used essentially chemical relaxation[35] and rheological[36] methods for measuring the relaxation time associated to the scission/fusion of wormlike micelles.

Among the chemical relaxation techniques, the temperature jump (T-jump) with light scattering monitoring of the relaxation proved to be most useful for obtaining the desired information. Indeed, the micelle size depends on temperature, and a jump of temperature generally results in a detectable variation of the scattered intensity with time reflecting the relaxation of the system. In the dilute regime, at concentration below the micelle overlap concentration, C^*, the scattered light intensity is proportional to the weight average micelle aggregation number, N_W. Thus a temperature jump will induce a perturbation of the system of wormlike micelles and will provide a *direct* measure of the relaxation time τ_2 for reversible scission. In the semidilute regime the elongated micelles form a network and the scattered intensity varies with the temperature-dependent mesh size of the network.[37] A temperature jump then results in a complex, two-steps relaxation of the system at concentration in the range between C^* and, say, $10C^*$. One step is characterized by a wave vector (q)-dependent relaxation time that reflects the rearrangement of the network, while the other time is q-independent and is associated to the reversible scission. A study of the effect of q on the measured relaxation times permits an easy assignment of the observed relaxations. At $C > 10\ C^*$ the relative amplitude associated to the chemical process becomes too small to be measured accurately.[37] Note that bond interchange reactions and end interchange reactions (see Figure 10.2) are not perturbed by a temperature jump as they are characterized by a zero enthalpy of reaction. They therefore cannot be investigated by the T-jump technique.

The relaxation time for reversible scission is obtained in an indirect manner from rheological measurements of the frequency dependence of the complex viscosity of wormlike micelle solutions. In the experimental situation where the linear stress relaxation function is found to be single exponential, the terminal viscoelastic relaxation time τ_{visco} is a geometric average of τ_{rep}, relaxation time characterizing the reptation of the average wormlike micelle if it were unbreakable, and of τ_{break}, the time characterizing the reversible scission of a wormlike micelle.[29,30]

$$\tau_{visco} = (\tau_{break}\tau_{rep})^{1/2} \qquad (10.36)$$

In Equation 10.36, $\tau_{break} = 2\tau_2$.

The rheological determination of τ_{break} therefore requires the measurement of τ_{visco} and τ_{rep}. As is shown below, the experimental conditions in T-jump and rheological measurements are often quite different. This led us to separate the review of the experimental results according to the method of measurement.

The T-jump setup with light scattering detection has been described.[37,38] The apparatuses used in rheological studies were in most instances commercial devices.

10.4.2 T-Jump Studies

The first investigations of the relaxation in a solution of elongated micelles were performed by Hoffmann et al.[39,40] They were followed by several studies of other systems[34,37,38,41–49] listed in Table 10.1 together with the parameters involved and the range of values of the relaxation time τ_2 for the micelle scission/fusion. It is striking that for the systems 1 through 8 the values of τ_2 all fall in a relatively narrow range, say, between 0.01 and 0.1 s at temperatures between 25 and 35°C, when measured in the absence of additives. Recall that studies of micelle kinetics

TABLE 10.1
Summary of the T-Jump Studies of the Scission/Fusion Process

	System[a]	Effect Investigated	τ_2 Range(s)	Ref.
1	TPySal + TTASal + 25 mM NaBr at 20°C (total [surfactant] = 25 mM)	Mixture Composition	0.01–1	39,40
2	10 mM TTABr + 10 mM NaSal + Pentanol at 25°C	[Pentanol]	0.2 s at 45°C and 2 s at 25°C (no pentanol)	41
3	TDMAO, 25°C	[Surfactant]	0.001–0.01	42
4	CTABr + KBr	T, [Surfactant], [KBr]	0.001–1	37,38
5	SDS + NaCl at 35°C	T, [Surfactant], [NaCl]	0.01–0.4	43
6	CTABr + KBr + Pentanol	[KBr], [Pentanol]	0.002–0.1	45
7	CPyCl + NaSal + 0.5 M NaCl [NaSal]/[CPyCl] = 0.5	T, [Surfactant]	0.1–1	46
8	CTABr + C18F17 + KBr	T, [C8F17]/[CTABr], [Surfactant]	0.002–1	47
9	12-2-12 + NaF	T, [Surfactant], [NaF]	0.1–1000	44
10	$C_{12}EO_6$	T, [Surfactant]	0.0002–0.005	48,49
11	$C_{14}EO_8$	T, [Surfactant]	0.0001–0.0006	34
	$C_{16}EO_8$	T, [Surfactant]	0.005–0.05	34

[a]TPySal = tetradecylpyridinium salicylate; TTASal = tetradecyltrimethylammonium salicylate; TTABr and CTABr = tetradecyl and cetyl trimethylammonium bromide; NaSal = sodium salicylate; TDMAO = tetradecyldimethylaminoxide; SDS = sodium dodecylsulfate; CPyCl = cetylpyridinium chloride; C18F17 = perfluorooctylbutane trimethylammonium bromide; 12-2-12 = gemini surfactant ethanediyl-1,2-bis(dodecyldimethylammonium bromide); $C_{12}EO_6$ = hexaethyleneglycol monododecyl ether; $C_{14}EO_8$ and $C_{16}EO_8$ = octaethyleneglycol mono tetradecyl and hexadecyl ether.

have shown that the values of τ_2 can vary by several orders of magnitude when changing the experimental parameters.[5] Systems 9 and 10 show τ_2 values rather different from the other systems. System 9 refers to a gemini surfactant (two alkyl chains-two head groups) that starts forming wormlike micelles at a much lower concentration than the corresponding conventional surfactant,[3] owing to a higher end-cap energy.[44] This explains the much larger τ_2 values found for this surfactant. System 10 concerns the nonionic surfactant $C_{12}EO_6$ which forms micelles that grow elongated as the temperature increases, beginning approximately 35°C below the cloud temperature.[48–50] However, up to a temperature of about 20°C, the $C_{12}EO_6$ micelles are not very long[50] and the system may be in the Kahlweit case (see Section 10.3.5), corresponding to fairly short times for reversible scission. The same may be true for the $C_{14}EO_8$ micelles (system 11). The difference between the τ_2 values for the $C_{14}EO_8$ and $C_{16}EO_8$ systems indicates a considerable increase of the micelle lifetime with the surfactant alkyl chain length, a well-known result in the case where micelle formation/breakdown only involves exchange reactions.[5]

An in-depth comparative discussion of most of the studies listed in Table 10.1 is difficult as they correspond to systems with different surfactants at different temperature and ionic strength. Besides, in most instances two parameters at a time were varied, most often the surfactant concentration and the ionic strength. Indeed, several of the listed investigations were performed at a salt (or surfactant) concentration not high enough for the ionic strength to remain constant when increasing the surfactant (or salt) concentration. As a result, the height of the potential barrier varied in such experiments and the reported variations of the relaxation time were due to changes of micelle size (through the end-cap energy) and also of height of the potential barrier with the surfactant or salt concentration.

Take for instance the study of growth of CTABr micelles in the presence of KBr.[37] At constant C the micelle length increases with the concentration of added KBr and, theoretically, the relaxation time should have also increased because the free energy change associated with micelle scission becomes larger. In fact, the results show a decrease of relaxation time at KBr concentration between 0.1 and 0.4 M that indicates that the effect due to the decrease of potential barrier is larger than that due to the increase of micelle size. Another example is the effect of addition of pentanol to CTABr solutions that result in a large increase of micelle size and decrease of τ_2 (see Figure 10.6).[45] The size variation indicates an increase of end-cap energy upon insertion of the short chain pentanol molecules between surfactant head groups whereas the decrease of τ_2 reveals a decrease of the potential barrier perhaps because an accumulation of pentanol molecules at a given location of the micelle cylindrical part can greatly facilitate scission at this location.

The results for SDS in the presence of NaCl are the only ones that closely correspond to experiments where the surfactant concentration was increased at constant ionic strength.[43] For this system the transition from spherical to elongated micelles is clearly observed at high [NaCl].[25] The non-monotonous variation of $1/\tau_2$ in Figure 10.7a at [NaCl] = 0.6 M shows that the so-called ladder model of

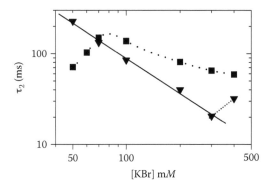

FIGURE 10.6 Variation of the relaxation time τ_2 in a 100 mM CTABr solution at 35°C with the concentration of KBr, in the absence of 1-pentanol (■) and in the presence of 50 mM 1-pentanol (▼). Reproduced from Ref. 45, with permission of the American Chemical Society.

micelle growth used by Missel et al.[25] for the interpretation of light scattering data for SDS solutions in the presence of NaCl is not an adequate description. The values of $-\log F(s)$ which reflect the end-cap energy for elongated micelles with $N > 300 \cong 5N_S$ for SDS, obtained from the fitting of the variation of τ_2 in Figure 10.7a are 12.6 and 17.4 k_BT at [NaCl] = 0.6 and 1 M, respectively. These

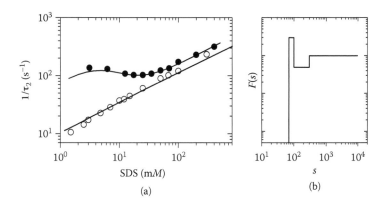

FIGURE 10. 7 (a) Variation of τ_2 with the SDS concentration in the presence of 0.6 M NaCl (●) and 1 M NaCl (○) at 35°C. The solid lines fitting the experimental results have been obtained by a simulation that used the distribution of equilibrium constant $F(s)$ represented in (b). The observed variations imply that SDS has a lower free energy in a spherical micelle than in an end-cap (see Figure 10.5). A higher free energy has been assumed for micelles with an aggregation number between that for spherical micelles and for elongated micelles and correspond to the trough in the plot of $F(s)$ vs. s. Reproduced from Ref. 43, with permission of the American Chemical Society.

values are very close to those calculated in Ref. 43, from the light scattering data reported in Ref. 25: 12.2 and 18 $k_B T$, respectively. However, because the variation of τ_2 is proportional to the derivative dN/dC whereas the intensity of scattered light depends on N_W, the departure from the model is more evident in relaxation data at the beginning of growth. Figure 10.7b shows the distribution of the equilibrium constants $F(s)$ used for the analysis of the T-jump results. The observed relaxation corresponds to the case reported in Figure 10.5b, plot 2. A more accurate distribution curve of $F(s)$ values cannot be obtained from T-jump experiments because for nearly spherical micelles the variation of size with temperature is too small to give rise to detectable relaxation signals. Nevertheless, the observed variation of τ_2 indicates clearly that the free energy of a surfactant is lower in short micelles than in end-caps. The values of the potential barrier for larger micelles estimated from these data are about 11 $k_B T$ and 8 $k_B T$ for [NaCl] = 0.6 and 1 M, respectively. This decrease of 3 $k_B T$ whereas the end-cap energy increases by nearly 5 $k_B T$ leads overall to a small decrease of $1/\tau_2$ upon increasing [NaCl] at constant surfactant concentration. At [NaCl] = 1 M the plot is nearly linear with a slope of 0.58 that is, larger than the value of 0.5 predicted from the scaling of N with C. Slope values above 0.5 were also reported in studies of the nonionic surfactants $C_{14}EO_8$ and $C_{16}EO_8$ for which micelle growth is not complicated by electrostatic effects. Similarly large exponents have been found in the rheological studies of the dynamics of wormlike micelles reviewed in the next section.

10.4.3 RHEOLOGICAL STUDIES

Rheological studies of wormlike micelle solutions are numerous (see Chapters 4, 7, 8, and 9). However, few of these studies report values of the time, τ_{break}, for the scission of wormlike micelles. Indeed, obtaining this time from measurements of complex viscosity, using Equation 10.36, is not straightforward.[51] The method strictly applies to systems where the stress relaxation function is single exponential or near so and the analysis of the results involves the full theory of the rheology of wormlike micelle solutions developed by Cates et al.[29,52] This section reviews the systems for which values of τ_{break} have been reported and whenever possible compares T-jump and rheological values. The systems investigated are listed in Table 10.2 together with the parameters whose effect has been investigated and the range of values measured for τ_{break}.[39,40,42,46,51,53–58]

The first study that showed that the rheology of wormlike micelle solutions can be controlled by the micelle scission/fusion kinetics was reported by Hoffmann et al.[39] who coined the expression of *kinetically controlled viscosities*.[40] The authors observed that the orientation relaxation time obtained using the extinction angle of flow birefringence for solutions of wormlike micelles was very close to the relaxation time τ_2 obtained by T-jump for the same system (Table 10.2, system 1, and Figure 10.8). As the micelles were very long, the rotation times should have been much longer and also stretched over a wide range for a given system. That such was not the case indicated that the time τ_{break} was much

TABLE 10.2
Summary of the Rheological Studies of the Scission/Fusion Process

	System[a]	Effect Investigated	τ_{break} Range(s)	Ref.
1	TPySal + TTASal + 25 mM NaBr (total [surfactant] = 25 mM) at 20°C	Mixture Composition	0.01–1	39,40
2	TDMAO, 25°C	[Surfactant]	0.0001–0.01	42
3	CTABr + KBr at 30–35°C	[Surfactant]	0.01–1	51
		[Surfactant], [KBr], T	0.04–1	54
4	12-2-12	[Surfactant]	0.1–6	55
		[NaCl]	0.01–4	56
5	CPyCl + NaSal + 0.5 M NaCl [NaSal]/[CPyCl] = 0.5	T, [Surfactant]	0.1–1	46
6	CTA-Tos at 30°C	[Surfactant]	0.1–1	53
7	CTA-C$_7$H$_{15}$SO$_3$	[Surfactant]	0.01–44	57
8	Er(bisEtOH)MeACl blended with 2-propanol + 400 mM KCl	T	10 at 40°C 100 at 10°C	58

[a] Same abbreviations as in Table 10.1; C$_7$H$_{15}$SO$_3$ = n-heptylsulfonate ion; CTA-Tos: cetyltrimethylammonium tosylate; Er(bisEtOH)MeACl = erucyl-bis(hydroxyethyl)methylammonium chloride.

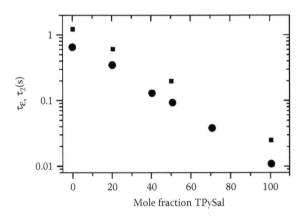

FIGURE 10.8 Mixtures of TPySal and TTASal (total surfactant concentration $C = 25$ mM, at 20°C) in the presence of 25 mM NaBr. Variation of the relaxation time τ_2 (■) obtained by T-jump and of the orientation relaxation time τ_ε (●) obtained from the dependence of the extinction angle on shear rate using flow birefringence, with the TPySal mole fraction in the mixture. Reproduced from Ref. 39, with permission of Verlag Chemie GmbH.

shorter than the rotation time of the micelle of average length. The situation is very different for the wormlike micelles of tetradecyldimethylaminoxide (Table 10.2, system 2).[42] In this system the structural relaxation time was shorter than the time for scission/fusion, and the rheological behavior of the system was not controlled by the micelle dynamics.

The measurements of the complex viscosity of wormlike micelle solutions of CTABr were analyzed on the basis of the theory of Turner and Cates[52] and yielded the first truly rheological values of τ_{break} (Table 10.2, system 3).[51,54] Although the results were not obtained at exactly the same KBr content and temperature as in the T-jump study of the same system,[37] the values of τ_{break} are in the same range as those of τ_2. This is in line with the observation of a single time constant for the stress relaxation function in these systems.

The rheological study of the wormlike micelles of the gemini surfactant 12-2-12 yielded values of τ_{break} that decreased rapidly upon increasing surfactant and NaCl concentration,[55,56] as expected theoretically (Table 10.2, system 4). The values of τ_{break} were reported to be of several seconds in the absence of salt at a surfactant concentration around 100 mM. In Ref. 56, NaCl was used as added salt because micelles bind less chloride ions than bromide ions and it was hoped that NaCl additions would affect only little the ionization of the wormlike micelles and, in turn, their end-cap energy. In a later study,[44] NaCl was replaced by NaF, as the fluoride ion is bound by micelles even less than the chloride ion.

The wormlike micelles in the solutions of CPyCl + NaSal at [NaSal]/[CPyCl] = 0.5 in the presence of 0.5 M NaCl were investigated by T-jump and rheological methods (Table 10.2, system 5).[46] Figure 10.9 shows that the values of $2\tau_2$ and of τ_{break} fall on a single line when plotted against the surfactant volume fraction ϕ (proportional to the concentration C). The scaling equation $\tau_{break} \propto \phi^{-0.67}$, provided a good fit to the data but the exponent 0.67 is in disagreement with that

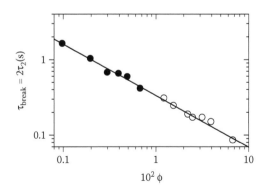

FIGURE 10.9 System CPyCl + NaSal in the presence of 0.5 M NaCl at 20°C ([NaSal]/[CPyCl] = 0.5). Variations of $2\tau_2$ (●) measured by T-jump and of τ_{break} (○) obtained from rheological measurements with the surfactant volume fraction. Reproduced from Ref. 46, with permission of the American Chemical Society.

expected on the basis of the scaling law for micelle growth ($N \propto \phi^{0.5}$). Similar differences were observed between scaling laws for τ_{break} and N for CTAB[38] and SDS[43] wormlike micelles. These differences cannot be attributed to a variation of ionic strength for the CTAB solutions or to the vicinity of the spherical-to-elongated shape transition for the SDS solutions. They can be explained only on the basis of a dependence of the potential barrier on micelle length. It is also noteworthy that Figure 10.9 shows no discontinuity or peculiarity in the variation of the relaxation time with ϕ at the volume fraction which separates the T-jump values measured in the dilute regime where $N \propto \phi^{0.5}$ from the rheological data obtained in the semidilute regime, where $N \propto \phi^{0.6}$. Similarly, T-jump experiments that covered both the dilute and semidilute regimes of other systems showed no peculiarity in the vicinity of C^*. For example, for SDS in the presence of 1 M NaCl, Figure 10.7 indicates $C^* \cong 43$ mM and $\tau_2 \propto \phi^{-0.58}$. Irrespective of the theoretical model, if the average scission rate constant is proportional to N and the scaling law for the scission/fusion relaxation time has an exponent that differs from 0.5, the detailed balance leads to an average fusion rate constant which depends on micelle size. Note that for CTA-Tos solutions (Table 10.2, system 6) $\tau_{break} \propto \phi^{-0.82}$.[53] The absence of added electrolyte which results in an increase of ionic strength with the surfactant concentration may be responsible for the large scaling exponent.

Two more studies report values of τ_{break} obtained by using the same method of analysis of the viscosity data. The first one concerns wormlike micelles of cetyltrimethylammonium-based surfactants with n-alkylsulfonate counterions.[57] The values of τ_{break} reported for the cetyltrimethylammonium n-heptylsulfonate (Table 10.2, system 7) decreased rapidly from about 44 to 0.1 s as the concentration was increased from 0.01 to 0.03 wt%. Unfortunately this study was not performed at constant ionic strength and the variation of τ_{break} is due to changes of micelle size and of the potential energy barrier. The last study concerns a very complex system based on the surfactant erucyl-bis(hydroxyethyl)methylammonium chloride blended with 2-propanol (at constant ratio [surfactant]/[2-propanol]) in the presence of 400 mM KCl (Table 10.2, system 8).[58] Little can be said about the reported values of τ_{break} for this system because of the presence of 2-propanol. Indeed it is likely that the partition of this alcohol between micelles and bulk phase varied with the surfactant concentration and temperature, possibly resulting in micelles of varying composition.

10.5 CONCLUSIONS

Much progress has been made in understanding the relaxation behavior of systems of wormlike micelles in recent years. Nevertheless this understanding has not yet reached a level comparable to that for spherical or nearly spherical micelles. For instance there has been no systematic study of the effect of the surfactant chain length on the scission/fusion kinetics of these micelles. Also in their overwhelming majority the studies concerned the kinetics of wormlike micelles of ionic surfactants. It would be interesting to extend the studies to more nonionic surfactants than listed in Table 10.1 because micelle scission/fusion appears to occur at relatively low concentration in a relatively narrow range of temperature, slightly below the cloud temperature.

As this chapter showed (see Section 10.4.2) the rheological behavior of wormlike micelle systems can be controlled by the kinetics of scission/fusion of these micelles under appropriate experimental conditions of temperature, concentration, and presence of specifically bound counterions. This of course will determine the uses of these systems. It may also guide the user in the handling of the surfactant for a given application. Chapters 15 to 18 in this volume all mention the importance of the dynamics of wormlike micelles in the applications of these systems to drag reduction, shampoos, oil recovery, and household products. A recent review[59] emphasized the importance of micelle kinetics in relation to various technological processes such as foaming, wetting, emulsification, solubilization and detergency. Results for sodium dodecylsulfate solutions showed a peculiar behavior of all of these properties at a concentration of about 200 mM, which is also the concentration where the relaxation time for micelle formation/breakdown goes through a maximum. Recall that this maximum is an indication that the contribution of scission/fusion processes of long micelles starts becoming important. Also the dynamics of wormlike micelles has been shown to be important in the analysis of results in studies of polymerization of wormlike micelles.[60]

REFERENCES

1. Almgren, M., Löfroth, J.-E., Rydholm, R. *Chem. Phys. Lett.* 1979, *63*, 265.
2. Majhi, P. R., Dubin, P. L., Feng, X., Guo, X., Leermakers, F. A. M., Tribet, C. *J. Phys. Chem. B* 2004, *108*, 5980.
3. Bernheim-Groswasser, A., Zana, R., Talmon, Y. *J. Phys. Chem. B* 2000, *104*, 4005.
4. Bernheim-Groswasser, A., Wachtel, E., Talmon, Y. *Langmuir* 2000, *16*, 4131.
5. Zana, R. In *Dynamics of Surfactant Self-Assemblies. Micelles, Microemulsions, Vesicles and Lyotropic Phases*, Zana, R., Ed., CRC Press, Boca Raton, 2005, Chapter 3, p. 75.
6. Zana, R. In *Dynamics of Surfactant Self-Assemblies. Micelles, Microemulsions, Vesicles and Lyotropic Phases*, Zana, R., Ed., CRC Press, Boca Raton, 2005, Chapter 4, p. 161.
7. Folger, R., Hoffmann, H., Ulbricht, W. *Ber. Bunsenges. Phys. Chem.* 1974, *78*, 986.
8. Muller, N. In *Reaction Kinetics in Micelles*, Cordes, E., Ed., Plenum Press, New York, 1973; p. 1.
9. Nakagawa, T. *Colloid Polym. Sci.* 1974, *252*, 56.
10. Aniansson, E. A. G., Wall, S. N. *J. Phys. Chem.* 1974, *78*, 1024 and 1975, *79*, 857.
11. Aniansson, E. A. G., Wall, S. N., Almgren, M., Hoffmann, H., Kielmann, I., Ulbricht, W., Zana, R., Lang, J., Tondre, C. *J. Phys. Chem.* 1976, *80*, 905.
12. Lessner, E., Teubner, M., Kahlweit, M. *J. Phys. Chem.* 1981, *85*, 1529.
13. Hall, D. G. *J. Chem. Soc. Faraday Trans.* 2 1981, *77*, 1973.
14. Lessner, E., Teubner, M., Kahlweit, M. *J. Phys. Chem.* 1981, *85*, 3167.
15. Israelachvili, J. N., Mitchell, J. D., Ninham, B. W. *J. Chem. Soc., Faraday Trans.*, 2 1976, *72*, 1525.

16. Aniansson, E. A. G. *Ber. Bunsenges. Phys. Chem.* 1978, *82*, 981 and *J. Phys. Chem.* 1978, *82*, 2805.
17. Aniansson, E. A. G. *Prog. Colloid Polym. Sci.* 1985, *70*, 2.
18. Kahlweit, M. *Pure Appl. Chem.* 1981, *53*, 2069 and *J. Colloid Interface Sci.* 1982, *90*, 92.
19. Kahlweit, M. In *Physics of Amphiphiles: Micelles, Vesicles and Microemulsions*, Degiorgio, V., Corti, M., Eds., North-Holland, Amsterdam, 1985.
20. Marques, C. M., Turner, M. S., Cates, M. E. *J. Chem. Phys.* 1993, *99*, 7260.
21. Dubbledam, J. L. A., van der Schoot, P. *J. Chem. Phys.* 2005, *123*, 14492 and references therein.
22. Eriksson, J. C., Ljunggren, S. *J. Chem. Soc., Faraday Trans.*, 2 1985, *81*, 1209.
23. Eriksson, J. C., Ljunggren, S. *Langmuir* 1990, *6*, 895.
24. Bauer, A., Woelki, S., Kohler, H.-H. *J. Phys. Chem.* B 2004, *108*, 2028 and references therein.
25. Missel, P. J., Mazer, N. A., Carey, M. C., Benedek, G. B. *J. Phys. Chem.* 1989, *93*, 1044 and references therein.
26. Porte, G., Poggi, Y., Appell, J., Maret, G. *J. Phys. Chem.* 1984, *88*, 5713.
27. Shikata, T., Hirata, H., Kotaka, T. *Langmuir* 1988, *4*, 354.
28. Turner, M. S., Marques, C., Cates, M. E. *Langmuir* 1993, *9*, 695.
29. Cates, M. E. Macromolecules 1988, 20, 2289 and *J. Phys. France* 1988, *49*, 1593.
30. Turner, M. S., Cates, M. E. *Europhys. Lett.* 1990, *11*, 681.
31. Waton, G. *J. Phys. Chem.* 1997, *101*, 9727.
32. Geng, Y., Ahmed, F., Bhasin, N., Discher, D. E. *J. Phys. Chem.* B 2005, *109*, 3772.
33. Kuni, F. M., Shechkin, A. K., Rusanov, A. I., Grinin, A. P. *Langmuir* 2006, *22*, 1534.
34. Ilgenfritz, G., Schneider, R., Grell, E., Lewitzki, E., Ruf, H. *Langmuir* 2004, *20*, 1620.
35. Zana, R. In *Dynamics of Surfactant Self-Assemblies. Micelles, Microemulsions, Vesicles and Lyotropic Phases*, Zana, R., Ed., CRC Press, Boca Raton, 2005, Chapter 2, p. 37.
36. Rehage, H. In *Dynamics of Surfactant Self-Assemblies. Micelles, Microemulsions, Vesicles and Lyotropic Phases*, Zana, R., Ed., CRC Press, Boca Raton, 2005, Chapter 9, p. 419.
37. Candau, S. J., Merikhi, F., Waton, G., Lemaréchal, P. *J. Phys. France* 1990, *51*, 977.
38. Faetibold, E., Waton, G. *Langmuir* 1995, *11*, 1972.
39. Löbl, H., Thurn, H., Hoffmann, H. *Ber. Bunsenges. Phys. Chem.* 1984, *88*, 1102.
40. Hoffmann, H., Löbl, H., Rehage, H., Wunderlich, I. *Tenside Detergents* 1985, *22*, 290.
41. Ohlendorf, D., Inherthal, W., Hoffmann, H. *Rheologica Acta* 1986, *25*, 468.
42. Hoffmann, H., Oetter, G., Schwandner, B. *Prog. Colloid Polym Sci.* 1987, *73*, 95.
43. Michels, B., Waton, G. *J. Phys. Chem.* B 2000, *104*, 228.
44. Oelschlaeger, C., Waton, G., Candau, S. J., Cates, M. E. *Langmuir* 2002, *18*, 7265.
45. Michels, B., Waton, G. *J. Phys. Chem.* B 2003, *107*, 1133.
46. Oelschlaeger, C., Waton, G., Candau, S. J. *Langmuir* 2003, *19*, 10495.
47. Buhler, E., Oelschlaeger, C., Waton, G., Candau, S. J. *J. Phys. Chem.* B 2004, *108*, 11236
48. Strey, R., Pakusch, A. In *Surfactants in Solution*, Mittal, K. L., Bothorel, P., Eds., Plenum Press, New York, 1986, Vol. 4, p. 465.
49. Zilman, A., Safran, S. A., Scottmann, T., Strey, R. *Langmuir* 224, *20*, 2199.

50. Zana, R., Weill, C. *J. Phys. Lett.* 1985, *46*, L-953.

51. Kern, F., Lemaréchal, P., Candau, S. J., Cates, M. E. *Langmuir* 1992, *8*, 437.

52. Turner, M. S., Cates, M. E. *J. Phys. (Paris)* 1990, 51, 307; *Langmuir* 1991, *7*, 1590.

53. Soltero, J. F. A., Puig, J. E., Manero, O. *Langmuir* 1996, *12*, 2654.

54. Khatory, A., Lequeux, F., Kern, F., Candau, S. J. *Langmuir* 1993, *9*, 1456.

55. Kern, F., Lequeux, F., Zana, R., Candau, S. J. *Langmuir* 1994, *10*, 1714.

56. Candau, S. J., Hebraud, P., Schmitt, V., Lequeux, F., Kern, F., Zana, R., *Nuovo Cimento* D 1994, *16*, 1401.

57. Oda, R., Narayanan, J., Hassan, P. A., Manohar, C., Salkar, R. A., Kern, F., Candau, S. J. *Langmuir* 1998, *14*, 4364.

58. Couillet, I., Hughes, T., Maitland, G., Candau, F., Candau, S. J. *Langmuir* 2004, *20*, 9541.

59. Patist, A., Kanicky, J. R., Shukla, P. K., Shah, D. O. *J. Colloid Interface Sci.* 2002, 245, 1.

60. Zhu, Z., Gonzalez, Y. I., Xu, H., Kaler, E. W., Liu, S. *Langmuir* 2006, *22*, 9494.

11 Giant Micelles at and near Surfaces

Rob Atkin, Annabelle Blom, and Gregory G. Warr

CONTENTS

11.1 INTRODUCTION

One of the most remarkable discoveries of the last decade is the widespread existence of giant micellar aggregates adsorbed to solid-solution interfaces. Long-standing debates about adsorbed layer morphology — micelle-like spheres[1-3] versus a laterally unstructured bilayer — on hydrophilic substrates, typified by mica and silica, were rekindled when atomic force microscopy (AFM) revealed that many cationic surfactants formed elongated cylindrical aggregates on mica.[4,5] The discovery of long, straight, hemicylindrical adsorbed aggregates on graphite sent a similar tremor through studies of hydrophobic substrates, but this was ultimately shown to be a substrate-specific effect.[6] Although mica is not a typical hydrophilic substrate for surfactant adsorption,[7] it is a widely used model surface. Giant micelles have not only been found to be almost ubiquitous on mica, but to occur on many other surfaces as well.

The formation of adsorbed giant micelles influences the equilibrium and dynamic properties of the adsorbed layer in several ways. The forces between surfactant-coated surfaces are highly sensitive to adsorbed layer morphology. Unlike bilayers, adsorbed micelles do not undergo hemifusion,[8,9] which affects their ability to act as a lubricant.[10,11] The growth of highly oriented mesoporous silicate films at interfaces[12,13] is only possible in systems where discrete aggregates form, as such structures require the condensation or adsorption of multiple layers of micelles with silicate precursors as counterions.

In addition to giant micelles on surfaces, AFM and scattering techniques have revealed near-surface structures consisting of multilayered arrays of organized micelles.[14,15] These may be aligned by flow, or organized by confinement between surfaces, or both. This effect may be observed in the drainage of soap films, or

---

TABLE 11.1

Representative Structures of Surfactants Forming Giant Micelles at or Near Surfaces

DTA+ (C$_{12}$), TTA+ (C$_{14}$), CTA+ (C$_{16}$)   C$_m$EO$_n$

12-s-12   DS−

TTeA+   HFDeP+

DDA+   18-3-1

occasionally in the preparation of samples for cryo-TEM.[16] Confined and near-surface micellar solutions thus may differ from bulk phases in their organization and sometimes their morphology.

Most conventional surfactants have by now been investigated by one of the two major techniques available for investigating adsorbed layer structure — AFM or neutron reflectometry. The surfactants discussed in this chapter are summarized in Table 11.1.

11.2 EXPERIMENTAL TECHNIQUES

11.2.1 ATOMIC FORCE MICROSCOPY

The *in situ* images of discrete adsorbed aggregates obtained using atomic force microscopy (AFM) have redefined our understanding of surfactant adsorption at the solid-liquid interface. AFM uses a sharp tip mounted on a springlike cantilever to probe the lateral morphology of an adsorbed surfactant layer. The substrate is mounted on top of a piezoelectric element, which is then raster scanned, that is, from side to side in horizontal lines and from top to bottom, beneath the tip. A laser is reflected off the cantilever into a position-sensitive detector, measuring the vertical deflection of the tip as the sample is moved laterally (Figure 11.1). The deflection at each position is then used to construct an image of the surface.

Generally AFM is used with the probe in hard contact with a solid substrate. However, hard contact imaging will destroy fragile adsorbed surfactant aggregates,

Laser

Position-
sensitive
detector

Tube
scanner

FIGURE 11.1 Schematic representation of the operation of an AFM. The surface is raster scanned in the *xy* plane beneath a tip upon which a laser is focussed. The *z*-deflection of the tip as it responds to underlying features on the surface is measured by the position-sensitive detector and an image generated.

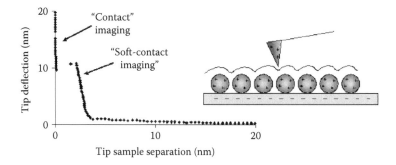

FIGURE 11.2 Representative force curve for a cationic surfactant solution adsorbed on a hydrophilic surface. When imaging in soft contact the tip is held above the adsorbed film by electrostatic or steric repulsion. An image of the underlying morphology is produced. The separation between the tip and surface at film rupture provides a measure of the thickness of the surfactant layer.

rendering this method useless for the study of surface micelles. Fortunately, the sensitivity of AFM is such that it can sense repulsive forces generated by the adsorbed species. These forces may be electrostatic, due to adsorption of charged surfactant onto both the tip and the substrate, or steric in the case of nonionic surfactants. Force curves similar in form to that shown in Figure 11.2 are typical, with a steeply repulsive interaction detected at close separations. Using an imaging force just less than that which ruptures the adsorbed surfactant film, it is possible to image in *soft contact*,[17] with the AFM tip scanning about 1 nm *above* the adsorbed layer. Sufficient contrast is obtained to image the aggregate structure while minimizing the risk of aggregate deformation, as shown in Figure 11.2.

Soft contact imaging requires that the surfactant layer have headgroups facing the solution to produce the repulsive force. Thus AFM imaging experiments are usually performed at concentrations greater than the critical micelle concentration (cmc). Image contrast results from variation in the force between the tip and adsorbed material. Contrast is maximized when forces just below that required to rupture the film are used. The tip radius is greater than the radius of curvature of most adsorbed micelles meaning the actual shape and size of highly curved adsorbed micelles cannot be accurately mapped. However, AFM is particularly well suited to detecting structures that are arranged periodically. From the lateral arrangement of surface aggregates, the shape can be inferred with confidence.

11.2.2 Neutron Reflectometry

The structure of surfactant films can be probed using the angular dependence of the intensity of an incident beam specularly reflected from an adsorbed layer on a surface (Figure 11.3). Neutron reflectivity has sufficient resolution to study the arrangement of surfactant molecules in the adsorbed layer, as different nuclei scatter neutrons with different amplitudes, and the combination of typical neutron

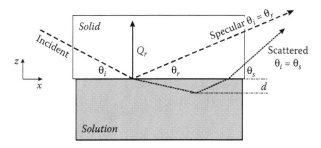

FIGURE 11.3 Schematic of neutron reflectometry or near-surface small-angle neutron scattering cell, showing incident and specularly reflected beams, as well as the transmitted and scattered (and refracted) beams.

wavelengths, λ, and incidence angles, θ, leads to experimental resolution at a nearly molecular scale. The resolution of a reflectivity experiment is $2\pi/Q_{max}$, where Q_r is the wave-vector of the reflection defined by

$$Q_r = \frac{4\pi}{\lambda} \sin\theta_r \qquad (11.1)$$

Because neutrons are highly penetrating, they are able to travel through thick samples of certain solids, making them especially useful for studying adsorbed layers on solid substrates. Quartz, sapphire, and oxidized silicon are among the most common substrates examined. Although x-rays have approximately the same wavelength and same resolution, their lack of penetration has seen them more widely employed in studies of layers at air/solution interfaces.

Critically, hydrogen and deuterium have vastly different scattering length densities, β, which allows contrast variation to be achieved through a combination of hydrogenated and deuterated surfactants and/or solvents. While isotopic substitution alters the reflectivity of the adsorbed layer, it does not affect morphology. Additionally, by careful adjustment of the hydrogen to deuterium ratio of the solvent, the contrast between the solvent and the substrate can be set to zero, resulting in a reflectivity profile that depends only on the adsorbed layer. By isotopically labeling a particular molecule (or part thereof) it is also possible to control its contribution to the reflected intensity and thus obtain details of the internal structure of adsorbed aggregates. This can be particularly useful in determining compositions of mixed adsorbed films.[18-20]

The major drawback of neutron reflectivity is that it is primarily sensitive to structure normal to the interface, z. Much of the interpretation of results depends on adopting a specific model for the adsorbed film that generates a scattering length density z-profile. Off-specular scattering has not, so far, been successfully separated from neutron scattering by the subphase, although this should be possible. Conventionally, the adsorbed layer has been regarded as a

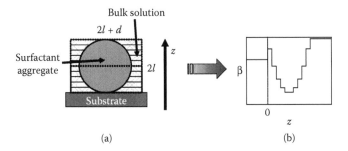

FIGURE 11.4 Schematic of the unit cell (a) used to calculate the scattering length density profile (b). The cell is divided into a number of layers and has an overall thickness equal to twice the length of the surfactant molecule, l. Coverage is altered by adjustment of the separation between structures, d, giving an overall dimension for the unit cell of $2l \times (2l + d)$.

laterally unstructured film consisting of one or more layers.[21] Recently, a model incorporating the discrete spherical or cylindrical aggregate morphologies observed in AFM experiments has been developed. This model correctly predicts the reduction in surface coverage as film curvature increases. A bilayer covers 100% of the substrate. Surface packing constraints dictate that cylinders and spheres can cover a maximum of 79% and 61% of the substrate, respectively.[22] This is an important observation, as the *patchy bilayer* or *island* interpretations suggested by many neutron reflectometry studies using a bilayer model are unphysical corrections made to account for the observed fractional surface coverages.

For models based on surface aggregates, the reflected interference fringes from the solid-layer and layer-solvent interfaces can be characterized by a thickness and a scattering length density profile $\beta(z)$ (Figure 11.4).[23] The composition z-profile is determined by dividing the adsorbed film into a series of layers. Each layer is characterized by its scattering length density, thickness, and if necessary, an interfacial roughness factor. The scattering length density of each layer is a weighted average contribution of the surfactant aggregate and the bulk solution. In spite of the morphological detail in this model, a particular profile is not necessarily a unique solution, that is, several models of adsorbed layer structure may fit an individual reflectivity curve, $R(Q)$. Several different contrasts, achieved by isotopic substitution, provide information that aids in identifying the most appropriate model, as do complementary AFM experiments.

Clearly, the combination of AFM and neutron reflectometry (NR) is very powerful, and allows the structure of an adsorbed layer to be ascertained with greater certainty than is possible using either technique in isolation. AFM provides a direct, visual image of the lateral structure of the film, whereas NR probes the structure of the layer normal to the substrate. Such complementary data allows the surface excess, layer thickness, and interaggregate spacing to be accurately determined.

In addition to adsorbed films, both neutron and x-ray reflectometry are particularly sensitive to the presence of multiple layers at an interface. This can give rise to Bragg peaks or other distinctive modulations in the reflectivity curve. In such cases, off-specular scattering can yield detailed information about near-surface structure.

11.2.3 SCATTERING FROM SURFACE AND NEAR-SURFACE STRUCTURES

Near-surface small-angle neutron scattering (NS-SANS), pioneered by Hamilton et al.,[15,24] employs a reflectometry-like geometry using a fixed incidence angle above the critical angle so that some of the beam is refracted into the solution. Unlike reflectometry, in which the intensity of the specularly reflected beam is collected as a function of angle of incidence, here a two-dimensional SANS detector is used to gather the scattered intensity from the incident beam that has penetrated into the solution (see Figure 11.3) at a fixed angle. The low angles used effectively serve to limit the depth sampled by the transmitted beam. This is because the path length that must be traversed in the solution in order to sample a scattering event increases dramatically with depth, d (Figure 11.3). This leads to attenuation of the beam and its scattered signal. For typical conditions the solution is sampled to a depth of around 50 µm,[25] so that differences between bulk and near-surface organization can be distinguished.

11.3 GIANT MICELLES AT THE SOLID/SOLUTION INTERFACE

11.3.1 GRAPHITE

The first images of surfactant aggregates adsorbed at the solid-liquid interface were of CTAB on the cleavage plane of highly ordered pyrolytic graphite, reported in 1994.[26] The basal plane of graphite is comprised of sheets of hexagonal rings of carbon atoms with a threefold axis of rotation, and is useful for AFM imaging because it is atomically smooth. The images showed parallel stripes with a periodicity of 4.2 nm which were aligned in one of three directions, due to surfactant adsorption being templated by the symmetry axes of the substrate. The period between the stripes was slighter greater than twice the length of a fully extended monomer, consistent with a hemicylindrical adsorbed structure. As this structure is very long in one direction, it is clearly a surface-adsorbed giant (hemi)micelle.

The strong attraction between surfactant tail groups and the graphite substrate is due to both hydrophobic and van der Waals interactions. This attraction, coupled with the large interaction area between the surface and the surfactant, allows graphite to exert the greatest influence over adsorbed aggregate structure of any substrate. The surfactant alkyl chain is adsorbed in an all trans configuration, with two methylene units fitting within one hexagonal ring of the lattice,[27] forming a zigzag line. This produces a head-to-head, tail-to-tail arrangement that templates subsequently adsorbed surfactant into very straight hemicylinders (Figure 11.5)

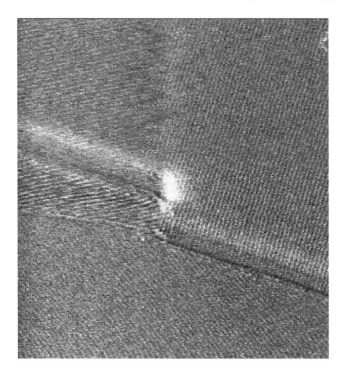

FIGURE 11.5 400×400 nm^2 AFM deflection image of $C_{12}EO_9$ adsorbed at the graphite–aqueous solution interface.[33] Very straight, hemicylindrical aggregates are aligned normal to each of the three symmetry axes of the underlying graphite lattice and meet at grain boundaries. The dark line about the middle of this image is a step in the graphite.

Hemicylindrical aggregation has been observed on graphite for ionic[4,26,28–30] (conventional and gemini), nonionic,[6,17,31–33] and zwitterionic[34] surfactants with hydrocarbon tail groups dodecyl or longer. Surfactants with tail groups decyl or shorter in length form a featureless monolayer on graphite, as the alkyl tail is too short to be strongly oriented by the substrate. Similar results have been observed for surfactants adsorbed onto graphite from the polar nonaqueous solvent formamide[35] and the room-temperature ionic liquid ethylammonum nitrate.[36] Due to the reduced driving force for surfactant self-assembly in such solvents, hemicylindrical adsorbed aggregates were only observed at very high concentrations and for longer alkyl chains than in water. In ethylammonium nitrate, at least 16 carbons are required to template hemicylinders.

Surfactants with geometries preferring low curvature, such as DDAB[37] and $C_{12}EO_3$, form lamellar mesophases in solution. When these surfactants are adsorbed to graphite, broad stripes with a repeat distance much greater than that for typical hemicylindrical aggregates are observed. These stripes are adsorbed surfactant bilayers oriented normal to the graphite surface. Such a structure is a compromise between the templating effect exerted by the substrate and monomer geometry, that is, the alkyl chains are sufficiently long to be templated by substrate, but the monomer curvature is too low to permit the formation of hemicylinders.

Several studies have examined surfactant adsorption on hydrophobic substrates where surface anisotropy is absent, typically self-assembled monolayers (SAMS). Hemicylindrical aggregates have not been observed, with the surfactant generally adopting a monolayer morphology.[38,39] However, a study using a rigid, amorphous hydrophobic substrate (silica with covalently attached trimethylchlorosilane groups) produced quite different results.[40] In the absence of anisotropic interactions, the geometry of the surfactant monomer dictates morphology rather than the substrate. Globular half-micelles were reported, with lateral periodicities that scale with micelle size. While these structures are clearly not giant micelles, this study does illustrate that the anisotropy of graphite is critical for producing giant micelles.

11.3.2 MICA

As crystalline mica is hydrophilic, the surfactant headgroups interact with the substrate rather than the tail, and thus, relative to graphite, the area of interaction per surfactant molecule is greatly reduced. This reduces the ability of mica to template adsorbed structure and leads to considerably different aggregate morphologies on mica compared to graphite. As a result, the geometry of the surfactant monomer plays a greater role in dictating surface structure on mica than for graphite.[4,17,41]

The density of charged sites on mica is high (one site per 0.48 nm²),[17] therefore monomeric surfactants are adsorbed with their headgroups in closer proximity than in bulk micelles (for comparison, the headgroup area for CTAB in a micelle is 0.64 nm²).[42] Thus, adsorbed aggregates have lower curvature than corresponding solution micelles, and elongated adsorbed micelles are favored. The negative surface groups on mica are arranged precisely with the surface lattice of the substrate. While this does influence aggregate morphologies, mica only weakly orientates the axial directions of the adsorbed structure.

DTAB and TTAB form giant cylindrical micelles on mica. The interaggregate distance is similar to the diameter of a solution micelle[4,43,44] and the long axis of the aggregates is aligned locally in one of three directions due to the crystallinity of the substrate. The formation of giant micelles on mica is hindered by the use of a weakly binding counterion, such as chloride. A weakly binding counterion does not sufficiently shield electrostatic repulsion between surfactant headgroups for cylindrical micelles and more curved structures are favored. This is illustrated by the fact that DTAC forms spherical structures on mica[5] despite the high surface charge. Spherical structures are also formed by cationic surfactants with two charged quaternary ammonium headgroups such as 18-3-1, 16-3-1, and 12-3-1.[28]

Gemini surfactants, denoted *m-s-m* (see Table 11.1), have proven useful for differentiating between the influence of surfactant geometry, and substrate effects, on adsorbed aggregate structure. The geometry of gemini surfactants can be

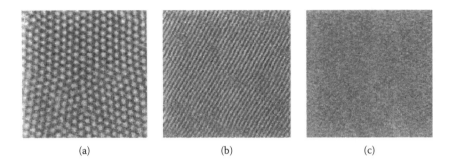

(a)   (b)   (c)

FIGURE 11.6 150×150 nm^2 AFM deflection images of (a) spherical aggregates arranged hexagonally on mica formed by the asymmetric surfactant 18-3-1, (b) giant cylindrical aggregates formed by the symmetric gemini surfactant 12-4-12, and (c) a bilayer formed by the symmetric gemini surfactant 12-2-12. Reprinted with permission from Ref. 28. © 1997 by American Chemical Society.

systematically altered by changing alkyl chain, m, and/or spacer length, s. On mica, a wide variety of surface morphologies have been observed, including giant micelles (Figure 11.6). As the headgroup size is systematically decreased, progressively flatter aggregate structures are favored. For example, 12-12-12 and 12-10-12 form globular aggregates, 12-8-12 forms short cylinders, 12-6-12 and 12-4-12 form long cylinders, and a reduction in the spacer length to 2 results in the formation of a featureless bilayer.[28,45] This aggregation pattern parallels that observed in solution, where the curvature of aggregates is lowered as the spacer length is decreased from 12 to 2.

Addition of electrolyte to a micellar solution generally reduces curvature by better screening electrostatic repulsions between headgroups. On the surface this progression has been observed to invert. The thermodynamically stable bilayer formed by CTAB can be transformed to cylinders with addition of KBr.[43] Increasing the electrolyte concentration resulted in the long axis of the cylinders becoming shorter, leading to the formation of discrete aggregates. The addition of H$^+$, which is known to be more strongly surface binding than K$^+$, produced a greater number of defects at the same concentration, supporting the idea of the ions competing with the surfactant for surface sites.

A corresponding effect was reported by Duval et al. in a study on the influence of CsCl on morphologies formed by a series of gemini surfactants.[45] Added caesium ions compete for charged sites with surfactant cations, lowering the density of adsorbed surfactant and increasing aggregate curvature. For 12-2-12, the addition of CsCl leads to the formation of an adsorbed mesh (Figure 11.7).[46]

An adsorbed mesh is a recently discovered structure at the solid-liquid interface and can be visualized as a film of densely branched cylinders.[46] It occurs at curvatures between those of cylinders and planar bilayers. Meshes are

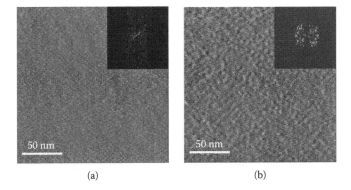

(a)   (b)

FIGURE 11.7 200×200 nm^2 AFM deflection images of (a) a bilayer formed by 12-2-12 and (b) mesh formed when 100mM CsCl is added to solution. The cesium cations reduce the negative charge on mica by competing for binding sites with the surfactant. Reprinted with permission from Ref. 45. © 2004 by American Chemical Society.

best identified by their place in a curvature progression as they are often similar in texture to a film of adsorbed globules.

Aromatic molecules and counterions have a strong influence on the adsorbed structure of TTeAB on mica.[47] The large headgroup of this surfactant leads to the formation of sperical structures in the absence of additives. When aromatic molecules such as naphthalene, 1-naphthol, and 2-naphthol are added to solution, they are preferentially solubilized in the palisade layer of the aggregate, lowering the film curvature. A structural transition from globules to cylinders—giant micelles—is observed.

A similar transition to giant cylindrical micelles was observed when sodium salicylate was added to films of globular micelles formed even by surfactants with large headgroups.[5] The effect is less pronounced when sodium benzoate is used to reduce the curvature of a TTAB film. In addition to the benzoate anion being solubilized in the surfactant layer, the sodium cation competes for surface binding sites. These two factors will change curvature in opposing directions and the net result was the formation of a mesh.[46]

When mixtures of DDAB and TTeAB were adsorbed on mica, the curvature of the film could be systematically tuned by varying the mole ratio of the two surfactants. This study was the first to report the mesh structure described above. By adding the bilayer forming DDAB to a spherical TTeAB film, the hydrocarbon volume was increased and a structural transition from spheres, to giant cylinders through to a mesh was induced (Figure 11.8).[46] A similar reduction in curvature from globular to giant micellar structure was observed in mixtures of the zwitterionic surfactant DDAPS with incorporation of DTAB.[44] The effect of partial miscibility of the alkyl tail in mixtures of the hydrocarbon surfactant tetradecyltriethylammonium chloride

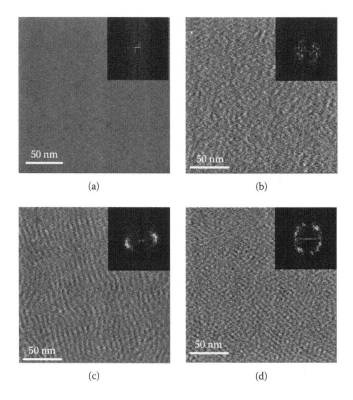

(a)   (b)

(c)   (d)

FIGURE 11.8 200×200 nm^2 AFM deflection images showing the increase in curvature of (a) a DDAB bilayer with sequential addition of the surfactant TTeAB forming more highly curved structures. At the bulk composition 0.035 mM DDAB, 4.10 mM TTeAB a mesh (b) was formed, giant cylindrical micelles (c) were formed at 0.023 mM DDAB, 4.89 mM TTeAB and (d) TTeAB forms a globular film at 6.3 mM. Reprinted with permission from Ref. 45. © 2004 by American Chemical Society.

(TTeAC) and fluorocarbon surfactant perfluorodecylpyridinium chloride (HFDePC) has been examined. The globular TTeAC micelles in the film were continuously transformed to giant cylinders despite the hydrocarbon surfactant being almost excluded from the surface once mixed.[48,49]

The spherical structures formed by TTeAB on mica can also be transformed into giant micelles through the addition of nonadsorbing polyethylene oxide monoalkylether (nonionic) surfactants.[50] These nonionic surfactants do not adsorb directly on mica, but coadsorb in the presence of cationic surfactants (due to attractive hydrophobic interactions between tail groups) to form a mixed adsorbed micelle. Nonionic surfactants with headgroups containing less than 8 ethylene oxide units lower the film curvature of the aggregate, promoting the formation of cylindrical aggregates at low nonionic surfactant concentrations. As the non-ionic surfactant concentration was increased these giant micelles branched to form a mesh structure and then a bilayer. This transition is a consequence of the increase in hydrocarbon volume of the adsorbed aggregate.

11.3.3 Mineral Oxide Surfaces—Silica

The charge on most mineral oxide surfaces is caused by protonation or deprotonation of surface hydroxyl groups. These amphoteric sites may be positive, negative, or neutral at any particular pH. The total surface charge is therefore determined by the relative population of each charged species at the surface. Two types of silica surfaces have been used for AFM and reflectance studies: crystalline quartz and amorphous silica.[51] Note that the properties of amorphous silica are dependent on the method of preparation.[52,53]

Silica or quartz surfaces have different charge densities and roughness compared to mica. A charge density of one site per 20 nm^2 at neutral pH is expected for amorphous silica surface (recall for mica there is one ionizable site per 0.48 nm^2). This will influence the ability of the surfactant molecules to pack closely, resulting in more curved aggregates on silica compared to mica. Nonionic surfactants do not adsorb on mica but can adsorb on silica through hydrogen bonding between the ethylene oxide headgroups and surface silanol groups.

Far fewer AFM imaging studies have employed silica than mica or graphite, primarily due to the roughness and amorphous nature of the surface making it more challenging to obtain clear images. As with other substrates, factors that affect aggregation in solution can produce corresponding structural effects at the silica-water interface.

In general, single chained cationic surfactants such as DTAB, TTAB, and CTAB have been reported to form spherical admicelles on both quartz and amorphous silica.[18,22,54–57] The saturation surface excess measured using neutron reflectometry on quartz increases with alkyl tail length from 5.9,[18] 6.5,[22] and 6.7[57] μmol m^{-2}, respectively. A corresponding trend is observed for these same surfactants on silica.[58]

On amorphous silica, the method of surface preparation is known to influence the surface excess[52,53] and adsorbed layer morphology. For CTAB, both giant wormlike admicelles[59] or spherical admicelles[55] have been reported under the same solution conditions but with substrates prepared using different methods (Figure 11.9). Wormlike admicelles form when the surface is soaked in strong basic solution prior to use.[59] Base treatment creates more negatively charged surface groups,[52,60] so surfactants adsorb with headgroups in closer proximity, favoring flatter structures. When the surface is pretreated with strong acid, the surface remains hydrophilic, but the density of charged sites is reduced. Surfactants adsorption is relatively diffuse, with greater distance between surface-adsorbed headgroups, and spherical structures form.[55]

It should be noted that the formation of giant admicelles could not be induced by pH variation of the bulk solution *after* the surfactant was adsorbed on either quartz or amorphous silica.[55,56] Increasing pH from 3 to 11 increased the ζ-potential

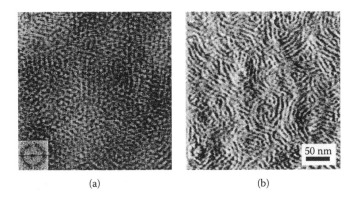

(a)   (b)

FIGURE 11.9 300×300 nm^2 AFM deflection images of 10 mM CTAB on silica after the surface being prepared using (a) strong acid[55] and (b) strong base.[59] Soaking the silica surface in a basic solution increases the negative charge[60] on the surface resulting in formation of giant wormlike micelles. Reprinted with permission from Refs. 55 and 59. © 2000 by American Chemical Society.

of the surface from below 20 mV to greater than 100 mV and reduced the interaggregate spacing of CTAB spheres from 10 nm to 6.3 nm.[56] The lack of a sphere to cylinder transition is due to the density of adsorption sites remaining insufficient for giant cylinders to form.

The adsorbed morphology of the gemini surfactants 12-2-12 and 12-3-12 at twice the cmc has been studied, with flattened ellipsoidal admicelles reported.[61] The structure of surfactants with spacer sizes s greater than 4 could not be determined due to proximal desorption of surfactant from the adsorbed layer as the AFM tip approached. This resulted in a change in the interaction between the tip and adsorbed layer, from repulsive to attractive, which is prohibitive for imaging under soft-contact conditions. Both 12-2-12 and 12-3-12 form bilayers on the more negatively charged mica surface,[28,46] and recently a mesh structure was observed in a study of 12-2-12 with added CsCl on mica.[45] As the addition of CsCl decreases the number of surface sites able to bind surfactant, this raises the possibility that the adsorbed morphology on silica is also a mesh. However, as the adsorbed morphology of surfactants with longer spacer groups on silica could not be determined, and the actual density of surface sites in the mica-CsCl system is unknown, it is impossible to state with certainty whether the flattened micelle or mesh structure forms for the 12-2-12–silica system.

Subramanian and Ducker investigated the adsorbed structure of CTA$^+$ with several different counterions.[55] Hard counterions, such as Cl$^-$, OAc$^-$, HSO$_3^-$, SO$_4^{2-}$, CO$_3^{2-}$, and SO$_3^{2-}$ interact strongly with water so are relatively unavailable for binding to surfactant ions. As a result, they do not alter the spherical morphology. Soft counterions, such as Br$^-$, HS$^-$, S$_2$O$_3^{2-}$, and CS$_3^{2-}$ interact weakly with water, so associate more readily with surfactant admicelles. This lowers the curvature of the aggregate by shielding the electrostatic repulsion between neighboring

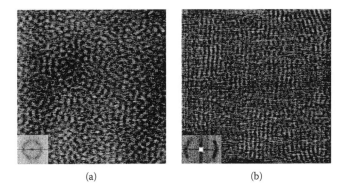

(a)   (b)

FIGURE 11.10 200×200 nm^2 AFM deflection images of (a) 10 mM CTAAc and (b) 10 mM CTAAc + 5 mM Na$_2$S$_2$O$_3$ on amorphous silica. The highly polarizable S$_2$O$_3$$^{2-}$ counterion partitioned to the micelle surface, shielding headgroup repulsion between neighboring surfactant molecules and induced formation of giant cylindrical micelles. Reprinted with permission from Ref. 55. © 2000 by American Chemical Society.

headgroups, producing a sphere to giant cylindrical adsorbed micelle transformation (Figure 11.10).[55]

In separate studies, similar results were found for CTAB with both NaBr and LiBr on amorphous silica,[55] and for TTAB with added NaBr on quartz.[22] Neutron reflectometry was used to show that the addition of electrolyte increases both surface excess and the fraction of surface covered with surfactant, as expected for a sphere to cylinder transition.[22] Similar increases in surface excess for CTAB on addition of KBr have been reported from optical reflectometry experiments.[53,62]

Surfactants with aromatic counterions such as salicylate (Sal) and 3,5-dichlorobenzoate (ClBz) are known to form long, entangled cylindrical micelles in solution.[63–65] On amorphous silica, CTASal was observed, using both AFM and neutron reflectometry, to form giant cylindrical micelles at the cmc. A bilayer was formed by CTAClBz as this counterion binds more strongly to the surfactant.[66]

Solubilization of the aromatic solute β-naphthol into a TTAB adsorbed layer of globular micelles on quartz produced giant cylinders at 2 mM and a bilayer at 2.5mM (Figure 11.11),[67] in accordance with results obtained for the same system on mica.[47] β-naphthol is understood to partition to the headgroup layer of the adsorbed aggregate, shielding repulsions between headgroups and reducing film curvature.[68]

Poly(oxyethylene) monoalkylether nonionic surfactants adsorb onto silica by hydrogen bonding with surface silanol groups. Aggregate curvature is determined by the relative sizes of the ethylene oxide headgroup and the alkyl tail, with headgroups in excess of 5 ethoxy units forming adsorbed globules at room temperature.[39] However, these globular structures can be transformed to giant cylindrical admicelles by increasing the temperature. This reduces the affinity of

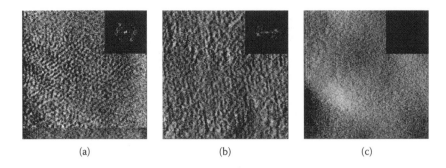

(a)   (b)   (c)

FIGURE 11.11 200×200 nm^2 AFM deflection images of (a) 2 cmc TTAB, (b) 2 cmc TTAB + 2 mM β-naphthol, and (c) 2 cmc TTAB + 2.5 mM β-naphthol on a quartz crystal. The aromatic solute partitions to the palisade layer of the adsorbed micelle, reducing interfacial curvature.

the headgroup for the aqueous medium, causing it to contract and the curvature to reduce, transforming globules into cylinders. Raising the temperature toward the cloud point of $C_{12}EO_5$, $C_{14}EO_6$, and $C_{16}EO_6$ caused these giant cylindrical admicelles to branch. Further increasing the temperature ultimately led to the formation of a mesh (Figure 11.12).[69]

On quartz, mixtures of DDAB, which forms a bilayer, and DTAB, which forms globules,[18] were shown to produce both giant cylindrical micelles and a branched-micelle mesh over a very narrow range of solution composition (Figure 11.13). In both systems, the mole fraction of DTAB in the bulk solution was in excess of 0.99. Neutron reflectometry using different contrasts was used to elucidate the composition of the mixed films for the cylindrical and mesh morphologies at the quartz interface. For both structures, the surface was found to be

(a)   (b)   (c)

FIGURE 11.12 400×400 nm^2 AFM deflection images of 0.12 mM $C_{16}EO_6$ at (a) 20°C, (b) 25°C, and (c) 30°C. The globular morphology observed at room temperatures transforms to giant cylindrical admicelles and then a network of branched cylindrical micelles at temperatures close to the cloud point. The same transitions were observed for $C_{12}EO_5$ and $C_{14}EO_6$. Reprinted with permission from Ref. 69. © 2005 by American Chemical Society.

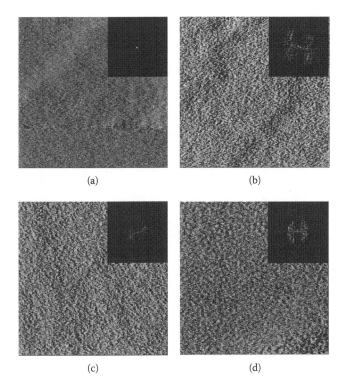

FIGURE 11.13 200×200 nm^2 AFM deflection images of (a) a laterally featureless bilayer formed by 1 mM DDAB, (b) a mesh structure formed by the mixture 0.07 mM DDAB and 8.1 mM DTAB (0.8 mole % DDAB), (c) rods formed by 0.06 mM DDAB and 9.8 mM DTAB (0.6mole % DDAB) and (d) globules formed by 26 mM DTAB on quartz.

almost exclusively composed of DDAB (mole fraction 0.98), despite the vast excess of DTAB in the bulk. That is, DDAB is preferentially adsorbed, most likely a consequence of its much greater hydrophobicity. A slight change in the bulk composition produced a small, but discernable, change in the adsorbed film composition, causing a structural evolution from bilayer to mesh, cylinders, and globules.

11.3.4 OTHER MINERAL OXIDE SURFACES

There are many other mineral oxide surfaces available for studying surfactant adsorption, such as rutile (TiO_2) and sapphire (Al_2O_3), for example. Single chained cationic and anionic surfactants typically form globular aggregates[56,57] on these substrates, whereas DDAB forms a bilayer, in accordance with results obtained for silica. Addition of sodium salicylate to an adsorbed TTAB film on sapphire produced the expected transition from adsorbed spheres to giant wormlike

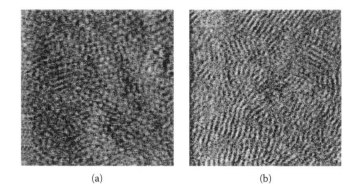

(a)   (b)

FIGURE 11.14 200×200 nm^2 AFM deflection images of (a) 2 cmc TTAB at pH 11.0 and (b) 2 cmc TTAB + salicylate on a sapphire platelet.

micelles (Figure 11.14).[57] Neutron reflectometry showed a corresponding increase in the fractional surface coverage from 50% to 68%, consistent with this change in structure.

11.4 GIANT MICELLES NEAR SURFACES

Perhaps the most striking difference between the structures of bulk solutions and adsorbed films of giant micelles is the high degree of alignment observed in AFM images. The presence of a macroscopic interface breaks the symmetry of the bulk solution and forces an alignment parallel to the surface. Strict confinement into an adsorbed surface film then imposes a preferred direction by parallel alignment among neighbors, and the high volume fraction of the adsorbed cylinders results in a regular spacing.

Near a macroscopic surface the orientation of micelles is anisotropic, although the confinement is less stringent and the micelle concentration much lower. Nevertheless this can be enough to give rise to distinct "near-surface" structures in some circumstances. Bilayer membranes of sponge phases align parallel to a planar interface yielding smectic order.[70]

Near-surface structures of giant micelles have been reported under a few specific circumstances: at high concentrations, near the L_1/H_1 phase boundary, and in dilute systems under shear.

Findenegg et al. have examined the $L_1 \rightarrow H_1$ phase transition in nonionic surfactants near hydrophobically modified silicon-solution and air-solution interfaces.[71] The adsorbed surfactant film is expected to be laterally unstructured at both interfaces.

Upon cooling a $C_{12}EO_5$ solution from L_1, the H_1 phase was found by neutron reflectivity to form with cylinders aligned parallel to the solid-liquid interface.

At the air-solution interface, grazing incidence x-ray diffraction also revealed a hexagonal phase comprised of highly aligned threadlike micelles. The higher resolution of x-ray diffraction made it possible to resolve the bulk and near-surface contributions to diffraction patterns from the H_1 phase, and to show that the melting point of the near-surface structure is slightly higher than the bulk phase.[72]

A similar effect was reported in AFM investigations of the surface structures of concentrated DTAB and DTAC solutions on graphite. At very high concentrations in the L_1 phase the force-separation curve displayed multiple oscillations as the AFM tip sensed the development of a layered near-surface structure above the elongated hemimicelles on the graphite substrate.[14] In DTAB this occurred close to the L_1/H_1 boundary, and the layered structure was concluded to consist of cylindrical micelles. As DTAC forms an I_1 discrete cubic phase, it was not possible to deduce whether the multilayers consisted of spherical or cylindrical micelles.

For these systems, alignment is a consequence of both surface and nearest-neighbor interactions, which should diminish as the micelle solution is diluted. Hamilton et al. have shown that it is possible to achieve a very high degree of order in giant micelle systems by coupling surface and shear effects.[15] A 0.02 M mixed CTAB + CTA 3,5-dichlorobenzoate forms a viscoelastic bulk solution of entangled, thread-like giant micelles. In quiescent solution this shows no structure by NS-SANS, but even very low flow rates give rise to a highly ordered near-surface state of hexagonally aligned rods. This state is more ordered than the shear-induced bulk state, and has dramatically different kinetics for both formation and relaxation after cessation of shear.[73] The existence of the highly aligned state is sensitive to solution composition, and to the flexibility of the giant micelles. Further examination of the effects of micellar structure and flow conditions on the induced surface state could impact on a variety of coating, flow, and lubrication applications, and enhance our understanding of the behavior of giant micelle solutions in porous media.

11.5 GIANT MICELLES CONFINED BETWEEN SURFACES

Many self-assembled surfactant phases undergo substantial structural modification when confined between two macroscopic surfaces. Just as a water droplet may nucleate from an organic solution confined between mica surfaces in a surface forces apparatus,[74] lamellar phases have been shown to condense from sponges[75] and microemulsions.[76] The condensed lamellar phase gives rise to a characteristic oscillatory force profile as a function of separation between the mica surfaces.

Such oscillatory forces also exist in confined micellar systems, and are reported in studies using the surface force apparatus as well as in studies of soap film drainage. In systems of spherical CTAB micelles a depletion attraction is reported at low concentrations, but this gives way to an oscillatory or structural repulsion at higher concentrations,[77] which can accurately be modelled as layers of charged micelles confined between charged walls.[78] Similar oscillatory forces occur between bubbles in static and dynamic film-drainage experiments containing spherical SDS micelles, but diminish upon addition of electrolyte.[79]

Confinement-induced structure has also been observed in giant micelle solutions. A weak oscillatory force attributed to "orientation depletion of the semi-flexible polymers aligning along the surfaces" has been reported for wormlike CTAB micelles in 0.1 M KBr using the surface forces apparatus.[80] More recently, oscillatory forces in semidilute solutions of giant micelles of 12-2-12 have been seen by both film drainage[81] and in the surface force apparatus.[82] The remarkable results in this system are thought to be due to the absence of added electrolyte leading to stronger and longer range correlations in the confined film. One might speculate that the effect may be enhanced in dynamic film drainage studies by flow-induced alignment effects observed in NS-SANS.

11.6   CONCLUSIONS

Many of the conditions that give rise to giant micelles in aqueous solution apply equally well to interfaces. Mica is almost unique in its capacity to drive adsorbed cationic surfactants towards cylindrical micelles. On mica, and other hydrophilic surfaces, strongly binding counterions favor lower curvature aggregates of cationic surfactants, but in adsorbed films this must be balanced against the possible ion-exchange effect of the cation. On silica surfaces, adsorbed nonionic surfactants exhibit transitions into elongated and branched micelles on heating toward their cloud point, also paralleling changes in bulk solution structure.

Strong, anisotropic interactions between graphite and surfactant alkyl chains make its ability to orient adsorbed films even more extreme. Long, straight, parallel, hemicylindrical hemimicelles are formed by many types of surfactants adsorbed on the graphite cleavage plane.

The high local concentration and confinement of surfactant in adsorbed films nevertheless generates highly ordered parallel arrays whenever conditions favor cylindrical adsorbed micelles. This arrangement can be mimicked in bulk solution near a surface when an additional confining or orienting field is applied by increasing concentration, by inducing flow, or by confining against a second interface.

ACKNOWLEDGMENTS

The authors acknowledge support from the Australian Research Council, and valuable discussions with Jamie Schulz, Simon Biggs, Vince Craig, and Erica Wanless.

REFERENCES

1. Zhu, B.-Y., Gu, T. *J. Chem. Faraday Trans. 1.* 1989, *85*, 3813.
2. Tiberg, F., Jensson, B., Tang, J., Lindman, B. *Langmuir* 1994, *10*, 2294.
3. Levitz, P., van Damme, H., Kervais, G. *J. Phys. Chem.* 1984, *88*, 2228.
4. Manne, S., Gaub, H.E. *Science* 1995, *270*, 1480.
5. Patrick, H.N., Warr, G.G., Manne, S., Aksay, I.A. *Langmuir* 1999, *15*, 1685.
6. Grant, L.M., Tiberg, F., Ducker, W.A. *J. Phys Chem. B.* 1998, *102*, 4288.
7. Schulz, J.C., Warr, G.G. *Langmuir* 2000, *16*, 2995.
8. Drummond, C., Israelachvili, J., Richetti, P. *Phys. Rev. E.* 2003, *67*, 066110.
9. Helm, C.A., Israelachvili, J.N., McGuiggan, P.M. *Biochemistry* 1992, *31*, 1794.
10. Drummond, C., Israelachvili, J., Richetti, P. *Europhys. Lett.* 2003, *58*, 503.
11. Richetti, P., Drummond, C., Israelachvili, J., In, M., Zana, R. *Europhys. Lett.* 2001, *55*, 653.
12. Aksay, I.A., Trau, M., Manne, S., Honma, I., Yao, N., Zhou, L., Fenter, P., Eisenberger, P.M., Gruner, S.M. *Science* 1996, *273*, 892.
13. Holt, S.A., Reynolds, P.A., White, J.W. *Phys. Chem. Chem. Phys.* 2000, *2*, 5667.
14. Fitzgerald, P.A., Warr, G. G. *Adv. Mat.* 2001, *13*, 967.
15. Hamilton, W.A., Butler, P.D., Baker, S.M., Smith, G.S., Hayter, J.B., Magid, L.J., Pynn, R. *Phys. Rev. Lett.* 1994, *72*, 2219.
16. Lin, Z., Mateo, A., Zheng, Y., Kesselman, E., Pancallo, E., Hart, D.J., Talmon, Y., David, H.T., Scriven, L.E., Zakin, J.L. *Rheol. Acta* 2002, *41*, 483.
17. Patrick, H.N., Warr, G.G., Manne, S., Aksay, I.A. *Langmuir* 1997, *13*, 4349.
18. Blom, A., Warr, G.G., Nelson, A. *Colloids Surf. A* submitted.
19. Penfold, J., Tucker, I., Thomas, R.K. *Langmuir* 2005, *21*, 6330.
20. Penfold, J., Staples, E., Tucker, I., Thomas, R.K. *Langmuir* 2000, *16*, 8879.
21. Zhou, X.L., Chen, S.H. *Phys. Rep.* 1995, *257*, 223.
22. Schulz, J.C., Warr, G.G., Butler, P.D., Hamilton, W.A. *Phys. Rev. E.* 2001, *63*, 0416041.
23. Rennie, A.R., Lee, E.M., Simister, E.A., Thomas, R.K. *Langmuir* 1990, *6*, 1031.
24. Baker, S.M., Smith, G.S., Pynn, R., Butler, P.D., Hayter, J.B., Hamilton, W.A., Magid, L.J. *Rev. Sci. Inst.* 1994, *65*, 412.
25. Butler, P.D. In *Supramolecular Structure in Confined Geometries*. Manne, S., Warr, G.G. Eds., ACS Symposium Series No. 736; Washington, DC, American Chemical Society, 1999.
26. Manne, S., Cleveland, J.P., Gaub, B.E., Stucky, G.D., Hansma, P.K. *Langmuir* 1994, *10*, 4409.
27. Cyr, D.M., Venkataraman, B., Flynn, G.W. *Chem. Mater.* 1996, *8*, 1600.
28. Manne, S., Schaffer, T.E., Huo, Q., Hansma, P.K., Morse, D.E., Stucky, G.D., Aksay, I.A. *Langmuir* 1997, *13*, 6382.
29. Wanless, E.J., Ducker, W.A. *J. Phys. Chem.* 1996, *100*, 3207.
30. Wanless, E.J., Ducker, W.A. *Langmuir* 1997, *13*, 1463.
31. Holland, N.B., Ruegsegger, M., Marchant, R.E. *Langmuir* 1998, *14*, 2790.
32. Patrick, H.N., Warr, G.G. *Colloids Surf. A.* 2000, *162*, 149.
33. Patrick, H.N. *Surfactant Self-Assembly at the Solid-Liquid Interface,* PhD thesis, The University of Sydney, 1998.
34. Ducker, W.A., Grant, L.M. *J. Phys. Chem.* 1996, *100*, 11507.
35. Duval, F.P., Warr, G.G. *Chem. Comm.* 2002, *19*, 2268.

36. Atkin, R., Warr, G.G. *J. Am. Chem. Soc.* 2005, *127*, 11940.

37. Fitzgerald, P.A., Warr, G. G. *Advanced Materials* 2001, *13*, 967.

38. Grant, L.M., Ederth, T., Tibert, F. *Langmuir* 2000, *16*, 2285.

39. Grant, L.M., Tibert, F., Ducker, W.A. *J. Phys. Chem.* 1998, *102*, 4288.

40. Wolgemuth, J.L., Workman, R.K., Manne, S. *Langmuir* 2000, *16*, 3077.

41. Hoffmann, H., Platz, G., Rehage, H., Schorr, W., Ulbricht, W. *Ber. Bunsen-Ges. Phys. Chem.* 1981, *85*, 255.

42. Israelachvili, J.N. *Intermolecular and Surface Forces, Part III*; London, Academic Press, 1992.

43. Ducker, W.A., Wanless, E.J. *Langmuir* 1999, *15*, 160.

44. Ducker, W.A., Wanless, E.J. *Langmuir* 1996, *12*, 5915.

45. Duval, F.P., Zana, R., Warr, G.G. *Langmuir* 2006, *22*, 1143.

46. Blom, A., Duval, F.P., Kovacs, L., Warr, G.G., Almgren, M., Kadi, M., Zana, R. *Langmuir* 2004, *20*, 1291.

47. Kovacs, L., Warr, G.G. *Langmuir* 2002, *18*, 4790.

48. Davey, T.W., Warr, G.G., Almgren, M., Asakawa, T. *Langmuir* 2001, *17*, 5283.

49. Davey, T.W., Warr, G.G., Asakawa, T. *Langmuir* 2003, *19*, 5266.

50. Blom, A., Warr, G.G. *Langmuir* 2006, *22*, 6787.

51. Iler, R.K. *The Chemistry of Silica*; New York, Wiley, 1979.

52. Chorro, M., Chorro, C., Dolladille, O., Partyka, S., Zana, R. *J. Colloid Interface Sci.* 1999, *210*, 134.

53. Atkin, R., Craig, V.S.J., Biggs, S. *Langmuir* 2000, *16*, 9374.

54. Liu, J.F., Min, G., Ducker, W.A. *Langmuir* 2001, *17*, 4895.

55. Subramanian, V., Ducker, W.A. *Langmuir* 2000, *16*, 4447.

56. Schulz, J.C., Warr, G.G. *Langmuir* 2002, *18*, 3191.

57. Schulz, J.C. *Interfacial Structures in Thin Surfactant Films*, PhD thesis, University of Sydney, 2000.

58. Atkin, R., Craig, V.S.J., Wanless, E.J., Biggs, S. *J. Colloid Interface Sci.* 2003, *266*, 236.

59. Velegol, S.B., Fleming, B.D., Biggs, S., Wanless, E.J., Tilton, R.D. *Langmuir* 2000, *16*, 2548.

60. Furst, E.M., Pagac, E.S., Tilton, R.D. *Ind. Eng. Chem. Res.* 1996, *35*, 1566.

61. Atkin, R., Craig, V.S.J., Wanless, E.J., Biggs, S. *J. Phys. Chem. B.* 2003, *107*, 2978.

62. Atkin, R., Craig, V.S.J., Biggs, S. *Langmuir* 2001, *17*, 6155.

63. Shikata, T., Sakaiguchi, Y., Uragami, H., Tamura, A., Hirata, H. *J. Colloid Interface Sci.* 1987, *119*, 291.

64. Carver, M., Smith, T. L., Gee. J.C., Delichere, A., Caponetti, E., Magid, L.J. *Langmuir* 1996, *12*, 691.

65. Bayer, O., Hoffmann, H., Ulbricht, W., Thurn, H. *Adv. Colloid Interface Sci.* 1986, *26*, 177.

66. Kleydish, J. *Near Surface Structure of Cationic Surfactants*, Honours Thesis, University of Sydney, 2005.

67. Schulz, J.C., Blom, A., Nelson, A., Warr, G.G. *Unpublished results*.

68. Almgren, M., Grieser, F., Thomas, J.K. *J. Am. Chem. Soc.* 1979, *101*, 279.

69. Blom, A., Warr, G.G. Wanless, E.J. *Langmuir* 2005, *21*, 11850.

70. Hamilton, W.A., Porcar, L., Butler, P.D., Warr, G.G. *J. Chem. Phys.* 2002, *116*, 8533.

71. Findenegg, G.H., Braun, C., Lang, P., Steitz, R. In *Supramolecular Structure in Confined Geometries*. Manne, S., Warr, G.G., Eds., ACS Sympsoium Series No. 736; Washington, DC, American Chemical Society, 1999.

72. Braun, C., Lang, P., Findenegg, G.H. *Langmuir* 1995, *11*, 764.

73. Hamilton, W.A., Butler, P.D., Magid, L.J., Han, J., Slawecki, T.M. *Phys. Rev. E.* 1999, *60*, 1146.

74. Christenson, H.K. *J. Colloid Interface Sci.* 1984, *104*, 234.

75. Antelmi, D.A., Kekicheff, P., Richetti, P. *J. Phys. II* 1995, *5*, 103.

76. Moreau-Biensan, L., Barois, P., Richetti, P. In *Supramolecular Structure in Confined Geometries*. Manne, S., Warr, G.G., Eds., ACS Symposium Series No. 736; Washington, DC, American Chemical Society, 1999.

77. Kekicheff, P., Richetti, P. *Phys. Rev. Lett.* 1992, *68*, 1951.

78. Pollard, M.L., Radke, C.J. *J. Chem. Phys.* 1994, *101*, 6979.

79. Bergeron, V., Radke, C.J. *Langmuir* 1992, *8*, 3020.

80. Kekicheff, P., Nallet, F., Richetti, P. *J. Phys. II. France* 1994, *4*, 735.

81. Espert, A., von Klitzing, R., Poulin, P., Colin, A., Zana, R., Langevin, D. *Langmuir* 1998, *14*, 4251.

82. Anthony, O., In, M., Marques, C.M., Zana, R., Richetti, P. Unpublished results.

12 Stimuli-Responsive Giant Micellar Systems

Frank Pierce Hubbard, Jr., and Nicholas L. Abbott

CONTENTS

12.1 INTRODUCTION

This chapter presents an overview of a recent series of studies that have investigated the response of giant micellar systems to external stimuli. These studies cover a diverse range of amphiphiles that are known to form giant micelles, including classical cationic, anionic, and nonionic surfactants, macromolecular amphiphiles, as well as more exotic surfactants that incorporate light-sensitive groups, redox-active groups, and biological motifs such as peptide-based head groups. The response of these giant micellar systems to changes in temperature, oxidation state (controlled electrochemically), illumination, and pH is discussed, with a focus directed to systems that demonstrate large changes in microstructure and solution properties. A remarkably diverse range of transitions in microstructure has been observed, including vesicle-to-giant micelle transitions, globular micelle-to-giant micelle transitions, changes in contour lengths of wormlike micelles, as well as branching of wormlike micelles. All of these transitions lead to significant changes in rheology and/or phase behavior, which suggests potential application of these systems in areas such as formulation of cosmetics and pharmaceutics (controlled release), biotechnology (scaffolds for growing bones), and micro-electromechanical systems (electro- and photo-rheological fluids).

This chapter is organized as follows. First, we review giant micellar systems that respond to changes in temperature. Next, we discuss the design of giant micellar systems that incorporate redox-active groups or light-sensitive groups that respond to changes in redox-potential and illumination, respectively, and cause large changes in solution properties. Finally, we discuss pH-sensitive systems, including giant micellar systems that undergo pH-triggered transformations to form networks suitable for use as scaffolds, for example, for growth of bone. In closing our introduction to this chapter, we mention here that this review does not attempt to present an exhaustive summary of all work in this area, but rather seeks to be selective in presenting studies from the recent literature that serve to illustrate progress and opportunity.

12.2 TEMPERATURE-SENSITIVE SYSTEMS

The first types of stimuli-responsive giant micellar systems that we address in this review are those systems that exhibit substantial responses to changes in temperature. Many past studies have reported on the existence of temperature-dependent properties in giant micellar systems, the majority reporting a substantial decrease in the viscosity of solutions containing giant micelles with increasing temperature. We refer the reader to Chapters 8 and 9 for a comprehensive discussion of these effects of temperature on giant micellar systems. In the interest of completeness, below we briefly comment on the above-mentioned temperature-dependent response of giant micellar systems, and then focus our discussion on systems that exhibit deviations from this classical behavior. We organize our discussion according to the type of surfactant (cationic, anionic, and nonionic).

12.2.1 CLASSICAL RESPONSE OF GIANT MICELLES TO CHANGES IN TEMPERATURE

A previous review by Cates and Candau[1] contains a useful summary of early studies that explored the rheological properties of wormlike micelles, including temperature-dependent behaviors. Many additional studies have been reported since this review,[2–13] as discussed in Chapters 8 and 9. The majority of these giant micellar systems (e.g., 40 mM erucyl bis-(hydroxyethyl)-methylammonium chloride (EHAC) and 450 mM sodium salicylate (NaSal)[14]) exhibit a zero-shear viscosity that decreases exponentially with increasing temperature.

Although the general trend of a decrease in viscosity with increase in temperature is common, it is interesting to note that the microscopic origin of this widely documented behavior may, in fact, vary significantly between surfactant systems. One model that leads to the above-described temperature-dependent behavior is based on temperature-dependent scission of the micelles.[1] As the contour length decreases, the viscosity is predicted to decrease due to a decrease in the extent of entanglement of micelles. It has also been proposed, however, that branching of wormlike micelles may lead to a decrease in viscosity with increasing temperature.[8,14–20] In particular, Lequeux has presented a simple theory that relates the viscosity of a branched network to a solution of an identical number density of entangled micelles.[20] This theory predicts that the viscosity of a branched network is less than the viscosity of an entangled network with no junction points. Branching of wormlike micelles is expected when the energy penalty to form branches becomes less than the energy penalty to form end-caps. Raghavan et. al. suggest that the high end-cap energy of the EHAC/NaSal system ($65\ k_BT$) may promote micellar branching, which in turn may lead to the decrease in viscosity with increasing temperature (and lower consolute point) that is measured in this system.[14]

Whereas the section above briefly addresses the classical effect of an increase in temperature on giant micellar systems (decrease in viscosity with increase in temperature), a number of recent experimental studies have revealed exceptions to this temperature-dependent behavior (e.g., increase in viscosity with increase in temperature). As we discuss next, these deviations appear to be driven by phenomena involving strong counterion binding to giant micelles. These studies, when combined, suggest that a diverse range of temperature-dependent responses in giant micellar systems can be engineered through the appropriate choice of the counterion.

12.2.2 CATIONIC SURFACTANT SYSTEMS

One of the early reports of an increase in viscosity with temperature in a giant micellar system was made by Manohar and co-workers.[21,22] These investigators mixed cetyltrimethylammonium bromide (CTAB) and sodium 3-hydroxynaptha-lene-2-carboxylate (SHNC) and removed the sodium and bromide counterions to prepare cetyltrimethylammonium hydroxynapthalenecarboxylate (CTAHNC). The surfactant CTAHNC was then dissolved in water and the viscosity of the resulting solution was recorded as a function of temperature between 30°C and 85°C (Figure 12.1a).[21] Inspection of Figure 12.1a reveals that the viscosity *increases* from 50 mPa·s at 30°C to a maximum value of 7000 mPa·s at approximately 60°C, and then decreases. The authors also reported that the solutions appeared turbid at low temperatures whereas upon heating they clarified (Figure 12.1c) and became strongly viscoelastic.

Manohar and co-workers used a variety of techniques, including nuclear magnetic resonance (NMR), light scattering, small angle neutron scattering (SANS), and freeze-fracture transmission electron microscopy (FF-TEM), to characterize the microstructures formed by CTAHNC at various temperatures. The combined results of these studies led to the conclusion that the surfactant in this system undergoes a temperature-dependent transition from a vesicular micro-structure (5 μm in size at 30°C, by FF-TEM) to a giant, wormlike microstructure at elevated temperatures. This transition was proposed to result from a "melting" of the head group region of the aggregates. Two pieces of evidence are cited in support of this proposition. First, the authors noted that at low temperatures, the

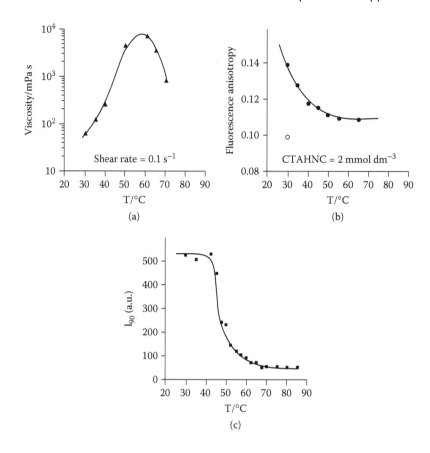

FIGURE 12.1 Temperature-dependent properties of aqueous CTAHNC (0.6% m/v). (a) Solution viscosity. (b) Fluorescence anisotropy. (c) Intensity of scattered light. Reprinted with permission from Ref. 21. © 1996 by The Royal Society of Chemistry.

NMR chemical shifts for HNC are broad and characteristic of a solid-like state. As the temperature was increased, the peaks were observed to sharpen, indicating a more fluid-like environment. Second, the authors used fluorescence anisotropy measurements (Figure 12.1b) to characterize the tendency of HNC to rotate when bound to the aggregates: the fluorescence anisotropy was high at lower temperatures, indicating that the movement of the HNC was restricted. As the temperature was increased, however, the anisotropy decreased, and approached that of HNC free in solution (open circle, Figure 12.1b), thus indicating an increase of counterion mobility.

A second example of an unusual change in viscosity with temperature in a giant micellar system has been reported by Raghavan and Kaler[15] using a mixture of 60 mM EHAC and 18 mM NaSal. As shown in Figure 12.2, the zero-shear viscosity of this system increased from 7,000 Pa·s at 25°C to 20,000 Pa·s at 40°C,

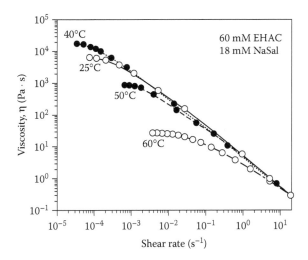

FIGURE 12.2 Viscosity of an aqueous solution of 60 mM EHAC and 18 mM NaSal, as a function of temperature and shear rate. Reprinted with permission from Ref. 15. © 2001 by American Chemical Society.

and then decreased with further increase in temperature. It is interesting to note that this system is comprised of the same components (surfactant and counterion) that lead to the classical temperature-dependent behavior. Classical behavior was observed in a system containing excess NaSal, whereas the solution used to obtain the results shown in Figure 12.2 contained excess EHAC. To provide insight into the temperature-dependent behavior shown in Figure 12.2, the authors reported measurements obtained by using dynamic rheology. Their measurements led them to propose that at low temperatures (25°C) either (1) not all of the surfactant in solution formed wormlike micelles, or (2) not all of the wormlike micelles were long enough to entangle. The level of incorporation of the surfactants and the salicylate counterions into the micelles was proposed to increase with temperature, thus increasing the number and/or the contour length of the micelles and hence the viscosity of the solution. The decrease in viscosity observed with increase in temperature above 40°C was attributed to scission of the wormlike micelles. The high end-cap energy of this surfactant system was proposed to keep the rate of scission low, however, thus providing an account for the unusually high viscosities observed at elevated temperatures.

Building from the work of Manohar and co-workers, Kalur et. al. have reported unusual temperature-dependent viscoelastic behaviors using wormlike micelles prepared from mixtures of EHAC and SHNC.[23] In contrast to the prior study by Manohar and co-workers, Kalur et al. did not remove the sodium and bromide counterions following the mixing of the two surfactants. Depending on the ratio of EHAC to SHNC, the latter authors measured a range of different rheological behaviors as a function of temperature. As shown in Figure 12.3,

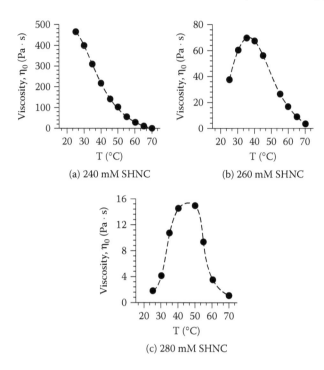

FIGURE 12.3 Temperature-dependent viscosities of aqueous solutions of EHAC and SHNC. The concentration of EHAC was 40 mM. Reprinted with permission from Ref. 23. © 2005 by American Chemical Society.

either a monotonic decrease in viscosity or a peak in viscosity was observed upon an increase in temperature. Using SANS, the authors observed an increase in scattering at low q values (where q denotes the magnitude of the scattering wave vector) as temperature was increased, consistent with the formation and growth of wormlike micelles upon heating.

Kalur et. al. discuss the origin of the temperature-dependent behaviors seen in Figure 12.3 with reference to the properties of the HNC and Sal counterions. The authors suggest that because HNC is less soluble in water than Sal (HNC has an additional phenyl ring as compared to Sal), it partitions more completely into the micelles. At low temperatures, the authors propose that most of the HNC ions partition into the micelles. This leads to a high negative surface charge and formation of micelles with high curvature. However, as temperature is increased, the authors propose that the solubility of HNC in water increases, resulting in desorption of some of the HNC counterions from the micelle surface. The decrease in negative surface charge promotes growth (elongation) of the micelles, and thus the measured increase in viscosity. The temperature-dependent dissociation of the HNC from the micelles is proposed to compete with the "classical" effects of increasing temperature (scission), leading to the maximum in viscosity seen in Figures 12.3b and 12.3c. The possibility of micellar branching, which might also provide an account for the decrease in viscosity at high temperatures, was not discussed in the paper.

12.2.3 ANIONIC SURFACTANT SYSTEMS

The temperature-dependent behaviors of giant micellar systems containing anionic surfactants have also been reported, although the number of reports is fewer than for the case of cationic surfactants.[24–26] In general, the viscosities of wormlike micelles formed from anionic surfactants tend to be lower than for cationic surfactants. For example, sodium dodecylsulfate (SDS) has been reported to form wormlike micelles with zero-shear viscosities of ~1 Pa·s (1,000 times that of water). This value is low compared to cationic systems where zero-shear viscosities of 10^4 Pa·s or more are typical. A recent study by Kalur and Raghavan describes zero-shear viscosities of the order 1–30 Pa·s for mixtures of sodium oleate (NaOA) with either potassium chloride (KCl) or triethylammonium hydrochloride (Et₃NHCl), consistent with the formation of wormlike micelles.[26] As shown in Figure 12.4, the authors reported that the zero-shear viscosity of this system exhibits an exponential decrease with increasing temperature, consistent with the classical response of wormlike micelles, and thus suggest that there is a reduction in micellar length as the temperature is increased. An interesting feature of this system, however, is the presence of a lower consolute point that is observed when NaOA is mixed with Et₃NHCl, but not when mixed with KCl. Small angle neutron scattering measurements revealed a temperature-dependent increase in scattering that the authors interpreted in terms of attractive interactions between micelles that drive phase separation. While the physical picture proposed by the authors is consistent with their measurements, we again mention here that it may also be possible to account for the SANS and rheological measurements as well as the

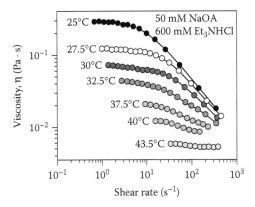

FIGURE 12.4 Change in viscosity with temperature for aqueous solution of NaOA and Et₃NHCl. Reprinted with permission from Ref. 26. © 2005 by American Chemical Society.

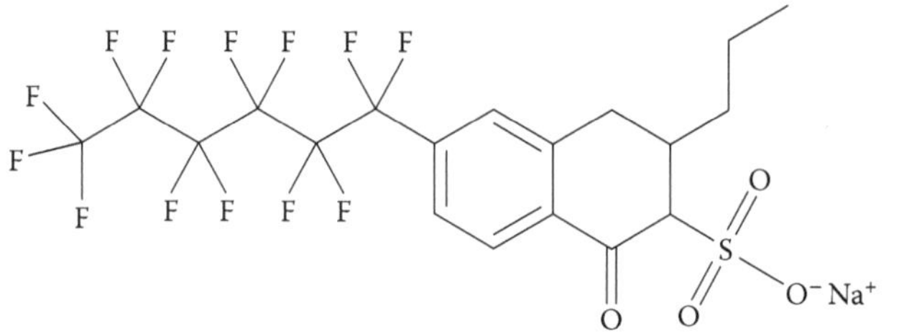

FIGURE 12.5 Structure of FC6-HC4, a fluorocarbon-hydrocarbon hybrid surfactant that displays unusual temperature-dependent rheological properties in aqueous solution.

phase behavior in terms of a temperature-dependent growth of the wormlike micelles into a branched network. Such a model could also offer an explanation as to why the zero-shear viscosity of mixtures of NaOA and KCl is an order of magnitude higher (30 Pa·s) than mixtures of NaOA and Et₃NHCl (0.8 Pa·s).

Unusual temperature-dependent rheological properties have also recently been reported for anionic surfactants that possess hybrid fluorocarbon-hydrocarbon tails. An example of the family of surfactants studied, sodium 1-oxo-1-[4-(tridecafluorohexyl)phenyl]-2-hexanesulfonate (FC6-HC4), is shown in Figure 12.5.[27,28] Aqueous solutions containing 10 wt% FC6-HC4 display large changes in solution viscosity with temperature, as depicted in Figure 12.6. Inspection of

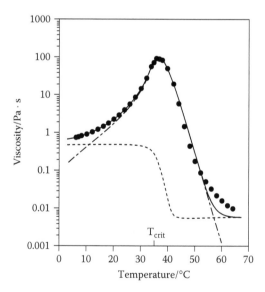

FIGURE 12.6 Viscosity as a function of temperature for a 10 wt% solution of FC6-HC4. Solid and dashed lines represent predictions of a model described in the text. Reprinted with permission from Ref. 28. © 1998 by American Chemical Society.

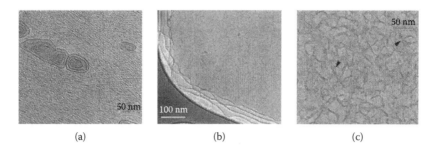

(a)   (b)   (c)

FIGURE 12.7 Cryo-transmission electron micrograph of a solution containing 10 wt% FC6-HC4 at different temperatures: (a) 20°C, (b) 40°C, (c) 60°C. Reprinted with permission from Ref. 29. © 2003 by Elsevier.

Figure 12.6 reveals that at 5°C, the solution viscosity is approximately 1 Pa·s. Upon increasing temperature to 35°C, the solution viscosity rises to 100 Pa·s, and then decreases with increasing temperature, dropping to a value of 0.01 Pa·s at 60°C.

In an effort to provide insight into the origin of the unusual temperature-dependent rheological properties of this system, the original investigators[27,28] and others[29] have attempted to characterize the microstructures formed in this system at different temperatures. In Figure 12.7, cryo-transmission electron micrographs are shown for the FC6-HC4 system at 20°C, 40°C, and 60°C.[29] The cryo-transmission electron micrographs reveal the presence of wormlike micelles (20°C), large bilayer structures that span the TEM grid (40°C), and an L_3 continuous lamellar phase (60°C). The authors caution, however, that the microstructures observed at 40°C may be artifacts of the sample preparation procedures as optical microscopy of samples prior to blotting did not reveal microstructures comparable in size to those seen in the cryo-transmission electron micrographs. Additional studies appear to be required in order to better understand the microstructures present in the FC6-HC4 system at temperatures that give rise to the maximum in viscosity.

Finally, it should be noted that the unusual temperature-dependent response of the above-described surfactant system appears to originate from a precarious balance of forces encountered under a particular set of solution conditions.[27,28] The authors report that an increase in concentration of FC6-HC4 to 20 wt% or 30 wt% results in little to no change in viscosity as a function of temperature. Likewise, if the length of either the hydrocarbon or fluorocarbon chain of the surfactant is changed, or the fluorocarbon chain is replaced with a second hydrocarbon chain, the temperature-dependent changes in viscosity also disappear.

12.2.4 Nonionic Surfactant Systems

It is well established that nonionic surfactants of the C_mEO_n type (where C_m denotes a saturated alkyl chain containing m carbon atoms, and EO_n denotes a head group composed of n ethylene oxide groups) exhibit an upper miscibility gap.[30,31] The complex, temperature-dependent behaviors of C_mEO_n surfactants are attributed to the balance of ethylene oxide-ethylene oxide and ethylene oxide water interactions. As ethylene oxide head group dehydrates with increase in temperature, it is thought to become more compact, which in turn decreases the curvature of the aggregate surface and promotes formation of wormlike aggregates or "pearl necklace" structures. Depending on the values of m and n, temperature-dependent growth can be substantial or minimal. For example, solutions of C_8EO_4[31] or $C_{12}EO_5$[30] show marked growth into rods or pearl necklace structures as temperature is increased, whereas solutions containing $C_{12}EO_8$[30] aggregate as small micelles over a wide range of temperatures.

Despite the numerous observations of phase separation in nonionic surfactant systems upon heating, the evolution of the microstructures in solution prior to phase separation is not fully understood. Historically, investigators have used NMR measurements in combination with static and dynamic light scattering to estimate quantities such as self-diffusion coefficients and radii of gyration or hydrodynamic radii to infer aggregate microstructure. More recently, however, cryo-transmission electron microscopy (cryo-TEM) has yielded interesting insights into the evolution of microstructure with increase in temperature in nonionic surfactant systems. Bernheim-Groswasser et. al., for example, have suggested that branched micellar networks form as the solution approaches criticality.[32] Whereas distinguishing between entangled and branched micellar networks is not possible using existing scattering techniques, these authors have found evidence of micellar networks in solutions of $C_{12}EO_5$ by using cryo-TEM. Figure 12.8 shows a cryo-TEM image at a temperature corresponding to the onset of micelle network formation. Inspection of the figure reveals the presence of threefold and fourfold micellar junctions ($T = 29°C$). Images taken at lower temperatures (8°C or 18°C, not shown) show no evidence of branching. The authors remark that micellar branching is both concentration- and temperature-dependent, which is consistent with theory.[33–35] They also suggest that these micellar networks are likely the precursor to phase separation that is observed upon heating of the nonionic surfactant systems.

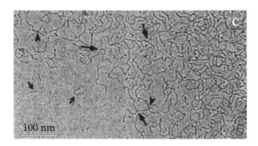

FIGURE 12.8 Cryo-TEM image of a 0.5 wt% $C_{12}EO_5$ solution. The arrows and arrowheads mark where threefold and fourfold micellar junctions have formed. Reprinted with permission from Ref. 32. © 2000 by American Chemical Society.

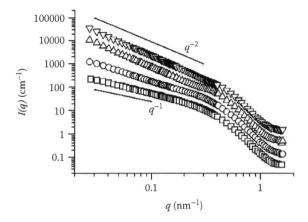

FIGURE 12.9 SANS spectra of aqueous solutions containing 1 wt% ChEO and 0.15 wt% $C_{12}EO_3$. Data is shown at four different temperatures. Bottom to top: 35°C, 40°C, 50°C and 60°C. Reprinted with permission from Ref. 36. © 2005 by American Chemical Society.

A recent study by Moitzi et al. also describes an interesting set of observations involving temperature-dependent microstructures seen with mixtures of $C_{12}EO_3$ and a cholesteryl surfactant possessing a poly(ethylene oxide) head group (abbreviated as ChEO).[36] The authors used a combination of dynamic light scattering (DLS) and small angle neutron scattering (SANS) to characterize the sizes and microstructures of the aggregates formed in solution. When using ChEO with either 10 or 15 EO units, the authors reported transitions from micelles to wormlike micelles or wormlike micelles to vesicles by either: (1) addition of $C_{12}EO_3$ to a solution containing ChEO, or (2) increasing the temperature of a solution containing ChEO with or without $C_{12}EO_3$. Figure 12.9 shows the SANS spectra for a sample containing 1 wt% ChEO and 0.15 wt% $C_{12}EO_3$ as a function of temperature. At lower temperatures, the solution contains wormlike micelles, as evidenced by the −1 slope plotted with the data. As temperature increases from 35°C to 60°C, the slope of the data changes from −1 to −2, consistent with formation of vesicular structures in solution. While the temperature-dependent microstructures in solution have been characterized, the accompanying changes in rheological properties do not yet appear to have been investigated.

12.3 REDOX-ACTIVE SYSTEMS

The second class of stimuli-responsive giant micellar systems that we discuss here are systems that contain surfactants incorporating redox-active groups. A particularly interesting class of redox-active surfactants utilizes the ferrocene moiety. Ferrocenyl surfactants have been widely studied in the past,[37–62] but only recently have giant micellar systems based on ferrocenyl surfactants of the type shown in Figure 12.10 been reported. As shown in Figure 12.10, oxidation of

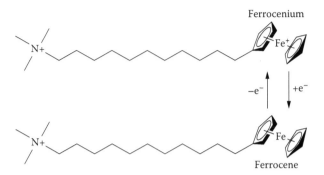

FIGURE 12.10 Structure of the redox-active surfactant FTMA.

ferrocene to ferrocenium leads to the generation of an additional charge on the surfactant. Past studies have demonstrated that this change in oxidation state can be effected reversibly by using electrochemical methods, thus providing access to approaches that provide both spatial and temporal control over the interfacial and bulk properties of these surfactant systems.

Recently, Tsuchiya and co-workers reported that mixtures of 11-ferrocenyl-undecyltrimethylammonium bromide (FTMA, Figure 12.10) and sodium salicylate (NaSal) form giant micellar systems.[45] In their work, the authors characterized solutions containing 50 mM FTMA in the presence of various concentrations of NaSal. Figure 12.11 provides a qualitative illustration of the dramatic effect that the oxidation state of FTMA has on the rheological properties of this system. Panels (a) and (b) of Figure 12.11 show the behavior of solutions of FTMA without NaSal or with 10 mM NaSal, respectively. These solutions readily flowed in a manner qualitatively similar to water. In contrast, addition of 20 mM NaSal led to formation of a highly viscous solution that flowed very slowly, as depicted in Figure 12.11c. Electrochemical oxidation of FTMA to FTMA$^+$ (see Figure 12.10) in the presence of 20 mM NaSal resulted in the reappearance of flow behaviors comparable to those seen in the absence of NaSal (Figure 12.11d).

The qualitative observations described above were accompanied by rheological measurements, as a function of concentration of NaSal and oxidation state of FTMA. When 20 mM NaSal was added to a solution containing reduced FTMA, a viscosity of 15 Pa·s was measured. Electrochemical oxidation of FTMA to FTMA$^+$ resulted in a dramatic decrease in viscosity (four orders of magnitude) to 0.0025 Pa·s.

The authors characterized the microstructures formed in solutions containing reduced and oxidized FTMA. As shown in Figure 12.12, by using freeze-fracture electron microscopy, they observed long wormlike micelles in solutions containing 50 mM reduced FTMA and 20 mM NaSal.[45] Micrographs of solutions containing oxidized FTMA and NaSal (not shown) displayed no evidence of aggregation, consistent with other reports that oxidized FTMA aggregates weakly, if at all, in solution.[59] Attempts to recover the initial solution viscosity by electrochemically

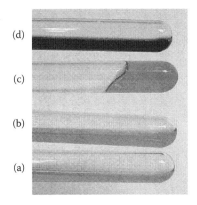

FIGURE 12.11 Photographs of aqueous solutions of 50 mM reduced FTMA containing (a) no added NaSal, (b) 10 mM NaSal, and (c) 20 mM NaSal. (d) Image of solution containing 50 mM oxidized FTMA and 20 mM NaSal. Reprinted with permission from Ref. 45. © 2004 by American Chemical Society.

reducing a solution containing FTMA$^+$ back to FTMA were hindered by gel formation near the electrode surface, which dramatically reduced the diffusion rate of surfactant to the electrode. Although the processes of electrochemical oxidation and reduction leading to the large changes in rheological properties were relatively slow (performed over 24 hr), such transformations in microscale systems, where mass transport to the electrodes is fast due to the small diffusion lengths, would likely yield more rapid changes in properties.

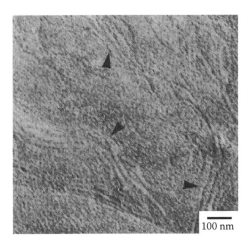

FIGURE 12.12 Freeze-fracture electron micrograph of a solution containing 50 mM reduced FTMA and 20 mM NaSal. Arrowheads point to wormlike micelles formed in solution. Reprinted with permission from Ref. 45. © 2004 by American Chemical Society.

12.4 LIGHT-SENSITIVE SYSTEMS

The third type of stimuli-responsive giant micellar system that we address in this chapter are systems that contain a surfactant with a light-sensitive group. Wolff and co-workers were among the first to report that the addition of substituted polyaromatic compounds (such as anthracenes,[63] acridizinium salts,[64] or coumarins[65]) to solutions containing simple surfactants (such as CTAB) could lead to the formation of light-responsive giant micellar systems with viscoelastic properties. In their studies, they found that photodimerization of the polyaromatic compounds led to changes (although modest in magnitude) of the rheological properties of these solutions. More recently, several research groups have incorporated different light-sensitive groups into surfactants, including azobenzene,[66–83] stilbene,[84–87] and spiropyrans,[88,89] to create surfactant systems that respond to light. The majority of these past studies focus on the azobenzene group, because of its robust and relatively simple photochemistry. Illuminating a molecule containing azobenzene with ultraviolet (UV) light causes a *trans* to *cis* isomerization, as shown in Figure 12.13. The reverse isomerization (*cis* to *trans*) can be driven by illumination with visible light. Isomerization of an azobenzene group within a surfactant leads to a change in the shape of the surfactant, which is generally viewed as the cause of the resulting changes in aggregate structure. Light-sensitive surfactants thus provide interesting opportunities to create stimuli-responsive giant micellar systems because illumination of a solution can be easily patterned and modulated.

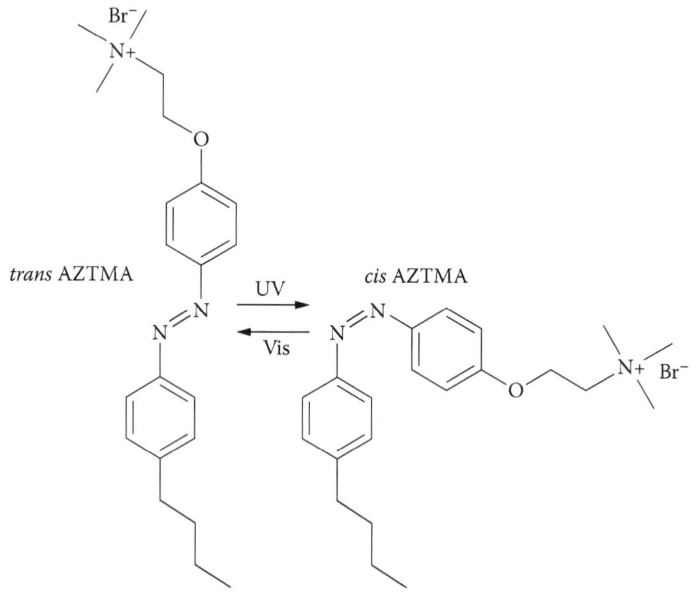

FIGURE 12.13 Structure of light-sensitive surfactant AZTMA, shown in the *trans* and *cis* states.

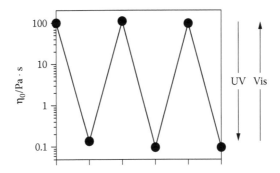

FIGURE 12.14 Changes in solution viscosity upon sequential illumination of an aqueous solution containing 50 mM CTAB, 50 mM NaSal, and 10 mM AZTMA with UV and visible light. Reprinted with permission from Ref. 68. © 2005 by American Chemical Society.

In a recent study, Sakai and co-workers have reported that addition of the light-sensitive surfactant 4-butylazobenzene-4'-(oxyethyl)-trimethylammonium bromide (AZTMA, Figure 12.13) to a solution containing wormlike micelles allows for light-triggered changes in solution viscosity.[68] A mixture of 50 mM CTAB, 10 mM AZTMA, and 50 mM NaSal produced a solution containing wormlike micelles with a viscosity of 100 Pa·s. As shown in Figure 12.14, illumination of the viscous solution with UV light for 2 hr led to a decrease in solution viscosity to 0.1 Pa·s (3 orders of magnitude). Subsequent illumination with visible light recovered the initial solution viscosity. Changes in viscosity with illumination were reversible and repeatable over multiple photocycles.

Characterization of the microstructure of aggregates formed in solution using mixtures of CTAB, AZTMA, and NaSal, and the connection between the microstructures and rheological properties, is the subject of ongoing work.[68] A series of interesting observations reported by these authors do, however, provide some useful insights into the mechanism by which AZTMA influences the microstructure and rheological properties of this system. First, the authors report that addition of *trans*-AZTMA to a solution of 50 mM CTAB and 50 mM NaSal causes only a modest increase in viscosity (from 60 Pa·s to 100 Pa·s). Also, they report that mixtures of *trans*-AZTMA and NaSal, in the absence of CTAB, do not form solutions with high viscosities. These results indicate that *trans*-AZTMA does not substantially promote the formation of wormlike micelles in the CTAB, NaSal, *trans*-AZTMA system, but rather the presence of the wormlike micelles is due to the CTAB and NaSal. The authors speculate that the linear shape of *trans*-AZTMA allows it to be easily incorporated into wormlike micelles formed from CTAB and NaSal. Second, the authors report that addition of the *cis*-isomer of AZTMA to the mixture of CTAB and NaSal leads to a reduction in the viscosity from 60 Pa.s to 0.1 Pa.s. Thus, the principal means by which AZTMA leads to photo-dependent rheological properties is through the *cis*-isomer of AZTMA disrupting the wormlike micelles formed by CTAB and NaSal. It appears likely that the *cis*-AZTMA disrupts the entangled network of wormlike micelles by lowering the end-cap energy of the wormlike micelles, thus causing the formation of short rodlike aggregates in solution.[68]

12.5 pH-SENSITIVE SYSTEMS

The final type of stimuli-responsive giant micellar system that we describe here are those systems that exhibit pronounced changes in properties with pH. We focus on the pH-dependent behavior of surfactants with peptide-based head groups, as they appear to exhibit large changes in microstructure with pH and present interesting opportunities in regenerative medicine. The pH sensitivity of these systems is derived from the incorporation of peptide residues into the head group that can be titrated with pH. The pH-triggered assembly of giant micelles has been shown to provide a route to the preparation of nanostructures that present biologically relevant chemical functionalities (e.g., binding domains for cells) on a densely packed aggregate surface.[90,91]

We illustrate the recent activity in this area by focusing our comments on a paper by Hartgerink and co-workers.[92] The authors synthesized a peptide-based surfactant with the goal of creating a giant micellar system that could be used in studies of the mineralization of bone. They utilized an amphiphile engineered with five specific regions, as depicted in Figure 12.15. The five

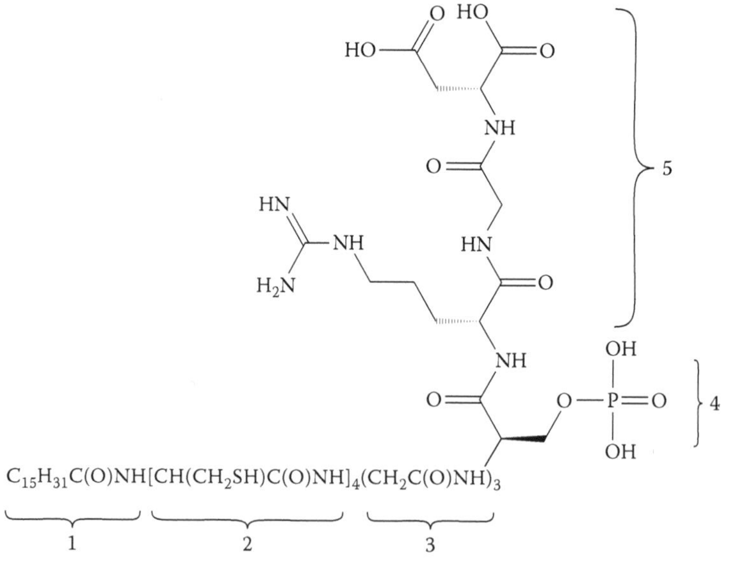

FIGURE 12.15 Structure of peptide-amphiphile that can be triggered to form giant micelles with a decrease in pH. The giant micelles can be used to direct the growth of hydroxyapatite. Reprinted with permission from Ref. 92. © 2001 by AAAS.

regions each served a particular function, as described below. Region 1 comprises a 16-carbon alkyl chain, which causes the molecule to self-associate in aqueous solution. Region 2 consists of four cysteine residues, which allows for disulfide cross-linking, making the giant micelles formed by this system structurally rigid. Region 3 is a flexible linker that provides conformational freedom to the peptides in the head group region of the surfactant. Region 4 presents a phosphorylated serine residue that functions to direct the mineralization of calcium phosphate. Finally, region 5 presents the Arg-Gly-Asp (RGD) peptide sequence, which serves to promote the adhesion of cells to the surface of the network of giant micelles.[92]

A solution of the peptide-based surfactant shown in Figure 12.15 was first prepared at pH 8, and treated with dithiothreitol (DTT) to reduce all the cysteine residues to free thiols. At this pH, the amphiphile does not self-assemble into giant micelles in solution. Acidifying the solution below pH 4, however, resulted in the formation of a self-supporting, birefringent gel. Examination of the gels by cryo-TEM revealed that giant micelles approximately 7 nm in diameter and more than a micrometer in length were formed in solution. Staining experiments demonstrated that the aliphatic tail of the amphiphile was buried in the core of the giant micelles, and that regions 4 and 5 of the amphiphile resided near the surface of the giant micelle. Addition of base to the gelled solution resulted in disassembly of the giant micellar network. However, if a giant micellar network was formed at acidic pH, and then oxidized with a solution of iodine, the network remained stable up to pH 8, due to cross-linking via formation of disulfide bonds. Treatment of the cross-linked network with dithiothreitol (a reducing agent) dispersed the network at basic pH. Thus, by manipulating both solution pH as well as the oxidation state of region 2 of the amphiphile (thiol/disulfide equilibrium), the self-assembly of the peptide-based surfactant into giant micelles could be reversibly manipulated.[92]

Hartgerink and co-workers also investigated the ability of the above-described networks of giant micelles to direct processes of mineralization relevant to formation of bone. A solution of the peptide-based amphiphile in Figure 12.15 was placed on a TEM grid, and then gelled by exposure to a vapor of HCl. The top of the gel was then exposed to a solution of $CaCl_2$ and the bottom of the gel was exposed to Na_2HPO_4. The intermixing of these salts within the giant micellar network lead to the formation of hydroxyapatite (HA). Interestingly, the authors also demonstrated that the crystallographic texture of HA was influenced by the morphology of the giant micellar network, with preferential growth of the HA crystals parallel to the giant micelles. The investigators performed control experiments in the absence of the giant micelles, or by using giant micelles formed from homologous peptide-based amphiphiles where the phosphoserine residue was replaced by serine. These studies lead the authors to conclude that networks of giant micelles formed from the peptide-based surfactant shown in Figure 12.15 are necessary to achieve oriented growth of HA.

12.6 CONCLUSIONS

We conclude this chapter with several comments. First, the examples described above demonstrate that a rich variety of transitions in microstructures, as well as rheological and phase behaviors, can be triggered in stimuli-responsive giant micellar systems. We point out, however, that the connection between microstructure, rheology, and phase behavior is not always obvious. In particular, the role of branching of giant micellar networks is considered in some studies but not others to potentially underlie changes in rheology and phase behavior. The factors driving phase separation (such as with increasing temperature) are also not fully elucidated. Systematic studies that combine cryo-TEM, small angle scattering, phase behavior, and rheological measurements, as a function of the relevant stimulus, appear warranted.

Second, we comment that a number of the unusual temperature-dependent rheological behaviors of giant micellar systems containing ionic surfactants are attributed to the partitioning of counterions between the water and micelle. Although these propositions lead to useful physical pictures, experimental measurements are need to test these propositions.

Third, in comparison to the number of studies that utilize changes in temperature to manipulate giant micellar systems, the number of studies utilizing other external stimuli such as redox-potential, illumination, and pH is small. The technological potential of these systems appears significant. Such systems also make possible types of experiments that cannot be performed with conventional surfactants, and thus they may also facilitate the development of a fuller understanding of the connection between microstructure, rheology, and phase behavior in giant micellar systems.

Fourth and finally, we conclude by noting that the formation of pH-dependent networks of giant micelles from peptide-based amphiphiles appears a fruitful direction of research. The multitude of different physical, chemical, and biological attributes that can be engineered into peptide-based amphiphiles makes them a rich area for discovery. Although the biological activities of some pH-dependent giant micellar systems have been characterized, many unresolved fundamental structure-property relationships exist and thus this area of research appears a particularly fertile one for the colloid and surfactant scientist.

REFERENCES

1. Cates, M. E., Candau, S. J. *J. Phys: Conden. Mat.* 1990, *2*, 6869.
2. Candau, S. J., Hirsch, E., Zana, R., Delsanti, M. *Langmuir* 1989, *5*, 1225.
3. Kern, F., Zana, R., Candau, S. J. *Langmuir* 1991, *7*, 1344.
4. Fischer, P., Rehage, H. *Langmuir* 1997, *13*, 7012.
5. Makhloufi, R., Cressely, R. *Colloid Polym. Sci.* 1992, *270*, 1035.
6. Berret, J.-F., Porte, G., Decruppe, J.-P. *Phys. Rev. E* 1997, *55*, 1668.
7. Ponton, A., Schott, C., Quemada, D. *Colloid Surface A* 1998, *145*, 37.
8. Kern, F., Lequeux, F., Zana, R., Candau, S. J. *Langmuir* 1994, *10*, 1714.

9. Soltero, J. F. A., Puig, J. E., Manero, O. *Langmuir* 1996, *12*, 2654.
10. Hassan, P. A., Valaulikar, B. S., Manohar, C., Kern, F., Bourdieu, L., Candau, S. J. *Langmuir* 1996, *12*, 4350.
11. Hassan, P. A., Candau, S. J., Kern, F., Manohar, C. *Langmuir* 1998, *14*, 6025.
12. Oda, R., Narayanan, J., Hassan, P. A., Manohar, C., Salkar, R. A., Kern, F., Candau, S. J. *Langmuir* 1998, *14*, 4364.
13. Narayanan, J., Manohar, C., Kern, F., Lequeux, F., Candau, S. J. *Langmuir* 1997, *13*, 5235.
14. Raghavan, S. R., Edlund, H., Kaler, E. W. *Langmuir* 2002, *18*, 1056.
15. Raghavan, S. R., Kaler, E. W. *Langmuir* 2001, *17*, 300.
16. Magid, L. J. *J. Phys. Chem. B* 1998, *102*, 4064.
17. Croce, V., Cosgrove, T., Maitland, G., Hughes, T., Karlsson, G. *Langmuir* 2003, *19*, 8536.
18. Gonzalez, Y. L., Kaler, E. W. *Curr. Opin. Coll. Int. Sci.* 2005, *10*, 256.
19. Koehler, R. D., Raghavan, S. R., Kaler, E. W. *J. Phys. Chem. B* 2000, *104*, 11035.
20. Lequeux, F. *Europhys. Lett.* 1992, *19*, 675.
21. Salkar, R. A., Hassan, P. A., Samant, S. D., Valaulikar, B. S., Kumar, V. V., Kern, F., Candau, S. J., Manohar, C. *Chem. Commun.* 1996, *10*, 1223.
22. Narayanan, J., Mendes, E., Manohar, C. *Int. J. Mod. Phys. B* 2002, *16*, 375.
23. Kalur, G. C., Frounfelker, B. D., Cipriano, B. H., Norman, A. I., Raghavan, S. R. *Langmuir* 2005, *21*, 10998.
24. Magid, L. J., Li, Z., Butler, P. D. *Langmuir* 2000, *16*, 10028.
25. Hassan, P. A., Raghavan, S. R., Kaler, E. W. *Langmuir* 2002, *18*, 2543.
26. Kalur, G. C., Raghavan, S. R. *J. Phys. Chem. B* 2005, *109*, 8599.
27. Tobita, K., Sakai, H., Kondo, Y., Yoshino, N., Iwahashi, M., Momozawa, N., Abe, M. *Langmuir* 1997, *13*, 5054.
28. Tobita, K., Sakai, H., Kondo, Y., Yoshino, N., Kamogawa, K., Momozawa, N., Abe, M. *Langmuir* 1998, *14*, 4753.
29. Danino, D., Weihs, D., Zana, R., Orädd, G., Lindblom, G., Abe, M., Talmon, Y. *J. Colloid Interface Sci.* 2003, *259*, 382.
30. Nilsson, P.-G., Wennerström, H., Lindman, B. *J. Phys. Chem.* 1983, *87*, 1377.
31. Strunk, H., Lang, P., Findenegg, G. H. *J. Phys. Chem.* 1994, *98*, 11557.
32. Bernheim-Groswasser, A., Wachtel, E., Talmon, Y. *Langmuir* 2000, *16*, 4131.
33. Drye, T. J., Cates, M. E. *J. Phys. Chem.* 1992, *96*, 1367.
34. Bohbot, Y., Ben-Shaul, A., Granek, R., Gelbart, W. M. *J. Phys. Chem.* 1995, *103*, 8764.
35. Tlusty, T., Safran, S. A., Strey, R. *Phys. Rev. Lett.* 2000, *84*, 1224.
36. Moitzi, C., Norbert, F., Glatter, O. *J. Phys. Chem. B* 2005, *109*, 16161.
37. Hays, M. E., Abbott, N. L. *Langmuir* 2005, *21*, 12007.
38. Hattori, T., Kato, M., Tanaka, S., Hara, M. *Electroanalysis* 1997, *9*, 722.
39. Hattori, T., Nakayama, M. *Electroanalysis* 2005, *17*, 613.
40. Imamura, H., Tsuchiya, K., Kondo, Y., Yoshino, N., Ohkubo, T., Sakai, H., Abe, M. *J. Oleo Sci.* 2005, *54*, 125.
41. Bai, G., Graham, M. D., Abbott, N. L. *Langmuir* 2005, *21*, 2235.
42. Swearingen, C., Wu, J., Stucki, J., Fitch, A. *Environ. Sci. Technol.* 2004, *38*, 5598.
43. Tsuchiya, K., Sakai, H., Kwon, K., Takei, T., Abe, M. *J. Oleo Sci.* 2002, *51*, 133.
44. Tsuchiya, K. Sakai. H. Saji, T., Abe, M. *Langmuir* 2003, *19*, 9343.
45. Tsuchiya, K., Orihara, Y., Kondo, Y., Yoshino, N., Ohkubo, T., Sakai, H., Abe, M. *J. Am. Chem. Soc.* 2004, *126*, 12282.

46. Sakai, H., Imamura, H., Kondo, Y., Yoshino, N., Abe, M. *Colloid Surf. A* 2004, *232*, 221.

47. Brake, J. M., Mezera, A. D., Abbott, N. L. *Langmuir* 2003, *19*, 8629.

48. Datwani, S. S., Truskett, V. N., Rosslee, C. A., Abbott, N. L., Stebe, K. J. *Langmuir* 2003, *19*, 8292.

49. Aydogan, N., Gallardo, B. S., Abbott, N. L. *Langmuir* 1999, *15*, 722.

50. Aydogan, N., Abbott, N. L. *Langmuir* 2001, *17*, 5703.

51. Aydogan, N., Rosslee, C. A., Abbott, N. L. *Colloids Surf. A* 2001, *201*, 101.

52. Aydogan, N., Abbott, N. L. *Langmuir* 2002, *18*, 7826.

53. Kakizawa, Y., Sakai, H., Nishiyama, K., Abe, M., Shouji, H., Kondo, Y., Yoshino, N. *Langmuir* 1996, *12*, 921.

54. Kakizawa, Y., Sakai, H., Yamaguchi, A., Kondo, Y., Yoshino, N., Abe, M. *Langmuir* 2001, *17*, 8044.

55. Kakizawa, Y., Sakai, H., Abe, M., Kondo, Y., Yoshino, N. *Mater. Technol.* 2001, *19*, 259.

56. Rosslee, C. A., Abbott, N. L. *Anal. Chem.* 2001, *73*, 4808.

57. Takei, T., Sakai, H., Kondo, Y., Yoshino, N., Abe, M. *Colloids Surf. A* 2001, *183*, 757.

58. Gallardo, B. S., Hwa, M. J., Abbott, N. L. *Langmuir* 1995, *11*, 4209.

59. Gallardo, B. S., Metcalfe, K. L., Abbott, N. L. *Langmuir* 1996, *12*, 4116.

60. Gallardo, B. S., Abbott, N. L. *Langmuir* 1997, *13*, 203.

61. Gallardo, B. S., Gupta, V. K., Eagerton, F. D., Jong, L. I., Craig, T. V. S., Shah, R. R., Abbott, N. L. *Science* 1999, *283*, 57.

62. Bennet, D. E., Gallardo, B. S., Abbott, N. L. *J. Am. Chem. Soc.* 1996, *118*, 6499.

63. Wolff, T., Kerperin, K. J. *J. Colloid Interface Sci.* 1993, *157*, 185.

64. Lehnberger, C., Wolff, T. *J. Colloid Interface Sci.* 1999, *213*, 187.

65. Yu, X., Wolff, T. *Langmuir* 2003, *19*, 9672.

66. Shang, T., Smith, K. A., Hatton, T. A. *Langmuir* 2006, *22*, 1436.

67. Shang, T., Smith, K. A., Hatton, T. A. *Langmuir* 2003, *19*, 10764.

68. Sakai, H., Orihara, Y., Kodashima, H., Matsumura, A., Ohkubo, T., Tsuchiya, K., Abe, M. *J. Am. Chem. Soc.* 2005, *127*, 13454.

69. Sakai, H., Matsumura, A., Yokoyama, S., Saji, T., Abe, M. *J. Phys. Chem. B* 1999, *103*, 10737.

70. Bonini, M., Berti, D., Di Meglio, J. M., Almgren, M., Teixeria, J., Baglioni, P. *Soft Matter* 2005, *1*, 444.

71. Hubbard Jr., F. P., Santonicola, G., Kaler, E. W., Abbott, N. L. *Langmuir* 2005, *21*, 6131.

72. Faure, D., Gravier, J., Labrot, T., Desbat, B., Oda, R., Bassani, D. M. *Chem. Commun.* 2005, *9*, 1167.

73. Lee, C. T., Smith, K. A., Hatton, T. A. *Macromolecules* 2004, *37*, 5397.

74. Lee, C. T., Smith, K. A., Hatton, T. A. *Biochemistry* 2005, *44*, 524.

75. Buwalda, R. T., Stuart, M. C. A., Engberts, J. B. F. N. *Langmuir* 2002, *18*, 6507.

76. Zou, B., Qiu, D., Hou, X., Wu, L., Zhang, X., Chi, L., Fuchs, H. *Langmuir* 2002, *18*, 8006.

77. Orihara, Y., Matsumura, A., Saito, Y., Ogawa, N., Saji, T., Yamaguchi, A., Sakai, H., Abe, K. *Langmuir* 2001, *17*, 6072.

78. Kang, H.-C., Lee, B. M., Yoon, J., Yoon, M. *J. Colloid Interface Sci.* 2000, *231*, 255.

79. Shin, J. Y., Abbott, N. L. *Langmuir* 1999, *15*, 4404.

80. Eastoe, J., Dominguez, M. S., Cumber, H., Burnett, G., Wyatt, P., Heenan, R. K. *Langmuir* 2003, *19*, 6579.

81. Eastoe, J., Dominguez, M. S., Cumber, H., Wyatt, P., Heenan, R. K. *Langmuir* 2004, *20*, 1120.

82. Eastoe, J., Vesperinas, A., Donnewirth, A.-C., Wyatt, P., Grillo, I., Heenan, R. K., Davis, S. *Langmuir* 2006, *22*, 851.

83. Bradley, M., Vincent, B., Warren, N. *Langmuir* 2006, *22*, 101.

84. Eastoe, J., Dominguez, M. S., Wyatt, P., Beeby, A., Heenan, R. K. *Langmuir* 2002, *18*, 7837.

85. Eastoe, J., Dominguez, M. S., Wyatt, P., Heenan, R. K. *Langmuir* 2004, *20*, 6120.

86. Kozlecki, T., Wilk, K. A. *J. Phys. Org. Chem.* 1996, *9*, 645.

87. Kozlecki, T., Wilk, K. A., Syper, L. *Prog. Colloid. Poly. Sci.* 1998, *110*, 193.

88. Sun, C., Arimitsu, K., Abe, K., Ohkubo, T., Yamashita, T., Sakai, H., Abe, M. *Mater. Technol.* 2004, *22*, 229.

89. Liu, S., Fujihira, M., Saji, T. *J. Chem. Soc., Chem. Comm.* 1994, *16*, 1855.

90. Niece, K. L., Hartgerink, J. D., Donners, J. J. J. M., Stupp, S. I. *J. Am. Chem. Soc.* 2003, *125*, 7146.

91. Claussen, R. C., Rabatic, B. M., Stupp, S. I. *J. Am. Chem. Soc.* 2003, *125*, 12680.

92. Hartgerink, J. D., Beniash, E., Stupp, S. I. *Science* 2001, *294*, 1684.

13 Hydrogen Bonded Supramolecular Polymers versus Wormlike Micelles: Similarities and Specificities

Laurent Bouteiller

CONTENTS

13.1 INTRODUCTION

Supramolecular polymers are linear chains of low molecular weight monomers held together by reversible and highly directional noncovalent interactions (Figure 13.1). Because of their macromolecular architecture, they display polymer-like properties both in solution and in the bulk. This concept has been developed in the last 15 years, mainly by chemists with the ambition to bring new properties and new functions to the field of polymer science. Although the analogy between supramolecular polymers and wormlike micelles is obvious, there has been little cross-fertilization between the two fields. (The most notable exception being the early recognition that the theoretical framework developed for wormlike micelles described rather well the rheological properties of supramolecular polymers.[1]) Thus, in this book devoted to wormlike micelles, my aim is to stress the similar-ities and specificities of supramolecular polymers in comparison to wormlike micelles. Given that excellent reviews on supramolecular polymers already exist,[2–6] I have chosen not to give an exhaustive description, but rather to focus on hydrogen bonded systems which have been characterized in solution, with the hope that this will both constitute a coherent subset and afford an easy comparison with wormlike micelles. The first two parts of this contribution are devoted to two specific systems which have been the most thoroughly characterized by rheology in solution. The last part is an attempt to show the versatility of supramo-lecular polymers.

13.2 BENZENE- AND CYCLOHEXANE-
TRICARBOXAMIDE

C3-symmetrical molecules are highly attractive building blocks for the formation of elongated supramolecular architectures, as long as reasonably strong interac-tions (here hydrogen bonds) are directed out of the plane of the molecule, in a cooperative fashion. The compounds considered here belong to two subgroups, depending on the exact structure of their core (Figure 13.2). It is particularly interesting to compare them because although their chemical structures are very similar, their properties are strikingly different.

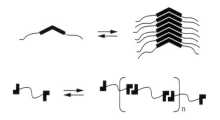

FIGURE 13.1 Schematic representation of supramolecular polymers.

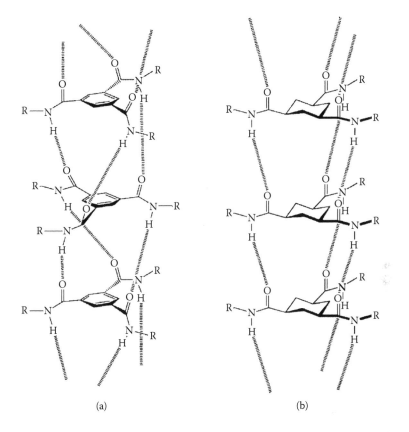

(a)   (b)

FIGURE 13.2 Self-assembled structure of BTC (a) and CTC (b) supramolecular polymers (R = alkyl group). Adapted with permission from Ref. 18. © 2006 by American Chemical Society.

13.2.1 BENZENE-TRICARBOXAMIDE (BTC)

In nonpolar solvents such as n-alcanes, cyclohexane, or toluene, various BTC derivatives have been shown to form viscoelastic solutions.[7] Based on the crystalline structure of a model compound,[8] a supramolecular structure has been proposed (Figure 13.2a). In this proposed structure, monomers are stacked onto each other and are held both by the formation of three hydrogen bonds between the amide groups and by π-stacking between the aromatic groups. Because aromatic and amide groups tend to favor a coplanar conformation, the hydrogen bonds do not lie parallel to the column axis, but are tilted. Thus, the hydrogen bond pattern should be helicoidal. This is confirmed by CD spectroscopy: the addition of a small amount of monomer with chiral side-chains to a nonchiral monomer solution[9,10] or to a racemic solution[11] induces a strong Cotton effect. This cooperative chiral induction is in agreement with the fact that the nonchiral

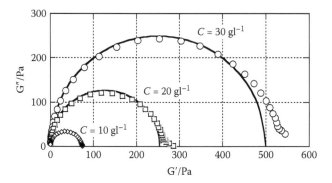

FIGURE 13.3 Cole–Cole plot for BTC solutions in decane, at 20°C. Reprinted with permission from Ref. 12. © 2006 by American Chemical Society.

monomer forms equal amounts of left and right handed helices, because the insertion of a chiral seed induces a strong preference of one helicity over the other. However, the presence of a significant fraction of free NH groups detected by FTIR spectroscopy suggests that many defects are present in this helical hydrogen bonding pattern.[11]

This organization at the molecular level has strong consequences on macroscopic properties of the solutions. Figure 13.3 shows that BTC solutions in decane form viscoelastic fluids with a nearly perfect Maxwellian behavior.[12,13] The fact that there is a single relaxation time (τ) is in agreement with a fast breaking regime of supramolecular chains. Moreover, the concentration dependence of the plateau modulus ($G \sim C^{2.0}$) is in agreement with the entanglement of flexible chains. However, the relaxation time (τ) is a slowly decreasing function of concentration, which is clearly in disagreement with the usual living polymer model for which $\tau \sim C^{1.25}$ is expected in the fast breaking regime of flexible chains (see Chapter 4). Consequently, reptation and chain scission are not the only mechanisms controlling entanglement release in this system. The authors propose the so-called phantom crossing model,[12] in which two chains make an entanglement point of lifetime τ_1 and then "go through each other," because of a local rearrangement of the chains. The hydrogen bonding defects mentioned above could be involved in such local rearrangements.

13.2.2 Cyclohexane-Tricarboxamide (CTC)

CTC derivatives also form viscoelastic solutions in nonpolar solvents,[14] but the rheological signature is clearly different from the case of BTC (Figure 13.4). The frequency dependence of the storage and loss moduli cannot be described by a single relaxation time.[15] Two relaxation times are necessary to adequately fit the data. The fast relaxation shows a similar behavior as in the case of BTC ($G_{fast} \sim C^{2.0}$), but the slow relaxation shows a concentration dependence ($G_{slow} \sim C^{1.3}$) which possibly indicates the presence of rigid rodlike species. Consequently, the

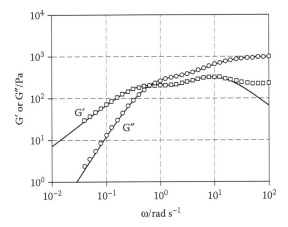

FIGURE 13.4 Frequency dependence of storage and loss moduli for a 30 gL^{-1} CTC solution in decane, at 20°C. Reproduced from Ref. 15, with permission of the Society of Rheology, Japan.

authors propose a model where the supramolecular polymer would present alternative sequences of rigid rodlike parts and more flexible parts. The increased rigidity of CTC compared to BTC could be due to its particular hydrogen bonding pattern. Indeed, because of the lack of π-stacking interaction and the lack of conjugation between the amides and the cyclohexane ring, the hydrogen bonds are believed to be parallel to the column axis (Figure 13.2b). This is supported by x-ray crystallography of a model compound.[16] The straight hydrogen bonding pattern of CTC may then lead to fewer defects (and thus more rigidity) than the helical pattern of BTC, because no helix reversal defects are expected. Unfortunately, no scattering data are available at the moment to confirm this hypothesis.

Interestingly, the parallel orientation of amide functions along the supramolecular columns is responsible for the buildup of large macrodipoles, as characterized by dielectric spectroscopy.[17]

Finally, it is worth mentioning the synthesis of a monomer bearing two such CTC moieties.[18] This compound self-associates strongly in chloroform.

13.3 BIS-UREAS

Because the urea moiety can form stronger hydrogen bonds than the amide moiety, many compounds bearing only two urea functions have been shown to form strong supramolecular architectures. If a parallel or antiparallel orientation of the two ureas is enforced by the spacer connecting them, then long one-dimensional supramolecular assemblies should be expected. Depending on the exact nature of the spacer and the lateral substituents, it is possible to tune the structure, but also the dynamic character of the assemblies. With symmetrical

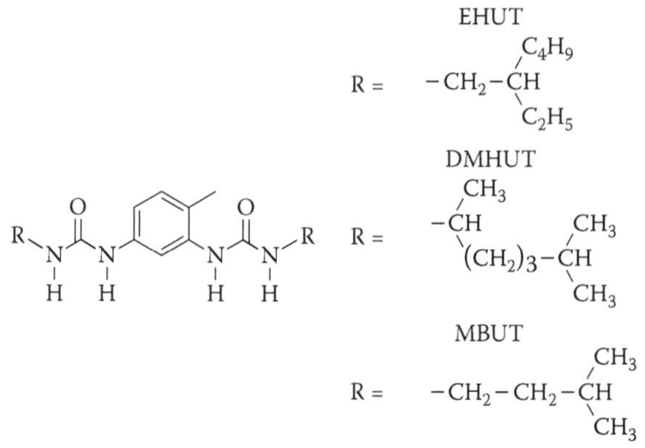

EHUT
$$R = \quad -CH_2-\overset{\overset{\displaystyle C_4H_9}{|}}{\underset{\underset{\displaystyle C_2H_5}{|}}{CH}}$$

DMHUT
$$R = \quad -\overset{\overset{\displaystyle CH_3}{|}}{\underset{\underset{\displaystyle (CH_2)_3-\overset{\overset{\displaystyle CH_3}{}}{\underset{\underset{\displaystyle CH_3}{|}}{CH}}}{|}}{CH}}$$

MBUT
$$R = \quad -CH_2-CH_2-\overset{\overset{\displaystyle CH_3}{|}}{\underset{\underset{\displaystyle CH_3}{|}}{CH}}$$

FIGURE 13.5 Structure of some bis-urea supramolecular polymers.

spacers and regular substituents, crystallization of the bis-urea is favored, so that organogelators can be obtained.[19–21] These compounds are soluble at high temperature in a particular solvent, but at lower temperature, highly anisotropic crystalline fibers are formed, which finally entrap the solvent. The strong gels obtained bear no real similarity to wormlike micelles, because they are metastable and there is no dynamic exchange between the fibers at room temperature. However, using an unsymmetrical spacer and/or branched substituents, one can hope to destabilize competing crystalline structures and stabilize dynamic one-dimensional filaments. These filaments should be maintained by hydrogen bonds along their axis and should show no tendency for lateral packing. Bis-ureas with a 2,4-toluene spacer (Figure 13.5) indeed form dynamic supramolecular polymers in nonpolar solvents.[22]

13.3.1 PSEUDOPHASE DIAGRAM OF EHUT

The pseudophase diagram of EHUT in toluene is shown on Figure 13.6.[23] It is likely that other (not necessarily dynamic) supramolecular structures exist at lower temperatures or at higher concentrations, but the remarkable feature about this bis-urea is that it forms two distinct supramolecular architectures, which are stable over a wide range of concentrations and temperatures. These two structures are in dynamic exchange with the monomer. Of course, the lines on this diagram are not true phase transitions, but limit the domains where each structure is the most abundant. For both supramolecular structures, FTIR spectroscopy indicates that all urea groups are hydrogen bonded. Moreover, small angle neutron scattering (SANS) shows that both structures are long and fibrillar (Figure 13.7), the high temperature structure being thinner than the low temperature structure. Based on the SANS derived dimensions, on molecular simulation and on the structure of a monolayer probed by STM (Figure 13.8),[24] a ladderlike supramolecular

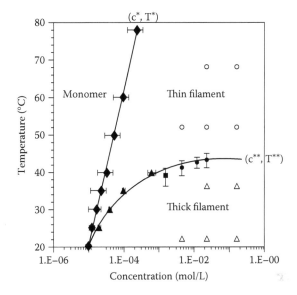

FIGURE 13.6 Pseudophase diagram for EHUT solutions in toluene. Transition between monomers and thin supramolecular filaments determined by calorimetry (ITC) (◆). Transition between thin filaments and thick tubes determined by ITC (▲), viscosimetry (■), and FTIR (●). SANS characterization of the thin filaments (○) and thick tubes (△). Reprinted with permission from Ref. 23. © 2006 by American Chemical Society.

arrangement has been proposed for the high temperature, thin filament structure (Figure 13.9a).[23] Similarly, a thick tubular arrangement has been proposed for the low temperature structure (Figure 13.9b).[25]

The intriguing aspect of the tubular structure is that it can be expected to be stable only if the inner cavity is filled with solvent. Consequently, a very strong solvent effect is expected, with solvents of large molecular dimensions destabilizing the tubular structure. This effect was indeed demonstrated (Figure 13.10) with a series of aromatic solvents of similar dielectric constants and solvating power.[25] For instance, the transition temperature between the thin and the thick structure is more than 50°C lower in bulky 1,3,5-triisopropylbenzene than in toluene.

13.3.2 PROPERTIES OF THE THIN FILAMENT STRUCTURE

Starting from the free monomer, the growth of the thin filament (Figure 13.11) has been characterized by FTIR spectroscopy[26] and isothermal titration calorimetry.[27] Both studies are in agreement both qualitatively and quantitatively, and show that the self-association of EHUT is highly cooperative, which means that the association constant for the formation of long oligomers ($K_n = K$, n > 2) is much larger than the association constant for the formation of dimers (K_2). This

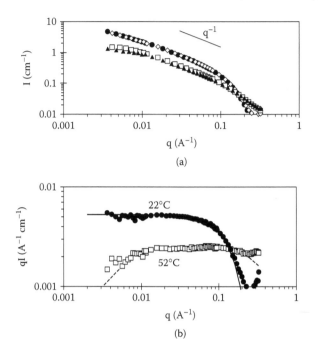

(a)

(b)

FIGURE 13.7 SANS curves for solutions of EHUT in d_8-toluene at several temperatures (22°C (●); 36°C (◇); 52°C (□); 68°C (▲)). (a) Intensity (I) versus momentum transfer (q), for a 22.9 mM solution. (b) qI versus q representation, for a 4.6 mM solution. The plain curve is a fit according to a model for infinitely long rigid filaments (diameter 2r = 26 Å and linear density n_L = 0.55 Å^{-1}). The dotted curve is a fit for short and rigid filaments (diameter 2r = 13 Å, linear density n_L = 0.25 Å^{-1} and length 2H = 400 Å). Reprinted with permission from Ref. 23. © 2006 by American Chemical Society.

FIGURE 13.8 High resolution STM image of an EHUT monolayer on Au(111) (5 × 10 nm^2, −0.4 V, 1.9 nA), with insets of a space filling model of EHUT. Reprinted with permission from Ref. 24. © 2006 by American Physical Society.

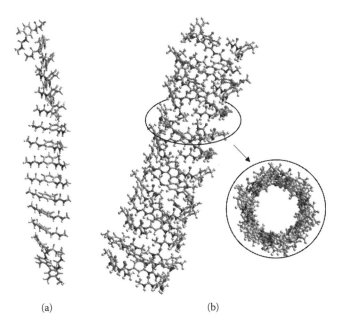

(a)   (b)

FIGURE 13.9 Tentative supramolecular structures proposed for the thin filaments (a) and the tubes (b).

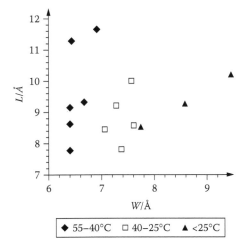

FIGURE 13.10 Transition temperature (T**, value represented by the different symbols) between the thin filaments and tubes for EHUT solutions in aromatic solvents, versus length (L) and width (W) of the solvent molecules. The length (L), width (W), and thickness (Th) are defined as the respective dimensions of the smallest right-angled parallelepiped containing the molecule, such that $L > W > Th$. Reproduced from Ref. 25 with permission of Wiley Interscience.

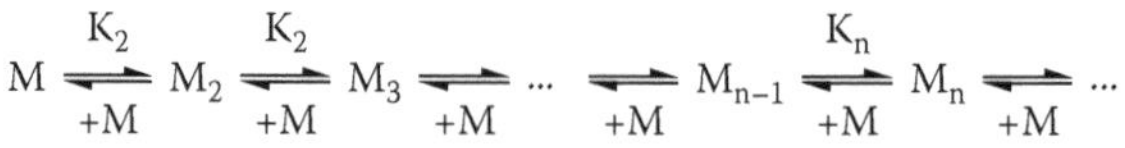

$$M \underset{+M}{\overset{K_2}{\rightleftharpoons}} M_2 \underset{+M}{\overset{K_2}{\rightleftharpoons}} M_3 \underset{+M}{\rightleftharpoons} \cdots \underset{+M}{\rightleftharpoons} M_{n-1} \underset{+M}{\overset{K_n}{\rightleftharpoons}} M_n \underset{+M}{\rightleftharpoons} \cdots$$

FIGURE 13.11 Association equilibria involved in the formation of a supramolecular polymer (M: monomer, Mn: chain of degree of polymerization n).

cooperativity results from an electronic effect,[28] and possibly also from a conformational effect.[29]

The knowledge of the association constants makes it possible to compute the whole length distribution of the filaments. For example, Figure 13.12 shows their number average length versus concentration in two solvents. The critical concentration below which no filament is present, is characteristic of the cooperativity of the growth process. Of course, the filaments are longer in less polar solvents and they can become extremely long. This results in a strong increase of viscosity with concentration. However, the solutions are not viscoelastic, even at concentrations well above the overlap concentration.[30] Consequently, the relaxation of entanglements, probably by chain scission, is very fast ($\tau < 0.01$ s).

13.3.3 PROPERTIES OF THE THICK FILAMENT STRUCTURE

In contrast, the thick filament structure yields strongly viscoelastic solutions in the semidilute regime.[31,32] Figure 13.13 shows a Cole–Cole plot for an EHUT solution in dodecane ($C^* = 0.1$ g/L). Experimental data can be fitted at low frequencies with a Maxwell model, in agreement with the release of entanglements through a scission and recombination mechanism. In the framework of

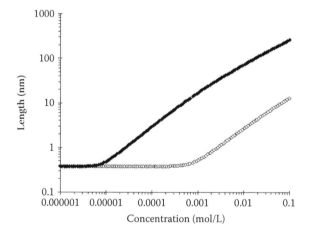

FIGURE 13.12 Number average curvilinear length (L_n) of the thin filaments, versus EHUT concentration in chloroform ($\bigcirc$) and 1,3,5-trimethylbenzene ($\blacklozenge$), at 30°C.

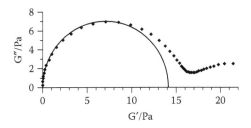

FIGURE 13.13 Cole–Cole plot for a 5 gL^{-1} EHUT solution in dodecane, at 25°C.

Cates theory (see Chapter 4), the concentration dependence of the elastic plateau modulus (G ~ $C^{2.2}$), the zero shear viscosity (η_0 ~ $C^{3.5}$), and the relaxation time (τ ~ $C^{0.8}$) could be analyzed to mean that the system is in a fast breaking regime of flexible filaments. However, this is in disagreement with two facts. First, the departure from monoexponentiality at higher frequencies (Figure 13.13) is an indication that the scission-recombination of the filaments may not be much faster than their reptation. Moreover, a static light scattering study on EHUT/cyclohexane solutions has shown that the persistence length of the EHUT filaments is at least 100 nm.[33] As will be confirmed in the next section, the rheological characteristics of EHUT solutions can in fact be explained better by the presence of semiflexible filaments for which the breaking and reptation times are of the same order of magnitude.

In the nonlinear regime also, the EHUT solutions display some similarities with wormlike micelles. Figure 13.14 shows a stress-strain curve typical of shear banding (Chapter 9, Section 9.5).[32]

However, a major difference with wormlike micelles is that in the case of EHUT solutions, the viscoelastic behavior (at low temperatures) can be switched to the purely viscous behavior (at high temperatures). Moreover, the transition

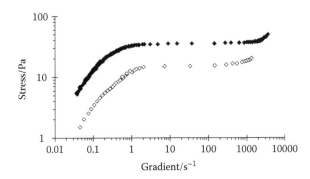

FIGURE 13.14 Flow curves for 4 gL^{-1} (○) and 10 gL^{-1} (◆) EHUT solutions in dodecane, at 25°C.

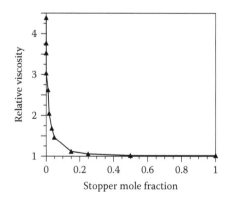

FIGURE 13.15 Structure of chain stopper DBUT.

has been shown to be fast, reversible (without hysteresis), and extremely cooperative: the conversion of thin filaments into thick filaments occurs within a temperature range of 5°C only.[23] This transition can be triggered by temperature, but also by a change in the solvent composition or by a change of the monomer composition (see below).

13.3.4 TUNING WITH CHAIN STOPPERS

The fact that supramolecular polymers are assembled by highly directional interactions allows one to design chain stoppers, that is, molecules able to interact specifically with the chain ends of the filaments. In the case of bis-ureas, such a chain stopper can, *a priori*, be obtained by blocking the NH groups, so that only hydrogen bond accepting functions are left. Figure 13.15 shows the structure of such a chain stopper (DBUT). Addition of low amounts of this compound to EHUT solutions reduces its viscosity significantly (Figure 13.16).[34] More interestingly, the use of chain stoppers means that the independent variation of the length and the concentration of the filaments becomes possible.[35,36] Indeed, if the association constant of the chain stopper is large enough, nearly all chain ends are occupied by a chain stopper, which means that the length of the filament is

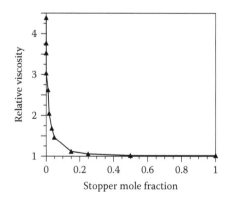

FIGURE 13.16 Relative viscosity of mixtures of EHUT and chain stopper DBUT in CCl₄ vs. stopper mole fraction. The overall concentration is the same for all experiments ([EHUT] + [DBUT] = 7.2 mM).

inversely proportional to the chain stopper fraction, and independent of the EHUT concentration. With this idea in mind, rheological experiments were performed to determine the chain length ($<N>$) and the concentration (C) dependencies of the zero shear viscosity of EHUT/DBUT solutions in cyclohexane: $\eta_0 \sim <N>^{2.7}C^{1.8}$.[35] Since $<N> \sim C^{0.5}$, the overall concentration scaling is 3.2, which (as mentioned above) could be in agreement with the exponent expected for fast breaking of flexible filaments (3.3). However, now that the chain length and the concentration dependencies can be decoupled, it is clear that the experimental results are not in agreement with the model for fast breaking of flexible filaments, for which $\eta_0 \sim <N>^{0.5}C^3$ is expected. In fact, the experimental exponents lie in between the limits for flexible chains and rigid rods, and also between the limits for fast and slow breaking.[35] Unfortunately, the scaling exponents of the relaxation time ($\tau \sim <N>^{2.3}C^0$) cannot be explained similarly. More experiments are needed to gain a better understanding of all the underlying mechanisms. In particular, it is possible that the chain stoppers not only reduce the length of the filaments, but also create defects within the trimolecular structure of the thick filaments.

13.3.5 Engineering

All the results presented so far were obtained with a particular monomer (EHUT), but are in fact typical of a family of compounds sharing the same 2,4-bis-ureidotoluene core (Figure 13.5).[26] As long as the monomer is soluble in the solvent of interest, the exact nature of the lateral substituent R does not change the global picture, but influences the boundary between the thin and thick filament structures. For instance, the transition temperatures of EHUT and DMHUT solutions in toluene are 20°C apart. Moreover, it is possible to finely tune the transition temperature of the solution by simply mixing the two monomers.[23]

The selective synthesis of bis-ureas with two different lateral substituents is also possible.[37] In particular, this opens up the possibility to graft bis-urea stickers onto macromolecules or to prepare multifunctional compounds.[38]

13.4 VERSATILITY OF SUPRAMOLECULAR POLYMERS

In suitable conditions, both wormlike micelles and supramolecular polymers form long reversible chains with similar rheological properties, but there is a most significant difference between them. Supramolecular polymers self-assemble because the monomer contains specific and directional complementary associating groups. In fact, the structure of the monomer can be understood as made of two independent parts: the associating groups, which are responsible for the self-assembly process, and the rest of the molecule, which can be altered nearly at will without compromising the self-assembly (as long as no interfering hydrogen bonding groups are introduced). Consequently, it is possible to tune the properties and to add functionality through chemical design. In contrast, wormlike micelles form because of the overall structure of the monomer (surfactant): the relevant parameter is the packing parameter measuring the balance between the hydrophobic tail, the polar head group, and the counterion. Introducing a new chemical group in one part of the surfactant may tilt the delicate balance and thus completely change the topology of the micelle. The following characteristics of supramolecular polymers illustrate this point. Some of the properties described here can also be obtained in the case of wormlike micelles, but the specificity of supramolecular polymers is the direct relationship between the chemical structure and the properties.

13.4.1 TUNING OF THE SOLUBILITY

Because the self-assembly of wormlike micelles is due to solvophobic forces, their study is often limited to a single solvent (usually water). In contrast, supramolecular polymers are often soluble in a range of organic solvents, and their solubility can be tuned by adding suitable substituents. Of course, the lower the polarity of the solvent, the stronger the association of hydrogen bonded supramolecular polymers. To increase the solubility of supramolecular polymers in apolar solvents, it is possible to add lipophilic alkyl chains at suitable positions. For instance, supramolecular polymer [AA3:BB] (Figure 13.17) is not soluble in decane. It can be dissolved in toluene, but it precipitates after some time. Replacing the butanoyl groups by longer decanoyl groups ([AA9:BB]) affords stable decane solutions, with reversible properties under mechanical and temperature cycling.[39]

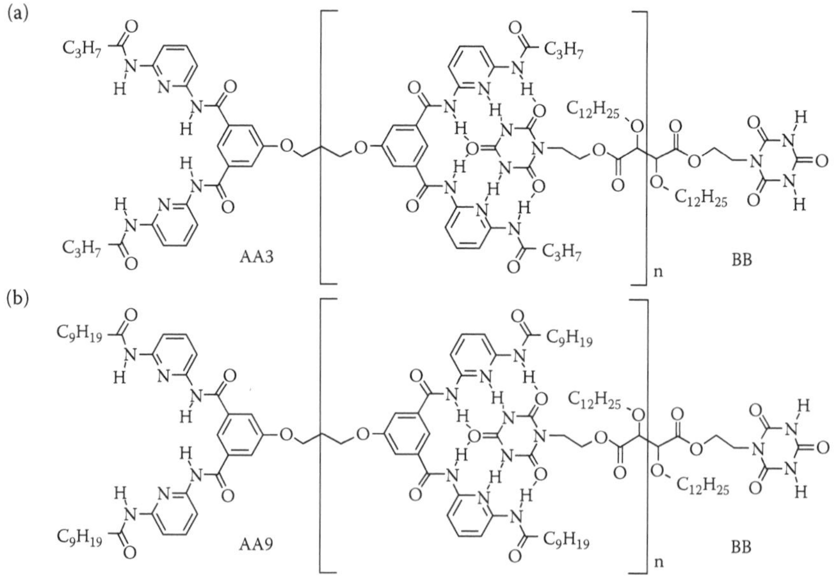

FIGURE 13.17 Hydrogen bonded supramolecular polymers [AA3:BB] (a) and [AA9:BB] (b).

13.4.2 SHIFTING RING-CHAIN EQUILIBRIUM

Ring-chain equilibrium in supramolecular polymers has been extensively studied both experimentally[40–44] and theoretically.[45–47] For strongly associating systems, a threshold concentration is expected below which only cyclic species are present, and above which the amount of cyclic species remains constant. It has been shown that the value of this threshold concentration, and thus the amount of cyclic species, is a strong function of the length[40] and the conformation[41,42,45] of the spacer connecting the associating groups. In the particular case of supramolecular polymers formed by the assembly of two complementary monomers, the cyclic content is minimized if the length of the two monomers are mismatched.[43] Generally, a careful choice of the connecting spacer thus makes it possible to design supramolecular polymers with a given cyclization tendency. This in turn influences the properties, mainly because of the lower molecular weight of cyclic species compared to linear chains.

13.4.3 INTRODUCING BRANCHES OR CROSSLINKS

The use of monomers bearing more than two associating groups is a straightforward way to introduce a controlled amount of branches or crosslinks in a supramolecular polymer structure.[1,48–52] The improvement of the mechanical properties in the case of concentrated[48] or solventless[51,52] systems is spectacular. However, no thorough rheological study of the influence of such a multifunctional monomer on the properties of semidilute supramolecular polymer solutions has been published yet.

13.4.4 CHIRALITY

Decorating a monomer structure with a chiral side-chain can lead to a transfer of chirality at the supramolecular level. Indeed, several examples show that disklike molecules, designed to pile up into long reversible columns, form in fact helical columns, driven by the favorable packing of the chiral lateral substituents.[9,11,53,54] This chiral packing effect is strong enough to induce a chiral amplification: a low amount of chiral monomer mixed with a nonchiral monomer can still drive the formation of helical columns. With a different design, where chirality is directly built in the hydrogen bond pattern, Aida et al. have demonstrated that it is possible to enforce a homochiral supramolecular polymerization.[55] In this case, a mixture of L and D monomers exclusively forms supramolecular chains of polyL and polyD homopolymers instead of copolymers.

13.4.5 COUPLING ELECTROOPTICAL PROPERTIES

Functional supramolecular polymers can be designed by introducing suitable chromophores, such as quinacridone,[56] perylenes,[57,58] or oligophenylenevinylenes,[59,60] in the monomer structure. The self-assembly process, by altering the distance between the chromophores, can then be coupled to a particular electrooptical property (fluorescence, energy transfer). This could be of interest in

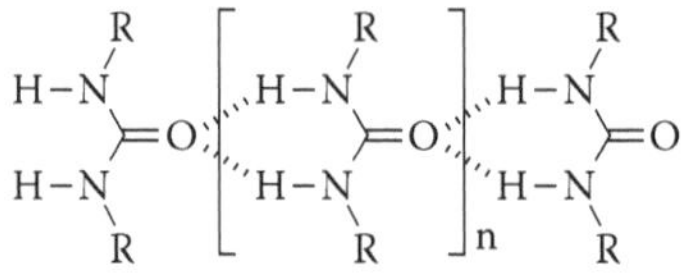

FIGURE 13.18 Hydrogen bonding pattern of dialkyl ureas.

the fields of light harvesting and long-range vectorial transport of excitation energy.

13.4.6 POLARITY OF THE CHAIN

Designing a wormlike micelle which would present two different end-caps (by analogy with the pointed and barbed ends of an actin filament[61]) is certainly a difficult task. In the case of supramolecular polymers, this is feasible. Dialkyl ureas (Figure 13.18) are very simple monomers where this breaking of symmetry directly results from the structure of the monomer: one extremity presents a free carbonyl group to the solvent, while the other extremity presents hydrogen giving groups.[62,63] This feature is potentially useful in the context of surface grafting of supramolecular polymers (see below). In more strongly associated systems like BTC and CTC (see Section 13.2), the symmetry breaking along the chain is responsible for the build up of large macrodipoles that may be useful for electrooptical or electromechanical devices.[17,64]

13.4.7 ENGINEERING OF SUPRAMOLECULAR CHAIN ENDS WITH CHAIN STOPPERS

The use of monofunctional compounds designed to interact specifically with extremities of supramolecular polymer chains (chain stoppers), has already been mentioned in the section devoted to bis-ureas (Section 13.3).[34–36] This concept had previously been applied to other hydrogen bonded supramolecular polymers.[1,65,66] Chain stoppers introduced in varying amounts are often used simply to reduce the length of the supramolecular chains. However, if the chain stopper is introduced in a fixed proportion relative to the monomer, its effect is then to block the concentration dependence of the length of the supramolecular polymer.[35,36] Finally, chain stoppers can be exploited to decorate the chain ends with particular functional groups or labels.[66]

13.4.8 GRAFTING ON SURFACES

One of the most interesting developments of the chain stopper concept is the use as a surface anchor. Indeed, covalent grafting of a chain stopper on a surface yields a supramolecular polymer brush, if the surface is immersed in a supramolecular polymer solution. Such brushes have been realized experimentally with ureidopyrimidinone[67] or oligonucleotide[68,69] based monomers. The properties of the brushes (thickness, adhesion) have been studied by AFM, with chain stopper grafted tips. The conclusion of these preliminary studies is that specific molecular recognition mediates direct bridging and thus adhesion between the surface and the tip. Moreover, the average length of the grafted chains seems to be shorter than the chains in the surrounding solution. Finally, the case of a polar supramolecular polymer chain with two complementary but different chain ends (A and B) is worth considering. If such a system is brought into contact with a surface grafted with an anchoring group bearing only A functions, then a theoretical model shows that the supramolecular brushes formed should exert repulsive forces between surfaces.[70,71]

13.4.9 Control of the Supramolecular Chains by External Triggers

Responsive supramolecular polymers can result from various functional elements introduced within the monomer structure. For instance, light controlled supramolecular polymers have been obtained by incorporating a photoresponsive chromophore between the two self-associating groups of the monomer. Thus, the light can trigger a conformational switch[72,73] or the formation of a reversible bond within the monomer,[74] which then leads to a change of the length of the chains. Alternatively, it is possible to use a chain stopper bearing a photocleavable protecting group.[75] The deprotection improves the efficiency of the chain stopper, so that the viscosity of the solution decreases upon irradiation.

Another approach is to introduce a chemically cleavable linker in between the two hydrogen bonding groups of the monomer. For example, this can be an organometallic complex cleaved by addition of a suitable ligand,[76,77] or a reversible covalent bond,[78] or an ionic interaction controlled by the presence of CO_2.[79,80]

13.4.10 Covalent Capture

Covalent capture of supramolecular assemblies can be tricky, because the energy involved in the covalent bond formation is large compared to the stabilizing energy of the self-assembling process.[81–83] Consequently, the covalent reaction should be very carefully designed to avoid disrupting the supramolecular structure. Moreover, in the case of supramolecular polymers, interchain crosslinking should be avoided. However, Meijer et al. have successfully polymerized columnar stacks of BTC derivatives bearing a photopolymerizable sorbate group.[84,85]

13.5 CONCLUSION

We have shown through two particular examples (benzene-tricarboxamides (BTC) and bis-ureas), that it is possible to design hydrogen bonded supramolecular polymers with rheological properties very similar to those of wormlike micelles. These systems self-assemble to form long semirigid filaments, which can break and recombine in a dynamic fashion. The most significant difference between these systems and wormlike micelles is that the former are maintained by the assembly of predictable and directional associating groups. Consequently, it is possible to tune the properties and to add functionality through chemical design. For instance, this opens the possibility to rationally tune the cyclic and branched species content of supramolecular polymers, or to graft supramolecular polymers on surfaces.

REFERENCES

1. Sijbesma, R.P., Beijer, F.H., Brunsveld, L., Folmer, B.J.B., Hirschberg, J.H.K.K., Lange, R.F.M., Lowe, J.K.L., Meijer, E.W. *Science* 1997, *278*, 1601.
2. Zimmerman, N., Moore, J.S., Zimmerman, S.C. *Chem. Ind.* 1998, 604.
3. Brunsveld, L., Folmer, B.J.B., Meijer, E.W., Sijbesma, R.P. *Chem. Rev.* 2001, *101*, 4071.
4. ten Cate, A.T., Sijbesma, R.P. *Macromol. Rapid Commun.* 2002, *23*, 1094.
5. Lehn, J.-M. *Polym. Int.* 2002, *51*, 825.
6. *Supramolecular Polymers*, Ciferri, A., Ed., Marcel Dekker, New York, 2005.
7. Hanabusa, K., Koto, C., Kimura, M., Shirai, H., Kakeki, A., *Chem. Lett.* 1997, 429.
8. Lightfoot, M.P., Mair, F.S., Pritchard, R.G., Warren, J.E. *Chem. Commun.* 1999, 1945.
9. Brunsveld, L., Schenning, A.P.H.J., Broeren, M.A.C., Janssen, H.M., Vekemans, J.A.J.M., Meijer, E.W. *Chem. Lett.* 2000, 292.
10. van Gorp, J.J., Vekemans, J.A.J.M., Meijer, E.W. *J. Am. Chem. Soc.* 2002, *124*, 14759.
11. Ogata, D., Shikata, T., Hanabusa, K. *J. Phys. Chem. B* 2004, *108*, 15503.
12. Shikata, T., Ogata, D., Hanabusa, K. *J. Phys. Chem. B* 2004, *108*, 508.
13. The Maxwell model has also been shown to be in agreement with creep and creep recovery experiments. Shikata, T., Sakamoto, A., Hanabusa, K. *Nihon Reoroji Gakkaishi* 2004, *32*, 203.
14. Hanabusa, K., Kawakami, A., Kimura, M., Shirai, H. *Chem. Lett.* 1997, 191.
15. Shikata, T., Ogata, D., Hanabusa, K. *Nihon Reoroji Gakkaishi* 2003, *31*, 229.
16. Fan, E., Yang, J., Geib, S.J., Stoner, T.C., Hopkins, M.D., Hamilton, A.D. *J. Chem. Soc., Chem. Commun.* 1995, 1251.
17. Sakamoto, A., Ogata, D., Shikata, T., Hanabusa, K. *Macromolecules* 2005, *38*, 8983.
18. Jang, W.-D., Aida, T. *Macromolecules* 2004, *37*, 7325.
19. van Esch, J., Schoonbeek, F., de Loos, M., Kooijman, H., Spek, A.L., Kellogg, R.M., Feringa, B.L. *Chem. Eur. J.* 1999, *5*, 937.
20. Terech, P., Weiss, R.G. *Chem. Rev.* 1997, *97*, 3133.
21. *Molecular Gels: Materials with self-assembled fibrillar networks*, Terech, P., Weiss, R.G., Eds., Kluwer, Dordrecht, The Netherlands, 2005.
22. Boileau, S., Bouteiller, L., Lauprêtre, F., Lortie, F. *New J. Chem.* 2000, *24*, 845.
23. Bouteiller, L., Colombani, O., Lortie, F., Terech, P. *J. Am. Chem. Soc.* 2005, *127*, 8893.
24. Vonau, F., Suhr, D., Aubel, D., Bouteiller, L., Reiter, G., Simon, L. *Phys. Rev. Lett.* 2005, *94*, 066103.

25. Pinault, T., Isare, B., Bouteiller, L. *Chem. Phys. Chem.* 2006, *7*, 816.
26. Simic, V., Bouteiller, L., Jalabert, M. *J. Am. Chem. Soc.* 2003, *125*, 13148.
27. Arnaud, A., Bouteiller, L. *Langmuir* 2004, *20*, 6858.
28. The urea moiety is polarized by the formation of a first hydrogen bond (in a dimer), which favors the formation of its next hydrogen bond (in a trimer).
29. The dihedral angles between the urea moieties and the toluene spacer are not identical in the free monomer and in the thin filament structure. This introduces an entropic penalty for the dimer formation.
30. Ducouret, G. private communication.
31. Lortie, F., Boileau, S., Bouteiller, L., Chassenieux, C., Demé, B., Ducouret, G., Jalabert, M., Lauprêtre, F., Terech, P. *Langmuir* 2002, *18*, 7218.
32. Ducouret, G., Chassenieux, C., Martins, S., Lequeux, F., Bouteiller, L. *J. Colloid Interface Sci.* 2007.
33. van der Gucht, J., Besseling, N.A.M., Knoben, W., Bouteiller, L., Cohen Stuart, M.A. *Phys. Rev. E* 2003, *67*, 051106.
34. Lortie, F., Boileau, S., Bouteiller, L., Chassenieux, C., Laupretre, F. *Macromolecules* 2005, *38*, 5283.
35. Knoben, W., Besseling, N.A.M., Bouteiller, L., Cohen Stuart, M.A. *Phys. Chem. Chem. Phys.* 2005, *7*, 2390.
36. Knoben, W., Besseling, N.A.M., Cohen Stuart, M.A. *Macromolecules* 2006, *39*, 2643.
37. Colombani, O., Bouteiller, L. *New J. Chem.* 2004, *28*, 1373.
38. Colombani, O., Barioz, C., Bouteiller, L., Chanéac, C., Fompérie, L., Lortie, F., Montès, H. *Macromolecules* 2005, *38*, 1752.
39. Kolomiets, E, Buhler, E., Candau, S.J., Lehn, J.-M. *Macromolecules* 2006, *39*, 1173.
40. Abed, S., Boileau, S., Bouteiller, L. *Macromolecules* 2000, *33*, 8479.
41. Folmer, B.J.B., Sijbesma, R.P., Meijer, E.W. *J. Am. Chem. Soc.* 2001, *123*, 2093.
42. ten Cate, A.T., Kooijman, H., Spek, A.L., Sijbesma, R.P., Meijer, E.W. *J. Am. Chem. Soc.* 2004, *126*, 3801.
43. Yamaguchi, N., Gibson, H.W. *Chem. Commun.* 1999, 789.
44. Söntjens, S.H.M., Sijbesma, R.P., van Genderen, M.H.P., Meijer, E.W. *Macromolecules* 2001, *34*, 3815.
45. Chen, C.-C., Dormidontova, E.E. *Macromolecules* 2004, *37*, 3905.
46. Ercolani, G., Mandolini, L., Mencarelli, P., Roelens, S. *J. Am. Chem. Soc.* 1993, *115*, 3901.
47. Ercolani, G. *J. Chem. Phys. B* 1998, *102*, 5699.
48. Castellano, R.K., Clark, R., Craig, S.L., Nuckolls, C., Rebek, J., Jr. *Proc. Natl. Acad. Sci. USA* 2000, *97*, 12418.
49. Castellano, R.K., Rebek, J., Jr. *J. Am. Chem. Soc.* 1998, *120*, 3657.
50. Berl, V., Schmutz, M., Krische, M.J., Khoury, R.G., Lehn, J.-M. *Chem. Eur. J.* 2002, *8*, 1227.
51. St. Pourcain, C.B., Griffin, A.C. *Macromolecules* 1995, *28*, 4116.
52. Lange, R.F.M., Van Gurp, M., Meijer, E.W. *J. Polym. Sci. Part A: Polym. Chem.* 1999, *37*, 3657.
53. Hirschberg, J.H.K.K., Brunsveld, L., Ramzi, A., Vekemans, J.A.J.M., Sijbesma, R.P., Meijer, E.W. *Nature* 2000, *407*, 167.
54. Hirschberg, J.H.K.K., Koevoets, R.A., Sijbesma, R.P., Meijer, E.W. *Chem. Eur. J.* 2003, *9*, 4222.

55. Ishida, Y., Aida, T. *J. Am. Chem. Soc.* 2002, *124*, 14017.

56. Keller, U., Müllen, K., De Feyter, S., De Schryver, F.C. *Adv. Mater.* 1996, *8*, 490.

57. Würthner, F., Thalacker, C., Sautter, A. *Adv. Mater.* 1999, *11*, 754.

58. Würthner, F., Thalacker, C., Sautter, A., Schärtl, W., Ibach, W., Hollricher, O. *Chem. Eur. J.* 2000, *6*, 3871.

59. El-ghayoury, A., Schenning, A.P.H.J., van Hal, P.A., van Duren, J.K.J., Janssen, R.A.J., Meijer, E.W. *Angew. Chem. Int. Ed.* 2001, *40*, 3660.

60. Varghese, R., George, S.J., Ajayaghosh, A. *Chem. Commun.* 2005, 593.

61. Korn, E.D., Carlier, M.-F., Pantaloni, D. *Science* 1987, *238*, 638.

62. Jadzyn, J., Stockhausen, M., Zywucki, B. *J. Phys. Chem.* 1987, *91*, 754.

63. Lortie, F., Boileau, S., Bouteiller, L. *Chem. Eur. J.* 2003, *9*, 3008.

64. Sakamoto, A., Ogata, D., Shikata, T., Urukawa, O., Hanabusa, K. *Polymer* 2006, *47*, 956.

65. Castellano, R.K., Nuckolls, C., Rebek, J., Jr. *Polymer News* 2000, *25*, 44.

66. Hirschberg, J.H.K.K., Ramzi, A., Sijbesma, R.P., Meijer, E.W. *Macromolecules* 2003, *36*, 1429.

67. Zou, S., Schönherr, H., Vancso, G.J. *Angew. Chem. Int. Ed.* 2005, *44*, 956.

68. Kersey, F.R., Lee, G., Marszalek, P., Craig, S.L. *J. Am. Chem. Soc.* 2004, *126*, 3038.

69. Kim, J., Liu, Y., Ahn, S.J., Zauscher, S., Karty, J.M., Yamanaka, Y., Craig, S.L. *Adv. Mater.* 2005, *17*, 1749.

70. van der Gucht, J., Besseling, N.A.M., Cohen Stuart, M.A. *J. Am. Chem. Soc.* 2002, *124*, 6202.

71. van der Gucht, J., Besseling, N.A.M., Fleer, G.J. *J. Chem. Phys.* 2003, *119*, 8175.

72. Lucas, N.L., van Esch, J., Kellogg, R.M., Feringa, B.L. *Chem. Commun.* 2001, 759.

73. Takeshita, M., Hayashi, M., Kadota, S., Mohammed, K.H., Yamato, T. *Chem. Commun.* 2005, 761.

74. Ikegami, M., Ohshiro, I., Arai, T. *Chem. Commun.* 2003, 1566.

75. Folmer, B.J.B., Cavini, E., Sijbesma, R.P., Meijer, E.W. *Chem. Commun.* 1998, 1847.

76. Hofmeier, H., El-ghayoury, A., Schenning, A.P.H.J., Schubert, U.S. *Chem. Commun.* 2004, 318.

77. Hofmeier, H., Hoogenboom, R., Wouters, M.E.L., Schubert, U.S. *J. Am. Chem. Soc.* 2005, *127*, 2913.

78. Kolomiets, E., Lehn, J.-M. *Chem. Commun.* 2005, 1519.

79. Xu, H., Hampe, E.M., Rudkevich, D.M. *Chem. Commun.* 2003, 2828.

80. Xu, H., Rudkevich, D.M. *Chem. Eur. J.* 2004, *10*, 5432.

81. Clark, T.D., Kobayashi, K., Ghadiri, M.R. *Chem. Eur. J.* 1999, *5*, 782.

82. Clark, T.D., Ghadiri, M.R. *J. Am. Chem. Soc.* 1995, *117*, 12364.

83. Bassani, D.M., Darcos, V., Mahony, S., Desvergne, J.P. *J. Am. Chem. Soc.* 2000, *122*, 8796.

84. Masuda, M., Jonkheijm, P., Sijbesma, R.P., Meijer, E.W. *J. Am. Chem. Soc.* 2003, *125*, 15935.

85. Wilson, A.J., Masuda, M., Sijbesma, R.P., Meijer, E.W. *Angew. Chem. Int. Ed.* 2005, 44, 2275.

14 Nonionic Block Copolymer Wormlike Micelles

You-Yeon Won and Frank S. Bates

CONTENTS

14.1 INTRODUCTION

Block copolymers are macromolecules composed of chemically distinct blocks that are covalently linked together. Molecular architectures that can be produced using the simplest combination of two types of monomers (A and B) include diblock (A-B) and triblock (A-B-A) copolymers. Chemical dissimilarity between the A and B blocks often confers an amphiphilic character to this class of materials. Particularly, there is a subset of block copolymers that contain both hydrophilic and hydrophobic blocks, and these compounds can be regarded as macromolecular analogs of conventional small molecule surfactants. A typical hydrophobic segment of a small molecule surfactant contains a linear hydrocarbon chain of oligomeric methylene (-CH$_2$-) of length usually less than twenty repeat units. Ethylene oxide (-CH$_2$-CH$_2$-O-) moieties commonly form the hydrophilic portion of many nonionic surfactants including the family referred to as C$_m$EO$_n$ (oligo(ethylene oxide) monoalkyl ether), where the subscripts m and n denote the numbers of methylene and ethylene oxide repeat units, respectively. The past decade has witnessed great advances in the investigation and application of polymeric surfactants thanks to the development of synthetic methods for the preparation of model poly(ethylene oxide)-poly(hydrocarbon) materials such as poly(ethylene oxide)-poly(styrene) (PEO-PS),[1] poly(ethylene oxide)-poly(ethyl ethylene) (PEO-PEE),[2] poly(ethylene oxide)-poly(isoprene) (PEO-PI),[3] and poly(ethylene oxide)-poly(butadiene) (PEO-PB).[4] Advances in the field of polymer chemistry, in fact, allow a wealth of other monomers to be used for the preparation of polymeric surfactants; see Refs. 5 and 6 for detailed reviews on the synthesis of macromolecular surfactants. Poly(acrylic acid)-poly(styrene) (PAA-PS)[7] and poly(acrylic acid)-poly(butadiene) (PAA-PB)[8] are two examples among many available in the category of ion-containing polymeric surfactants. Commercial products include PEO-based diblock and triblock copolymers having poly(propylene oxide) (PPO) as the hydrophobic component, which were developed more than 50 years ago.[9] New polymeric surfactants continue to emerge, expanding the arsenal of amphiphiles available to researchers and industrialists working in this field.

When dispersed in a selective solvent, typically water, which is compatible with one block but not with the other, asymmetric thermodynamic interactions lead to the formation of nanodomains with dimensions comparable to the molecular dimensions. Compared to small surfactants (i.e., C$_m$EO$_n$, lipids and soaps), the polymeric nature of macromolecular surfactants (in particular, the high molecular weight of the hydrophobic moieties) has the important consequence of enhancing the nanophase separation, as elucidated in the following discussion. According to Flory[10] and Huggins,[11] the free energy change per monomer G associated with molecular mixing of polymer chains (p) with a solvent (s) is given by

$$\frac{\Delta G}{k_B T} = \frac{\phi_p}{N_p}\ln\phi_p + \phi_s\ln\phi_s + \chi_{ps}\phi_p\phi_s \qquad (14.1)$$

where ϕ_p and ϕ_s are the polymer and solvent volume fractions, respectively, N_p is the number of repeat units in the polymer chain, and χ_{ps} is the dimensionless Flory-Huggins interaction parameter that is proportional to the enthalpy of mixing between the polymer segments and solvent species. The Flory-Huggins parameter has a critical value of 1/2 for solubilization of very long polymers. Recognizing that the first (entropic) term scales as $1/N_p$, while the third (enthalpic) term scales as χ_{ps} in the polymer-solvent mixture, one can expect on the same grounds that the hydrophobic effect, which causes self-assembly of the surfactant in water, will be enhanced with increasing molecular weight ($\sim N_p$) at a given choice of hydrophobic monomer (that is, at fixed χ_{ps}). As demonstrated throughout this chapter, large hydrophobic blocks, relative to the short oligomeric hydrophobes found in normal surfactants, are fundamentally responsible for many of the unique properties associated with polymeric surfactants.

Surfactant solutions offer a wealth of fascinating states of self-assembly at intermediate and low concentrations in water. Disordered phases, which occur at low surfactant concentrations have attracted the interest of many researchers, due in part to the occurrence of viscoelastic micellar phases, which find applications in a variety of industries. The viscoelastic surfactant phase usually contains long flexible entangled cylindrical aggregates often referred to as wormlike, threadlike, rodlike, or giant micelles. The polymer-like configuration of the micelles gives rise to unique rheological properties,[12,13] as is discussed in detail in Chapters 3, 4, and 8 to 10. Speculation regarding wormlike micelles in surfactant solutions dates as far back as the early 1950s.[14] A wormlike micelle morphology also has been associated with block copolymers in nonaqueous (organic) solvents for nearly three decades.[15-18] However, only within the last 11 years has the formation of nonspherical (elongated) micelles been demonstrated with water-compatible block copolymers,[4,7,19,20] and today a great deal of effort is being directed at the aqueous-based systems. In this chapter, we provide an overview of the state of knowledge concerning the properties and applications of polymeric wormlike micelles. Our emphasis is on wormlike micelles formed in water, as opposed to those occurring in organic solvents. These water-based systems are relevant to many advanced technologies; including several applications[21] described in Chapters 15 to 18. This chapter first reviews the basic micelle geometries commonly observed in dilute block copolymer solutions (which parallel those previously established with conventional surfactants) and the underlying principles that govern the micelle morphologies (Section 14.2). Then we discuss the attributes and properties of block copolymer micelles, while drawing distinctions between the behavior of macromolecular and conventional surfactants, at both single molecular and supramolecular levels (Sections 14.3–14.5). Topics that are discussed include conformations of the block copolymer molecules within the micelles, and nonergodicity in the micelle solutions due to the absence of molecular exchange between the micelles (Section 14.3); static and dynamic attributes of the wormlike micelles that occur as a consequence (Section 14.4); defect structures (including branches and swollen end caps) that further reflect the truly polymeric nature of the system (Section 14.5). Finally, this chapter summarizes recent examples of proofs of concepts for new applications of block copolymer-based wormlike micelles (Section 14.6). All these examples suggest a great potential for expanding the use of polymeric wormlike micelles along various technological fronts.

14.2 MICELLAR POLYMORPHISM IN BLOCK COPOLYMER SOLUTIONS

Amphiphilic self-assembly results in a variety of nanostructures in surfactant solutions: micelles and vesicles at low concentrations, and various liquid-crystalline morphologies in the high-concentration regime. Three basic structures can be created in the dilute limit: spheres, cylinders, and bilayers, dictated primarily by the ratio of the sizes of the hydrophobic and hydrophilic parts of the molecule. For instance, in the homologous series of $C_{12}EO_n$ surfactants, as the number n of ethylene oxide groups is systematically varied, the preferred micelle shape changes correspondingly; at room temperature, lamellar phases occur in dilute solutions of $C_{12}EO_3$ and $C_{12}EO_4$,[22] whereas $C_{12}EO_5$[23,24] and $C_{12}EO_6$[25-27] favor cylindrical micelles, and $C_{12}EO_7$ and $C_{12}EO_8$ form spherical micelles.[24] This well-documented sequence of morphological transitions from bilayer to cylinder to sphere is understood to be a result of the increase in the preferred interfacial curvature, which tends to reconcile the increased asymmetry between the excluded volumes of the hydrophilic and hydrophobic segments with increasing hydrophilic composition. Similar trends are seen for many other surfactants, regardless of whether the surfactants are nonionic or ionic; see Refs. 28–30 for an example of micelle shape variations among members of a set of homologous quaternized gemini surfactants. These basic micelle morphologies can be modeled using simple geometrical concepts that correlate molecular structure with interfacial curvature. By extending the molecular packing consideration of Tanford,[31] Israelachvili and co-workers developed a phenomenological description of the optimal micelle geometry in terms of the packing parameter defined as $v/l_o a$ where v is the volume of the hydrophobic chain, l_o is the maximum effective length of the hydrophobic chain, and a is the measured interfacial area per chain.[32] The preferred geometries are spheres for $v/l_o a \leq 1/3$, cylinders for $1/3 \leq v/l_o a \leq 1/2$, bilayers for $1/2 \leq v/l_o a \leq 1$, and inverted structures for $v/l_o a > 1$.[33]

A useful physical picture that serves as the basis for models including those of Tanford and Israelachvili involves a competition among various opposing forces such as interfacial tension, steric and/or electrostatic repulsions between adjacent hydrophilic head groups, and the entropic penalty for hydrophobic chain deformation necessary for uniform occupation of space.[34] Analogous considerations have been advanced to analyze the micellar aggregation behavior of non-charged block copolymers in selective solvents.[35] Three competing forces contribute to a free-energy balance: an interfacial free energy $F_{interface}$, which is an increasing function of the interfacial area per chain a, and free energies associated with the core and corona deformation (F_{core} and F_{corona}, respectively). Minimizing entropically unfavorable chain stretching favors an increase in a; this simple optimization leads to the selection of a specific morphological length scale and shape. Large polymer blocks actually simplifies the theoretical modeling of self-assembly due to statistical averaging over well-established conformational states that produce universal scaling relationships and asymptotic thermodynamic properties in the long chain limit.[36] Along this line, scaling theories with varying levels of simplifying assumptions have been put forward to evaluate the relationship between the size of the constituent block copolymer and the size (or the aggregation number) of the resulting micelle with a given choice of the micelle geometry, for example, sphere.[37–40] Scaling predictions are in reasonable agreement with experiments.[41] However, only recently have these theories been quantitatively extended to describe the thermodynamic mechanisms that discriminate between micellar geometries in block copolymer solutions. Notably, a semiempirical free-energy-balance argument of Rubinstein and co-workers[42] provides accurate predictions regarding the experimentally observed trends in the interfacial area per chain (a) (which increases as the micelle morphology is systematically varied from bilayers to cylinders to spheres) and the degree of hydrophobic chain stretching (which increases with the changes in the micelle morphology from bilayers to cylinders to spheres).[43] Additional discussion of the molecular packing behavior of self-assembled block copolymers is provided in Section 14.3. These findings add detailed (molecular-level) insights into the roles that the hydrophobic and hydrophilic chain conformations play in the selection of an optimal geometry. For cylindrical block copolymer micelles, minimization of the overall free energy pits hydrophobic chain stretching (which favors less interfacial curvature or a bilayer geometry) against hydrophilic chain stretching (which favors more interfacial curvature or a sphere shape) in concert with creating the minimum interfacial area. Wormlike micelles represent a delicate compromise between these opposing tendencies, leading to a much narrower window of compositions over which this morphology is found relative to bilayers (vesicles) and spheres (see Figure 14.1). Balancing between these geometric limits also likely contributes to a susceptibility to defect formation, particularly as the molecular weight increases, as discussed in Section 14.5.

Recently there have been a number of experimental reports exploring the relationship between the composition of the block copolymer and the structure of the resulting micelles. Occurrence of the traditional sequence of bilayers to cylinders to spheres with increasing the soluble block composition has been confirmed with a range of examples, including aqueous solutions of PEO-PB,[43,44] PEO-PEE,[43] PEO-PS[1] and PAA-PS,[19,45] and nonaqueous solutions of PI-PS in heptane (which is a selective solvent for PI).[42] In particular, the PEO-PB and PEO-PEE systems have been studied in detail by the present authors with the perspective that these materials can be regarded as direct polymeric extensions of the well-studied C_mEO_n surfactants. Figure 14.1 summarizes the micelle morphologies observed from two sets of PEO-poly(hydrocarbon) nonionic surfactants, each with systematically varying PEO molar masses at a fixed hydrocarbon chain size; these are macromolecular analogous of $C_{12}EO_n$ when n is varied between 3 and 8. Overall, the polymeric surfactants exhibit the three basic dispersed structures found with $C_{12}EO_n$ along with a few intriguing features that distinguish their behavior from the small surfactants. As displayed in Figure 14.1, in the polymer limit large micelle coexistence regions occur in two ranges of PEO compositions, one between the sphere and cylinder regimes, and the other between the cylinder and bilayer regimes. Furthermore, the bilayer-cylinder

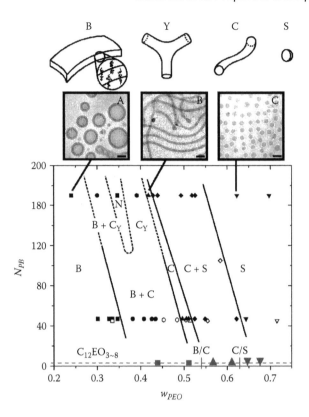

FIGURE 14.1 Summary of micelle morphologies observed using cryo-TEM with dilute aqueous solutions of PEO-PB diblock copolymers (reproduced with permission from Ref. 44). The regions of stability for spherical (S), cylindrical (C), branched (C_Y), network (N), and bilayer morphologies are given as functions of the degree of polymerization of the PB block (N_{PB}) and the weight fraction of the PEO block (w_{PEO}). The results were drawn from studies of mainly two sets of PEO-PB materials. In each set, the copolymers studied have different PEO compositions at a fixed PB molecular weight (i.e., $N_{PB} = 46$ or 170). The representative cryo-TEM images of the basic micelle geometries (S, C, and B) are shown in the insets (A through C); the scale bars correspond to 100 nm. For comparison, also shown are the data collected from the literature (cited in text), summarizing the room-temperature morphological behavior of the micellar solutions of $C_{12}EO_n$ surfactants containing varying numbers of ethylene oxide units (n = 3 to 8).

coexistence regime is characterized by complex hybrid morphologies that are stable over an extended period of time.[43] As discussed elsewhere,[43,44,46,47] such coexistence can be attributed to at least two factors. First, a lack of communication between micelles (referred to as nonergodicity) results in long-lived metastable structures. Second, polydispersity, an inevitable consequence of synthetic techniques, may contribute as well. However, the exact origins of the extended coexistence of different morphologies and the hybrid morphologies remain unresolved. An important clue comes from the fact that a huge energetic penalty is associated with exposing the hydrocarbon block to water (i.e., $\chi_{PB/water}$ ≈ 3.5 at 25°C[48]), which can completely inhibit the evolution of the morphological state of the system toward global equilibrium.[49] Another (perhaps the most) interesting manifestation of the far-from-equilibrium (nonergodic) nature of the PEO-PB micelles is the unexpected occurrence of complex network structures (i.e., the region denoted with an "N" on the morphology diagram) above a critical PEO-PB molecular weight,[44] which is discussed in more detail in Section 14.5. From Figure 14.1, it should also be noted that the bilayer-to-cylinder and cylinder-to-sphere boundaries are tilted toward lower values of the weight fraction of PEO (w_{PEO}) as the hydrophobic chain size (N_{PB}) is increased, and therefore one could observe a bilayer-to-cylinder or cylinder-to-sphere transition by increasing the overall molecular weight of the surfactant at a fixed PEO weight fraction, which is consistent with the picture that the hydrophilic (corona) chains are normally in a more extended conformation than the hydrophobic (core) chains.[35]

Many of the earlier studies performed by Eisenberg and co-workers regarding the micellar polymorphism achievable with polymeric surfactants involved water-compatible block copolymers containing PS as the hydrophobic component such as PEO-PS[1,50–52] and PAA-PS.[7,45,52,53] Unlike the PB and PEE materials discussed in the previous example, PS has a glass transition temperature (T_g) of about 100°C,[54] and the hydrophobic nanodomains formed by self-assembled PS chains are glassy (i.e., frozen) below that temperature;[55] note that the T_g's of PB and PEE are −12°C[56] and −20°C,[57] respectively. Advantageously, the mechanical stability of the glassy core makes the micelle structure nonlabile in response to removal of solvent and thus amenable to direct electron microscopic imaging under high vacuum. On the other hand, a glassy core makes it difficult to prepare a micelle dispersion by direct dissolution of the block copolymer in water. To circumvent this problem, the Eisenberg team has devised a processing method in which the copolymer is first dissolved in a cosolvent (i.e., a good solvent for both of the hydrophobic and hydrophilic segments such as N,N-dimethylformamide (DMF) or dioxane), and then nanoscale aggregation of the hydrophobic chains is induced by gradually replacing the original solvent with water. This solvent exchange procedure introduces additional factors (such as initial cosolvent/water ratio[58] and cosolvent chemistry[59]) which influence the resultant morphologies of the micellelike aggregates. In addition, in the cases involving ion-containing block copolymers such as PAA-PS or poly(4-vinyl pyridine)-poly(styrene) (PVP-PS), the charged character of the hydrophilic block enables various electrostatically relevant parameters to be used for controlling the aggregate morphology; demonstrated examples of such control parameters include pH[19,60] and ionic strength of the medium.[19,61]

14.3 STRUCTURES AND DYNAMICS OF BLOCK COPOLYMER MOLECULES IN MICELLE SOLUTIONS

14.3.1 MOLECULAR PACKING IN THE CORE DOMAIN

In surfactant solutions, micelle shape is determined by the value of the packing parameter $v/l_o a$, as introduced in Section 14.2. Consider that a similar ratio $v/R_c a$ based on the micelle core radius R_c numerically defines a micelle geometry (i.e., $v/R_c a = 1$ for bilayer, 1/2 for cylinder, and 1/3 for sphere). Then it is easy to see that the packing parameter $v/l_o a$ $(= (v/R_c a) \cdot (R_c/l_o))$ characterizes the preferred structure of the self-assembled amphiphilic molecule in terms of the closeness of its value to the geometric specification $(v/R_c a)$. Although the optimal interfacial area per chain a can only be determined empirically, the packing parameter $v/l_o a$ provides a useful tool for explaining *a posteriori* the observed interrelationship between micelle type and surfactant molecular structure. The critical values of the packing parameter that define the ranges of stability for the different micelle structures are derived purely from geometrical considerations of molecular packing structures (as discussed above), and therefore the packing criteria can readily be extended to explain the observed micelle geometries in block copolymer solutions. However, because of the coil-like conformation that the hydrophobic chain adopts in the core domain of a block copolymer micelle, the critical (stretched) hydrophobic chain length is an inappropriate reference length scale (l_o). Instead, the end-to-end distance of the hydrophobic chain in the unperturbed random-walk configuration (R_o), although somewhat arbitrary, is a more appropriate estimate of l_o within a numerical factor of the order of unity. The packing parameter can be equivalently expressed as $v/R_o a$ $(= (v/R_c a) \cdot (R_c/R_o))$ for block copolymers, and this parameter can be used for rationalizing the structure of the micelles.

There are limited data in the literature discussing the packing properties of the micellized block copolymers. Figure 14.2 presents a summary of experimental values of the packing parameter $(v/R_o a)$ and the degree of hydrophobic chain stretching (R_c/R_o) as functions of micelle morphology, extracted from cryogenic transmission electron microscopy (cryo-TEM) images of PEO-PB and PEO-PEE micelles.[43] We note that, due to the arbitrariness of the choice of the reference length scale (R_o), the ranges of the values of $v/R_o a$ for different micelle morphologies do not exactly match those of surfactants. The results summarized in Figure 14.2 provide valuable insight into how the hydrophobic block adjusts itself in response to changes in the hydrophilic chain length within and across the boundaries of the different morphologies. For instance, starting from the sphere regime, as the corona chain size is reduced, the core block becomes continually stretched in order to accommodate the decreased interfacial area per chain (a) under constant volume of the hydrophobic chain (v). This trend persists until the stretching ratio R_c/R_o reaches a threshold value (≈ 3). However, beyond this limit, the free-energy penalty associated with additional stretching of the hydrophobic chain exceeds the amount of free-energy reduction obtainable by a corresponding

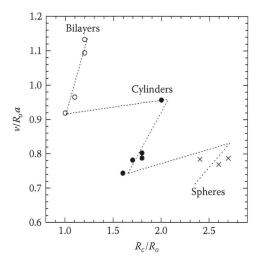

FIGURE 14.2 Summary of the observed interrelationships between the packing parameter (v/R_oa where v is the volume of the hydrophobic chain, R_o is the end-to-end distance of the hydrophobic chain in the unperturbed random-walk configuration, and a is the measured interfacial area per chain) and the degree of hydrophobic chain stretching (R_c/R_o where R_c is the micelle core radius) for the three basic micelle morphologies (i.e., sphere, cylinder, and bilayer) in dilute aqueous solutions of PEO-PB and PEO-PEE block copolymer surfactants. This figure was generated by modifying Figure 11 in Ref. 43. As explained in this paper, the values of v/R_oa and R_c/R_o were calculated on the basis of the micelle core dimensions determined from the cryo-TEM images of the block copolymer micelles. The dotted lines represent hypothetical paths that correspond to changes in the values of v/R_oa and R_c/R_o, within and across the boundaries of the different morphologies, as one reduces the corona block sizes at a fixed core block size starting from the sphere regime (or vice versa).

decrease of the interfacial area per chain. Further reduction of the corona molecular weight results in a shift of the preferred micelle geometry from sphere to cylinder, which significantly lessens the core block stretching for a given interfacial area per chain because of the reduced interfacial curvature. The same set of trends with continued decrease of the corona size repeats again within the cylinder region and also through the cylinder-to-bilayer transition. These observations regarding the core-block packing properties in block copolymer micelles are supported by similar findings in other systems (such as PI-PS in heptane[42] and PAA-PS in dioxane/water mixture[62]), and are also consistent with theoretical scaling explanations[42] and Monte Carlo simulation results.[63]

14.3.2 CORONA CHAIN CONFORMATIONS

The conformational properties of the corona chains are a long-studied subject in polymer and colloid science. Many of these studies have been performed in the context of polymer brushes attached to a surface having a preset geometry. In micelle solutions of block copolymers, the corona conformations are also fundamental to the macroscopic behavior of the micelles. A good illustrative example is the dramatic difference in the ordering behavior of the crew-cut versus starlike micelles at high concentrations; the former with shorter and steeper interactions forms face-centered cubic arrays, while the latter favors body-centered cubic structures.[64,65] When dissolved in a good solvent, polymer chains undergo global swelling, and the overall chain dimension scales as $N^{3/5}$ (that is, $R \sim N^{3/5}$ where R is the root-mean-square end-to-end distance of the chain, and N is the number of repeat units in the chain),[66] which is different from the one-half power law that corresponds to the random-walk configuration that the chains would assume in the solvent-free state.[67] When the solvated polymer chains are attached at one end to an impenetrable interface at high grafting density, the chains are further stretched to accommodate the imposed geometric constraints. The effects of the dense chain-end immobilization are usually considered in terms of the two contrary tendencies: osmotic stretching (due to excluded-volume interactions), and entropic resistance to the stretching.[35,68] Various theoretical methodologies have been developed to evaluate the structure of polymer brushes that would occur as a result of the balance of these forces. In connection with block copolymer micelles, rough estimates of how the conformation of the corona chains is influenced by the curvature of the grafting interface can, for instance, be achieved through the simple scaling arguments based on the concept of the correlation *blob*.[38,39,69] More exact treatments of this question on the basis of the self-consistent field (SCF) models are also possible and yield detailed information about the brush structures.[64,68,70–73] Predictions from these theories converge on an idea that polymer chains attached to planar surfaces (i.e., planar brushes) show parabolic segment-density profiles, while cylindrical and spherical brushes have more rapidly decaying (hyperbolic) density profiles. The theoretical predictions regarding the properties of planar brushes have received numerous confirmations from experiments (such as those based on neutron[74] and x-ray[75] reflectivity and surface force measurements[76]) and computer simulations.[77] As model systems for studying curved brushes, block copolymer micelles have been extensively investigated via small-angle neutron and x-ray scattering (SANS and SAXS, respectively); see, for example, Ref. 78 for a recent review on this subject. These results establish reasonable, albeit mostly qualitative, agreement between theory and experiment on the diffusely decaying character of the radial segment density distribution in spherical brushes[79–82]; however, little experimental data is available on the structure of cylindrical brushes except for the SANS study by the present authors,[81] which is discussed below in the context of the peculiarity shown by PEO brushes.

The most important example of a polymer commonly encountered in the form of a polymer brush is poly(ethylene oxide) (PEO) (also referred to as poly(ethylene glycol) (PEG) or poly(oxy ethylene) (POE)). When tethered to a surface as a brush,

PEO exhibits a unique ability to protect the surface from unwanted adsorption of other macromolecules (such as proteins). For this reason, surface modification by PEO grafting has become a common practice for creating bioinert interfaces for pharmaceutical compounds and biomedical implants.[83–85] Scientifically, exploring the possible role that the conformation of PEO plays in creating this useful property has been a subject of great interest to the polymer community. In the unbound state, coarse-grained trajectories of monomers along the PEO chain in water exhibit a self-avoiding random-walk statistics (characteristic of a swollen chain in a good solvent).[86] On the local length scale, however, the dissolved state of PEO involves a configurational rearrangement of the polymer segments in coordination with the surrounding water molecules.[87,88] The structural (i.e., counter-entropic) nature of the hydrophilic interaction that PEO has with water is well manifested in the lower-critical-solution-temperature (LCST)-type phase behavior of PEO in water; that is, PEO becomes insoluble in water at temperatures above the LCST, that is, when the thermal energy disrupts the specific configuration of the polymer, which enables hydrophilic interaction with water.[89,90] Another related property of PEO, which has a fundamental influence on the structure and functioning of PEO especially under the brush-like arrangements, is amphiphilicity. The effect of the amphiphilic character of PEO has long been evidenced by the adsorption of the polymer at the air-water interface, with both homopolymer[91,92] and block-copolymer[93,94] forms of this macromolecule. Similarly, with micellar PEO brushes the amphiphilicity of the ethylene oxide segments can cause the end-grafted PEO chains to (partially) collapse from the aqueous solution toward hydrophobic surfaces. A water ^{17}O spin relaxation study on the state of water in micelle solutions of $C_{12}EO_8$ surfactants first provided indirect evidence for the formation of a compact layer of the oligo(ethylene oxide) surrounding the micelle core,[95] and this observation inspired new thermodynamic models based on the assumption of dualistic (i.e., hydrophilic vs. hydrophobic) states of the ethylene oxide units toward explaining the various phenomena originating from the amphiphilicity of PEO.[96] As predicted by an SCF theoretical study, the interfacial adsorption of the end-grafted PEO has an important ramification to, for instance, how the molecular weight of the PEO impacts its protein adsorption resistance.[97] Experimental data that give direct confirmation of the adsorption of the coronal PEO chains at the micelle interfaces have been reported more recently. Combined techniques of selective deuterium labeling and contrast-matching SANS analysis have been applied to the determination of the segment density distribution of the coronal PEO brushes of the micelles formed from PEO-PB diblock copolymers, and the results indicate that the conformational behavior of the micellar PEO brushes deviates over the entire corona length scales from what would be anticipated for equivalent brushes without attractive interactions with grafting surfaces (Figure 14.3).[81] The corona domain of these micelles (in which the PEO segments are expected to be highly accumulated toward the hydrophobic interface) has even been directly visualized using cryo-TEM (Figure 14.4).[43,98]

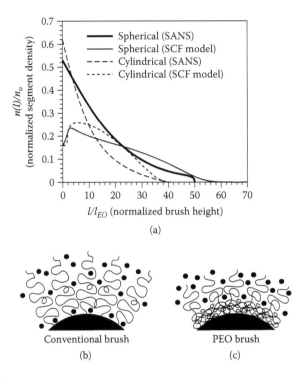

FIGURE 14.3 (a) Segment distributions of the micellar brushes of PEO attached to the hydrophobic surfaces of the spherical and cylindrical micelles formed in water by PEO-PB diblock copolymers (reproduced with permission from Ref. 81). Here, $n(l)$ denotes the position-dependent number density of PEO segments, and n_o is the melt density of PEO under solvent-free condition at room temperature (≈ 15.4 nm^{-3}). The distributions are displayed as functions of the normalized distance (l/l_{EO}) where l is the distance from the surface of the micelle core, and l_{EO} is the size of the ethylene oxide (EO) repeat unit (≈ 3.6 Å). In the figure, the density profiles determined from SANS measurements (thick curves) are compared to the chain conformations that nonadsorbing PEO brushes would assume under comparable brush conditions (thin curves), as predicted from a self-consistent field (SCF) numerical analysis. (b and c) Schematic demonstration of the difference in the brush conformation between the two stated cases.

14.3.3 MOLECULAR EXCHANGE BETWEEN MICELLES[99]

Micelles are dynamic entities that continuously undergo thermally driven changes in physical makeup. After formation, the micelles can disintegrate into smaller units or even into unimeric molecules, and during the lifetime of a micelle the constituent molecules constantly enter and escape from the micelle structure.[100] Therefore, the time scales associated with these dynamic processes are essential parameters in defining the overall characteristics of a micellar system. Two distinct mechanisms for molecular exchange between micelles are documented

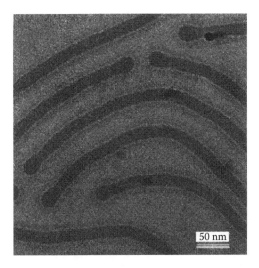

FIGURE 14.4 A representative cryo-TEM image recorded from a 1 wt% aqueous solution of a PEO-PB diblock copolymer (with number-average block molecular weights of $\bar{M}_{n,PEO}$ = 7,500 g/mol and $\bar{M}_{n,PB}$ = 5,600 g/mol), revealing two interesting morphological details: clearly resolved core-corona structures and spherical end caps (reproduced with permission from Ref. 98).

in the surfactant literature: (i) unimeric exit and entry of the molecules ($M_i \leftrightarrow M_{i-1} + M_1$ where M_i designates a micellar aggregate containing i surfactant molecules)[101–103] and (ii) micelle fusion and fission ($M_i + M_j \leftrightarrow M_{i+j}$).[104,105] These ideas (originally developed in the surfactant field) also served as the inspiration for numerous theoretical and experimental explorations of analogous mechanisms for the component exchange of block copolymer micelles.[106] A scaling theory predicts that, in dilute micelle solutions of block copolymers, the activation free energy is lower for the unimolecular exchange mechanism.[107] Many experimental kinetic data (e.g., those obtained from fluorescence nonradiative energy transfer (NRET) probing[55,108–110] and temperature-jump results[111]) have been successfully analyzed on the basis of the models derived from the premise of unimeric exchange.[112] However, it has also been reported that, in micelle solutions of PEO-PS in methanol/water mixture, for instance, the two exchange mechanisms (i.e., unimer exchange and micelle fusion/fission) are both operative in combination.[110,113]

These experiments studying the exchange kinetics in polymeric micelles were typically based on certain sets of choices for the polymer/solvent chemistries, which are adjusted to make the exchange processes observable within the experimentally accessible time scales. However, in many other cases, block copolymer micelles exist in the mutually isolated state in which the intermicellar redistribution of molecules is practically absent. One such situation occurs when a high-T_g polymer is used as the core-forming block. Although little is known about how the nanoscale confinement imposed by the micellar aggregation would quantitatively impact the otherwise glassy character of the core block, evidence is abundant that, at ambient temperature, no chain exchange occurs with micelle cores composed of high-T_g materials such as PS[114] ($T_g \approx 100°C$), PMMA[115] ($T_g \approx 100°C$), poly(t-butyl acrylate) (PtBA)[116] ($T_g \approx 45°C$), and poly(DL-lactic acid)[117]

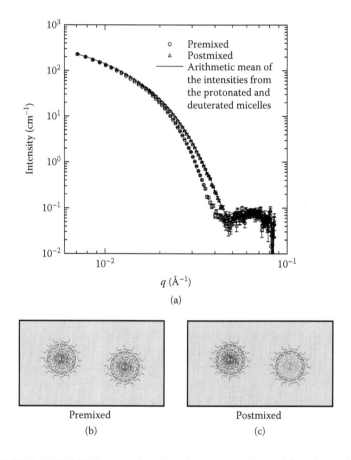

Premixed   Postmixed
(b)   (c)

FIGURE 14.5 (a) SANS data corroborating the absence of material exchange between individual PEO-PB wormlike micelles (reproduced with permission from Ref. 49). For this study, aqueous solutions containing equimolar mixtures between a pair of hydrogenated and deuterated PEO-PB diblocks (having an identical molecular weight and composition) were prepared in two different ways: mixing of the two polymers before ("premixed") or after ("postmixed") hydration. The SANS profile taken from an 8-day-old postmixed solution exactly coincides with the intensity profile constructed by taking an average between the intensities obtained separately from the protonated and deuterated micelles. Also shown are schematics illustrating the states of mixing between two diblock copolymers having normal versus deuterated core blocks for the (b) premixed versus (c) postmixed micelles in the absence of intermicellar molecular redistribution.

($T_g \approx 60°C$). In other situations when the block copolymer has a negligible solubility in a solvent (i.e., when the critical micellization concentration (cmc) is nearly zero), the micelle exchange can also be completely inhibited (in this case, thermodynamically) even with strictly nonglassy cores. Within the monomolecular exchange picture of Aniansson and Wall[101,102] the residence time of a surfactant molecule within a micelle is estimated to scale exponentially with the negative of the free-energy change of the micellization (see Chapter 10, Equation 10.10). Remember from Equation (14.1) that, even with a marginal incompatibility between the monomer and the solvent (i.e., a small value of the χ_{ps} parameter), the material in the form of polymer may become insoluble in the solvent when the molecular weight (N_p) is sufficiently large. In the extreme case of the polymeric equivalent of the C_mEO_n surfactants, where the molecules are highly amphiphilic with the enthalpy change of hydration of the methylene segment (-CH$_2$-) being a few k_BT at room temperature,[48] the nonsoluble behavior can become effective at relatively low molecular weight. From the literature values for the standard free energies of micelle formation (ΔG_{mic}(-CH$_2$-) $\approx -1.15k_BT$, and ΔG_{mic}(-CH$_2$-CH$_2$-O-) $\approx +0.32k_BT$)[118] and the residence time of C_8EO_8 in a micelle ($\tau \approx 1.0$ µs at 25°C),[119] one can extrapolate that, once micellized, PEO-PB with 66 ethylene oxide and 46 butadiene units, for instance, is likely to be locked in within the confine of a micelle indefinitely. In this strongly amphiphilic limit, the unimolecular dispersion of block copolymer molecules is quite possible. Formation of unimolecular micelles might occur through collapse of individual hydrophobic blocks into globules that are surrounded by a corona layer of the soluble block.[120] However, studies suggest that, even under the presence of unimeric micelles, the insolubility of the core-forming block normally defies any micelle exchange mechanisms.[121] As shown in Figure 14.5, time-resolved SANS experiments have confirmed that the PEO-PB molecules having the aforementioned size and composition indeed become (permanently) confined within the individual micelles separated by water; similar observations also have been reported for other types of hydrophobic block chemistries (e.g., poly(ethylene propylene) (PEP)[122] and plasticized poly(methyl methacrylate) (PMMA)[123]). Yet it remains a question how far a micellar dispersion produced by the simple (mechanical) mixing of the block copolymer and the solvent would be from thermodynamic equilibrium. In systems of kinetically and/or thermodynamically frozen micelles, the micelle structure and properties become critically dependent upon the sample preparation procedures as amply demonstrated by the work of Eisenberg and co-workers.[1,7,8,19,45,58–61]

14.4 STRUCTURAL AND MECHANICAL PROPERTIES OF BLOCK COPOLYMER WORMLIKE MICELLES

The static appearance of the wormlike micelles in amphiphilic solutions closely resembles that of semiflexible polymers in good solvents. Just like the solutions of molecular polymers, the macroscopic properties (such as rheological properties[124,125] and lyotropic phase behavior[126,127]) of the solutions containing polymer-like micelles are mainly influenced by a few key parameters that statistically characterize the structural and mechanical properties of the polymer-like objects, such as average size, size distribution, and stiffness, which are the topics of this section.

In surfactant wormlike micelles, the continuous random breakage and reformation of the polymer-like structures (and also possibly the condensation and evaporation of unimers to and out of the micelles) give rise to an inherent nonuniformity of the micelle lengths. Under the thermal equilibrium assumption, a mean-field theory predicts a single-exponential decay profile in form for the micelle length distribution: that is, $c(\tilde{L}) \propto \exp(-\tilde{L}/\tilde{L}_o)$ where $c(\tilde{L})$ is the number concentration of micelles of dimensionless length $\tilde{L}$, and $\tilde{L}_o$ is a reference length defined as $\tilde{L}_o \approx \phi^{1/2} \exp(E/2k_BT)$ where ϕ the total volume fraction of the micelles, and E is the scission energy of the cylindrical micelle (as needed for creating two new hemispherical micelle ends).[12] In many block copolymer wormlike micelles, the micelle sizes and size distributions are expected to be heavily influenced by the procedures adopted for the sample preparation due to the "nonlivingness" of the micelles formed. A recent study with wormlike micelles produced via film rehydration of a PEO-PB diblock copolymer indicates that the resultant micelle size distribution is reasonably represented by the above-mentioned exponential function, and the fitting of the measured distribution to the theoretical profile leads to an estimate of the scission energy $E \approx 26\ k_BT$.[128] This value is lower by more than an order of magnitude than the energy associated with generating the additional interfaces as a result of the scission, that is, $E \approx \gamma \cdot (4\pi/3)R_c^2 = 1.3 \times 10^3\ k_BT$, estimated on the basis of the experimental values of the interfacial tension ($\gamma \approx 0.027$ J/m^2)[129] and the radius of the micelle core ($R_c \approx 7$ nm).[4] The populations of different contour lengths in the PEO-PB wormlike micelles are far from their equilibrium states; the measured micelle sizes (which were likely quenched during the early stages of the film hydration process) generally are much shorter than what would be expected when the systems is in thermodynamic equilibrium.

The stiffness characteristics of long chain-like objects (such as polymers and wormlike micelles) in solution are typically quantified in terms of the persistence length l_p, defined as the orientational correlation length between two unit tangent vectors along the contour of the chain.[130] Using this parameter, one can write the root-mean-square end-to-end distance (R) of a semiflexible (Kratky-Porod) chain as

$$R = \sqrt{2l_p^2(L/l_p - 1 + \exp(-L/l_p))} \tag{14.2}$$

where L is the chain contour length (i.e., $L = Nb$ in the polymer analogy where N is the number of monomers and b is the monomer size).[130] For our discussion later in this section of the determination of the persistence length by scattering, it is useful to note that in this equation there are two asymptotic limits of the mass scaling of the coil size: when $L \gg l_p$ (i.e., in the ideal chain limit), $R \approx \sqrt{2l_pL} \sim N^{1/2}$, whereas $R \approx L \sim N$, when $L \ll l_p$ (i.e., in the rodlike limit). That is, for length scales smaller than l_p, the local segment of a semiflexible chain can be described as a rigid rod, and the global chain conformation on length scales much greater than l_p follows the random-walk statistics. Conceptually, how long the orientational correlation persists along the contour line depends on the inherent rigidity of the elongated structure against local bending caused by the thermal fluctuation, and therefore the persistence length can be related to the bending rigidity modulus (κ) by the equation $l_p = \kappa/k_B T$.[130]

There are a few different methods that have been demonstrated for experimentally determining the persistence length of wormlike micelles. In fact, the flexible nature of surfactant wormlike micelles was first conjectured on the basis of the intrinsic viscosity data (obtained from aqueous solutions of dodecyl ammonium chloride with added NaCl) which showed a micelle size dependence of the intrinsic viscosity that is different from those anticipated for rigid prolate and oblate ellipsoids.[131] In principle, one can apply the numerical/semiempirical formalism derived by Yamakawa and co-workers for the intrinsic viscosity ($[\eta]$), hydrodynamic radius (R_h), and gyration radius (R_g) of semiflexible chains in a theta solvent[132,133] to the determination of the persistence length of wormlike micelles from combined intrinsic viscosity (or dynamic light scattering (DLS)) and static light scattering (SLS) measurements. In practice, however, the uncertainty associated with the polydispersity in the size of the surfactant wormlike micelles prohibits any accurate determination of the persistence length through analysis of the interrelationships among the measured average quantities of $[\eta]$, R_h, R_g, and $\bar{M}_w$, as has been pointed out by Schmidt.[134] Also, the concentration-dependent variation of the micelle size complicates the interpretation of experimental data from the combination of intrinsic viscosity and static light scattering measurements to obtain estimates of $[\eta]$ and $\bar{M}_w$, which requires an extrapolation of the corresponding data toward zero concentration. Nevertheless, these methods are expected to have great utility for accurate evaluation of the persistence length of nonliving wormlike micelles such as those formed in water by strongly amphiphilic polymer surfactants (e.g., PEO-PB), similar to what has been demonstrated with monodisperse supramolecular cylindrical brushes prepared by the vulcanization of the PI core domain of the nanofibril composed of PS-PI diblock copolymers and subsequent isolation of a fraction of the nanofiber material by centrifugation.[135,136]

A more precise measurement of the micelle stiffness is possible using small-angle scattering techniques. In dilute micelle solutions in which the individual micellar objects are on average far apart from each other, the isotropic scattering intensity ($I(q)$) measured as a function of the magnitude of the scattering wave vector q (defined as $q \equiv (4\pi/\lambda)\sin(\theta/2)$ where λ is the wavelength of the radiation, and θ is the scattering angle) over the q range $1/R < q < 1/b$ (where R represents the chain end-to-end distance, and b is the monomer size) is dominated by intrachain interference effects. In the ideal chain limit of a semiflexible chain in which $l_p < r < R$ (i.e., $1/R < q < 1/l_p$ due to the conjugate nature of momentum transfer (q) and real-space distance (r)), the mean-field monomer-monomer pair correlation ($g(r)$) within a volume ($\sim r^3$) containing n monomers scales as $g(r) \sim n/r^3 \sim 1/r$, and therefore the Fourier transform of $g(r)$ which is equivalent to the scattered intensity due to the monomer-monomer scattering interference follows the scaling behavior $I(q) \sim q^{-2}$.[66] On the other hand, on local length scales corresponding to the rodlike limit where $b < r < l_p$ (i.e., $1/l_p < q < 1/b$), the scaling relations become $g(r) \sim n/r^3 \sim 1/r^2$ and $I(q) \sim q^{-1}$. Therefore, by analyzing the scattering data collected at the appropriate (i.e., intermediate) q range, one can determine the persistence length l_p which inversely corresponds to the value of q at which the q^{-2} to q^{-1} scaling crossover occurs. It should be noted that, as opposed to the intrinsic viscosity/DLS and SLS (which typically requires experiments to be performed in the Guinier (i.e., low-q) scattering regime) combination, a great advantage of using the intermediate-q scattering for probing the micelle stiffness is that, in this q range on the order of $\sim 1/l_p$, the intensity profile becomes largely insensitive to the overall size of the micelle, and therefore the measurement becomes unaffected by polydispersity. The analysis of scattering in the medium q range normally involves the fitting of the scattering data using theoretical models for the intensity profile such as the full numerical model proposed by Yoshizaki and Yamakawa.[137]

Wormlike micelles of various surfactants have been investigated, in particular, using SANS (the use of deuterium labeling makes it possible to measure scattering from dilute solution samples with reasonable statistics), and the typical l_p values obtained from these measurements range from 10 to 40 nm: for example, $l_p \approx 17$ nm for nonionic $C_{16}EO_6$ in water,[138] 28 nm for sodium 1-hexadecane sulfonate $NaC_{16}SO_3$ in water with no added salt,[138] 19 to 34 nm for mixed cetyltrimethylammonium 2,6-dichlorobenzoate/cetyltrimethylammonium chloride in water in the presence of sodium 2,6-dichlorobenzoate/NaCl,[139] 18 nm for cetylpyridinium bromide (CPyBr) in 0.8 M NaBr brine,[140] 11 to 21 nm for sodium dodecyl sulfate in water with added NaCl,[141] 12 nm for lecithin in cyclohexane,[142] and 15 nm for lecithin in isooctane.[143] In the case that the stiffness of self-assembled wormlike micelles is dictated by interfacial tension (γ), dimensional analysis suggests that the persistence length of a cylindrical micelle has the following dependence on the diameter of the micelle core (d): $l_p = c\gamma d^{4-D}/k_B T$ where c is a numerical constant, and D is the dimensionality of the geometry (i.e., $D = 1$ for cylinder),[128] suggesting that, from a comparison with the data obtained from $C_{16}EO_6$ (i.e., $l_p \approx 17$ nm for $d \approx 3$ nm),[138] the wormlike micelles of PEO-PB having $d \approx 15$ nm (i.e., such as those shown in Figure 14.4) are expected to be stiffer than their surfactant analogs by about two orders of magnitude. This calculation indicates that the transition in the scattering from the ideal chain to the rodlike behavior occurs in the q range between 10^{-4} and 10^{-3} nm^{-1} ($\sim 1/l_p$), which in fact lies beyond the typical q range of 0.05 to 1 nm^{-1} accessible in conventional SANS.[144] Recently, the ultra small-angle neutron scattering (USANS) technique[145] has been used to probe the conformation of the wormlike micelles formed in water from the PEO-PB diblock copolymer, and a least-squares analysis of the resultant USANS profile

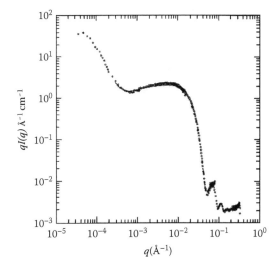

FIGURE 14.6 Combined USANS and SANS data taken from 1 wt% PEO-PB ($\bar{M}_{n,PEO}$ = 2,500 g/mol and $\bar{M}_{n,PB}$ = 2,400 g/mol) wormlike micelles in D$_2$O (reproduced with permission from Ref. 81). The data presented in the bending plot (i.e., $\log(q \cdot I(q))$ vs. $\log(q)$) format display the transition between the coil-like (i.e., $I(q) \sim q^2$) to rodlike (i.e., $I(q) \sim q^1$) scaling behavior. The scattering data were analyzed using the Yoshizaki-Yamakawa model over the intermediate q regime (i.e., $1/R < q < 1/b$ where R is the end-to-end distance of the chain, and b is the monomer size), yielding an estimate for the persistence length of $l_p \approx 570$ nm; the solid curve represents the model fit to the data.

using the Yoshizaki-Yamakawa model (Figure 14.6) gives an estimate of $l_p \approx$ 570 nm,[81] which is about thirty times that of the C$_{16}$EO$_6$ wormlike micelles. The exact same system has been examined via fluorescence imaging of the PEO-PB wormlike micelles confined to a 1-μm gap between two glass slides.[128] In strictly two-dimensional (2-D) space, correction for taking into account the difference in dimensionality (i.e., $l_{p,2-D} \approx 2l_p$)[146] yields the 2-D equivalent Kratky-Porod relation for the root-mean-square end-to-end distance:[147]

$$R_{2D} \approx \sqrt{4l_p^2(L/l_p - 2 + 2\exp(-L/l_p))} \qquad (14.3)$$

By comparing the time average of the square of distances between micelle ends measured from time-resolved video images (such as those shown in Figure 14.7) to the above relation, the persistence length of the PEO-PB wormlike micelles is estimated to be $l_p \approx 500$ nm,[128] which is in excellent agreement with the USANS result (i.e., $l_p \approx 570$ nm).

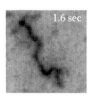 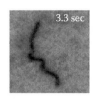

FIGURE 14.7 Time-resolved fluorescence images of the wormlike micelles of PEO-PB ($\bar{M}_{n,PEO}$ = 2,500 g/mol and $\bar{M}_{n,PB}$ = 2,400 g/mol) confined between two glass plates that are separated by ~1 μm (reproduced with permission from Ref. 128). Comparison of the average of the end-to-end distances of the wormlike micelle fluctuating in time (in relation to its contour length) to the 2-*D* Kratky-Porod relation (Equation 14.3) leads to an estimation of $l_p \approx 500$ nm for the persistence length. Scale bar corresponds to 5 μm.

14.5 DEFECTIVE WORMLIKE MICELLES

Branching and network formation were first proposed as a possible mechanism that enables the phase separation of wormlike micelles in ionic surfactant/brine mixtures.[148] The same idea was also adopted in explaining the observed concentration dependences of linear viscoelastic responses in some systems, such as cetylpyridinium chlorate/sodium chlorate brine (CPyClO$_3$/ClO$_3$Na/H$_2$O)[149] and hexadecyltrimethylammonium bromide/potassium chloride/water (CTAB/KCl/H$_2$O)[150] in the high salt concentration limit, which significantly deviate from what is expected from the known dynamic behavior of linear living wormlike micelles in the semidilute concentration regime.[151] To date, many cryo-TEM images have been reported in the literature that confirm the occurrence of branched wormlike micelles in both mixed[152–154] and single-component[23,155] surfactant solutions. Purely on the grounds of surfactant molecular packing, the saturated Y-junction morphology is estimated to be energetically stable as an intermediate structure between the simple cylinder and perforated lamellar phases.[148] A full thermodynamic description of the branching phenomenon has been presented more recently, which explains that the branching (i.e., the welding between a micelle end and the cylinder body to form a Y-junction) is favored when the energy lowering associated with removing an end defect at the expense of generating a new branch junction exceeds the corresponding entropic loss.[156,157] (See also Chapter 2, Section 2.6.) This conceptual framework reveals that micelle branching has interesting parallels to the conventional phase transition from gas (cylinder end) to liquid (Y-junction) in molecular systems,[157] providing an explanation for the formation of phase-separated network structures in analogy to the condensed liquid phase coexisting with the gas phase under subcritical conditions. Network formation and phase separation have been confirmed experimentally in both micellar solutions of surfactants[23] and ternary water/oil/surfactant microemulsions.[158,159]

In many cases involving block copolymers, the above-described thermodynamic explanation regarding the branched micelle formation loses its applicability because of the absence of molecular exchange in block copolymer micelles.

Within the isolated (nonergodic) micelle structure, however, the redistribution of constituent block copolymer molecules is still feasible, and this allows the individual micelles to achieve interfacial geometries that optimally accommodate the specific molecular structures. Parallel to the equilibrium gyroid morphology occurring in block copolymer melts (in which the three-dimensional cubic networks of the minority domain contain threefold connectors with an interfacial curvature intermediate to those of the cylindrical and lamellar structures),[160] the Y-junction morphology has also been reported in micellar solutions of block copolymers. For example, short-armed Y-junction structures have been documented with the crew-cut micelle-like aggregates of PEO-PS formed in water/DMF mixtures via solvent exchange,[52] and long-branched cylinders with multiple three-armed junctions have been visualized by TEM in studies of methanol solutions of poly(glyceryl methacrylate)-poly(2-cinnamoyloxyethyl methacrylate)-poly(allyl methacrylate) (PGMA-PCEMA-PAMA)[161] and poly(t-butyl acrylate)-poly(2-cinnamoyloxyethyl methacrylate)-poly(isoprene) (PtBA-PCEMA-PI)[162] triblock copolymers. Recently, systematic investigations of the micelle morphologies in a set of model PEO-PB diblock copolymers with varying PEO compositions at a constant PB molecular weight of 9,200 g/mol revealed that the threefold junction indeed appears as a stable structural element in the PEO composition range between the cylinder and bilayer regimes (Figure 14.1).[44] As shown in Figure 14.8, the detailed structural features of the branched cylindrical micelles, such as planar Y-junction cores and swollen spherical end caps, are strikingly similar to the morphologies predicted for microemulsion networks formed by ternary water/oil/surfactant microemulsions; a major contributing factor to the surfactant interfacial geometry is the Helfrich's curvature energy.[163] As displayed in the micelle morphology diagram for the PEO-PB diblock copolymers

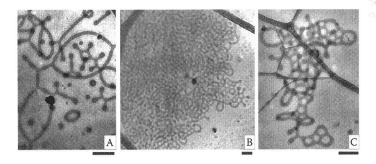

FIGURE 14.8 Representative cryo-TEM micrographs taken from aqueous solutions containing 1 wt% PEO-PB diblock copolymers with (A) $\bar{M}_{n,PEO}$ = 5,900 g/mol and $\bar{M}_{n,PB}$ = 9,200 g/mol (w_{PEO} = 0.39) and (B and C) $\bar{M}_{n,PEO}$ = 4,700 g/mol and $\bar{M}_{n,PB}$ = 9,200 g/mol (w_{PEO} = 0.34) (reproduced with permission from Ref. 44). (A) Branched wormlike micelles with Y-junctions, swollen end caps, loops, and undulating cylindrical structures. (B and C) Dense network micelles which undergo macroscopic phase separation from the aqueous phase. Scale bars are 200 nm.

(Figure 14.1), the preferred interfacial curvature decreases with decreasing PEO composition, leading to a network that resembles interconnected cylinders at the center of the cylinder-to-bilayer transition region. This sequence agrees with recent theoretical predictions for the sequence of morphologies around the network state in microemulsions.[159] In block copolymer network micelles (Figure 14.8), tightly arrayed Y-junction clusters give rise to small regular cylindrical loops that are often arranged locally in a threefold symmetric pattern (reminiscent of the metastable hexagonally ordered perforated lamellar morphology observed in block copolymer melts[164]). Interestingly, when the overall size of PEO-PB is reduced by a factor of about 4 (i.e., in the morphology sequence observed with the lower molar mass series of PEO-PB diblocks in Figure 14.1), the compound branched morphologies disappear throughout the cylinder-bilayer transition, suggesting that, below a certain molecular weight, the conformational degree of freedom of the core block chain becomes insufficient to overcome the molecular packing stress that would be imposed by the Y-junction formation. It should also be noted that, unlike the equilibrium branching in surfactant wormlike micelles, the absence of micelle recombination in the PEO-PB-based systems excludes any thermodynamically driven post-micellization branching processes, which require fusion between a cylinder end and the body of an already formed cylindrical micelle.

14.6 UNIQUE APPLICATIONS OF BLOCK COPOLYMER WORMLIKE MICELLES

14.6.1 DRUG DELIVERY

In recent years, block copolymer micelles have received expanded attention from scholars in pharmaceutical sciences; detailed discussions of current developments in this field can be found in many recent review articles.[165-169] Incorporation of bioactive substances (such as drugs or nutrients) into the hydrophobic core domain of the micelle offers a convenient means of obtaining stable aqueous dispersions of the therapeutic compounds that are otherwise nondispersible in water. In particular, block copolymer micelles possess many advantageous qualities that favor their use over surfactant micelles as carriers of pharmaceutical agents, summarized as follows (see also Chapter 18, Section 18.5):

1. As discussed in Section 14.3, high molar-mass block copolymers are normally incapable of escaping from the micelle structure, even at infinite dilution, and are therefore safe and inert in a biological setting. In contrast, ordinary (small) surfactants can freely move away from the micelles as unimers and interact, for instance, with proteins and lipids in the surrounding medium, at times resulting in unwanted protein denaturation,[170-172] membrane disruption and cellular lysis;[173-175]
2. A wide shell of the swollen corona chain (typically PEO) provides the polymeric drug carriers with stability against interactions with charged

cell/tissue surfaces and also resistance to nonspecific adsorption of proteins, enabling prolonged circulation and release of the encapsulated drugs;[83,176-178]

3. Larger core volumes have correspondingly higher payload capacities for solubilization of hydrophobic drugs and imaging dyes than their surfactant analogs;

4. With widened choices for the core block chemistries, the polymeric approach affords new strategies for fine-tuning the kinetics of release of encapsulated drugs. Polymer-specific properties that can be readily incorporated for modification of the release characteristics include biodegradability (which is achievable with polymers containing depolymerizable moieties along the backbone chain such as poly(D,L-lactic acid-co-glycolic acid) (PLGA),[179] poly(caprolactone) (PCL),[180] and poly(lactic acid) (PLA) in both semicrystalline poly(L-lactic acid)[181] and glassy poly(D,L-lactic acid)[182] forms), thermo- and/or pH-responsive cores (for which poly(propylene oxide) (PPO), poly(N-isopropylacrylamide) (PNIPAAM), and also the above-mentioned polymers (i.e., PLA, PLGA and PCL) are well-documented examples),[183,184] and internally phase-segregated hydrophobic nanodomains (of which the basic principles have recently been demonstrated through the use of multicomponent block copolymers comprising at least two hydrophobic blocks that are mutually immiscible).[65,185]

Along these lines, most of the prior research has concentrated on demonstrating the applications of spherical micelles. Recently these studies have been extended to exploring the utility of wormlike micelles of block copolymers in drug delivery technologies.[186,187] Although the capacity for drug loading per mass of the carrier is unlikely to be much affected by the geometry of the delivery vehicle, a possible benefit of using wormlike micelles (as opposed to spherical micelles) is that this elongated structure may provide an additional mode of controlling the biodistribution and release profile of the delivery systems. For example, biodegradable cores have been demonstrated for allowing a time-dependent shortening of long wormlike micelles, which in turn greatly impacts the dynamics of the release of the loaded material.[187] However, it remains to be further investigated whether the wormlike micelles (which are typically many micrometers long in the as-hydrated state with a broad distribution of sizes)[4,43] can pass through or beyond the smallest capillaries of the body's circulation system to arrive at the desired target sites; there are references to the minimum capillary diameters of a few microns for humans and other mammals,[188,189] but unhindered navigation through other parts of the body (for example, the passage across the blood-brain barrier) requires the drug carriers to be as small as a few nanometers in diameter.[190,191] Conversely, a cylindrical geometry may actually facilitate transport through microcapillaries due to localization along the center streamlines thereby reducing the probability of entrapment in stagnant regions of the lumen.[128] Moreover, functionalized wormlike micelles should display a greatly enhanced tendency to bind to tissue.

14.6.2 Organic Nanofibrils via Crosslinking

Using the combined procedures of amphiphilic self-assembly (to form a desired nanostructure) and subsequent chemical treatment of the self-assembled object (to convert it into a crosslinked gigantic single molecular entity with the structural features preserved from its original self-assembled precursor) is an attractive method for creating stable organic nanomaterials with prescribed sizes and shapes. Earlier examples involved the efforts of organic chemists to develop "polymerizable" lipid building blocks (which contain reactive moieties in the molecular structures) for creating biomimetic membrane structures with enhanced stability and biological functionalities (such as surface recognition and stimuli-responsive release capabilities).[192,193] More recently, with expanded access to the nanoscale morphologies via block copolymer self-assembly, research has been active in the direction of developing block copolymer-based chemistries and processing strategies for constructing chemically stabilized organic nanostructures; for instance, see Ref. 194 for an extensive up-to-date review of the chemical methods available for the covalent stabilization of the self-assembled morphologies in block copolymer solutions. In fact, there are specific advantages to using block copolymers as building block materials for producing crosslinked nanoscopic assemblies. In many demonstrated situations, the block copolymers employed contain a large number of potentially reactive sites (such as double bonds for free radical crosslinking, or tertiary amine or carboxylic acid groups for condensation reaction with added crosslinking agents),[194,195] typically one per repeat unit in the crosslinkable block, and therefore only a small conversion of the reactive functional groups is required to achieve complete covalent integration of the constituent molecules. When the number of reactive sites equals the degree of polymerization of the crosslinkable block (DP), gelation (defined as the point at which the incipient gel first appears, i.e., $\bar{M}_w \to \infty$) occurs at a critical monomer conversion of $\pi_i = 1/(DP - 1)$, and the entire mass within the self-assembled structure becomes incorporated into a single covalently connected entity (i.e., $\bar{M}_n \to \infty$) at a conversion of $\pi_f = 2/DP$.[196] This implies that, for example, with $DP = 100$ the original self-assembled structure would be completely locked in at a monomer conversion as low as 1% to 2%, which is likely to be well below the limit of the degree of crosslinking above which the crosslinking-induced structural perturbations[197] may cause rearrangement of the precursor morphology. Further, the stationary nature of the block copolymer micelles with infinitely long micelle lifetime (as discussed in Section 14.3) makes it more tractable to perform the crosslinking with a wider range of reaction conditions and time scales. These benefits are reflected in the great number of available examples that illustrate successful applications of the polymer-based assembly/crosslinking approach for generating various crosslinked nanostructures.

With interest in the general theme of self-assembly based manufacture of one-dimensional nanostructures, the paradigm of the chemical crosslinking of self-assembled block copolymers has been adopted in many studies directed toward developing methodologies for creating organic nanorods (as potential templates for fabricating inorganic nanomaterials). The first documented demonstration of chemically stitched block copolymer nanofibrils involved the preparation of the hexagonally packed cylindrical phase in the melt of poly(styrene)-poly(2-cinnamoylethyl methacrylate) (PS-PCEMA) diblock copolymer, followed by selective covalent crosslinking of the PCEMA chains in the cylindrical core; the individual crosslinked nanostructures retained their cylindrical shape even after dissolution of the crosslinked material in tetrahydrofuran (THF) (which is a good solvent for both blocks), confirming the complete chemical fixation of the parent morphology.[198] A proof of concept has also been reported that demonstrates the mass fabrication of nanometers-thin rubber threads via free radical crosslinking of the cores of the gigantic cylindrical micelles derived by bulk dispersion of PEO-PB diblock copolymers in water (Figure 14.9).[4] Crosslinking of the corona layer of the rodlike micelle formed by a poly(acrylic acid)-poly(methyl acrylate)-poly(styrene) (PAA-PMA-PS) triblock copolymer in a water/THF mixture through condensation reaction between PAA and diamine species has also been proven to be useful for preserving the morphological state in the dilute concentration limit.[199] More recently, many significant variations that bear upon the basic theme of nanorod fabrication via self-assembly and crosslinking of block copolymers have been realized, particularly creation of organic

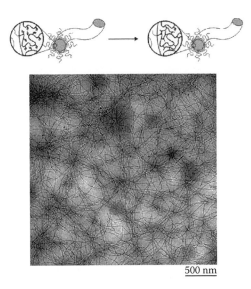

500 nm

FIGURE 14.9 A sketch illustrating the chemical crosslinking of the double bonds in the PB core of the wormlike micelles created by the self-assembly of PEO-PB copolymers in water. A representative cryo-TEM image of a 0.05 wt% crosslinked solution (shown below) confirms that the original wormlike micelle morphology is preserved without disruption through the crosslinking reaction. These figures are reproduced with permission from Ref. 4.

hollow nanotubules through the utilization of multicomponent block copolymers, which is discussed in the next subsection.

14.6.3 TEMPLATING OF INORGANIC NANOWIRES

Nanometer-thin metal/semiconductor wires have attracted much research interest as an enabling component in the miniaturization of various electronic-based devices.[200–203] Diverse approaches for the fabrication of nanowires have been developed which range from anisotropic crystal growth to template-guided synthesis to self-assembly of nanospheres.[203] In the block copolymer community, many researchers have recognized the potential usefulness of the elongated micelles of block copolymers either as direct building blocks or as templates for producing functional inorganic nanomaterials;[204–206] here, the great stability of the block copolymer micelles in fact makes this class of materials uniquely suited for templating applications which require the original self-assembled structure to be unaffected by any further chemical/physical treatments.

Recent developments in this area include the utilization of novel organic-inorganic hybridized block copolymers for creating the nanofibril morphologies; a notable example is the cylindrical micelles of various organometallic poly(ferrocenyldimethylsilane) (PFDMS)-based block copolymers. PFDMS contains ferrocene (i.e., bis(η^5-cyclopentadienyl)iron (II)) moieties which are redox active (for example, in response to the electric field),[207] and especially under oxidized conditions, this polymer may give rise to many useful properties, including semiconductivity[208] and catalytic activity.[209] Also, nanostructures made from PFDMS can be converted into ferromagnetic-ceramic composite (i.e., Fe/SiC/Si-containing) materials via pyrolysis.[210] The formation of cylindrical micelles having the PFDMS chains incorporated in the core domain has been demonstrated with a few types of soluble block chemistry; water-dispersible nanorods have been obtained from PFDMS-based amphiphilic block copolymers with examples in both nonionic (PEO-PFDMS with the PEO weight fraction (w_{PEO}) of about 0.46)[211] and ionic (poly(2-dimethylaminoethyl methacrylate)-poly(ferrocenyldimethylsilane) (PDMAEMA-PFDMS) with $w_{PDEAEMA} \approx 0.76$)[212] categories. Studies using poly(dimethylsiloxane)-poly(ferrocenyldimethylsilane) (PDMS-PFDMS) diblock copolymers have shown that the elongated micelle morphologies (such as rods ($w_{PDMS} \approx 0.65$)[213] and tubes ($w_{PDMS} \approx 0.78$)[214]) can also be created in hydrocarbon media. Because of the semicrystalline nature of the PFDMS material (of which the melting temperature is about 120 to 145°C),[215] the micelles of PFDMS-based block copolymers are expected to be mechanically stable at ambient temperature. Chemically stabilized PFDMS nanorods can also be generated, for example, through the use of block copolymers containing reactive groups in the soluble block (e.g., PI-PFDMS) for the chemical crosslinking of the coronal layer of the resultant micelle.[216]

The utilization of polymeric wormlike micelles to create nanoscale organic templates is another important paradigm for the synthesis of metallic and metal oxide nanowires. A major breakthrough in this area came from the development of the cylindrical micelle morphology composed of three coaxial tubular compartments from ABC-type triblock copolymers.[162] Such a morphology, when derived on the basis of proper design of the triblock chemistries, can provide a convenient route to construction of a hollow organic nanotube which can then be directly utilized as a molding cavity for inorganic material. Specifically, it has been demonstrated that the tri-layered cylindrical micelles prepared in methanol from poly(t-butyl acrylate)-poly(2-cinnamoyloxyethyl methacrylate)-poly(isoprene) (PtBA-PCEMA-PI) triblock copolymer can be chemically fixed through the photoinitiated crosslinking of the double bonds within the PCEMA shell, and the PI chains located in the innermost core domain subsequently can be removed by ozonolysis, resulting in a crosslinked nanotube of PtBA-PCEMA with a cavity inside.[162] The same strategy has also been demonstrated for the fabrication of water-dispersible polymeric nanotubes for which a water-soluble segment (such as poly(glyceryl methacrylate) (PGMA)) was incorporated in the triblock precursor.[217,218] After rendering them water-compatible, the polymer nanotubes are capable of serving as templates for growing nanowires via impregnation with aqueous metal precursors. For this purpose, the chemical properties of the inner surface of the nanotube can be further modified to facilitate the deposition of the metal or metal oxide material.[217,218] Examples are now abundant in the literature illustrating the versatility of the above-summarized combination of principles and procedures for producing various nanowires and tubes formulated with conducting (e.g., Ag,[219] Pd,[217] and Pd/Ni[217,218]) and semiconducting (e.g., γ-Fe$_2$O$_3$[220] and CdS[221]) compounds.

Another example of unique opportunities offered by polymer-based cylindrical micelles is the use of amphiphilic block copolymers as encasing agents for premade nanomaterials such as carbon nanotubes or metal/ceramic-based one-dimensional nanostructures. Recently, it has been reported that, when the micellization is induced (by solvent exchange) in the presence of codispersed carbon nanotubes, PAA-PS diblock copolymers (having molecular compositions that favor the formation of spherical micelles with no added nanotubes) form cylindrical micelles with carbon-nanotube cores encapsulated within the micelle shells.[222] Further, this self-assembled PAA-PS coating can be permanently fixed by carbodiimide-activated crosslinking of the PAA corona chains with added diamine coupling agents, leading to chemically stable polymer/nanotube composite materials that are dispersible in a wide range of solvents.[222] We envision that this micelle-encapsulation method can be extended to create multilayered polymer coatings on nanotubes and nanowires (by utilizing multicomponent block copolymers), which can serve as templates for assembling other inorganic layers thereupon.

14.7 SUMMARY AND OUTLOOK

The present chapter was designed to provide an overview of recent developments in the studies of block copolymer micelles, in particular, the wormlike micelles formed by block copolymers in solution. Over the past decade, much new knowledge and information has been created regarding the fundamental properties of block copolymer wormlike micelles and the unique technological opportunities associated with this form of self-assembly. These topics are discussed in this chapter, placed in the perspective of the conventional low molecular weight analogs. Giant block copolymer micelles exhibit rich morphological and dynamic behavior on both single molecular and micellar levels (Sections 14.2–14.4), including micellar polymorphism (Section 14.2) and well-defined molecular conditions for the creation of particular morphologies (Section 14.3), closely paralleling the behavior of conventional surfactants. On the other hand, other aspects of self-assembly make block copolymer thermodynamics and dynamics qualitatively different from that of small molecule surfactants, most significantly the absence of molecular exchange (i.e., nonergodicity) between individual micelles (Section 14.3). This "nonlivingness" is manifested in many ways, including the nonequilibrium distribution of sizes of the wormlike micelles (Section 14.4) and the great stability of the self-assembled structure in various environments (Section 14.6). In the high molecular-weight limit, enhanced conformational freedom spawns various defective morphologies unknown in conventional surfactants (Section 14.5). As reflected in our presentation, the scientific development of this aspect of block copolymer thermodynamics and dynamics has greatly benefited from the knowledge and methodology developed previously in the field of surfactant science.

A vast reservoir of knowledge accumulated in the field of polymer chemistry awaits exploitation in the creation of new block copolymer micelles for application in advanced technologies (Section 14.6). This includes the formation of multiblock micelles with unlimited control over the geometry and subdivision of innumerable categories of dispersed structures. This review has been restricted in limited scope, and accordingly we have not discussed many recent discoveries that extend well beyond conventional wormlike structures. Noteworthy examples include segmented micelles,[65,223] and rings and tori.[224,225]

Finally, we have not intended that this review be exhaustive. Our selection and coverage of topics are naturally biased by our own interests and expertise. We preemptively apologize to anyone who has contributed to the field but whose work is not adequately represented here.

ACKNOWLEDGMENT

The authors are grateful for financial support from Purdue University and from the National Science Foundation through the Materials Research Science and Engineering Center (MRSEC) at the University of Minnesota.

REFERENCES

1. Yu, K., Eisenberg, A. *Macromolecules* 1996, *29*, 6359.
2. Hillmyer, M. A., Bates, F. S. *Macromolecules* 1996, *29*, 6994.
3. Allgaier, J., Poppe, A., Willner, L., Richter, D. *Macromolecules* 1997, *30*, 1582.

4. Won, Y.-Y., Davis, H. T., Bates, F. S. *Science* 1999, *283*, 960.
5. Forster, S., Antonietti, M. *Adv. Mater.* 1998, *10*, 195.
6. Liu, S., Armes, S. P. *Curr. Opin. Colloid Interface Sci.* 2001, *6*, 249.
7. Zhang, L., Eisenberg, A. *Science* 1995, *268*, 1728.
8. Yu, K., Zhang, L., Eisenberg, A. *Langmuir* 1997, *13*, 2578.
9. Vaughn, T. H., Suter, H. R., Lundsted, L. G., Kramer, M. K. *J. Am. Oil Chem. Soc.* 1951, *28*, 294.
10. Flory, P. J. *J. Chem. Phys.* 1941, *9*, 660.
11. Huggins, M. L. *J. Am. Chem. Soc.* 1942, *64*, 1712.
12. Cates, M. E., Candau, S. J. *J. Phys. Condens. Mat.* 1990, *2*, 6869.
13. Herb, C. A., Prud'homme, R. K., Eds. *Structure and Flow in Surfactant Solutions*, ACS Symp. Ser. Vol. 578, American Chemical Society, Washington, 1994.
14. Debye, P., Anacker, E. W. *J. Phys. Colloid Chem.* 1951, *55*, 644.
15. Brown, R. A., Masters, A. J., Price, C., Yuan, X. F. in *Comprehensive Polymer Science*, Booth, C., Price, C., Eds., Pergamon Press, Oxford, U.K., 1989.
16. Canham, P. A., Lally, T. P., Price, C., Stubbersfield, R. B. *J. Chem. Soc. Farad. Trans. I* 1980, *76*, 1857.
17. Tuzar, Z., Sikora, A., Petrus, V., Kratochvil, P. *Makromol. Chem.* 1977, *178*, 2743.
18. LaRue, I., Adam, M., Sheiko, S. S., Rubinstein, M. *Macromolecules* 2004, *37*, 5002.
19. Zhang, L., Yu, K., Eisenberg, A. *Science* 1996, *272*, 1777.
20. Mortensen, K., Talmon, Y. *Macromolecules* 1997, *30*, 6764.
21. Yang, J. *Curr. Opin. Colloid Interface Sci.* 2002, *7*, 276.
22. Mitchell, D. J., Tiddy, G. J. T., Waring, L., Bostock, T., McDonald, M. P. *J. Chem. Soc. Faraday Trans. I* 1983, *79*, 975.
23. Bernheim-Groswasser, A., Wachtel, E., Talmon, Y. *Langmuir* 2000, *16*, 4131.
24. Brown, W., Pu, Z., Rymden, R. *J. Phys. Chem.* 1988, *92*, 6086.
25. Lin, Z. *Microstructure and Phase Behavior of Surfactants in Solution*, Ph.D. Thesis, University of Minnesota, Minneapolis, MN, 1993.
26. Brown, W., Johnson, R., Stilbs, P., Lindman, B. *J. Phys. Chem.* 1983, *87*, 4548.
27. Yoshimura, S., Shirai, S., Einaga, Y. *J. Phys. Chem. B* 2004, *208*, 15477.
28. Danino, D., Talmon, Y., Zana, R. *Langmuir* 1995, *11*, 1448.
29. Karaborni, S., Esselink, K., Hilbers, P. A. J., Smit, B., Karthauser, J., van Os, N. M., Zana, R. *Science* 1994, *266*, 254.
30. Manne, S., Schaffer, T. E., Huo, Q., Hansma, P. K., Morse, D. E., Stucky, G. D., Aksay, I. A. *Langmuir* 1997, *13*, 6382.
31. Tanford, C. *The Hydrophobic Effect*, Wiley, New York, 1973.
32. Israelachvili, J. N., Mitchell, D. J., Ninham, B. W. *J. Chem. Soc. Faraday Trans. 2* 1976, *72*, 1525.
33. Israelachvili, J. N. *Intermolecular and Surface Forces*, 2nd ed, Academic Press, San Diego, 1992.
34. Ben-Shaul, A., Gelbart, W. M. in *Micelles, Membranes, Microemulsions, and Monolayers*, Gelbart, W. M., Ben-Shaul, A., Roux, D., Eds., Springer-Verlag, New York, 1994.
35. Halperin, A., Tirrell, M., Lodge, T. P. *Adv. Polymer Sci.* 1992, *100*, 31.
36. de Gennes, P.-G. in *Liquid Crystals*, Liebert, L., Ed., Academic Press, New York, 1978.
37. Noolandi, J., Hong, K. M. *Macromolecules* 1983, *16*, 1443.
38. Halperin, A. *Macromolecules* 1987, *20*, 2943.

39. Birshtein, T. M., Zhulina, E. B. *Polymer* 1989, *30*, 170.
40. Izzo, D., Marques, C. M. *Macromolecules* 1993, *26*, 7189.
41. Hamley, I. W. *The Physics of Block Copolymers*, Oxford University Press, Oxford, U.K., 1998.
42. Zhulina, E. B., Adam, M., LaRue, I., Sheiko, S. S., Rubinstein, M. *Macromolecules* 2005, *38*, 5330.
43. Won, Y.-Y., Brannan, A. K., Davis, H. T., Bates, F. S. *J. Phys. Chem. B* 2002, *106*, 3354.
44. Jain, S., Bates, F. S. *Science* 2003, *300*, 460.
45. Zhang, L., Eisenberg, A. *J. Am. Chem. Soc.* 1996, *118*, 3168.
46. Jain, S., Bates, F. S. *Macromolecules* 2004, *37*, 1511.
47. Won, Y.-Y. *Korean J. Chem. Eng.* 2004, *21*, 296.
48. Barton, A. F. M., Ed. *CRC Handbook of Polymer-Liquid Interaction Parameters and Solubility Parameters*, CRC Press, Boston, 1990.
49. Won, Y.-Y., Davis, H. T., Bates, F. S. *Macromolecules* 2003, *36*, 953.
50. Yu, K., Bartels, C., Eisenberg, A. *Macromolecules* 1998, *31*, 9399.
51. Yu, K., Eisenberg, A. *Macromolecules* 1998, *31*, 3509.
52. Yu, K., Zhang, L., Eisenberg, A. *Langmuir* 1996, *12*, 5980.
53. Shen, H., Zhang, L., Eisenberg, A. *J. Phys. Chem. B* 1997, *101*, 4697.
54. Brandrup, J., Immergut, E. H., Eds. *Polymer Handbook*, 3rd ed., Wiley, New York, 1989.
55. Wang, Y., Balaji, R., Quirk, R. P., Mattice, W. L. *Polymer Bulletin (Berlin)* 1992, *28*, 333.
56. Ferry, J. D. *Viscoelastic Properties of Polymers*, Wiley, New York, 1980.
57. Khandpur, A. K., Macosko, C. W., Bates, F. S. *J. Polym. Sci. Part B: Polym. Phys.* 1995, *33*, 247.
58. Shen, H., Eisenberg, A. *J. Phys. Chem. B* 1999, *103*, 9473.
59. Yu, Y., Zhang, L., Eisenberg, A. *Macromolecules* 1998, *31*, 1144.
60. Shen, H., Zhang, L., Eisenberg, A. *J. Am. Chem. Soc.* 1999, *121*, 2728.
61. Zhang, L., Eisenberg, A. *Macromolecules* 1996, *29*, 8805.
62. Burke, S. E., Eisenberg, A. *Langmuir* 2001, *17*, 6705.
63. Sun, P., Yin, Y., Li, B., Chen, T., Jin, Q., Ding, D., Shi, A.-C. *J. Chem. Phys.* 2005, *122*, 204905.
64. Lin, E. K., Gast, A. P. *Macromolecules* 1996, *29*, 390.
65. Li, Z., Kesselman, E., Talmon, Y., Hillmyer, M. A., Lodge, T. P. *Science* 2004, *306*, 98.
66. de Gennes, P.-G. *Scaling Concepts in Polymer Physics*, Cornell University Press, Ithaca, 1979.
67. Flory, P. J. *Principles of Polymer Chemistry*, Cornell University Press, Ithaca, 1953.
68. Milner, S. T., Witten, T. A., Cates, M. E. *Macromolecules* 1988, *21*, 2610.
69. Daoud, M., Cotton, J. P. *J. Phys. (Paris)* 1982, *43*, 531.
70. Dan, N., Tirrell, M. *Macromolecules* 1992, *25*, 2890.
71. Netz, R. R., Schick, M. *Macromolecules* 1988, *31*, 5105.
72. Wijmans, C. M., Zhulina, E. B. *Macromolecules* 1993, *26*, 7214.
73. Zhulina, E. B., Borisov, O. V., Priamitsyn, V. A. *J. Colloid Interface Sci.* 1990, *137*, 495.
74. Majewski, J., Kuhl, T. L., Gerstenberg, M. C., Israelachvili, J. N., Smith, G. S. *J. Phys. Chem. B* 1997, *101*, 3122.

75. Majewski, J., Kuhl, T. L., Kjaer, K., Gerstenberg, M. C., Als-Nielsen, J., Israelach-vili, J. N., Smith, G. S. *J. Am. Chem. Soc.* 1998, *120*, 1469.
76. Kuhl, T. L., Leckband, D. E., Lasic, D. D., Israelachvili, J. N. *Biophys. J.* 1994, *66*, 1479.
77. Rex, S., Zuckermann, M. J., Lafleur, M., Silvius, J. R. *Biophys. J.* 1998, *75*, 2900.
78. Hamley, I. W., Castelletto, V. *Prog. Polym. Sci.* 2004, *29*, 909.
79. Forster, S., Wenz, E., Lindner, P. *Phys. Rev. Lett.* 1996, *77*, 95.
80. Liu, Y., Chen, S.-H., Huang, J. S. *Macromolecules* 1998, *31*, 2236.
81. Won, Y.-Y., Davis, H. T., Bates, F. S., Agamalian, M., Wignall, G. D. *J. Phys. Chem. B* 2000, *104*, 7134.
82. Willner, L., Poppe, A., Allgaier, J., Monkenbusch, M., Lindner, P., Richter, D. *Europhys. Lett.* 2000, *51*, 628.
83. Harris, J. M., Zalipsky, S., Eds. *Poly(ethylene glycol): Chemistry and Biological Applications*, American Chemical Society, Washington, 1997.
84. Kodera, Y., Matsushima, A., Hiroto, M., Nishimura, H., Iishi, A., Ueno, T., Inada, Y. *Prog. Polym. Sci.* 1998, *23*, 1233.
85. Veronese, F. M. *Biomaterials* 2001, *22*, 405.
86. Kawaguchi, S., Imai, G., Suzuki, J., Miyahara, A., Kitano, T. *Polymer* 1997, *38*, 2885.
87. Connor, T. M., McLauchlan, K. A. *J. Phys. Chem.* 1965, *69*, 1888.
88. Tasaki, K. *J. Am. Chem. Soc.* 1996, *118*, 8459.
89. Karlstrom, G. *J. Phys. Chem.* 1985, *89*, 4962.
90. Bjorling, M., Karlstrom, G., Linse, P. *J. Phys. Chem.* 1991, *95*, 6706.
91. Bailey, F. E., Callard, R. W. *Poly(ethylene oxide)*, Academic Press, New York, 1976.
92. Cao, B. H., Kim, M. W. *Faraday Discuss.* 1994, *98*, 245.
93. Bijsterbosch, H. D., de Haan, V. O., de Graaf, A. W., Mellema, M., Leermakers, F. A. M., Cohen Stuart, M. A., van Well, A. A. *Langmuir* 1995, *11*, 4467.
94. Faure, M. C., Bassereau, P., Carignano, M. A., Szleifer, I., Gallot, Y., Andelman, D. *Eur. Phys. J. B* 1998, *3*, 365.
95. Carlstrom, G., Halle, B. *J. Chem. Soc. Farad. Trans. I* 1989, *85*, 1049.
96. Bjorling, M., Linse, P., Karlstrom, G. *J. Phys. Chem.* 1990, *94*, 471.
97. Szleifer, I. *Biophys. J.* 1997, *72*, 595.
98. Zheng, Y., Won, Y.-Y., Bates, F. S., Davis, H. T., Scriven, L. E., Talmon, Y. *J. Phys. Chem. B* 1999, *103*, 10331.
99. Zana, R. in *Dynamics of Surfactant Self-Assemblies: Micelles, Microemulsions, Vesicles and Lyotropic Phases*, Zana, R., Ed., CRC Press, Boca Raton, 2005, p. 161.
100. Evans, D. F., Wennerstrom, H. *The Colloidal Domain: Where Physics, Chemistry, Biology, and Technology Meet*, VCH, New York, 1994.
101. Aniansson, E. A. G., Wall, S. N. *J. Phys. Chem.* 1974, *78*, 1024.
102. Aniansson, E. A. G., Wall, S. N. *J. Phys. Chem.* 1975, *79*, 857.
103. Aniansson, E. A. G., Wall, S. N., Almgren, M., Hoffmann, H., Kielmann, I., Ulbricht, W., Zana, R., Lang, J., Tondre, C. *J. Phys. Chem.* 1976, *80*, 905.
104. Kahlweit, M. *J. Colloid Interface Sci.* 1982, *90*, 92.
105. Rharbi, Y., Li, M., Winnik, M. A., Hahn Jr., K. G. *J. Am. Chem. Soc.* 2000, *122*, 6242.
106. Zana, R. in *Amphiphilic Block Copolymers: Self-Assembly and Applications*, Alexandridis, P., Lindman, B., Eds., Elsevier Science B.V., Amsterdam, The Netherlands, 2000.

107. Halperin, A., Alexander, S. *Macromolecules* 1989, *22*, 2403.
108. Prochazka, K., Bednar, B., Mukhtar, E., Svoboda, P., Trnena, J., Almgren, M. *J. Phys. Chem.* 1991, *95*, 4563.
109. Smith, C. K., Liu, G. *Macromolecules* 1996, *29*, 2060.
110. Wang, Y., Kausch, C. M., Chun, M., Quirk, R. P., Mattice, W. L. *Macromolecules* 1995, *28*, 904.
111. Hecht, E., Hoffmann, H. *Colloids Surf. A: Physicochem. Eng. Asp.* 1995, *96*, 181.
112. Cantu, L., Corti, M., Salina, P. *J. Phys. Chem.* 1991, *95*, 5981.
113. Haliloglu, T., Bahar, I., Erman, B., Mattice, W. L. *Macromolecules* 1996, *29*, 4764.
114. Selb, J., Gallot, Y. *Makromol. Chem.* 1980, *181*, 809.
115. Baines, F. L., Billingham, N. C., Armes, S. P. *Macromolecules* 1996, *29*, 3416.
116. Prochazka, K., Martin, T. J., Munk, P., Webber, S. E. *Macromolecules* 1996, *29*, 6518.
117. Tanodekaew, S., Pannu, R., Heatley, F., Attwood, D., Booth, C. *Macromol. Chem. Phys.* 1997, *198*, 3385.
118. Meguro, K., Ueno, M., Esumi, K. in *Nonionic Surfactants: Physical Chemistry*, Schick, M. J., Ed., Marcel Dekker, New York, 1987.
119. Frindi, M., Michels, B., Zana, R. *J. Phys. Chem.* 1992, *96*, 6095.
120. Tuzar, Z., Kratochvil, P. *Adv. Colloid Interface Sci.* 1976, *6*, 201.
121. Chu, B. *Langmuir* 1995, *11*, 414.
122. Willner, L., Poppe, A., Allgaier, J., Monkenbusch, M., Richter, D. *Europhys. Lett.* 2001, *55*, 667.
123. Rager, T., Meyer, W. H., Wegner, G. *Macromol. Chem. Phys.* 1999, *200*, 1672.
124. Morse, D. C. *Macromolecules* 1998, *31*, 7030.
125. Morse, D. C. *Macromolecules* 1998, *31*, 7044.
126. Khokhlov, A. R., Semenov, A. N. *Physica* 1981, *108A*, 546.
127. Vroege, G. J., Odijk, T. *Macromolecules* 1988, *21*, 2848.
128. Dalhaimer, P., Bates, F. S., Discher, D. E. *Macromolecules* 2003, *36*, 6873.
129. Discher, B. M., Bermudez, H., Hammer, D. A., Discher, D. E., Won, Y.-Y., Bates, F. S. *J. Phys. Chem. B* 2002, *106*, 2848.
130. Landau, L. D., Lifshitz, E. M. *Statistical Physics. Part I*, 3rd ed, Elsevier, Oxford, U.K., 1980.
131. Stigter, D. *J. Phys. Chem.* 1966, *70*, 1323.
132. Yamakawa, H., Fujii, M. *Macromolecules* 1974, *7*, 128.
133. Yamakawa, H., Yoshizaki, T. *Macromolecules* 1980, *13*, 633.
134. Schmidt, M. *Macromolecules* 1984, *17*, 553.
135. Liu, G., Yan, X., Duncan, S. *Macromolecules* 2002, *35*, 9788.
136. Liu, G., Yan, X., Duncan, S. *Macromolecules* 2003, *36*, 2049.
137. Yoshizaki, T., Yamakawa, H. *Macromolecules* 1980, *13*, 1518.
138. Jerke, G., Pedersen, J. S., Egelhaaf, S. U., Schurtenberger, P. *Langmuir* 1998, *14*, 6013.
139. Magid, L. J., Han, Z., Li, Z., Butler, P. D. *J. Phys. Chem.* 2000, *104*, 6717.
140. Marignan, J., Appell, J., Bassereau, P., Porte, G., May, R. P. *J. Phys. (Paris)* 1989, *50*, 3553.
141. Magid, L. J., Li, Z., Butler, P. D. *Langmuir* 2000, *16*, 10028.
142. Schurtenberger, P., Magid, L. J., King, S. M., Lindner, P. *J. Phys. Chem.* 1991, *95*, 4173.
143. Jerke, G., Pedersen, J. S., Egelhaaf, S. U., Schurtenberger, P. *Phys. Rev. E* 1997, *56*, 5772.

144. Lodge, T. P. *Mikrochim. Acta* 1994, *116*, 1.
145. Agamalian, M., Wignall, G. D., Triolo, R. *J. Appl. Cryst.* 1997, *30*, 345.
146. Hendricks, J., Kawakatsu, T., Kawasaki, K., Zimmermann, W. *Phys. Rev. E* 1995, *51*, 2658.
147. Rivetti, C., Guthold, M., Bustamante, C. *J. Mol. Biol.* 1996, *264*, 919.
148. Porte, G., Gomati, R., El Haitamy, O., Appell, J., Marignan, J. *J. Phys. Chem.* 1986, *90*, 5746.
149. Appell, J., Porte, G., Khatory, A., Kern, F., Candau, S. J. *J. Phys. II* 1992, *2*, 1045.
150. Khatory, A., Lequeux, F., Kern, F., Candau, S. J. *Langmuir* 1993, *9*, 1456.
151. Cates, M. E. *J. Phys. (Paris)* 1988, *49*, 1593.
152. Harwigsson, I., Soderman, O., Regev, O. *Langmuir* 1994, *10*, 4731.
153. Lin, Z. *Langmuir* 1996, *12*, 1729.
154. Silvander, M., Karlsson, G., Edwards, K. *J. Colloid Interface Sci.* 1996, *179*, 104.
155. Danino, D., Talmon, Y., Levy, H., Beinert, G., Zana, R. *Science* 1995, *269*, 1420.
156. Drye, T. J., Cates, M. E. *J. Chem. Phys.* 1992, *96*, 1367.
157. Tlusty, T., Safran, S. A. *Science* 2000, *290*, 1328.
158. Bernheim-Groswasser, A., Tlusty, T., Safran, S. A., Talmon, Y. *Langmuir* 1999, *15*, 5448.
159. Tlusty, T., Safran, S. A., Strey, R. *Phys. Rev. Lett.* 2000, *84*, 1244.
160. Bates, F. S., Fredrickson, G. H. *Physics Today* 1999, *52*, 32.
161. Liu, F., Liu, G. *Macromolecules* 2001, *34*, 1302.
162. Stewart, S., Liu, G. *Angew. Chem. Int. Ed.* 2000, *39*, 340.
163. Tlusty, T., Safran, S. A. *J. Phys.: Condens. Matter* 2000, *12*, A253.
164. Khandpur, A. K., Forster, S., Bates, F. S., Hamley, I. W., Ryan, A. J., Bras, W., Almdal, K., Mortensen, K. *Macromolecules* 1995, *28*, 8796.
165. Allen, C., Maysinger, D., Eisenberg, A. *Colloids Surf. B: Biointerfaces* 1999, *16*, 3.
166. Rosler, A., Vandermeulen, G. W. M., Klok, H. A. *Adv. Drug Deliver. Rev.* 2001, *53*, 95.
167. Kataoka, K., Harada, A., Nagasaki, Y. *Adv. Drug Deliver. Rev.* 2001, *47*, 113.
168. Gaucher, G., Dufresne, M. H., Sant, V. P., Kang, N., Maysinger, D., Leroux, J. C. *J. Control. Release* 2005, *109*, 169.
169. Torchilin, V. P. *Expert Opin. Ther. Pat.* 2005, *15*, 63.
170. Hayashi, T., Itagaki, H., Fukuda, T., Tamura, U., Sato, Y., Suzuki, Y. *Biol. Pharm. Bull.* 1995, *18*, 540.
171. Otzen, D. E. *Biophys. J.* 2002, *83*, 2219.
172. Chi, E. Y., Krishnan, S., Randolph, T. W., Carpenter, J. F. *Pharma. Res.* 2003, *20*, 1325.
173. Kondo, T., Tomizawa, M. *J. Pharm. Sci.* 1968, *57*, 1246.
174. Gustafsson, J., Oradd, G., Almgren, M. *Langmuir* 1997, *13*, 6956.
175. Groot, R. D., Rabone, K. L. *Biophys. J.* 2001, *81*, 725.
176. Malmsten, M. in *Amphiphilic Block Copolymers: Self-Assembly and Applications*, Alexandridis, P., Lindman, B., Eds., Elsevier Science B.V., Amsterdam, The Netherlands, 2000.
177. Otsuka, H., Nagasaki, Y., Kataoka, K. *Curr. Opin. Colloid Interface Sci.* 2001, *6*, 3.
178. Bhadra, D., Bhadra, S., Jain, P., Jain, N. K. *Pharmazie* 2002, *57*, 5.
179. Jeong, B., Bae, Y. H., Kim, S. W. *Macromolecules* 1999, *32*, 7064.
180. Hwang, M. J., Suh, J. M., Bae, Y. H., Kim, S. W., Jeong, B. *Biomacromolecules* 2005, *6*, 885.
181. Fujiwara, T., Kimura, Y. *Macromol. Biosci.* 2002, *2*, 11.

182. Jule, E., Nagasaki, Y., Kataoka, K. *Langmuir* 2002, *18*, 10334.
183. Jeong, B., Kim, S. W., Bae, Y. H. *Adv. Drug Deliver. Rev.* 2002, *54*, 37.
184. Gil, E. S., Hudson, S. A. *Prog. Polym. Sci.* 2004, *29*, 1173.
185. Brannan, A. K., Bates, F. S. *Macromolecules* 2004, *37*, 8816.
186. Dalhaimer, P., Engler, A. J., Parthasarathy, R., Discher, D. E. *Biomacromolecules* 2004, *5*, 1714.
187. Kim, Y., Dalhaimer, P., Christian, D. A., Discher, D. E. *Nanotechnology* 2005, *16*, S484.
188. Whitmore, R. L. *Rheology of the Circulation*, Pergamon Press, Oxford, U.K., 1968.
189. Fung, Y. C. *Biomechanics: Mechanical Properties of Living Tissues*, 2nd ed, Springer-Verlag, New York, 1993.
190. Huwyler, J., Wu, D., Pardridge, W. M. *Proc. Nat. Acad. Sci. USA* 1996, *93*, 14164.
191. Kabanov, A. V., Batrakova, E. V. *Curr. Pharm. Design* 2004, *10*, 1355.
192. O'Brien, D. F., Whitesides, T. H., Kilingbiel, R. T. *J. Polym. Sci. C Polym. Lett.* 1981, *19*, 95.
193. Ringsdorf, H., Schlarb, B., Venzmer, J. *Angew. Chem. Int. Ed.* 1988, *27*, 113.
194. Rodriguez-Hernandez, J., Checot, F., Gnanou, Y., Lecommandoux, S. *Prog. Polym. Sci.* 2005, *30*, 691.
195. Wooley, K. L. *J. Polym. Sci. B Polym. Chem.* 2000, *38*, 1397.
196. Dotson, N. A., Galvan, R., Laurence, R. L., Tirrell, M. *Polymerization Process Modeling*, Wiley-VCH, New York, 1996.
197. Nielsen, L. E. *J. Macromol. Sci. Rev. Macromol. Chem. C* 1969, *3*, 69.
198. Liu, G., Qiao, L., Guo, A. *Macromolecules* 1996, *29*, 5508.
199. Ma, Q., Remsen, E. E., Clark Jr., C. G., Kowalewski, T., Wooley, K. L. *Proc. Nat. Acad. Sci. USA* 2002, *99*, 5058.
200. Hulteen, J. C., Martin, C. R. *J. Mater. Chem.* 1997, *7*, 1075.
201. Lieber, C. M. *Solid State Commun.* 1998, *107*, 607.
202. Lieber, C. M. *MRS Bull.* 2003, *28*, 486.
203. Xia, Y., Yang, P., Sun, Y., Wu, Y., Mayers, B., Gates, B., Yin, Y., Kim, F., Yan, H. *Adv. Mater.* 2003, *15*, 353.
204. Forster, S. *Topics Curr. Chem.* 2003, *226*, 1.
205. Lazzari, M., Lopez-Quintela, M. A. *Adv. Mater.* 2003, *15*, 1583.
206. Hamley, I. W. *Soft Matter* 2005, *1*, 36.
207. Togni, A., Hayashi, T., Eds. *Ferrocenes: Homogeneous Catalysis/Organic Synthesis/Materials Science*, VCH, Weinheim, Germany, 1995, 1995.
208. Manners, I. *Angew. Chem. Int. Ed.* 1998, *35*, 1603.
209. Atkinson, R. C. J., Gibson, V. C., Long, N. J. *Chem. Soc. Rev.* 2004, *33*, 313.
210. MacLachlan, M. J., Aroca, P., Coombs, N., Manners, I., Ozin, G. A. *Adv. Mater.* 1998, *10*, 144.
211. Gohy, J. F., Lohmeijer, B. G. G., Alexeev, A., Wang, X. S., Manners, I., Winnik, M. A., Schubert, U. S. *Chem. Eur. J.* 2004, *10*, 4315.
212. Wang, X. S., Winnik, M. A., Manners, I. *Macromolecules* 2005, *38*, 1928.
213. Massey, J., Power, K. N., Manners, I., Winnik, M. A. *J. Am. Chem. Soc.* 1998, *120*, 9533.
214. Raez, J., Manners, I., Winnik, M. A. *J. Am. Chem. Soc.* 2002, *124*, 10381.
215. Massey, J. A., Temple, K., Cao, L., Rharbi, Y., Raez, J., Winnik, M. A., Manners, I. *J. Am. Chem. Soc.* 2000, *122*, 11577.

216. Wang, X. S., Arsenault, A., Ozin, G. A., Winnik, M. A., Manners, I. *J. Am. Chem. Soc.* 2003, *125*, 12686.
217. Li, Z., Liu, G. *Langmuir* 2003, *19*, 10480.
218. Yan, X., Liu, G., Haeussler, M., Tang, B. Z. *Chem. Mater.* 2005, *17*, 6053.
219. Wang, X. S., Wang, H., Coombs, N., Winnik, M. A., Manners, I. *J. Am. Chem. Soc.* 2005, *127*, 8924.
220. Yan, X., Liu, G., Liu, F., Tang, B. Z., Peng, H., Pakhomov, A. B., Wong, C. Y. *Angew. Chem. Int. Ed.* 2001, *40*, 3593.
221. Duxin, N., Liu, F. T., Vali, H., Eisenberg, A. *J. Am. Chem. Soc.* 2005, *127*, 10063.
222. Kang, Y., Taton, T. A. *J. Am. Chem. Soc.* 2003, *125*, 5650.
223. Zhu, J., Jiang, W. *Macromolecules* 2005, *38*, 9315.
224. Pochan, D. J., Chen, Z., Cui, H., Hales, K., Qi, K., Wooley, K. L. *Science* 2004, *306*, 94.
225. Zhu, J., Liao, Y., Jiang, W. *Langmuir* 2004, *20*, 3809.

15 Oilfield Applications of Giant Micelles

Phil Sullivan, Erik B. Nelson, Valerie Anderson, and Trevor Hughes

CONTENTS

15.1 INTRODUCTION

Viscoelastic surfactant systems based on wormlike micelles have generated considerable scientific interest and have been used in applications ranging from industrial cleaning, detergents, hydraulic fluids, and slurry transport fluids. The oil and gas industry has also employed viscoelastic surfactant (VES) fluids in a variety of well completion and stimulation applications. Unlike conventional polymer fluids that rely on entangled and crosslinked polymer chains to build viscosity, VES systems exploit the self-assembly of surfactant molecules into large wormlike micelles that entangle to form a viscoelastic structure.[1] This wormlike micellar structure is dynamic in nature and can be controlled by the

453

local fluid environment. For example, the structure can be destroyed upon exposure to hydrocarbon fluids, or the fluid viscosity controlled by adjusting the brine salinity. This responsiveness makes VES fluids attractive candidates for numerous oilfield operations where removal of the treatment fluid is important. The unique properties of this "living polymer" network give rise to numerous advantages in many applications.

This chapter begins with a brief discussion of applications and requirements of fluids used in oilfield operations. Next, the chapter describes the self-assembly process of VES fluids and some of the surfactant chemistries used in oilfield operations. Key properties of a successful VES fluid are explained, and important advantages of VES gels are detailed. Where applicable, experimental data are presented to contrast VES technology with polymer counterparts. Finally, well completion operations involving VES fluids are explained.

15.2 INTRODUCTION TO THE WELL SITE

The purpose of drilling oil and gas wells is to produce hydrocarbons from formations beneath the earth's surface. The borehole provides a conduit for the flow of fluids from the hydrocarbon-bearing formations to the surface. Certain equipment must be placed in the wellbore, and various procedures must be employed to initiate and control the fluid flow. The equipment and procedures are collectively known as a *well completion.* Many well completion procedures involve pumping fluids into the borehole and the producing formation. The two principal types of rock, or *matrix,* in which the hydrocarbon is found are sandstones and carbonates.

In most well completions, steel casing is lowered into the borehole and cemented in place. The cemented casing plays several roles: preventing wellbore collapse, providing hydraulic isolation between the rock layers, and controlling which layers are connected to the wellbore. The latter is especially important if there are water zones. The cemented casing not only reduces the amount of water produced with the oil, but also prevents contamination of fresh-water aquifers. Fluid flow is established by creating holes or perforations that extend beyond the casing and the cement sheath, thereby connecting the reservoir to the wellbore (Figure 15.1).

Well completion operations can cause *formation damage,* a loss of reservoir permeability resulting from treatment-fluid invasion and interactions between treatment fluids and reservoir fluids. The interactions can cause precipitates or emulsions to form, hindering fluid flow in the rock. During well production, fluids flow radially through the rock towards the wellbore, and there are additional obstacles that must be overcome. The permeability of the rock matrix may be very low. Some rocks contain fine sand that can migrate and block flow. Salty water produced with hydrocarbon can form scales in the rock pores or inside the casing. Paraffins or asphaltenes in the hydrocarbon fluid may precipitate and damage the rock matrix or plug the casing.

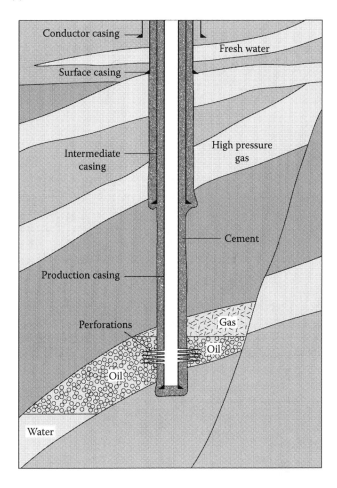

FIGURE 15.1 Diagram of a wellbore. Well depths can reach 30,000 ft [9,144 m]. The diameter of a typical wellbore varies from about 14 in. [35.6 cm] at surface to about 2 in. [5.1 cm] at the bottom of the well.

All of these problems can be mitigated by several techniques that either prevent loss of well productivity or reestablish production. In this chapter we will discuss four: gravel packing, coiled-tubing cleanout, matrix acidizing, and hydraulic fracturing. Gravel packing is a preventive technique, used to stop or slow the movement of fine sand from the formation into the wellbore, where it might block production or damage equipment further up the well. Coiled-tubing cleanout is a remedial technique to remove sand or other materials that have accumulated in the wellbore. Matrix acidizing and hydraulic fracturing are *stimulation* techniques that remove or bypass formation damage, and increase hydrocarbon production to levels far exceeding the rates that would be possible under natural flow conditions. Matrix acidizing involves removing formation damage close to the wellbore. Hydraulic fracturing involves creating a permeable fracture that extends past near-wellbore damage, and provides a pathway through which hydrocarbons may flow to the wellbore. Each of these services requires fluids with specific chemical and rheological characteristics, and VES fluids demonstrate clear advantages compared to traditional technologies.

Logistics is another important well service parameter. The wellsite may be offshore, in the middle of a desert or in a jungle. Transportation of chemicals and equipment to mix and pump the fluids can be a major challenge. The volumes of fluid required vary from about 1,000 [3,790 L] gallons for a tubing cleanout treatment to hundreds of thousands of gallons for a hydraulic fracturing treatment. Depending on the treatment, equipment must be capable of pumping fluids at rates between about 10 to 2,500 gal/min [38 to 9,500 L/min]. Mix water is usually sourced locally; therefore, caution must be exercised to ensure that contaminants do not adversely affect fluid performance. To avoid environmental damage, waste disposal is a major consideration. In many cases, wastes must be collected and treated far away from the wellsite.

Traditionally, polymers such as guar or hydroxyethyl cellulose have been used as fluid thickeners in oilfield applications. They are inexpensive, readily available, and relatively environmentally friendly. However, polymer-fluid preparation is complicated. The polymer must hydrate in the mix water, and the resulting fluid requires several chemical additives to adjust performance.

Surfactants are more expensive than polymers; however, surfactant-base fluids are easier to prepare. A simple dilution of the surfactant, with agitation, is all that is required (Figure 15.2). Such ease of preparation is a major advantage, and

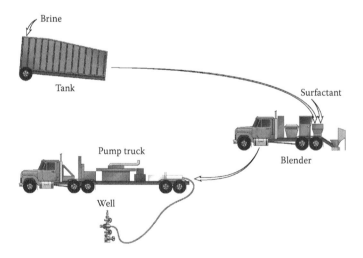

FIGURE 15.2 Field mixing procedure for VES fluids. Unlike polymer-base fluids, VES fluids are prepared by a simple dilution of the surfactant in brine. Reproduced from Ref. 2 with permission of Schlumberger.

saves considerable time at the wellsite. Costs are reduced and more treatments can be performed each day. In addition, fewer additives are required for performance adjustments.[2]

15.3 SELF-ASSEMBLY AND ENTANGLEMENT OF WORMLIKE MICELLES

Under certain conditions, surfactant molecules can self-assemble and form wormlike micelles (see Chapter 5). The principal conditions affecting such behavior are surfactant concentration, solvent characteristics, and temperature. The long micelles act as "living" polymers, increasing the viscosity of the solution in the same manner as conventional polymers.

Unlike polymer strands, wormlike micelles can reform if they are broken—in fact, they are continually breaking and recombining (see Chapters 4, 8, 9, and 10). During well service treatments, fluids can be exposed to shear rates exceeding 5,000 s^{-1} (e.g., when the fluid enters rock pores at high pumping rates). Polymers degrade at such high shear rates, negatively affecting fluid performance. Because wormlike micelles can recover after exposure to shear, VES fluids are able to maintain the appropriate fluid properties throughout the treatment.

15.4 SURFACTANT CHEMISTRIES USED FOR WELLBORE OPERATIONS

Three classes of viscoelastic surfactants are useful in oilfield applications—cationic, anionic, and zwitterionic. Each has advantages and disadvantages that influence the manner in which they are used.

15.4.1 CATIONIC

The most common cationic VES technology is based on quaternary ammonium salts of fatty acids. This choice was derived from work by Rose et al., in which such surfactants were used as bleach thickeners.[3,4] The basic raw material is rapeseed oil, which is mainly comprised of saturated fatty acids with carbon chain lengths ranging from C_{18} to C_{22}. An early quaternary ammonium salt surfactant for oilfield applications was prepared from raw rapeseed oil. The pure surfactant is a waxy solid at ambient temperature; therefore, a solvent is added to prepare a pourable product that can be easily handled at the wellsite.

The viscosity of a brine containing rapeseed oil quaternary ammonium salt versus temperature is shown in Figure 15.3. At a shear rate of 100 s^{-1}, a viscosity of at least 100 cp is generally required to initiate and propagate a fracture. Using this criterion, the maximum practical fluid temperature for rapeseed quaternary ammonium salt systems is about 130°F (54°C).

A system with improved temperature stability employs an ethoxylated quaternary ammonium salt based solely on erucic acid. Inspection of Figure 15.3 reveals that this surfactant is effective at temperatures up to about 175°F (79°C).

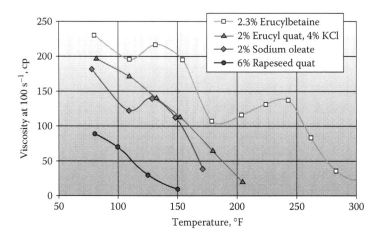

FIGURE 15.3 Fluid viscosity versus temperature of systems containing various visco-elastic surfactants.

The fluid range can be extended to 200°F (93°C) by adding sodium salicylate as a counterion to enhance micelle growth.[5,6]

For fracturing and gravel packing applications, the surfactant concentrate is added to brine to form a viscous fluid. An example of the viscosity/shear-rate curve for a typical fluid is shown in Figure 15.4. The zero shear-rate viscosity is very temperature dependent, but viscosities at higher shear rates are less so.

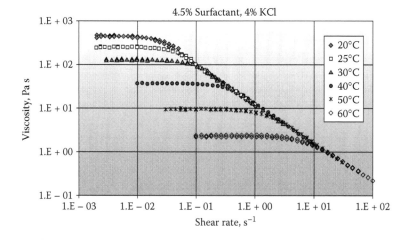

FIGURE 15.4 Viscosity of cationic VES solution versus shear rate and fluid temperature. Fluid contains 4.5 vol% erucyl quaternary ammonium salt surfactant and 4 wt% KCl.

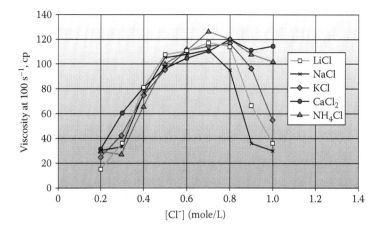

FIGURE 15.5 Effect of chloride-ion concentration on the viscosity of a 2 vol% solution of erucyl quaternary ammonium salt surfactant.

Figure 15.5 shows the variation of the viscosity (at a shear rate of 100 s¹) as a function of the chloride-ion concentration. The viscosity is more dependent on the chloride concentration than on the nature of the cation. The optimal chloride-ion concentration in the brine is 0.6–0.8 mole/L.

A major advantage of cationic viscoelastic surfactants is their low sensitivity to variations in water quality, divalent cations in particular. A major disadvantage is their high cost. Another problem with cationic surfactants is their propensity to form emulsions with some crude oils, particularly those containing paraffins and asphaltenes. Emulsion formation in the producing reservoir may hinder cleanup and lower the hydrocarbon production rate.

15.4.2 ANIONIC

An anionic viscoelastic surfactant, based on oleic acid, addresses the cost disadvantage of the cationic surfactant. The surfactant is prepared at the wellsite by neutralizing the fatty acid with sodium hydroxide, and is immediately pumped downhole.

The performance of the anionic surfactant as a thickener is shown in Figure 15.3. The maximum practical fluid temperature is about 130°F (54°C). Unlike the quaternary ammonium salts, anionic surfactants are sensitive to water hardness. Calcium and magnesium in the mix water will react with the surfactant to form an insoluble precipitate. The fluid viscosity is affected, and the precipitate may damage the producing reservoir. When hard mix water is the only option, chelating agents such as EDTA are added to prevent precipitate formation. Another important advantage of the anionic system is its lower propensity to form emulsions.

15.4.3 ZWITTERIONIC

The third type of viscoelastic surfactant for oilfield use is zwitterionic. One principal advantage of such surfactants is their efficacy at high fluid temperatures—up to about 250°F (121°C) (Figure 15.3).

15.5 TYPICAL PROPERTIES AND KEY ADVANTAGES OF VES FLUIDS

15.5.1 RHEOLOGY AND PARTICLE TRANSPORT

In many oilfield applications, fluid rheology is a key parameter because it governs particle transport, pumping requirements, and flow through porous media. The rheology of viscoelastic surfactant fluids is discussed elsewhere in this volume, but it is worth mentioning key features of the fluids' rheological properties that are important for oilfield applications, and comparing the performance of VES fluids with that of polymer fluids that have historically been used in the same applications.

Figure 15.6 presents rheograms of fluids viscosified by a viscoelastic surfactant and by hydroxyethylcellulose (HEC).[7] The two fluids were designed to have comparable viscosities at shear rates in the neighborhood of 100 s^{-1}. However,

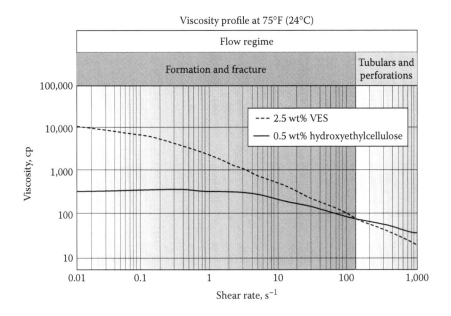

FIGURE 15.6 Constant-shear viscosity profiles of fluids viscosified with VES and hydroxyethylcellulose. Reproduced from Ref. 7 with permission of Schlumberger.

at other shear rates, the fluids' behavior is substantially different. The zero-shear viscosity of the VES fluid is more than an order of magnitude greater than the HEC counterpart. The VES fluid also demonstrates a much stronger shear thinning effect, with a nearly constant stress plateau over almost three orders of magnitude of shear rate.

An important advantage of VES fluids is their ability to fully regain viscosity after exposure to high shear rates. This occurs when fluids are pumped at high rates through tubing, or when the fluid passes through casing perforations. High shear rates can permanently degrade the performance of polymer-base fluids. Shear can physically break down the polymer or destroy irreversible crosslinks (e.g., with zirconium crosslinked guar fluids). Shear stability is necessary to ensure that the fluid will retain the designed properties throughout the treatment.

It is often advantageous to use foamed VES fluids. Foams reduce water contact with sensitive formations, and their compressible nature promotes better fluid cleanup after the treatment. In addition, foamed VES fluids are more economical because less surfactant is required per unit volume. Nitrogen is usually the gas used in the foam and, depending on the application, the foam quality varies from 30 to 80%. Figure 15.7 illustrates the rheological behavior of foamed VES fluids at various qualities, measured with a capillary foam rheometer. Maximum viscosity is achieved at foam qualities between 75 and 80%.

Many completion operations involve the hydraulic transport of solid particulates (typically sand particles) either within a wellbore or within a reservoir fracture. Because of horsepower and pump-rate limitations, particle transport must take place in laminar flow without the benefit of turbulent eddies. Field results and laboratory investigations have shown that VES fluids can efficiently

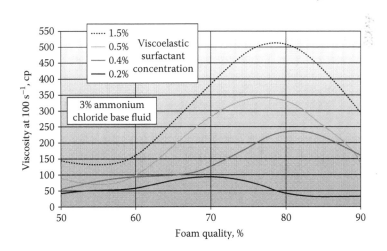

FIGURE 15.7 Viscosity of foamed VES fluids versus foam quality. Reproduced from Ref. 2 with permission of Schlumberger.

suspend and transport particles in laminar flow. This phenomenon is not well understood and remains an area of investigation. Velocity profile measurements of VES fluids flowing in pipes suggest that the constant-stress rheology plateau gives rise to a virtual slip layer near the pipe wall, with the bulk of the fluid experiencing very little shear. This observation suggests that the zero-shear viscosity of the VES gel may be the dominant factor governing particle transport. A similar effect occurs with polymer fluids, but the effect is more extreme for VES solutions because of their profound shear thinning behavior. Elasticity and normal force differences that cause particles to migrate to the center of the flow may also play a role in the transport process.

15.5.2 FRICTION PRESSURE REDUCTION

Efficient drag reduction has been a distinguishing characteristic of viscoelastic surfactants since their early use in district heating systems (see Chapter 16).[8] In oilfield operations, this drag reduction provides an advantage for reduced pressure and horsepower when pumping fluids through wellbores and long tubulars. Figure 15.8 provides an example comparison between a VES fluid and a guar polymer counterpart for use in hydraulic fracturing treatments. Both datasets are measurements for fluids pumped through 3.5-in. internal diameter tubing. The VES fluid imparts approximately 30% lower friction pressure at a flow rate of 10 barrels per minute (1.6 m³/min). Such a reduction allows treatments to be performed with less pumping equipment at the wellsite.

15.5.3 CLEAN-UP

When exposed to liquid hydrocarbons, the wormlike micelles in VES fluids collapse into spheres and the fluid viscosity drops dramatically. This is important

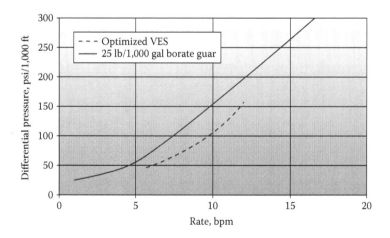

FIGURE 15.8 Friction pressure comparison between VES and crosslinked guar fluids.

because, at the conclusion of a treatment, the fluid must leave the producing formation and the wellbore, and clear the way for hydrocarbon production.

Polymer fluids do not lose viscosity upon contact with hydrocarbons; instead, chemicals known as *breakers* are added that degrade the polymers, reducing fluid viscosity. Common breakers include oxidizers such as ammonium persulfate and sodium bromate, and enzyme formulations that attack polymers at a more fundamental level. This process is not as efficient as micellar collapse, and frequently leaves insoluble polymer residue in the flow path that hinders production.

Nevertheless, there are situations when breakers are necessary with VES fluids. For example, many wells produce *dry gas* that contains a small amount of liquid hydrocarbons. Under these circumstances, salts are added to alter the ionic environment of the fluid and hasten fluid-viscosity reduction. To prevent premature fluid-viscosity collapse, the salts are encapsulated by a polymer coating. The coating ruptures when exposed to an elevated temperature, or to chemical and physical stresses during or after a treatment.[9,10] Release of the breaker profoundly decreases VES fluid viscosity, particularly at low shear rates.

15.6 WELL COMPLETION TECHNIQUES

This section provides more details about the use of VES fluids in gravel packing, cleanout, and stimulation applications.

15.6.1 GRAVEL PACKING

Some hydrocarbon producing formations are composed of poorly consolidated sands that can follow the hydrocarbons into the wellbore. Negative consequences include wellbore plugging and fouling of valves and other equipment at the surface. Sand control techniques keep the formation sand at bay, allowing the well to produce without hindrance.

The most common sand control technique is to create a *gravel pack* that acts as a size-exclusion filter between the formation sand and the wellbore.[11] The gravel size is selected such that the interstices of the pack exclude the formation sand, yet allow hydrocarbons to flow freely. Typical sand sizes range from 12/20 mesh to 40/60 mesh (approximately 0.25 to 2.0 mm).

A gravel pack screen is centralized inside the casing (Figure 15.9), and gravel is placed in the annulus between the casing and the screen and inside the perforations. The placement technique involves pumping a gravel slurry down the casing/screen annulus. As the gravel is deposited, carrier fluid invades the sand formation and also escapes to the interior of the screen and back up to the surface. The rest of the carrier fluid must break to make way for hydrocarbon production.

Polymers such as HEC and xanthan gum have traditionally been used as viscosifiers for gravel pack fluids. During the operation, as the fluid escapes into the formation, insoluble residue is left behind as a filter cake on the formation wall. The filter cake reduces the permeability of the formation/wellbore interface,

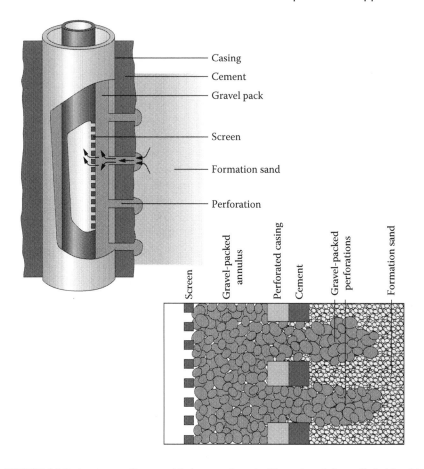

FIGURE 15.9 Anatomy of a cased-hole gravel pack. Reproduced from Ref. 11 with permission of Schlumberger.

hindering hydrocarbon production. VES fluids do not contain insoluble residues, resulting in less formation damage and higher well productivity.[12]

15.6.2 COILED-TUBING CLEANOUT

Accumulation of sand or other particulates in a wellbore can be extremely detrimental to well productivity. When this occurs, remedial operations are performed to remove the particulates. The most common technique involves placing coiled tubing into the wellbore to form an annulus (Figure 15.10). Particulate material is removed by pumping a fluid down the tubing string, whereupon it mixes with the particles and carries them up the annulus to the surface.

The inside diameter of coiled tubing can be as small as 1 in. (2.54 cm); therefore, drag reduction is important to attain the high flow rates necessary to mobilize the particulates. In many cases, the wellbore diameter is large (e.g., >

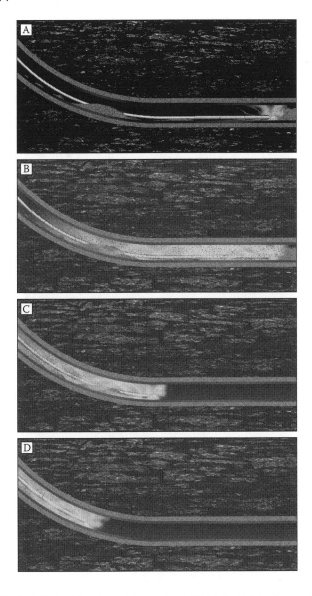

FIGURE 15.10 Coiled-tubing cleanout procedure. (A) Coiled tubing lowered into casing; (B) cleanout fluid mobilizes and suspends solids; (C and D) tubing is pulled out of the hole as solids are removed from casing. Reproduced from Ref. 13 with permission of Schlumberger.

9 in. [22.9 cm]) and turbulent flow is not possible. Cleanout fluids with good particle suspension capabilities are essential to achieve success. VES fluids provide good drag reduction and particle suspension ability, and are useful for some of the most demanding applications.[13]

15.6.3 ACIDIZING AND DIVERSION

In instances where formation damage extends only a few feet from the wellbore, the well may be *acidized* to dissolve or remove the damage. Acidizing treatments are often categorized by formation rock type (sandstone or carbonate). Such treatments are performed through the naturally existing flow channels in the rock, at pumping rates that are sufficiently low to avoid creating fractures in the formation.[14]

Carbonate acidizing is the reaction of hydrochloric acid on calcium carbonate (calcite) or calcium magnesium carbonate (dolomite). As the acid flows through the perforations into the formation, and carbonate is dissolved, *wormholes* form in the formation. Wormholes are highly conductive channels radiating from the point of acid injection (Figure 15.11). The length, direction, and number of wormholes depend on the formation reactivity and the rate at which acid leaks off to the matrix. Once formed, the wormholes carry virtually all of the flow. As shown in Figure 15.11, the wormholes form a dendritic network.[15]

FIGURE 15.11 Wormholes formed during laboratory-scale matrix acidizing treatment.

For efficient stimulation, the wormhole network should uniformly cover the formation interval; unfortunately, the acid has a natural tendency to form a limited number of wormholes that become larger and larger, ignoring less permeable regions. Matrix acidizing is even more complex when there are multiple intervals having significantly different permeabilities. High-permeability zones preferentially take the acid and leave zones with lower permeability untreated. These untreated intervals mean less production and lost reserves. Various mechanical and chemical methods have been developed to force acid away from the primary wormholes and create new ones, or to ensure treatment of lower permeability intervals. This process is called *diversion*.

Conventional chemical diversion methods include nitrogen foam, bridging agents like benzoic acid, and crosslinked polymer gels. These methods temporarily plug high-permeability carbonate zones to effectively divert the treatment fluids to zones of lower permeability. Chemical diversion methods vary in effectiveness. Sometimes temporary plugs become permanent, and the reservoir that was meant to be stimulated becomes damaged, diminishing well productivity.

Specially designed acidic polymer gel systems initially have a low viscosity to allow easy pumping, but once the fluid enters a carbonate formation and the acid spends, the polymer crosslinks when the pH rises to 2, increasing fluid viscosity. The viscosity increase restricts further flow of new acid through the wormholes, thereby diverting fresh acid to zones with lower permeabilities. As the acid continues to dissolve the rock the pH increases further. Once the pH reaches about 3.5, the gelled acid breaks, reducing the viscosity and enabling fluids to flow back and clean up. Polymer-base acid systems have several drawbacks. Because of the narrow pH window, the crosslinking and breaking phenomenon can be difficult to control. Moreover, the stability of polymer systems degrades as the bottomhole temperature increases. This instability hinders proper diversion or, at worst, permanently damages the formation to the point of preventing flow.

VES-base acidizing systems can overcome the drawbacks of polymer gels.[7,16,17] The gel system is a blend of HCl, one or more surfactants, and common additives required for acid treatment. At HCl concentrations above about 10 wt%, the fluid viscosity is low (Figure 15.12). As the acid is consumed through the reaction with calcite or dolomite, the fluid gels and diversion commences (Figure 15.13). Two factors trigger the gelation process. When the acid spends, the increased pH allows the surfactant molecules to form wormlike micelles. The carbonate dissolution also results in the formation of calcium chloride brine, further stabilizing the wormlike micelles. This system is easier to control than polymer gels because the surfactant gel breaks upon contact with produced hydrocarbon and does not rely on pH. The absence of insoluble residue further improves formation permeability after the treatment. This acid system can be used to stimulate wells that have bottomhole temperatures up to 300°F [149°C].

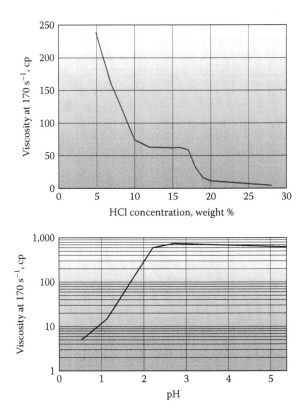

FIGURE 15.12 Viscosity response of VES-base acid fluids. The HCl concentration determines the fluid viscosity as it is pumped downhole (top). The reaction of HCl with the carbonate formation increases pH and produces $CaCl_2$ brine as a reaction product. As a result, wormlike micelles form and fluid viscosity increases (bottom). Reproduced from Ref. 7 with permission of Schlumberger.

15.6.4 Hydraulic Fracturing

Hydraulic fracturing is a stimulation technique that creates a fracture extending beyond the damage area.[18,19] Often, the fracture extends hundreds of feet from the borehole, greatly increasing the surface area of productive formation connected to the wellbore; as a result, the hydrocarbon production rate increases significantly (Figure 15.14). Note that the proportional production-rate increase is greater at lower formation permeabilities. Without hydraulic fracturing, many wells would not produce hydrocarbons at an economically viable rate.

Hydraulic fracturing treatments comprise two basic fluid stages. The first stage, or *pad*, is pumped through casing perforations at a rate and pressure sufficient to break down the formation and create a fracture. The fracture should

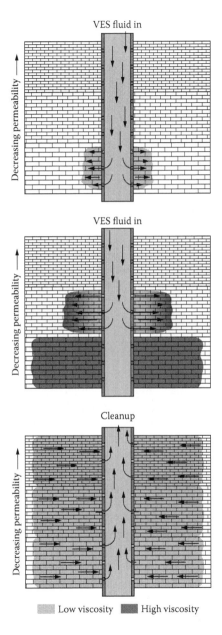

FIGURE 15.13 Matrix acidizing with in-situ diversion. The acid containing VES maintains a low viscosity as it is pumped down the well (top). It enters the most permeable zone (light gray) first. As the acid begins to react, the fluid viscosity increases and flow is diverted to a more permeable zone (center). Upon contact with hydrocarbons, the wormlike micelles in the fluid collapse into spheres and cleanup begins (bottom). Reproduced from Ref. 7 with permission of Schlumberger.

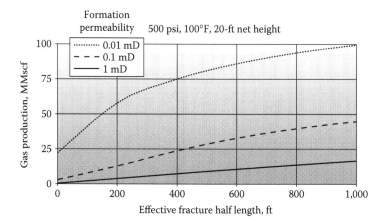

FIGURE 15.14 Effect of fracture length on well productivity for various formation permeabilities. The left side of the graph shows the production rate that would be achieved without stimulation. It is important to note that, proportionally, the benefit of fracturing increases as formation permeability decreases.

remain within the *pay zone* that contains hydrocarbons; otherwise, the efficiency of treatment is not optimal. The fracture geometry is a function of the fracturing-fluid viscosity.

The second stage contains spherical particles called *proppant*, which are carried by the fluid through the perforations into the open fracture. The proppant fills the fracture as the fluid leaks off into the formation (Figure 15.15). When pumping ceases, the fracture closes onto the proppant, thereby preserving the open fracture. Most proppant is made of sand; however, for deeper wells with higher formation pressures, stronger ceramic or bauxite proppants are used. Like gravel for sand control, proppant is available in a variety of sizes.

Proppant transport in fracturing fluids is governed by a complex combination of parameters, including particle size and density, fracture dimensions, and base-fluid rheological properties. Fluid viscosity is particularly important because it provides resistance to gravitational settling, and helps transport the proppant along a fracture. Conventional fracturing fluids employ polymers such as guar and HEC to impart viscosity. As mentioned earlier, field experience and laboratory studies with polymeric fluids have converged on one overarching guideline—for adequate proppant transport, the fluid viscosity should be at least 100 cp at a shear rate of 100 s^{-1}.

After proppant deposition and fracture closure, the fracturing fluid viscosity must be lowered to allow recovery from the well and to open a conductive pathway for hydrocarbon production. As discussed earlier, VES fluids will break spontaneously upon contact with liquid hydrocarbons, leaving little or no residue. If the well produces dry gas, encapsulated salts are added to the proppant slurry. When the fracture closes after pumping ceases, point loading causes the capsules to

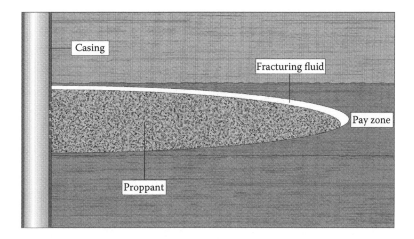

FIGURE 15.15 Hydraulic fracture after proppant placement. Reproduced from Ref. 19 with permission of Schlumberger.

rupture. The salt changes the ionic environment in the fluid, the wormlike micelles collapse into spheres and cleanup commences.

Today, almost all VES fluids used for hydraulic fracturing are foamed.[20] This is done for three principal reasons. Foams allow the use of less surfactant to achieve a given fluid viscosity, reducing the cost of the treatment fluid. The two-phase flow behavior of foams also reduces the rate at which fluid leaks off into the formation. This reduces the volume of fluid required to create a fracture, again reducing the cost of the treatment. Finally, upon cleanup the volume of waste fluid at the surface is substantially reduced. This has both environmental and economic benefits.

15.7 CONCLUSIONS

Viscoelastic surfactant fluids have found application in a number of oilfield operations. These operations include gravel packing, coiled-tubing cleanouts, matrix acidizing, and hydraulic fracturing.

The self-assembly of surfactants into long wormlike micelles gives these systems unique rheological properties and distinct advantages over traditional fluids viscosified with polysaccharide polymer chains. These advantages include:

- Superior logistics and simplicity of mixing
- Minimal formation damage
- Effective drag reduction for flow through wellbore tubulars
- Excellent particle transport abilities for moving solids into and out of the wellbore
- Spontaneous viscosity reduction upon contact with liquid hydrocarbons, improving cleanup after the treatment

Research and development is in progress to improve VES fluid systems. The principal goals are to reduce the fluid cost and increase the practical temperature range.

REFERENCES

1. Raghavan, S.R., Kaler, E.W. *Langmuir* 2001, *17*, 300.
2. Chase, B., Chmilowski, W., Marcinew, R., Mitchell, C., Dang, Y., Krauss, K., Nelson, E., Lantz, T., Parham, C., Plummer, J. *Oilfield Review* 1997, *9*, 20.
3. Rose, G.D., Teot, A.S., Foster, K.L. U.S. Patent 4800036, 1989.
4. Rose, G.D., Teot, A.S. In *Structure and Flow in Surfactant Solutions*, Herb, C.A., Prud'homme, R.K., Eds., ACS Symposium Series No. 578, American Chemical Society, Washington, DC, 1994, Chapter 25, p.352.
5. Brown, J.E., Card, R.J., Nelson, E.B. U.S. Patent 5964295, 1999.
6. Marczak, K.M., Clark, P.E. *Proc. Soc. Pet. Eng.*, paper SPE 63240, 2000.
7. Kefi, S. Lee, J., Pope, T.L., Sullivan, P., Nelson, E., Hernandez, A.N., Olsen, T., Parlar, M., Powers, B., Roy, A., Wilson, A., Twynam, A. *Oilfield Review* 2005, *16*, 10.
8. De Groot, M.C., Kievit, E.A. Tech. Univ. Delft, Publication EV-1670, 1994.
9. Nelson, E.B., Lungwitz, B., Dismuke, K., Samuel, M., Salamat, G., Hughes, T., Lee, J., Fletcher, P., Fu, D., Hutchins, R., Parris, M., Tustin, G.J. U.S. Patent 6881709, 2005.
10. Lee, J.C., Nelson, E.B. U.S. Patent 6908888, 2005.
11. Carlson, J., Gurley, D., King, G., Price-Smith, C., Waters, F. *Oilfield Review* 1992, *4*, 41.
12. Parlar, M., Nelson, E.B., Walton, I.C., Park, E., DeBonis, V.M. *Proc. Soc. Pet. Eng.*, paper SPE 30458, 1995.
13. Ali, A., Blount, C.G., Hill, S., Pokhiral, J., Weng, X., Loveland, M.J., Mokhtar, S., Pedota, J., Rødsjø, M., Rolovic, R., Zhou, W. *Oilfield Review* 2005, *17*, 4.
14. Crowe C., Masmonteil, J., Touboul, E., Thomas, R. *Oilfield Review* 1992, *4*, 24.
15. Daccord, G., Lenormand, R. *Nature* 1987, *325*, 41.
16. Chang, F., Qu, Q., Frenier, W. *Proc. Soc. Pet. Eng.*, paper SPE 65033, 2001.
17. Al-Anzi, E., Al-Mutawa, M., Al-Habib, N., Al-Mumen, A., Nasr-El-Din, H., Alvarado, O., Brady, M., Davies, S., Fredd, C., Fu, D., Lungwitz, B. *Oilfield Review* 2003, *15*, 28.
18. Armstrong, K., Card, R., Navarette, R., Nelson, E., Nimerick, K., Samuelson, M., Collins, J., Dumont, G., Priaro, M., Wasylycia, N., Slusher, G. *Oilfield Review* 1995, *7*, 34.
19. Bivins, C., Boney, C., Fredd, C., Lassek, J., Sullivan, P., Engels, J., Fielder, E.O., Gorham, T., Judd, T., Sanchez Mogollon, A.E., Tabor, L., Muñoz, A.V., Willberg, D. *Oilfield Review* 2005, *17*, 34.
20. Hogelin, P.T., England, K.W., Tabor, L.A. *Proc. Soc. Pet. Eng.*, paper SPE 59377, 1999.

16 Drag Reduction by Surfactant Giant Micelles

Jacques L. Zakin, Ying Zhang, and Wu Ge

CONTENTS

16.1 INTRODUCTION

At concentrations above cmc (critical micelle concentration) surfactant molecules dissolved in aqueous solution assemble into colloidal aggregates such as micelles or vesicles, which vary in shape and size depending on system conditions. Among the variety of micelle structures in solution, wormlike micelles (WLMs), also denoted as cylindrical, rodlike, or threadlike micelles, which resemble the long chain molecules of high polymers, may reduce friction energy loss in turbulent flow by up to 90% at relatively low surfactant concentrations under appropriate flow and temperature conditions. This phenomenon is called drag reduction (DR) and it has significant potential impacts on fluid transport and on the environment.

While high polymers with flexible chains and molecular weight above 10^6 g/mole can be DR effective at concentrations as low as a few ppm, they lose their effectiveness permanently when exposed to extreme shear or extensional stresses which break primary bonds predominantly near the molecules' midpoints. Although WLMs also break up upon mechanical or thermal degradation causing loss of DR, they can self-repair in regions of low stress or at lower temperatures and regain DR effectiveness. The self-assembly of surfactant DR additives (DRAs) permits them to be used in recirculation applications such as in district heating/cooling systems. Since the phenomenon of surfactant DR involves the interactions of thermodynamics and kinetics of surfactant self-assembly and turbulent non-Newtonian fluid mechanics, which are not completely understood, proposed mechanisms are still speculative.[1]

The role of giant micelles in surfactant DR, their effects on turbulent structures, their microstructures, their rheological characteristics, and the variables affecting their DR effectiveness along with practical applications will be reviewed in this chapter.

16.2 TURBULENT DRAG REDUCTION AND EFFECTS ON TURBULENT STRUCTURE

Turbulent flow DR is observed when addition of small amounts of additives such as high polymers or surfactants to a solvent leads to significant reduction in friction energy loss in conduit flows compared to the Newtonian solvent.[2-4] DR in pipe flow is evaluated by comparing the turbulent pressure gradients for the test solution and the Newtonian solvent at the same flow rate. In flow through a pipe of diameter, D, the wall shear stress, σ_w, is related to the pressure drop (ΔP) per unit length (L) by:

$$\sigma_w = -\Delta P \cdot D/4L \tag{16.1}$$

The Fanning friction factor, f, is defined as:

$$f = 2\sigma_w/\rho U^2 \tag{16.2}$$

where U is the mean flow velocity and ρ the solution density, usually assumed to be the same as the solvent density for dilute solutions.

The friction factor of the Newtonian solvent (water, water/cosolvent mixture or hydrocarbon) in turbulent pipe flow, f_s, follows the von Karman (VK) equation:

$$1/\sqrt{f_s} = 4\log_{10}(\text{Re}\sqrt{f_s}) - 0.4 \tag{16.3}$$

which can be approximated by the Blasius equation:

$$f_s = 0.791 \text{Re}^{-1/4} \qquad (16.4)$$

where Re is the solvent Reynolds number defined by:

$$\text{Re} = \rho_s UD/\eta_s \qquad (16.5)$$

where ρ_s is the solvent density and η_s the solvent dynamic viscosity.

The %DR in internal flows is calculated as:

$$\%DR = 100 \times (f_s - f)/f_s \qquad (16.6)$$

Reductions in heat and mass transport in turbulent flows parallel DR (reduced momentum transport).

When the pipe flow Re exceeds a critical value, the flow becomes turbulent, which is characterized by constant generation, dissipation, and regeneration of irregular disturbances such as eddies and vortices. While the motion of these turbulent structures results in increased convection of mass, momentum, and heat in the flow, it also consumes significant amounts of energy and leads to higher pressure drops than would prevail for laminar flow at the same Re. While laminar flow velocity can be described by a well-defined mean velocity profile, instantaneous axial turbulent velocity profiles are a combination of a local mean velocity and a time-dependent local velocity fluctuation, u'.

For Newtonian fluids, the turbulent time-averaged velocity profile is divided into three regions: the viscous sublayer, the buffer layer, and the turbulent core, with local mean velocity increasing with distance from the wall. In DR flows, the buffer layer is thickened and the local mean velocities in the turbulent core are higher than those predicted for Newtonian solvents. While turbulent intensities in the radial and the tangential directions are always lower than those for Newtonian solvents,[5–11] turbulent intensities for DR surfactant solutions in the streamwise direction are lower at low Re and higher at high Re.[5]

Friction factors for DR surfactant solutions may approach a maximum drag reducing asymptote (MDRA):[12]

$$f = 0.32 \text{Re}^{-0.55} \qquad (16.7)$$

This surfactant MDRA lies more than 90% lower than that for Newtonian solutions at high Re (see Figure 16.1). It lies more than 30% lower than Virk's MDRA[13] for high polymers:

$$f = 0.58 \text{Re}^{-0.58} \quad (\text{Re} = 4{,}000 - 40{,}000) \qquad (16.8)$$

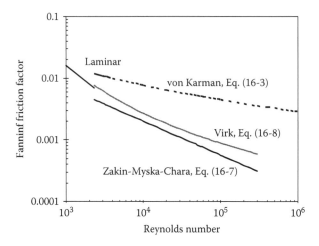

FIGURE 16.1 Maximum drag reduction asymptotes. Reproduced from Ref. 66.

The slope of the ultimate mean velocity profile for MDRA surfactant solutions is steeper than that for MDRA polymer solutions.[12]

16.3 DRAG REDUCING SURFACTANT ADDITIVES

Among various types of surfactant systems studied for DR, soap-type anionic surfactants such as potassium oleate are effective DRAs with good mechanical stability.[14,15] However, their tendency to precipitate in the presence of Ca^{2+} and Mg^{2+} ions in tap and sea water has limited their use as DRAs.[16–18] Both nonionic and cationic surfactants are less sensitive to metal cations. Large effective surfactant microstructures form in nonionic surfactants at their coacervation temperatures (cloud points) and they are DR in a narrow temperature range around the cloud point. Cationics, particularly quaternary ammonium salts with organic counterions, can have broad effective DR temperature ranges and have been the most widely studied surfactant DRAs. They are, however, mildly toxic with poor biodegradability, so there is a need to develop more environmentally friendly surfactant DRAs such as zwitterionic surfactants which biodegrade more rapidly. Desired DR properties can also be obtained by utilizing synergistic effects that may arise when two surfactant species are mixed. Mixed surfactant systems studied for DR include cationic surfactants of mixed alkyl chain lengths, cationic/anionic, nonionic/nonionic, nonionic/anionic and zwitterionic/anionic surfactant mixtures in aqueous solutions and in water/cosolvent systems (see also Section 16.3.6).

Surfactant DR effectiveness depends on the concentrations and molecular structures of surfactants and additives, temperature, pH, cosolvents, and shear, all of which affect micelle microstructure. DR in turbulent pipe flow usually

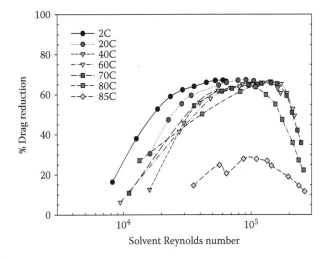

FIGURE 16.2 Drag reduction vs. solvent Reynolds number for 5 mM Ethoquad O12 with 12.5 mM NaSal in water in 6 mm diameter tube. Reproduced from Ref. 79.

increases with Re, reaches a maximum, then at a critical wall shear stress, σ_{wc}, decreases with further rise in flow rate due to breakup of the WLMs by the high stresses in the flow field (see Figure 16.2). DR lost at high flow rates can be recovered when the stresses are reduced allowing the micelles to reassemble more rapidly than they break up. The σ_{wc} values of cationic and other surfactant DR solutions generally show a maximum at some temperature and then decrease at higher temperatures.[19,20]

Chou concluded that WLMs must be long enough to give DR.[19] The three important measures of surfactant DR effectiveness: maximum DR ability, effective temperature range and σ_{wc} depend on the concentration, flexibility and strength of the WLMs.[21] At high temperatures, the WLMs shrink in size with decreasing DR effectiveness. The temperature above which the effective WLM length is too short to be DR effective is the upper limiting temperature for DR, T_{UC}. In cationic surfactant systems, organic counterions such as salicylate bind to the micelles and promote WLM formation. For the same surfactant concentration, one generally observes higher T_{UC} for higher counterion concentrations. The lower limiting temperature for DR for ionic surfactants, T_{LC}, is determined by surfactant solubility corresponding to the Krafft point, the temperature at which solubility equals the cmc. For poly(ethyleneglycol) monoalkylether nonionic surfactants, whose micelles grow in size with rising temperature, phase separation at their coacervation temperatures limits their DR at high temperatures. At low temperatures, nonionic micelles are too small to be DR effective. For both ionic and nonionic DRAs, the lost DR due to temperature changes is reversible as it can be regained when the system temperature returns to its DR effective temperature range.

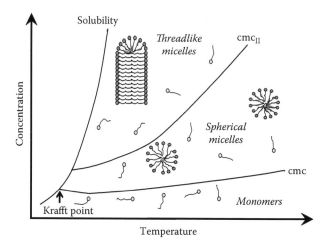

FIGURE 16.3 Schematic phase diagram for cationic surfactant solutions. Reproduced from Ref. 19.

16.3.1 MICROSTRUCTURES

Figure 16.3 is a schematic phase diagram of a dilute aqueous cationic surfactant showing temperature and concentration effects on its microstructures. When the temperature is lower than the Krafft point, the surfactant is partially in solid or gel form in the solution. At temperatures above the Krafft point and concentrations higher than the cmc, spherical micelles form in the surfactant solution. With further increase in concentration and/or upon addition of counterions, the micelles form cylindrical rods or threads or worms, with entangled wormlike and some-times branched wormlike structures. The contour length of WLMs can grow to a few microns while the cross-sectional diameter is invariant at roughly twice the length of the alkyl chain of the surfactant molecules. The concentration at which surfactants form wormlike micelles is called cmc_{II}.[22] While cmc is almost inde-pendent of temperature, cmc_{II} increases with temperature.

Increasing the ionic strength of the bulk media can shield the electrostatic repulsions between ionic surfactant headgroups, reducing the effective headgroup area and promoting WLM formation. Typical DR effective cationic surfactant solutions are dilute with concentrations of a few mM. Cationic surfactant DRAs require the addition of strongly binding organic counterions such as salicylate to form WLMs.[23–28] Oppositely charged surfactants or uncharged small compounds like alcohols may also induce sphere-to-WLM transition.[25] The presence of these additives gives a smaller interfacial charge density on the micelle, shielding headgroup repulsions and allowing denser packing of surfactants in the micelles.[29]

Although WLMs are believed to be necessary microstructures for effective DR,[18,21] vesicles can transform to WLMs under shear as observed by Mendes et al. in their small angle neutron scattering (SANS) studies[30,31] and by Zheng et al.

in their cryogenic transmission electron microscopy (cryo-TEM) studies of surfactant solutions.[32] This explains the surprising observation that a surfactant solution with vesicle structures in the quiescent state showed very good DR ability because they could form DR WLMs in turbulent flow fields.[32]

16.3.2 CORRELATIONS AMONG DR, RHEOLOGICAL PROPERTIES, AND MICELLE MICROSTRUCTURES

DR surfactant systems may contain classic WLMs, branched WLM networks, or vesicles, each of which can exhibit a variety of rheological properties and DR effectiveness. In this section, we classify DR surfactant systems into three categories according to their quiescent microstructures and for each category, correlate DR with rheological properties.

(a) Most surfactant DR systems form WLMs in the quiescent state, which can easily align in the flow direction under shear, giving high %DR and high σ_{wc}. Similar to semidilute WLM solutions ("living" polymers), DR solutions with WLMs generally have distinctive rheological properties such as high zero shear viscosities, shear thinning behavior with increasing shear rate followed by a local rise in shear viscosity caused by the formation of a shear-induced structure (SIS), high viscoelasticity with large first normal stress, N_1, quick recoil and stress overshoot and high extensional viscosity to shear viscosity ratio, θ (above ~100).[18] After reviewing the rheological behaviors of many DR surfactant solutions, Qi and Zakin concluded that SIS and viscoelasticity are not always observed in DR surfactant solutions while high θ values may be a requirement for surfactant solutions to be DR.[21]

(b) Some surfactant DR systems form WLM networks with branching points, that is, three-way joints of WLMs, with fewer WLM ends than those in (a). Branching occurs when the end-cap energy for the formation of semispheroidal WLM end-caps is high enough to overcome the energy associated with the formation of saddle-like branched joints.[33] Porte et al.,[34] Hassan et al.,[35] and Walker and Truong[36] suggested that as the end-cap energy increases significantly at high salt concentration, formation of intermicellar connections is favored and a branched micelle network may form. Branching points can slide freely along primary axes of the WLMs, leading to faster stress relaxation which also reduces the shear viscosity of the solution.[37] The formation of branched micelle networks at high salt concentration is well established from cryo-TEM imaging[38,39] (see Figure 16.4 and also Chapters 5 and 7). They have similar DR effectiveness as WLMs but give lower maximum %DR values, which may be attributed to their smaller persistence lengths compared to those of nonbranched WLMs. The branching points may relieve extra stresses by moving along the cylindrical body of WLMs giving greater flexibility to the branched WLM

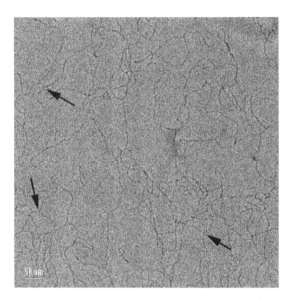

FIGURE 16.4 Branched WLMs (arrows) observed in 5 mM Ethoquad O12/12.5 mM NaSal water solution. Reproduced from Ref. 79.

networks and greater resistance to turbulent stresses than nonbranched WLMs. Therefore, solutions of branched WLM networks exhibit higher σ_{wc} than solutions in Section 16.3.2(a) and the complicated behaviors of branching points in flow give rise to complex rheological properties. Owing to the high flexibility of the microstructures, the solutions generally have low zero shear viscosity with SIS delayed to higher shear rates and low viscoelasticity with zero first normal stress N_I. However they can have either high or low extensional viscosities.

(c) A few DR surfactant solutions form vesicles or liquid crystals in the quiescent state which transform to WLMs in turbulent flow when a critical shear rate is exceeded, thus giving DR capability to their solutions.[32] The shear-induced transformation from non-DR microstructure to WLMs is analogous to the onset of DR in turbulent flow. However, onset of DR may not be observed if the critical shear rate for microstructure transformation is lower than the wall shear rate at the laminar-turbulent transition. In rheological measurements, such DR solutions may exhibit nearly Newtonian behavior, low shear viscosity with no SIS, zero N_I and low θ, although non-Newtonian behaviors might exist at higher shear rates than have been measured.[21]

16.3.3 CATIONIC SURFACTANT DRAs

Because of their relatively low cost and DR effectiveness, quaternary ammonium cationic surfactant DRAs with alkyl chain lengths mainly from C_{16} to C_{22} and binding counterions have been the most widely studied.[40] Concentrations as low as 50 ppm are DR effective.[41] Increasing concentration of cationic surfactant at the same counterion to surfactant molar ratio, ξ, generally enhances a system's DR ability.

16.3.3.1   Effects of Cationic Surfactant Structure on DR

16.3.3.1.1   Structure of the Alkyl Group

Increasing alkyl chain length increases T_{UC} for DR and σ_{wc} values.[19,42–44] Ohlendorf et al. reported an increase in T_{UC} of cationic surfactants of about 8.5°C for each additional $-CH_2-$ in the alkyl chain.[42] However, T_{LC} also increases with the alkyl chain length because of reduction in surfactant solubility at low temperatures.[19,43]

Lin et al. observed odd-even effects of cationic surfactant alkyl tails on the Krafft point and T_{LC} with alkyltrimethylammonium chloride/3-chlorobenzoate (counterion) systems.[45] C_{15} and C_{17} chains gave lower Krafft temperature and lower T_{LC} than expected from interpolation of even chain results, probably because of different packing arrangements inside the micelle cores.

Introducing a double bond on the surfactant alkyl tail increases the polarity of the alkyl chain, enhances low temperature solubility, and extends DR effectiveness to lower temperatures with little effect on T_{UC}.[43,19,46] Qi and Zakin studied the effects of cis vs. trans configurations of monounsaturated C_{18} trimethyl quaternary ammonium salt/sodium salicylate (NaSal) systems.[21] At $\xi = 1.0$ and 1.5, their effective DR temperature ranges were the same but, because of the kinked chain in the cis configuration and its larger volume giving a larger packing parameter, the resulting micelle microstructure gave a higher σ_{wc}. At $\xi = 2.5$, however, the cis surfactant system was insoluble below 50°C while the trans system was DR effective from 4 to 80°C.

16.3.3.1.2   Effects of Headgroup Structure

Rose and Foster[43] and Chou[19] noted that replacing the methyl (Me) groups in quaternary ammonium surfactants with hydroxyethyl (HE) groups enhanced their low-temperature solubility and improved their DR abilities at low temperatures. With the same alkyl group, surfactants with bulkier headgroups have smaller packing parameters than those with smaller headgroups and tend to form smaller aggregates. For example, substitution of a large ethyl group for a methyl on the headgroup (dimethylethylammonium) at 5mM surfactant and $\xi = 1.0$ and 2.5 reduced T_{UC}.[47] However, Chou found that 5mM saturated C_{18} surfactants (~2000ppm)/NaSal ($\xi = 2.5$) with headgroups of increasing size (increasing number of HE: 0, 2, and 3) all had the same DR effective temperature range but σ_{wc} increased with the number of HEs on the headgroup.[19] He attributed the increased σ_{wc} with headgroup HEs to the hydrogen bonding ability of HEs with both water and salicylate counterions, leading to a thick hydration layer at the micelle surface that stabilized and strengthened the micelles. In other experiments with oleyl cationic surfactants with NaSal of different concentrations and different ξ values, the order of DR effectiveness with HE substitutions for Me (0, 1, 2 or 3 methyl) varied.[48–52] The headgroup effect of HE on DR effectiveness apparently depends on both ξ and surfactant concentration.

16.3.3.2 Effects of Counterions

16.3.3.2.1 Counterion Structure Effects

Strongly binding organic counterions, such as salicylate, tosylate, certain mono and di chlorobenzoates, mono and dimethylbenzoates, and hydroxynaphthoates are among the most effective in promoting micellar growth in cationic surfactant solutions.[18,23,37,53] Counterion hydration and strength of dispersion interactions (van der Waals forces) between surfactant headgroups and counterions are important factors determining the binding ability of counterions to micelles.[54] These small molecules have organic portions that can penetrate into the lyophilic core of the micelle.[55] The delocalized π-electron cloud of the aromatic ring may interact with the positively charged surfactant head groups, promoting binding of aromatics to micelles.[56] Organic counterions with delocalized negative charges owing to their resonance structure and which form intramolecular hydrogen bonds[57,58] and bonds with water molecules[59,60] are especially effective in enhancing DR.[19] For example, systems with salicylate counterions usually have high T_{UC} for DR. Hydrogen bonds formed between the carboxylate and hydroxyl groups on salicylate counterions bound to the headgroups of the cationic surfactant micelles help to strengthen the micelle network structure and increase T_{UC}.[19] Figure 16.5 is a schematic showing a section of the elongated cylindrical micelle formed by an alkyltrimethylammonium chloride surfactant in the presence of sodium salicylate.

For aromatic counterions, the positions of the substituent groups influence the penetration ability of the counterions and, therefore, the micelle structures

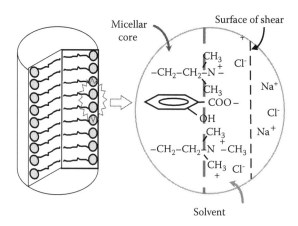

FIGURE 16.5 Schematic of WLM structure of quaternary alkyltrimethylammonium chloride with (sodium) salicylate counterion. Reproduced from Ref. 19.

as well as the properties of their solutions.[61–64] Lu et al. conducted DR and rheological tests of the three cetyltrimethylammonium chloride/sodium chlorobenzoate systems.[63] At ξ = 2.5, rheological and DR measurements and cryo-TEM images showed high viscoelasticity, high extensional viscosity, a well-developed wormlike network, and good DR over a wide temperature range for the 4-chlorobenzoate system. The 2-chlorobenzoate system, which formed spherical micelles with diameters around 5 nm, showed no viscoelasticity and no DR. The 3-chlorobenzoate system was viscoelastic, had high extensional viscosity, but its DR effective temperature only ranged from 30 to 50°C. Different cryo-TEM images of the 3-chlorobenzoate system showed both a wormlike network and large vesicles at 20°C. The microstructure is metastable at this temperature as precipitates were observed after shear in rheological measurements. Presumably, similar precipitation behavior occurred in high shear regions in the turbulent DR measurements at 20°C and no DR was observed, while a more stable, presumably WLM microstructure gave good DR at 30°C. Qualitatively, similar results were observed with the methylbenzoate system but the less hydrophobic methyl groups on the benzoate were less effective as counterions than the chlorobenzoates.[64]

Based on the ability of the 3-chlorobenzoate and 4-chlorobenzoate to induce system viscoelasticity, Bachofer and Turbitt[65] suggested that these counterions are intercalated between surfactant headgroups at the micelle surface, while 2-chlorobenzoate counterions do not intercalate between headgroups but are tilted with their loci tangential to the micellar interface. However, based on nuclear magnetic resonance (NMR) measurements of the cetyltrimethylammonium chloride-chlorobenzoate systems, Smith et al. proposed counterion binding arrangements in which the carboxyl group and the two adjacent (2 and 6) positions on the chlorobenzoates reside in the water phase and the remaining (3, 4, 5) positions on 2-chloro, 3-chloro, or 4-chloro reside in the hydrocarbon core of the micelles (see Figure 16.6).[64] Since chlorine is strongly hydrophobic, the 4-chlorobenzoate with the chlorine deeply embedded in the hydrocarbon core is bound most strongly, the 3-chloro is not so strongly bound, and the 2-chloro is very weakly bound as its chlorine is in the water phase.

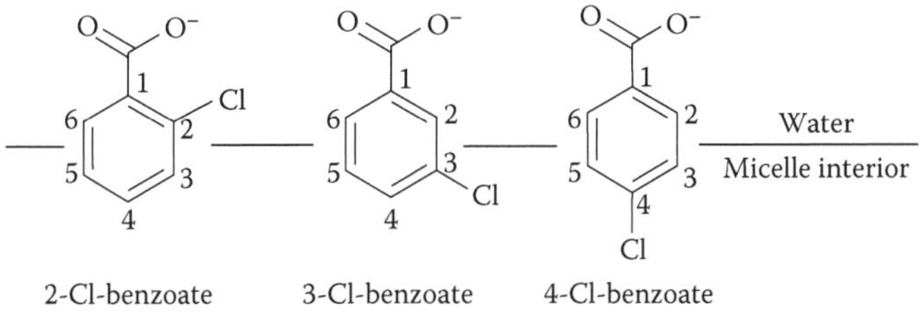

FIGURE 16.6 Orientation of 2-chlorobenzoate, 3-chlorobenzoate, and 4-chlorobenzoate counterions in micelles. Reproduced from Ref. 64 with permission of *Journal of Rheology*.

16.3.3.2.2   Effects of Counterion to Surfactant Molar Ratio, ξ

Increasing ξ up to the solubility limits promotes micelle growth and increases T_{UC} and σ_{wc} for DR.[19,43,66–68] Recent surfactant DR studies suggest that more flexible WLMs give better DR.[21] However, the effect of ξ on WLM microstructures is rather complicated and largely depends on the molecular structures of both the surfactant and the counterion species.

It is generally accepted that when ξ < 1, increasing ξ promotes micelle growth. At ξ = 1, there are free counterions and the cationic WLM surface remains positively charged. It takes ξ > 1 for 100% counterion binding and charge saturation of the micelle surface.

Zeta-potential measurements with tetradecyltrimethylammonium bromide/sodium salicylate at high ξ systems showed a charge reversal on the surface of measurement, which is the shear plane close to the micelle surface.[69] The charge profile on the micelle interface was probed by dye absorption measurements,[70] which showed that even a hundredfold increase in ξ does not lead to charge reversal on the micelle surface, although a high concentration of strongly binding counterions may create an anionic environment close to the micelle surface.

Magid et al. suggested that counterion binding cannot cause surface charge reversal.[23] Later SANS studies with cetyltrimethylammonium/2,6-dichlorobenzoate systems showed increasing WLM flexibility with increasing ξ from 1 to 20.[71] Clarifying the effects of ξ on WLM structures helps understand the effects of ξ on DR. However, one should not expect a universal rule for the effects of ξ on DR because micelle shape and size are determined by the overall geometric packing of specific systems. For example, Chou observed decreasing DR ability with increasing concentration (ξ increased from 1.5 to 2.5) of 3-hydroxy-2-naphthoate, a counterion which is very effective in promoting WLM formation and DR. Considering the large hydrophobic double ring structure of this counterion, he attributed the decrease in DR at high ξ values to increasing rigidity of the WLMs.[19]

Besides the effects on the shape and size of micelles, counterion concentration also modifies the kinetics of WLM networks. Shikata et al. suggested that the junctions of entangled WLMs can be cleaved with the aid of free excess counterions and the two threads can virtually pass through each other like a ghost.[72] Therefore, free counterions behave as catalysts for the disentangling relaxation of WLM networks. Odijk[73] and Hartmann and Cressely[53] proposed that excess free counterions facilitate the scission process of the WLMs, resulting in smaller micelles. Chou, on the other hand, suggested that excess counterions boost the reformation rate of broken micelles by rapidly binding to the micelle surface thus promoting micelle growth.[19]

16.3.4 NONIONIC SURFACTANT DRAS

Solutions of poly(ethylene glycol) monoalkylether nonionic surfactants are DR within a narrow temperature range around their cloud points which is caused by micelle growth at those temperatures. Cloud point determination provides a convenient method for estimating the effective temperature range. With rising temperatures, the dehydration of the poly(ethylene oxide) headgroups leads to reduced headgroup hydrophilicity, favoring micelle aggregation until phase separation occurs as the temperature reaches the cloud point.[74]

Nonionic surfactants such as ethoxylated alkylethers, ethoxylated fatty acid monoethanolamides and their mixtures have been reported to give DR. Chang and Zakin[75] found that mixtures of nonionic surfactants with the structure $C_{18}H_{35}(OCH_2CH_2)_xOH$ have cloud points intermediate between that of each component and are effective DRAs around their cloud temperatures with fairly high σ_{wc} values. The addition of sodium sulfate or phenol lowered the cloud temperature, increased the relative viscosity, and enhanced DR.

DeRoussel made DR measurements on a number of nonionic surfactants with saturated and unsaturated C_{12} to C_{18} alkyl groups and 4 to 23 ethylene oxide (EO) groups.[76] He observed maximum DR around their cloud temperatures. Decreasing alkyl chain length, increasing the number of EO and introducing chain unsaturation to the alkyl group all add hydrophilicity to the surfactant, limiting the size of aggregation and increasing the cloud temperature and thus shifting the DR effective temperature range to higher temperatures. Generally, T_{UC} for DR extends to about 5 to 10°C above the cloud point.[76] Therefore, the cloud temperature and DR effective temperature range can be tuned by selection of alkyl chain length and structure, number of EO groups, salt addition, and pairing of different nonionic surfactants.

Harwigsson and Hellsten studied the low temperature DR abilities of combinations of ethoxylated fatty acid monoethanolamides and alkylethers.[77] One combination of these nonionic surfactants with 5% fatty acid (anionic) was DR effective from 7 to 21°C but only with addition of 1mM $MgSO_4$. The solution lost its viscoelasticity at the cloud point (~12°C). The screening effect of Mg^{2+} ions on the carboxylate anionic surfactant headgroup reduces the electrostatic repulsions at the aggregate surface, promoting micelle growth. Replacing $MgSO_4$ with $NaNO_2$ did not give effective DR. For sea water or hard tap water to which large amounts of $NaNO_2$ are added as an anticorrosion agent, nonionic surfactants of the ethoxylated alkylether type are effective DRAs. Lu found that a mixture of ethoxylated fatty acid monoethanolamides was DR effective at 2°C at a concentration of about 2000 ppm (4 mM), whereas the 3500 ppm solution gave 70% DR from 2 to 12°C and the 5000 ppm solution was effective up to 15°C and to higher Re.[66]

16.3.5 Zwitterionic Surfactant DRAs

Zwitterionic surfactants have both positive and negative charges on their headgroups (zero net charge). They have good solubility in water and are relatively insensitive to salts and temperature. Although more expensive, their environmental friendliness has attracted increasing interest.[78] Among them, carboxylate betaine, amine oxide, aminimide, and their derivatives are effective as DRAs. Details of their DR effectiveness and that of their mixtures are given in Sections 16.3.6.3 and 16.3.7. Zhang studied a family of novel DR zwitterionic surfactants of the aminimide type. Oleyl trimethylaminimide, oleyl aminimide morpholine, and oleyl aminimide methoxyether showed good DR effectiveness in water from 5 to 30°C and can be used as DRAs at low temperatures for district cooling applications.[79]

16.3.6 Mixed Surfactant DR Systems

16.3.6.1 Mixtures of Cationic Surfactants with Different Chain Lengths

Small amounts of C_{12} surfactant added to a C_{22} cationic surfactant significantly reduced T_{LC} from 50 to 2°C but only slightly decreased its T_{UC}.[19,44,80] Incorporation of short chain cationic surfactants into the micelles of longer chain surfactants causes a less-ordered packing in the micelle core, lowers the Krafft temperature, and reduces T_{LC}.[19] Thus, the effective DR temperature range can be tuned by adjusting concentrations of surfactants of different chain lengths, but the mixed system loses DR ability at high C_{12} concentration.[44]

Tomašic et al. pointed out that the spacing between adjacent surfactants increases proportionally to the difference in the chain lengths of the mixed surfactant, that is, the asymmetric alkyl chains cannot pack as efficiently as symmetric alkyl chains.[81] Increase in the asymmetry of the hydrocarbon chain lengths usually results in increased tail transfer energy from the bulk solution to the micelle core.[82] Both explanations indicate weaker hydrophobic attraction between surfactant alkyl groups, which is the main driving force for micellization. Chain asymmetry results in the formation of smaller micelles, which explains the slight decrease observed in T_{UC} of the $C_{12} - C_{22}$ mixture.[19,44,80]

16.3.6.2 Cationic/Anionic Surfactant Mixtures

Studies of the microstructure, rheological, and DR properties of 10 mM solutions of cetyltrimethylammonium tosylate (CTAT) and sodium dodecylbenzene-sulfonate (SDBS) surfactant mixtures in the CTAT-rich region showed that increasing amounts of anionic surfactant SDBS added to CTAT with CTAT/SDBS mass ratios of 100/0, 99/1, 98/1, 97/3, and 96/4, extended T_{UC} and enhanced the solutions' viscosity and viscoelasticity.[83,79] Mixtures with sodium dodecyl sulfate showed similar behavior. Addition of excess sodium tosylate further increased shear and extensional viscosity and T_{UC}. Sodium chloride was less effective. Cryo-TEM images showed that excess sodium tosylate in the CTAT/SDBS (96/4) system induced the formation of branching points within the WLM networks. Qi concluded that the mixed surfactants are both more DR effective and less expensive than pure CTAT.[83]

16.3.6.3 Zwitterionic/Anionic Surfactant Mixtures

Pure C_{16} to C_{18}-sarcosinate zwitterionic surfactants are DR effective at rather high temperatures (60 to 100°C) at low concentrations (500 ppm).[77] Because the net charge of the zwitterionic aminoacid headgroup of sarcosinate surfactants becomes non-zero at both low and high pH, their DR abilities are pH sensitive. This, together with their tendency to crystallize at lower temperatures, limits their DR application potential. A mixture of C_{18}-betaine and branched SDBS at a molar ratio of ~6.5 was a good DRA between 50 and 85°C in the concentration range of 250 to 2000 ppm and was insensitive to pH changes from 7 to 10.5. At low temperature, the C_{16}-betaine/SDBS mixture, at a molar ratio of 12:1, was only weakly DR at rather low Re from 13 to 24°C. Mixtures of 2.5 mM alkylbetaine/0.5 mM SDBS had DR effective temperature ranges of 6 to 60°C for C_{16}-betaine/SDBS and 20 to 100°C for C_{18}-betaine/SDBS.[77]

Details of how zwitterionic and anionic surfactants interact with each other are still not well understood, however, in spite of extensive theoretical[84] and experimental studies.[85] Hoffmann et al. suggested that the large synergistic effects resulting from combining these two surfactants are caused by denser packing of surfactant molecules in micelles compared with individual components.[85] They noted that the electrostatic interaction between surfactant headgroups for such mixtures is usually fully developed with about 20% of anionic surfactant. With its high DR ability over a wide temperature range and good biodegradability, a mixture of C_{16} to C_{18} betaines and C_{11}-C_{13} alkylbenzenesulfonate (sodium salt) at a molar ratio of about 4:1 was developed as a possible commercial DRA by Akzo Nobel.

Amine oxide zwitterionic surfactants, oleyl dihydroxyethylamineoxide (OHAO) and saturated C_{22}, behenyl dihydroxyethylamineoxide (BMAO), are weakly DR effective at 5 to 15°C and 60 to 80°C, respectively. Addition of a betaine type zwitterionic surfactant, 2-lauryl (C_{12}) carboxymethyl hydroxyethyl imidazolium betaine at a molar ratio of amineoxide to betaine of about 1.5, significantly increased DR from 55% to 70% for OHAO at 15°C and from 10% to 83% for BMAO at 80°C.[86]

16.3.7 SOLVENT EFFECTS ON SURFACTANT DRAG REDUCTION

Since the freezing points of water/cosolvent systems are lower than 0°C, their enhanced low temperature capacity for heat removal in district cooling systems would reduce mass flow rate requirements. Further energy savings could be obtained with addition of DRAs.[52] The presence of polar nonaqueous solvent in water generally causes increase in cmc and decrease in aggregation number[87,88] with complicated effects on surfactant self-assembly and therefore on their DR abilities.

Cationic surfactants Ethoquad O12 (oleyl dihydroxymethylammonium chloride) and Ethoquad O13 were tested in water, 15%, 20%, and 28% ethylene glycol (EG) with excess salicylate counterion.[52] The addition of EG decreased T_{UC}, the maximum percentage DR and σ_{wc}. EG and other cosolvents also inhibited the formation of WLMs in these cationic/counterion surfactant systems, decreased relative shear viscosity and θ. Increased concentration of excess counterion promoted the formation of WLMs. The cosolvent effects diminished with decreasing temperature. At 0°C, in surfactant solutions in 20% EG/water the WLMs were better developed than at 25°C and, while not as effective as in water, gave significant low temperature DR.[52]

Commercially available alkylammonium carboxylate zwitterionic surfactants are good DRAs in water and EG/water systems when mixed with the anionic surfactant, SDBS, at a molar ratio of 4:1. However, newly synthesized oleyl aminimide zwitterionic surfactants showed better DR effectiveness in water and EG/water at low temperatures than the carboxylate/SDBS surfactant mixtures.[79] As in the cationic/counterion systems, an oleyl chain improves DR at low temperatures in zwitterionic systems.

Several mixtures of anionic and zwitterionic surfactants were found to be DR effective in pipe flow in a solvent containing up to 40% EG at both low temperature (5~15°C) and high temperature (60~80°C).[86] Hellsten and Oskarsson inferred the DR ability of zwitterionic/anionic surfactant mixtures in EG/water from 30~80°C based on their ability to inhibit vortex formation in a swirling flow.[89] A zwitterionic surfactant increased flow rate and boosted the heating capacity of 17% EG/water in oil production bundles in the North Sea.[90]

16.4 POTENTIAL APPLICATIONS

The temperature reversibility and mechanical recoverability of surfactant DRAs are particularly useful for the very promising application of DR additives in district heating/cooling systems, where hot or cold fluids are circulated in a loop between a heat source or a chiller to exchange heat with buildings in a district. By utilizing more efficient centralized combustors and/or waste heat, district heating provides more efficient use of energy sources, reduces environmental pollution by combustion products, centralizes maintenance, and frees up space in buildings. It is widely used in northern and eastern Europe and use in the US, Canada, Japan, and Korea is expanding.[18] District cooling is often used in the US, Canada, and Japan.

Successful large scale district heating field tests with cationic surfactants giving significant pipe flow energy savings have been carried out in Herning (Denmark),[91] Volklingen (Germany),[92] Prague (Czech Republic)[93] as well as district cooling tests at the University of California at Santa Barbara[94] and in Japan. Saeki et al. reported the application of cationic surfactant DRAs in practical air conditioning systems for energy saving using ice slurries.[95] Utilization of surfactant DRAs to prevent the agglomeration of ice slurries has been investigated.[96] The combination of the ice-dispersion ability and DR effectiveness was reported to improve the performance of ice slurry systems in advanced cold storage, transportation, and heat-exchanging systems.[97]

A novel application of surfactant DRAs lies in their effectiveness in preventing flow-induced localized corrosion (FILC).[98,99] Surfactant DRAs inhibit corrosion because they not only reduce wall shear stresses at a constant Re, but they also suppress the generation of turbulent eddies in the vicinity of the wall, which produce repeated impacts and cause intermittent stresses on the wall, leading to mechanical damage to the surface material.

Other promising novel potential applications of DR micellar systems are in hydraulic fracturing fluids and as anti-misting agents. Other suggested applications are for slurry pipelines, transportation of water for irrigation, reducing sewage line overflow after heavy rains, reducing cavitation, enhancing ship propulsion, firefighting, and jet cutting.[100]

REFERENCES

1. Gyr, A., Bewersdorff, H.-W. In *Fluid Mechanics and Its Applications*, Kluwer Academic Publishers, Dordrecht, Holland, 1995, p.170.
2. Toms, B. A. *Proc. Intern. Rheol. Congr.* 1949, *II; III*, 135.
3. Mysels, K. J. US Patent 1949, 2492173.
4. Agoston, G. A., Harte, W. H., Hottel, H. C., Klemm, W. A., Mysels, K. J., Pomeroy, H. H., Thompson, J. M. *J. Ind. Eng. Chem.* 1954, *46*, 1017.
5. Warholic, M. D., Schmidt, G. M., Hanratty, T. J. *J. Fluid Mech.* 1999, *388*, 1.
6. Gampert, B., Rensch, A. *Proc. ASME Fluids Eng. Div.* 1996, *237*, 129.
7. Chara, Z., Zakin, J. L., Severa, M., Myska, J. *Exp. Fluids* 1993, *16*, 36.
8. Kawaguchi, Y., Tawaraya, Y., Yabe, A., Hishida, K., Maeda, M. *Proc. ASME Fluids Eng. Div.* 1996, *237*, 47.
9. Myska, J., Zakin, J.L., L. Chara, Z. *Appl. Sci. Res.* 1996, *55*, 297.
10. Park, S. R., Suh, H. S., Moon, S. H., Yoon, H. K. *Proc. ASME Fluids Eng. Div.* 1996, *237*, 177.
11. Warholic, M. D., Massah, H., Hanratty, T. J. *Exp. Fluids* 1999, *27*, 461.
12. Zakin, J. L., Myska, J., Chara, Z. *AIChE J.* 1996, *42*, 3544.
13. Virk, P. S. *AIChE J.* 1975, *21*, 625.
14. Savins, J. G. *Rheol. Acta* 1968, *7*, 87.
15. Savins, J. G. US patent 1968, 3361213.
16. Zakin, J. L., Poreh, M., Brosh, A., Warshavsky, M. *Chem. Eng. Prog., Symp. Ser.* 1971, *67*, 85.
17. Shenoy, A. V. *Colloid Polym. Sci.* 1984, *262*, 319.
18. Zakin, J. L., Lu, B., Bewersdorff, H.-W. *Rev. Chem. Eng.* 1998, *14*, 253.
19. Chou, L. C. Ph.D dissertation 1991, The Ohio State University, Columbus, OH.
20. Steiff, W. A., Groth, S., Helmmig, J., Kleuker, H. H., Wocadlo, T., Weinspach, P. W. *Int. Symp. Fluids for District Heating*, 1991, 35, Copenhagen, Denmark.
21. Qi, Y., Zakin, J. L. *Ind. Eng. Chem. Res.* 2002, *41*, 6326.
22. Gyr, A., Bewersdorff, H.-W. in *Fluid Mechanics and Its Applications*, Kluwer Academic Publishers Dordrecht, Holland, 1995, p. 30.
23. Magid, L. J., Han, Z., Warr, G. G., Cassidy, M. A., Butler, P. D., Hamilton, W. A. *J. Phys. Chem. B* 1997, *101*, 7919.
24. Lin, Z., Cai, J. J., Scriven, L. E., Davis, H. T. *J. Phys. Chem.* 1994, *98*, 5984.
25. Rehage, H., Hoffmann, H. *Mol. Phys.* 1991, *74*, 933.
26. Cates, M. E., Candau, S. J. *J. Phys.: Condens. Matter* 1990, *2*, 6869.
27. Hoffmann, H., Schwandner, B., Ulbricht, W., Zana, R. In *Physics of Amphiphiles: Micelles, Vesicles, Microemulsions,* Degiorgio, V., Corti, M., Eds., North Holland Publ., Amsterdam, 1985, p. 261.
28. Oda, R., Narayanan, J., Hassan, P. A., Manohar, C., Salkar, R. A., Kern, F., Candau, S. J. *Langmuir* 1998, *14*, 4364.

29. Bewersdorff, H.-W. *Proc. ASME Fluids Eng. Div.* 1996, *237*, 25.
30. Mendes, E., Menon, S. V. G. *Chem. Phys. Lett.* 1997, *275*, 477.
31. Mendes, E., Narayanan, J., Oda, R., Kern, F., Candau, S. J., Manohar, C. *J. Phys. Chem. B* 1997, *101*, 2256.
32. Zheng, Y., Lin, Z., Zakin, J. L., Talmon, Y., Davis, H. T., Scriven, L. E. *J. Phys. Chem. B* 2000, *104*, 5263.
33. May, S., Bohbot, Y., Ben-Shaul, A. *J. Phys. Chem. B* 1997, *101*, 8648.
34. Porte, G., Gomati, R., El Haitamy, O., Appell, J., Marignan, J. *J. Phys. Chem.* 1986, *90*, 5746.
35. Hassan, P. A., Candau, S. J., Kern, F., Manohar, C. *Langmuir* 1998, *14*, 6025.
36. Walker, L. M., Truong, M. H. *Abstracts of Papers, 221st ACS National Meeting, San Diego, CA, US*, 2001, 18,
37. Ali, A. A., Makhloufi, R. *Colloid Polym. Sci.* 1999, *277*, 270.
38. Danino, D., Talmon, Y., Levy, H., Beinert, G., Zana, R. *Science* 1995, *269*, 1420.
39. Lin, Z. *Langmuir* 1996, *12*, 1729.
40. Chou, L. C., Christensen, R. N., Zakin, J. L. in *Drag Reduction in Fluid Flows*, Sellin, R. H. J., Moses, J. T., Eds., Ellis Horwood Ltd., Chichester, England, 1989, p.141.
41. Kawaguchi, Y., Yu, B., Wei, J., Feng, Z. *Proceedings of the ASME/JSME Joint Fluids Engineering Conference, 4th, Honolulu, HI, US* 2003, *2A*, 721.
42. Ohlendorf, D., Interthal, W., Hoffmann, H. *Rheol. Acta* 1986, *25*, 468.
43. Rose, G. D., Foster, K. L. *J. Non-Newtonian Fluid Mech.* 1989, *31*, 59.
44. Lin, Z., Chou, L. C., Lu, B., Zheng, Y., Davis, H. T., Scriven, L. E., Talmon, Y., Zakin, J. L. *Rheol. Acta* 2000, *39*, 354.
45. Lin, Z., Mateo, A., Zheng, Y., Kesselman, E., Pancallo, E., Hart, D. J., Talmon, Y., Davis, H. T., Scriven, L. E., Zakin, J. L. *Rheol. Acta* 2002, *41*, 483.
46. Laughlin, R. G. in *The Aqueous Phase Behavior of Surfactants*, Academic Press, Harcourt Brace & Company, London, 1996, p.558.
47. Zhang, Y., Qi, Y., Zakin, J. L. *Rheol. Acta* 2005, *45*, 42.
48. Horiuchi, T., Majima, T., Yoshii, T., Tamura, T. *Nippon Kagaku Kaishi* 2001, 423.
49. Horiuchi, T., Yoshii, T., Majima, T., Tamura, T., Sugawara, H. *Nippon Kagaku Kaishi* 2001, 415.
50. Sugawara, H., Yamauchi, M., Wakui, F., Usui, H., Suzuki, H. *Chem. Eng. Commun.* 2002, *189*, 1671.
51. Usui, H., Suzuki, H., Okunishi, T., Sugawara, H., Yamauchi, M. *Proceedings of the 13th International Congress on Rheology, Cambridge, United Kingdom, Aug. 20-25, 2000* 2000, *2*, 294.
52. Zhang, Y., Schmidt, J., Talmon, Y., Zakin, J. L. *J. Colloid Interface Sci.* 2005, *286*, 696.
53. Hartmann, V., Cressely, R. *Rheol. Acta* 1998, *37*, 115.
54. Magid, L. J., Han, Z., Li, Z., Butler, P. D. *J. Phys. Chem. B* 2000, *104*, 6717.
55. Prud'homme, R. K., Warr, G. G. *Langmuir* 1994, *10*, 3419.
56. Lindemuth, P. M., Bertrand, G. L. *J. Phys. Chem.* 1993, *97*, 7769.
57. Shapley, W. A., Bacskay, G. B., Warr, G. G. *J. Phys. Chem. B* 1998, *102*, 1938.
58. Rakitin, A. R., Pack, G. R. *Langmuir* 2005, *21*, 837.
59. Bijma, K., Rank, E., Engberts, J. B. F. N. *J. Colloid Interface Sci.* 1998, *205*, 245.
60. Underwood, A. L., Anacker, E. W. *J. Phys. Chem.* 1984, *88*, 2390.
61. Gravsholt, S. *J. Colloid Interface Sci.* 1976, *57*, 575.
62. Kreke, P. J., Magid, L. J., Gee, J. C. *Langmuir* 1996, *12*, 699.

63. Lu, B., Li, X., Scriven, L. E., Davis, H. T., Talmon, Y., Zakin, J. L. *Langmuir* 1998, *14*, 8.
64. Smith, B. C., Chou, L. C., Zakin, J. L. *J. Rheol.* 1994, *38*, 73.
65. Bachofer, S. J., Turbitt, R. M. *J. Colloid Interface Sci.* 1990, *135*, 325.
66. Lu, B. Ph.D. dissertation 1997, The Ohio State University, Columbus, Ohio.
67. Lin, Z., Lu, B., Zakin, J. L., Talmon, Y., Zheng, Y., Davis, H. T., Scriven, L. E. *J. Colloid Interface Sci.* 2001, *239*, 543.
68. Lu, B., Zheng, Y., Davis, H. T., Scriven, L. E., Talmon, Y., Zakin, J. L. *Rheol. Acta* 1998, *37*, 528.
69. Imae, T., Kohsaka, T. *J. Phys. Chem.* 1992, *96*, 10030.
70. Cassidy, M. A., Warr, G. G. *J. Phys. Chem.* 1996, *100*, 3237.
71. Magid, L. J., Li, Z., Butler, P. D. *Langmuir* 2000, *16*, 10028.
72. Shikata, T., Hirata, H., Kotaka, T. *Langmuir* 1988, *4*, 354.
73. Odijk, T. *J. Phys. Chem.* 1989, *93*, 3888.
74. Israelachvili, J. N. in *Intermolecular and Surface Forces*, Academic Press, San Diego, CA, 1991, p.291.
75. Chang, R. C., Zakin, J. L. *Influence Polym. Addit. Velocity Temp. Fields*, 1985, 61, Symposium Universitat-GH-Essen, Germany.
76. DeRoussel, P. B.S. honors thesis 1993, The Ohio State University, Columbus, Ohio.
77. Harwigsson, I., Hellsten, M. *J. Am. Oil Chem. Soc.* 1996, *73*, 921.
78. Li, F., Li, G. -Z., Chen, J.-B. *Colloids Surf., A* 1998, *145*, 167.
79. Zhang, Y. Ph.D. dissertation 2005, The Ohio State University, Columbus, Ohio.
80. Lin, Z. Ph.D. dissertation 2000, The Ohio State University, Columbus, Ohio.
81. Tomasic, V., Stefanic, I., Filipovic-Vincekovic, N. *Colloid Polym. Sci.* 1999, *277*, 153.
82. Yuet, P. K., Blankschtein, D. *Langmuir* 1996, *12*, 3819.
83. Qi, Y. Ph.D. dissertation 2002, The Ohio State University, Columbus, Ohio.
84. Shiloach, A., Blankschtein, D. *Langmuir* 1998, *14*, 1618.
85. Hofmann, S., Stern, P., Myska, J. *Rheol. Acta* 1994, *33*, 419.
86. Nobuchika, K., Nakata, T., Sato, K., Yamagishi, F., Tomiyama, S., Inaba, H., Horibe, A., Haruki, N. *Proceedings of Symposium on Energy Engineering in the 21st Century, Hong-Kong, China, Jan. 9-13, 2000* 2000, 2, 695.
87. Warnheim, T. *Curr. Opin. Colloid Interface Sci.* 1997, 2, 472.
88. Carnero Ruiz, C. *J. Colloid Interface Sci.* 2000, *221*, 262.
89. Hellsten, M., Oskarsson, H. *Application: WO Patent* 2002, 2002059228.
90. Sletfjerding, E., Gladso, A., Oskarsson, H., Elsborg, S. *Proc. 12th European Drag Reduction Working Meeting*, 2002, Herning, Denmark.
91. Hammer, F. *Inter. Symp. on Fluids for District Heating*, 1991, 139, Copenhagen, Denmark.
92. Kleuker, H. H., Althaus, W., Steiff, A., Weinspach, P. M., *Inter. Symp. Fluids for District Heating*, 1991, 123, Copenhagen, Denmark.
93. Pollert, J., Zakin, J. L., Myska, J., Kratochvil, P., *85th Int. District Heating and Cooling Assoc.*, 1994, 141, Seattle, Washington.
94. Gasljevic, K., Matthys, E. F. *Proc. ASME Fluids Eng. Div.* 1996, *237*, 249.
95. Saeki, T., Tokuhara, K., Matsumura, T., Yamamoto, S. *Nippon Kikai Gakkai Ronbunshu, B-hen* 2002, *68*, 1482.
96. Modak, P. R., Usui, H., Suzuki, H., *HVAC&R Res* 2002, *8*, 453.

97. Inaba, H., Inada, T., Horibe, A., Suzuki, H., Usui, H. *Int. J. Refrigeration* 2004, *28*, 20.

98. Schmitt, G. *Annali dell'Universita di Ferrara, Sezione 5: Chimica Pura ed Applicata, Supplemento* 2000, *11*, 1089.

99. Deslouis, C. *Electrochim. Acta* 2003, *48*, 3279.

100. Gyr, A., Bewersdorff, H.-W. In *Fluid Mechanics and Its Applications*, Kluwer Academic Publishers, Dordrecht, Holland, 1995, p. 202.

17 Giant Micelles and Shampoos

Luc Nicolas-Morgantini

CONTENTS

17.1 INTRODUCTION

Shampoos are the most important hair care products in terms of volume. More than 4 billion bottles, manufactured and marketed by several hundred companies, are sold annually around the world.

A shampoo is generally presented as a liquid, gel, or cream. Its formulation is based on surfactants that give it detergent, wetting, emulsifying, and foaming properties.

17.1.1 FUNCTIONS OF A SHAMPOO[1,2]

The primary function of a shampoo is to clean the hair and the scalp. This must be fulfilled, while respecting the following criteria:

- Remove dirt from hair quickly and efficiently without irritating the eyes, the scalp, or the hands
- Clean correctly with quantities that often vary widely from one user to another (from a few grams to a few tens of grams), at a temperature close to room temperature, irrespective of the level and nature of dirt, the type of hair (abundant or sparse, long or short, smooth or frizzy) and the water hardness
- Develop sufficient foam, which is associated with cleanliness

The expectations of consumers today largely go beyond these functions. That is, the shampoo is required to wash hair, but in the most pleasant way possible by:

- Having a pleasant appearance with appropriate rheology to flow easily from the container and be easily applied onto the hair, particularly in order to avoid the so-called "gelatinous" effect which characterizes discontinuous jelly-like dispensing
- Producing abundant, creamy foam with the right amount of scent but that is easily eliminated on rinsing

Moreover, it is demanded that the shampoo:

- Helps detangling and makes combing easier through wet hair
- Improves the feel of wet hair
- Provides for easy styling of the hair once dry
- Gives body, volume, and discipline to dried hair and makes the hair shine
- Reduces static electricity
- Avoids build up on repeated use of the product
- Fulfills specific needs such as clear from dandruff, keep the scalp in healthy condition, preserve the shade of colored hair, etc.

In other words, a modern shampoo must be attractive, well-scented, pleasant to use, and wash efficiently while providing a number of benefits according to the nature of hair and consumer expectations.

17.1.2 FORMULATION OF SHAMPOOS[1,2]

The primary function of shampoos rests on the use of synthetic detergents. Careful selection is essential because they set the washing power and for a large part the quality of the foam, as well as the level of eye and skin tolerance of the composition. They also condition the nature of the additives that can be used to obtain the sought-after "secondary" properties. At this time, most shampoos use aqueous solutions of sodium or ammonium salts of sulfated fatty alcohols from plants or synthetic, oxyethylenated or not, as the principal detergent, often associated with an amphoteric surfactant of the cocoamidopropyl betaine or cocoamphodiacetate type or sometimes a nonionic surfactant such as an alkylpolyglucoside.

The "secondary" functions of hair conditioning are provided nowadays in high performance shampoos by two major types of ingredients used alone or combined, that is, cationic polyelectrolytes and silicone oils.

Cationic polyelectrolytes are water-soluble polymers, which generally contain quaternary ammonium groups in their main or grafted chain. These groups give them a positive electric charge able to interact with anionic groups or sites. Cationic polyelectrolytes are easily adsorbed on the hair which has a negative surface charge in water, owing to the presence of anionic groups, the number of which strongly increases as a result of sun exposure and oxidative treatments. They have a high affinity with hair and high conditioning effects that are useful for all types of hair.

Silicones, endowed with high chemical inertia, having low surface tension, low glass transition temperature, weak cohesive forces and being insoluble in water, easily "wet" numerous materials, including hair, by covering them with a thin hydrophobic lubricant film.

Furthermore, a large number of shampoos have a pearlescent appearance. This optical effect is obtained by using certain fatty substances that form platelet-like particles, which produce multiple reflections and luminous interferences by crystallizing in the surfactant medium. Ethyleneglycol distearate is a typical pearlescent agent.

Shampoos can also contain thickening polymers such as cross-linked polyacrylic acids or certain polysaccharides.

Finally, shampoos also contain perfume and agents for bacteriological protection.

This chapter is organized as follows. The section that follows this introduction deals with the micelles present in solutions of surfactant mixtures that represent model shampoos. It is shown that they effectively contain wormlike micelles. The effect of several parameters linked to shampoo formulation on the rheological properties of model shampoos is presented in Section 17.3. Section 17.4 deals with real shampoos in the flask and when used. It is shown that their rheological properties are very similar to those of model shampoos. A short conclusion ends this chapter.

17.2 MICELLES IN MODEL SHAMPOOS

Most hydrophilic surfactants are endowed with the property of shifting from molecular solution to spherical micelles then to elongated (wormlike) micelles before reaching the domain of mesomorphic phases (hexagonal, lamellar, etc.) as their concentration is increased.

Shampoos are likely to undergo similar phase transitions as a function of the nature and concentration of surfactants they contain. Since the structure of these phases has a direct influence on the consistency and texture of the shampoos, it is of interest to be able to identify the micellar structure and most particularly to specify at which concentration the transition from spherical micelles to wormlike micelles may take place.

17.2.1 SPHERICAL TO WORMLIKE MICELLE TRANSITION STUDIES

Several techniques can be used to investigate changes in micelle size and shape and to characterize the transition from spherical to wormlike micelle.[3–7] Electrical conductivity, static and dynamic light scattering (SLS, DLS), small angle neutron scattering (SANS), transmission electron microscopy under cryogenic conditions (cryo-TEM), nuclear magnetic resonance (NMR), or differential scanning calorimetry (DSC) can be named.

Our studies have been carried out on simplified shampoos, such as anionic surfactant/amphoteric (zwitterionic at neutral pH) surfactant mixtures in the presence or not of salt, for example, sodium chloride, by using techniques that are sensitive to either the micelle mobility (electrical conductivity, DLS), the motions of surfactant molecules (NMR), or their organization (x-ray diffraction).

The following combinations of technical grade commercial surfactants were investigated:

*   Sodium alkyl diethyleneoxide sulfate (fatty alkyl chain: approximately 2/3 dodecyl and 1/3 tetradecyl; polar headgroup: 2 ethylene oxide units is an average value) also named sodium laurylether sulfate (SLES)
*   N-alkyl-N,N-dimethyl-N-carboxymethyl ammonium, inner salt (fatty alkyl chain: approximately 2/3 dodecyl and 1/3 tetradecyl) also named cocoyl betaine (CB)

In the mixtures, the SLES/CB weight ratio was set at 4/1 and the content of added NaCl at 1.5 wt%. Commercial surfactants may already contain some amount of salt; for instance, CB is commonly supplied as a 30% aqueous solution with 1% NaCl.

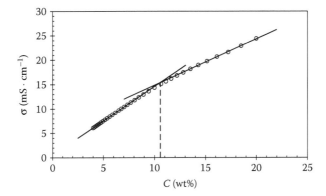

FIGURE 17.1 Variation of the electrical conductivity with the total surfactant concentration for the SLES/CB (4/1 wt/wt) combination.

17.2.1.1 Electrical Conductivity

Since the surfactants studied have ionic headgroups, it is possible to follow any change in the micelle organization by measuring the variations of the electrical conductivity as a function of surfactant concentration, C. Indeed, at a given temperature, the electric conductivity depends on micelle charge, that is, the number of surfactants per micelle and the degree of counterion binding, as well as the extent of mobility in the medium linked to the size and shape of the micelles. When micelles undergo a transition from spherical to cylindrical shape, their mobility decreases and a change in slope is observed on the plot representing the variation of conductivity with the surfactant concentration.

The variations of the conductivity, σ, with the surfactant concentration C, have been recorded at 25°C for SLES alone, CB alone, and SLES/CB 4/1 mixtures with or without added salt. The results are illustrated in Figure 17.1 for the SLES/CB mixture in the absence of NaCl.

It was concluded from this study that:

- A transition takes place when concentration reaches about 17% with SLES without salt and around 12–13% in the presence of salt.
- No transition can be detected with cocoylbetaine (the plot showed a continuous change of slope).
- The transition occurs at C around 10–11% with the SLES/CB mixtures. In the presence of NaCl this concentration drops to 6%.

17.2.1.2 Dynamic Light Scattering

The translation diffusion coefficient measured by dynamic light scattering depends on the shape and size of the micelles. It decreases as the micelle size increases.

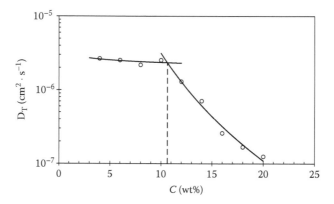

FIGURE 17.2 Variation of the diffusion coefficient with the total surfactant concentration for the SLES/CB (4/1 wt/wt) combination.

A series of measurements has been carried out at 25°C with SLES alone, CB alone, and SLES/CB 4/1 mixtures with or without added NaCl. The values of the average diffusion coefficient were obtained by the cumulant method at a scattering angle equal to 90 degrees.

Figure 17.2 shows the variation of the diffusion coefficient D_T with the surfactant concentration C for the SLES/CB mixture without added salt. The sharp decrease in translation diffusion coefficient indicates a change in micellar pattern at a concentration level between 10 and 11%.

The DLS measurements also reveal that:

- SLES solutions show a similar transition when C reaches around 16% without added salt and 13% in the presence of salt.
- CB solutions present a slow variation in the diffusion coefficient over a large concentration range from 2 to 25% with no pronounced change of slope.
- The SLES/CB mixtures show a transition at around 10–11%, which takes place at only 5–6% when adding NaCl.

17.2.1.3 X-Ray Diffraction

X-ray diffraction experiments have been carried out on SLES and SLES/CB solutions with or without added NaCl.

The transition from spherical to elongated micelles is identified on the diffraction spectrum by the appearance of a diffuse band at large angles when the surfactant concentration is increased (see Figure 17.3). This band, which corresponds to a spacing of 0.42 to 0.44 nm (i.e., a value very close to the diameter

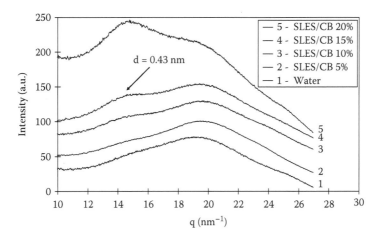

FIGURE 17.3 Wide-angle x-ray diffraction spectrum for the SLES/CB (4/1 wt/wt) combination. The insert gives the total surfactant concentration. q = wave vector.

of the paraffinic chain), is characteristic of an average lateral distance between surfactant molecules whose fatty chains are arranged in a parallel fashion along one spatial direction at least. It is present in hexagonal (H_1 and H_2) and lamellar (L_α) mesomorphic phases above the chain transition temperature (T_C). Below T_C (i.e., in the L_β phase) one or several sharp Bragg peaks reflect the interchain distance.

In wormlike micelles, surfactant fatty chains can take a parallel arrangement perpendicularly to the long micellar axis. In fact, the appearance of a diffuse band at the large angles is observed in a concentration range where wormlike micelles are likely to form as detected by the other techniques.

Figure 17.3 groups the large angle x-ray diffraction spectra recorded at 25°C for the SLES/CB combination at different concentrations (staggered curves). At $C > 10\%$, the diffuse band is observed at 0.43 nm. Its intensity increases with C up to 25%. From 30% upwards it is very intense but it then corresponds to the lateral distance (0.45 nm) between molecules arranged in a direct hexagonal structure (H_1) as confirmed by the small angle diffraction spectrum. This result is shown graphically in Figure 17.4.

From x-ray diffraction measurements the following conclusions can be drawn:

- With SLES solutions, the transition takes place from C around 15% without salt and at a lower concentration, around 12%, in the presence of NaCl.
- In SLES/CB solutions, the transition occurs beyond 10% without added salt and at around 7% after adding NaCl.

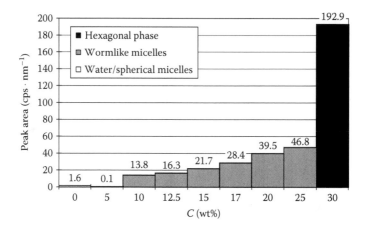

FIGURE 17.4 Intensity of the diffuse band at 0.43 nm as a function of total surfactant concentration for the SLES/CB (4/1 wt/wt) combination.

17.2.1.4 Nuclear Magnetic Resonance[7]

Proton NMR measurement of relaxation times T_1 and T_2 of the CH_2 protons of the aliphatic fatty chains allows the aggregation state of surfactants in aqueous solution to be characterized:

- T_1, longitudinal relaxation time, is characteristic of local rapid motions that is, molecular shifting of the surfactants in solution or in micelles.
- T_2, transversal relaxation time, is linked to slow motions, that is, shifting of the whole micelles.

The times T_1 and T_2 can be obtained by liquid-state NMR measurements. However, when used to follow the changes in size and shape of the micelles with the surfactant concentration, this technique soon shows limitations because when the concentration is increased, the viscosity increases and the mobility of the components is increasingly reduced. This leads to line broadening and prevents the analysis of the spectra. The magic-angle spinning (MAS) technique, a solid-state technique, produces higher resolution spectra with viscous systems and therefore makes it possible to get a precise measurement of the magnetic relaxation time.

Variations in the transversal relaxation rate ($1/T_2$) with the surfactant concentration and temperature have been measured on SLES and SLES/CB solutions with or without added salt. These variations can be attributed to a change in size or shape of components such as a spherical micelle/wormlike micelle transition.

It appears that:

- SLES solutions without salt added do not show changes over the investigated concentration range (1–20%). In contrast, in the presence of NaCl, $1/T_2$ greatly increases beyond 10%.
- SLES/CB mixture solutions show a strong increase of $1/T_2$ at a concentration between 10 and 15%. In the presence of NaCl, $1/T_2$ increases progressively and the transition may take place below the first concentration studied, which was 5%.

At a given temperature, T_1 hardly varies with C, which suggests that there are no significant changes in local motions of the surfactants during the change of micelle size or shape.

The variations of T_1 and T_2 with temperature reflect an increase in local and global motions with temperature.

17.2.2 COMPARISON OF THE DIFFERENT TECHNIQUES

Table 17.1 lists the values of the concentration at which the spherical micelle/wormlike micelle transition was detected by the different techniques used.

Fairly similar results were found with the four techniques as regards the concentration threshold for the spherical to wormlike micelle transition with the exception of SLES solutions in the absence of NaCl where NMR did not detect changes below 20%.

It appears that the micelle shape transition takes place at a lower surfactant concentration when anionic surfactant is combined with amphoteric surfactant than when used alone. The electrostatic interaction between the sulfate group of the anionic surfactant and the quaternary ammonium of the amphoteric betaine leads to an anionic compound with two fatty chains that favors the formation of wormlike micelles.

TABLE 17.1
Total Concentration Threshold (wt%) for Spherical/Wormlike Micelle Transition

	Electrical Conductivity	DLS	XRD	NMR
SLES	17	16	≥ 15	> 20
SLES + 1.5% NaCl	12–13	13	≥ 12	> 10
CB	?	?	—	—
SLES/CB (4/1)	10–11	10–11	≥ 10	> 10
SLES/CB (4/1) + 1.5% NaCl	6	5–6	≥ 7	< 5

When sodium chloride is added, the transition threshold decreases significantly in both types of solution, whether it contains the anionic surfactant SLES
only or the anionic/amphoteric combination SLES/CB.

Considering the nature of the mixtures of surfactants and the total surfactant
(12–18%) and salt (0.2–2%) concentrations generally used in shampoos, the
above results reveal that shampoos will always involve giant micelles.

17.3 RHEOLOGICAL PROPERTIES OF MODEL SHAMPOOS IN RELATION TO THEIR MICELLAR STRUCTURE

17.3.1 GENERAL OBSERVATIONS

The properties of solutions of giant micelles underlie the rheological properties
and texture of shampoos. These structures have been the subject of numerous
studies and have largely been described in the literature (see Chapters 5, 8, and
9).[8–12] A large part of the academic studies have been devoted to reference models
consisting of alkyl ammonium or pyridinium halides (bromide, chloride) in the
presence of particular salts (salicylate, tosylate), combinations of cationic and
anionic surfactants and also well-defined nonionic surfactants. Nevertheless, the
principal conclusions can be applied to industrial systems of interest to us for the
manufacturing of shampoos such as the mixtures of anionic surfactants/ampho-
teric surfactants.[13–15]

The sensitivity of ionic surfactants to the addition of salt that screens the
electrostatic repulsions between polar head groups, leads to the "salt curves"
where viscosity of the shampoo is plotted versus salt concentration and reaches
a peak well known to formulators (see results below).

In the increasing part of the salt curve, the spectacular thickening of shampoos
could be explained by the following sequence:

↗ salt concentration ⇒ ↘ headgroup repulsions ⇒ ↘ surfactant curvature ⇒
↗ scission energy ⇒ ↘ number of end caps of wormlike micelles ⇒ ↗ length of
micelles ⇒ ↗ entanglements ⇒ ↗ viscosity

In the decreasing part of the salt curve, at salinity higher than at the peak,
the decrease in viscosity is most probably related to connections between
micelles, also referred to as micelle branching[16] that reduce entanglements. Moreover, these junction points are likely to "slide" along the micelles, thereby
introducing a new relaxation process:

↗ salt concentration ⇒ ↘ surfactant curvature ⇒ ↗ inverse curvature ⇒ ↗ number
of connections ⇒ ↘ entanglements ⇒ ↘ viscosity

A further increase in salt concentration can bring about a situation where the
micelles form a totally connected network, in which no free micelle end cap can
exist. The notion of individual micelle is therefore no longer valid: one must
think in terms of average "mesh" size. The viscosity of the medium is no longer

ensured by the branches but by the knots of the network that slide along the length of the links.[16]

It should be noted that all of the previously described situations have been clearly observed by electron microscopy (see Chapter 5)[17] and are discussed in Chapters 4 and 8.

Thus, upon increasing salt concentration, micelles progressively turn from linear to branched then to a totally connected network. Finally, if an excess of salt is added (> 10 wt%), lamellar phases are produced which induce turbidity in the sample and a drop of viscosity.

The rheological behavior under flow and oscillation of a technical grade surfactant commonly used in the field of shampoos is illustrated by SLES at 15 wt% + NaCl 4.5 wt%. Such concentrations are well above those for the spherical/wormlike micelle transition and ensure that the solutions contain giant micelles. The measurements have been carried out using a stress-controlled rheometer (RS600 Thermo Electron) in cone and plate geometry. The spectra were determined in the linear viscoelastic regime. The pH of the solutions was set at 7.0.

The flow curve (Figure 17.5) is characterized by a plateau at low shear rate ($\eta_0 = 14.0$ Pa·s), followed by a region of strong fluidification beyond approximately 10 s^{-1}, which corresponds to a constant stress slip (slope -1), a typical feature of an alignment of the micelles under shear that is, of an isotropic/nematic phase transition.

The viscoelastic spectrum is shown in Figure 17.6. At low frequency, the loss modulus G″ is higher than the storage modulus G′, which indicates a liquid character. At low frequency, slopes of 2 and 1 are respectively observed for G′ and G″ as predicted by the polymer reptation theory.[18] At high frequency, G′ levels off ($G_0 = 230$ Pa) and becomes higher than G″, which reflects a quite solid character. The G′ and G″ plots cross at a frequency $f_c = 2.65$ Hz that represents the limit between the liquid and solid-state behavior. This frequency corresponds to a rheological relaxation time for the system: $\tau = 1/2\pi f_c = 60.0$ ms.

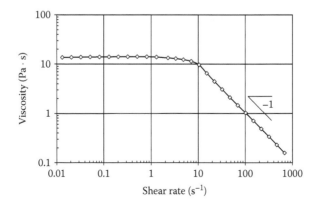

FIGURE 17.5 Flow curve of SLES 15% + NaCl 4.5% at 25°C.

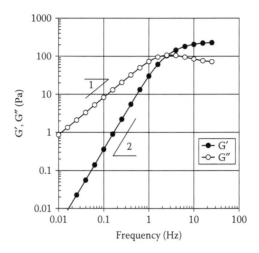

FIGURE 17.6 Viscoelastic spectrum of SLES 15% + NaCl 4.5% at 25°C.

Like many systems in which surfactants are assembled as wormlike micelles, this SLES solution shows the model behavior of a Maxwellian fluid[8,10] for the major part of the spectrum.

These viscoelastic fluids are characterized by a single relaxation time τ, linked to the zero shear viscosity and to the plateau modulus according to the relationship:

$$\eta_0 = G_0 \cdot \tau$$

Consequently, the knowledge of two of these three parameters allows the rheological behavior of the system to be fully characterized.

17.3.2 INFLUENCE OF FORMULATION PARAMETERS

The influence of formulation parameters has been studied on simplified shampoo bases.

17.3.2.1 Nature of the Surfactants

The rheological behavior of the two following surfactant combinations often used in shampoos has been investigated using flow and oscillation measurements:

- sodium laurylether sulfate (SLES)/cocoylamidopropylbetaine (CAPB) (12.5/2.5 wt%), an anionic/amphoteric surfactant combination
- ammonium laurylether sulfate (ALES)/ammonium lauryl sulfate (ALS) (10.0/5.0 wt%), a mixture of two anionic surfactants

in the presence of NaCl 3 wt%.

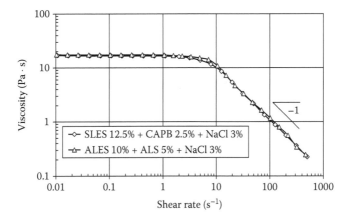

FIGURE 17.7 Flow curves of SLES/CAPB and ALES/ALS combinations with added salt at 25°C.

The two combinations behave similarly. As for SLES solutions, their flow curves (Figure 17.7) are characterized by a plateau at low shear rates, followed by a region of strong fluidification beyond approximately 10 s⁻¹. Besides the viscoelastic spectra shown in Figure 17.8 are quite similar and match those of wormlike micelle solutions. The values of η_0, G_0, and τ, extracted from these curves for the two combinations, are listed in Table 17.2. It is seen that the two

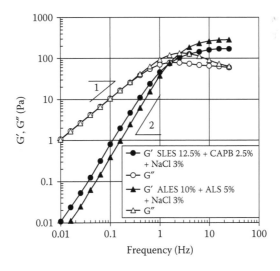

FIGURE 17.8 Viscoelastic spectra of SLES/CAPB and ALES/ALS combinations with added salt at 25°C.

TABLE 17.2

Values of η_0, G_0 , and τ for the Two Surfactant Combinations

	η_0 (Pa·s)	G_0 (Pa)	τ (ms)
SLES/CAPB (5/1) + 3% NaCl	16.6	171	94.0
ALES/ALS (2/1) + 3% NaCl	17.3	286	60.7

systems present very close viscosity plateaus but differ by their elastic modulus and relaxation time. The agreement with the Maxwell model is good with a difference less than or equal to 3%.

17.3.2.2 Nature of the Amphoteric Surfactant

The salt curves, η_0 versus [NaCl], have been determined for SLES 15 wt% and combinations of SLES 12.5 wt% with an amphoteric surfactant (CB or CAPB) at 2.5 wt%. The salt concentration takes into account the salt contained in the supplied surfactants.

Figure 17.9 shows the variation of the plateau viscosity values with the NaCl concentration. First of all, it is seen that the viscosity varies by approximately five orders of magnitude irrespective of the surfactant solution used. The very high peak value accounts for the thickening power of wormlike micelles. The maximum values of the viscosity are of the same order for all tested surfactant systems, between 50 Pa·s (SLES) and 95 Pa·s (SLES/CAPB).

In contrast, marked differences are observed regarding the sensitivity to salt. Thus, the curve obtained with the surfactant combination involving cocoylbetaine is sharply shifted toward lower salt content. For example, to reach a viscosity

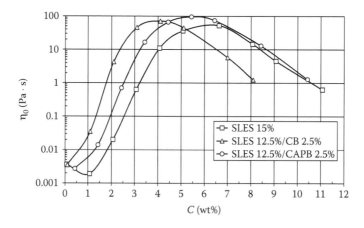

FIGURE 17.9 Effect of salt concentration on the plateau viscosity at 25°C as a function of the amphoteric surfactant nature.

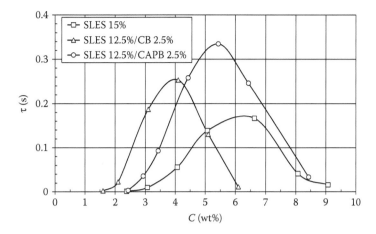

FIGURE 17.10 Effect of salt concentration on the relaxation time at 25°C according to the nature of amphoteric surfactant.

of 10 Pa·s, a total salt concentration of only 2.3 % is required instead of 3.2 % for the SLES/CAPB combination and 4.1% for SLES alone.

The variations of the relaxation time with the salt content are shown in Figure 17.10. They are linked to variations of the plateau viscosity η_0, which explains peak positions similar to those in Figure 17.9. Total salt concentration limits can be fixed in order to avoid the "gelatinous" effect that appears when the relaxation time is too high, typically superior to 60 ms. It is interesting to note that the same salt concentration values are thus obtained as those seen previously that give a viscosity of 10 Pa·s.

17.3.2.3 Total Surfactant Concentration

This factor has been studied with an anionic/amphoteric combination with a 5/1 weight ratio and a constant total NaCl content of 2.5 wt%. In Figure 17.11, the plateau viscosity η_0 is plotted versus the total surfactant concentration. As expected, the viscosity increases with C. However, a very strong increase is noticed as reflected by the slope of the curves that can reach a value of 11 in logarithmic coordinates. In fact, this value is much higher than the exponent n which generally appears in power laws that express the viscosity as a function of concentration: $\eta_0 \sim C^n$. In comparison, the exponent value is 3.5 for solutions of polymers (modified cellulose, polyoxyethylene, polyvinylpyrrolidone, etc.) in a good solvent. In fact, a value as high as 11 is reached here because the length of wormlike micelles is not constant over the investigated range of concentration; it increases rapidly with C.

It is also noticed that the curve for SLES/CB is clearly above that for SLES/CAPB, due to the fact that structural associations between SLES and CB need little salt to form (see Figure 17.9).

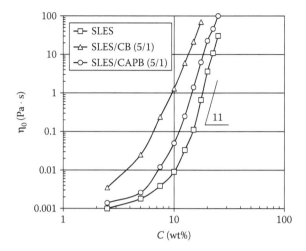

FIGURE 17.11 Variation of the plateau viscosity at 25°C as a function of total surfactant concentration at a given salt content (NaCl 2.5 wt%).

17.3.2.4 pH Variation

The vast majority of shampoos has a pH close to neutral, typically in the range 5.0 to 7.0. Lowering of pH can apparently be a source of problems when amphoteric surfactants are used. These include carboxylate groups that tend to lose their anionic character when getting close to their pKa, giving the surfactant a partially cationic nature: associations with anionic surfactants are then reinforced.

We have studied the impact of pH on the salt curve for the anionic/amphoteric SLES/CAPB combination. The results are given in Figure 17.12. The effect of

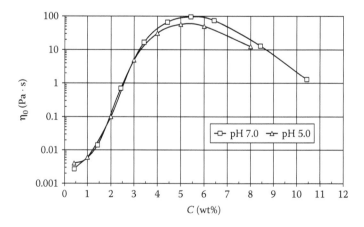

FIGURE 17.12 Effect of pH on plateau viscosity of SLES/CAPB combination versus salt concentration at 25°C.

the pH turns out to be very weak. Specifically it is nil in the usual range of shampoo formulation. A slight reduction in the salt effect is noted with an acid pH, the maximum viscosity values being slightly lower.

17.3.2.5 Type of Salt Added

The nature of counterion is very important in ionic surfactant systems. Indeed, the polar head must be taken in its totality, that is, with the corresponding counterion. The stronger the binding of the counterion to the micelles, the lower the micelle net charge and the stronger the modification of the spontaneous curvature of the surfactant. Since the rheological behavior of the system is controlled by this curvature, one may expect major changes in the texture depending on the type of salt added. In view of clarifying this aspect, we performed a study with SLES 12.5 wt%/CAPB 2.5 wt% shampoo base. The salts tested can be ranked according to the degree of binding of the counterion:

$$MEACl < NaCl \approx NH_4Cl \approx KCl < CaCl_2$$

MEA stands for monoethanolamine.

Figure 17.13 shows the corresponding plateau viscosity versus total salt concentration, expressed in eq/l in order to permit comparisons (this represents the total number of moles of charges present in solution, whatever the valence of the cation).

If one defines the "thickening" power of a salt as its ability to increase the viscosity at low concentration, the results clearly show a dependence on the nature of the cation and define the following sequence:

$$MEACl \ll NaCl < NH_4Cl < CaCl_2 \approx KCl$$

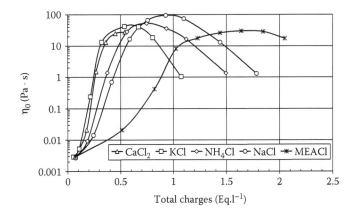

FIGURE 17.13 Effect of the concentration and type of salt on plateau viscosity of SLES/CAPB combination at 25°C.

It also appears that the maximum value of the viscosity depends on the nature of the cation. Thus, the highest viscosities were obtained with NaCl and the lowest with $CaCl_2$. Note that for $CaCl_2$, the decrease from the maximum of the salt curve could not be observed due to cloudiness of the solution, which was linked to the formation of lamellar phases.

Thickening of surfactant systems when salt is added results from an overall curvature reduction, due to reduced repulsion between polar heads. Thus, K^+ and Ca^{2+} are highly efficient in increasing viscosity at low concentration probably because they strongly reduce the net charge of polar heads, due to their higher binding and, thus, induce a stronger micellar growth. The presence of the ether group in SELS may introduce some specificity in the binding of counterions and may be responsible for a stronger binding of potassium than of other monovalent ions.

17.4 RHEOLOGICAL BEHAVIOR OF MARKETED SHAMPOOS

17.4.1 IN THE FLASK

17.4.1.1 Clear Shampoos

Clear shampoos are the closest products to the surfactant solutions described above. The flow curve at 25°C of a currently marketed clear shampoo (A) is represented in Figure 17.14. The curve comprises a plateau zone at low shear rate ($\eta_0 = 6.1$ Pa·s), followed by a region of high fluidification beyond around $30\ s^{-1}$. In fact, the plateau viscosity is representative of the behavior of the product under slow motion in the flask, for example, when the flask is tilted. The

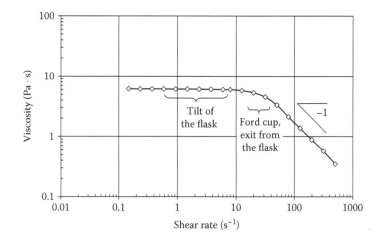

FIGURE 17.14 Flow curve of clear shampoo (A) at 25°C.

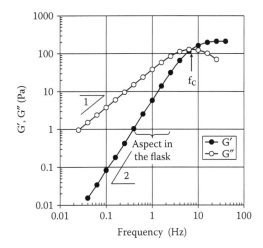

FIGURE 17.15 Viscoelastic spectrum of clear shampoo (A) at 25°C.

intermediate region near the start of the fluidification zone is related to Ford cup measurements[19] or exit of the product from the flask at a quicker flow. In practice, the zone of fluidification is rarely observed since it corresponds to a sudden expulsion of the product. It can be seen, for example if using an inadequate Ford cup, that is, with a too-large opening diameter.

The viscoelastic spectrum of this shampoo is shown in Figure 17.15. At low frequency, the loss modulus G'' is higher than the storage modulus G', which reflects a liquid nature. Slopes of 2 and 1 are found for G' and G'', respectively, as in the case of simple systems. In terms of application, this liquid nature concerns both the perception of the product in the flask ($f \approx 1$ Hz) and storage tests ($f \rightarrow 0$). At high frequencies, G' shows a plateau zone ($G_0 = 212$ Pa) and is higher than G''. The crossover frequency $f_c = 6.8$ Hz corresponds to a relaxation time of 23.4 ms.

The value of the relaxation time is crucial as it conditions the response of the shampoo at a frequency corresponding to its handling. Specifically, a too long relaxation time, typically $\tau > 60$ ms, will be directly responsible for a "gelatinous" effect that may be found unpleasant by the consumer. In terms of usage, this effect is reflected by an irregular flowing out of the flask, for example, running out in blobs followed by flopping back into the flask.

Clear shampoos can also contain cationic polyelectrolytes used as conditioning agents. Shampoo (B) whose rheological behavior in oscillation is shown in Figure 17.16 is representative of this type of product. This composition, which is more complex than the previous one, shows a behavior that departs still further from the Maxwell model. Specifically, in the low frequency region, the G' plot yields a slope below 2. In the high frequency area, after the crossover point, G' does not level out and continues to increase, while G'' hardly decreases. This

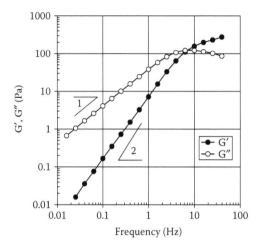

FIGURE 17.16 Viscoelastic spectrum of clear shampoo (B) at 25°C.

behavior reflects the existence of a wide spectrum of relaxation times linked to the modification of τ_{break} and $\tau_{reptation}$ (see Chapters 4, 8, and 10) by the interactions between polymer molecules and giant micelles.

17.4.1.2 Conditioning Shampoos

The rheological behavior of conditioning shampoos, which include a dispersion of silicon oil, a cationic polymer, and a pearlescent agent, is illustrated by shampoo (C) whose flow curve and viscoelastic spectrum, measured at 25°C, are shown in Figure 17.17 and Figure 17.18. The flow curve no longer shows a

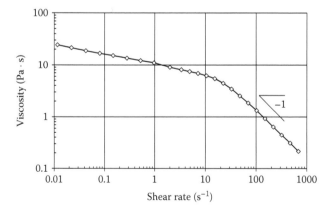

FIGURE 17.17 Flow curve of conditioning shampoo (C) at 25°C.

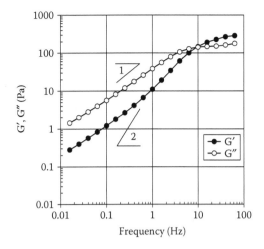

FIGURE 17.18 Viscoelastic spectrum of shampoo (C) at 25°C.

plateau at low shear rate, but a shear-thinning regime with a weak slope (−0.2). At higher shear rate, above approximately 20 s⁻¹, fluidification with the usual slope of 1 is observed. The viscoelastic spectra show large departures from the Maxwell model as in the previous example, but in a still more pronounced way with a rise in the G′ plot in the low frequency region.

Additional studies undertaken on simpler systems have shown the negligible role of the dispersion of silicon oil (average diameter of globules from 1 to 10 μm) on the rheological properties. Similarly, the presence of pearlescent agent, such as glycol distearate in moderate amounts (≤ 2 %), does not alter the rheological behavior either.

Conversely it should be stressed that the presence of dispersed phases of neutralized reticulated polyacrylic acid of microgel type leads away from the Maxwell model. Strictly speaking, these systems should be addressed in terms of granular media, which make their description extremely complicated.

17.4.2 DURING APPLICATION ONTO THE HAIR

When using shampoos, the level of viscosity and the viscoelastic nature of the product contribute to an easy holding in the hand and its easy spreading onto and through the hair. Application to wet hair leads to a threefold to tenfold dilution of the formulation depending on the length and thickness of the hair. As a result, the viscoelastic nature disappears, and the viscosity decreases significantly. For instance, the viscosity of clear shampoos after a threefold dilution drops to about 1.5 mPa·s. Such a low value allows foam to develop and active principles to be deposited.

17.5 CONCLUSIONS

Shampoos are viscoelastic materials owing to the presence of wormlike micelles. They behave more or less as Maxwell fluids depending on the complexity of the formulation.

Knowledge of the micellar structure of surfactant solutions and of the rheological properties of giant micelles is used to fine tune the formulation of shampoos, to improve usage qualities, and more adequately meet with a variety of consumer expectations.

ACKNOWLEDGMENTS

To Dr. F. Simonet for rheological measurements and interpretations; to Dr. F. Clément for light and x-ray scattering experiments; to both of them for fruitful discussions.

REFERENCES

1. Beauquey, B. In *The Science of Hair Care*, Bouillon, C., Wilkinson, J., Eds., Taylor & Francis, Boca Raton, 2005, 83.
2. Dubief, C., Nardello-Rataj, V. *Actualité Chimique* 2004, *274*, 4.
3. Yoshimura, S., Shirai, S., Einaga, Y. *J. Phys. Chem. B* 2004, *108*, 15477.
4. Bergstroem, L.M., Bastardo, L.A., Garamus, V.M. *J. Phys. Chem. B* 2005, *109*, 12387.
5. Liu, S., Gonzalez, Y.I., Damino, D., Kaler, E.W. *Macromolecules* 2005, *38*, 2482.
6. Grell, E., Lewitzki, E., Schneider, R., Ilgenfritz, G., Grillo, I., von Raumer, M. *J. Thermal Analysis and Calorimetry* 2002, *68*, 469.
7. Triba, M.N., Traïkia, M., Warschawski, D.E., Nicolas-Morgantini, L., Lety, A., Gilard, P., Devaux, P.F. *J. Colloid Interface Sci.* 2004, *274*, 341.
8. Cates, M.E., Candau, S.J. *J. Phys. Condens. Matter* 1990, *2*, 6869.
9. Khatory, A., Lequeux, F., Kern, F., Candau, S.J. *Langmuir* 1993, *9*, 1456.
10. Larson, R.G. In *The Structure and Rheology of Complex Fluids*, Oxford University Press, New York, 1999, 551.
11. Raghavan, S.R., Kaler, E.W. *Langmuir* 2001, *17*, 300.
12. Acharya, D.P., Kunieda, H. *J. Phys. Chem. B* 2003, *107*, 10168.
13. Clancy, S.F., Fuler, J.G., Scheidt, T., Paradies, H.H. *Z. Phys. Chem.* 2001, *215*, 905.
14. Blazer, D., Varwig, S., Weihrauch, M. *Colloids Surf. A* 1995, *99*, 233.
15. Penfield, K. *IFSCC Magazine* 2005, *8*, 115.
16. Lequeux, F. *Europhys. Lett.* 1992, *19*, 675.
17. Lin, Z., Lu, B., Zakin, J. L., Talmon, Y., Zheng, Y., Davis, H.T., Scriven, L.E. *J. Colloid Interface Sci.* 2001, *239*, 543.
18. de Gennes, P.-G. *J. Chem. Phys.* 1971, *55*, 149.
19. The Ford cup is a flowing device consisting of a cylindrical reservoir connected to a cone-shaped base with a calibrated cylindrical opening. The test consists in measuring, at constant temperature, the drainage time of a fixed quantity of liquid.

Shmaryahu Ezrahi, Eran Tuval, Abraham Aserin, and Nissim Garti

CONTENTS

18.1 INTRODUCTION

As their name implies, wormlike micelles have two salient features: (1) an elongated, cylindrical shape, and (2) flexibility (i.e., the ability to depart from the optimal curvature).[1] Three structural length scales are pertinent to the flexibility and mobility of these polymer-like aggregates:[2]

1. The cross-sectional radius, r_{cs}, that is independent of the surfactant concentration. Its value is typically a few nanometers.[3]
2. The overall (or contour) length, L. Its value may range in dimension from hundreds of nanometers to several micrometers.[3] The variation

of L with the surfactant volume fraction, ϕ, obeys a power law, which will not be discussed here.

3. The persistence length, l_p, which characterizes the rigidity of the worm-like micelle. Scattering methods (see Chapter 6) yield values of l_p that can vary by orders of magnitude but usually are of tens of nm.[3]

The relation between L and l_p determines the flexibility of elongated micelles. Clearly, when $l_p > L$, the micelles have a stiff, rodlike structure, whereas when $L \gg l_p$, the micelles are considered to be flexible, wormlike or threadlike.[3]

Undoubtedly, the most conspicuous property of wormlike micelles, which has received considerable attention from both experimental and theoretical points of view, is their flow behavior that underlies many of their applications. Wormlike micelle solutions are viscoelastic, that is, their response to mechanical perturbations involves both viscous flow and elastic deformation.[4] It should be noted that viscoelasticity is not unique to wormlike micelles and can be observed, for instance, in block copolymers,[5–7] biopolymers,[8] and plastic sulfur.[9] This property is commonly associated with a three-dimensional network microstructure formed by the entanglement of the wormlike micelles.[10]

Wormlike micelles can be formed from all types of surfactants. However, today consumers are becoming conscious of environmental considerations, and utilizing wormlike micelles made of environmentally friendly surfactants (such as sucrose esters that are biodegradable and biocompatible) is becoming more frequent.[11]

Wormlike micelles are currently being used rather extensively in applications as diverse as oilfield stimulation fluids (see Chapter 15),[12,13] drag reduction (see Chapter 16), and personal care products, shampoos in particular (see Chapter 17). This wide range of applications is based, at least partly, on the properties of wormlike micelles, which have been described in the preceding chapters. Especially, there is an increasing need for multifunctional products, which can transform their characteristics in response to physical and chemical triggers (see Chapter 12).[14]

In the following, three applications of wormlike micelles will be dealt with. Two of them, which can be described, broadly speaking, as small-scale cleaning and cosmetics, will be treated in detail. The third application is drug delivery by wormlike micelles. We realize that in view of the early stage of this technique and the fact that many pieces of information concerning the development of products based on drug-carrying wormlike micelles are kept confidential, the utilization of these micelles for therapeutic purposes cannot be discussed here in depth. Therefore, our review may not represent the state of the art in this spectacular area.

18.2 REQUIRED PROPERTIES OF WORMLIKE MICELLES

In many applications, wormlike micelles function as vehicles and containers for the active ingredients (for instance, a medication or a cleaning agent) of the product. However, the rheological properties of these micelles are perhaps more important. Thus, although home care and personal care products appear to be widely disparate consumer oriented products, their rheological behavior has some generic aspects. It should also be noted that the efficacy of such finished products depends upon the choice and concentration of active ingredients and expedients.[15] Yet, the rheological properties and the components that modify them determine many aesthetical and psychological facets of exploiting these consumer products. Thus, it affects many sensory perceptions induced by attributes such as the physical form, texture, flow behavior, and appearance of the product.[16] Moreover, as we will see in the following, they also influence some practical features such as the penetration of cosmetic actives into the stratum corneum[15] or the lengthening of contact time of cleaners on inclined and vertical surfaces.

In many surfactant-based products, relatively high viscosity is desired. The reasons for this requirement may be purely technical:

1. The maintenance of particles in suspension.[17]
2. Increasing residence time on contaminated surfaces.[18] Nonviscous, free-flowing liquids do not remain in contact with such surfaces long enough. Thickened cleansing agents are, for instance, dish-washing detergents, toilet bowl cleaners, wall cleaners, and shampoos.[18]
3. Easier dosing and easier application to the surface.[18]
4. Nonviscous compositions bring about problems such as misting the product and drift to unprotected surfaces.[19]
5. Undesired splashing and sputtering during application and use is minimized.[20,21]
6. Viscous formulations are compatible with a large variety of adjuvants such as suds boosters, abrasives, or disinfectants that enhance effectiveness.[18]

However, reasons directed at the user are also known:

1. Thickening of the product will confer a feeling of "quality" on the formulation.[15]
2. Many consumers have the impression that viscous compositions convey strength and aesthetic attributes.[18]
3. Thickened products can impart a particular sensory feel that is favored by the consumer.[17]
4. Viscous compositions can be dispensed not only by simple pouring but also by squeezing from tubes, for example.[18]
5. Viscous products are compatible with additives such as colorants, perfumes, emollients, and the like that enhance consumer appeal.[18]

Indeed, increased viscosity is just one aspect of rheological control. Rheological profiles will affect how easy the product is to manufacture, how the product appears in the packaging, how easy it is to pour or scoop from the packaging, how it is applied to the skin or hair, and how easy it is to rinse and remove after use.[15]

An obvious shortcoming of aqueous surfactant formulations is their low viscosities. Adding salts and polymers is a common solution, but the use of wormlike micelles that have relatively high viscosities at low shear rates is also considered in practice.[17] The variation of the viscosity of threadlike micelles as a function of shear rate is advantageous for many applications as will be shown below.

18.3 HOME-CARE PRODUCTS

The aforementioned properties of wormlike micelles underlie many applications of these micelles in home-care products as will be illustrated in the following examples.

18.3.1 Hard Surface Cleaning

In this category we include floors, tiles, toilet bowls, kitchen sinks, and the like. In products intended to clean hard surfaces and based on viscoelastic surfactants, the two conspicuous features of the amphiphiles, namely elasticity and viscosity, are manifested. The obvious advantages and the appeal of viscous products have been already discussed. In formulations intended for hard surface cleaning (as well as in herbicide formulations)[22] the drifting of mist droplets (i.e., gas-suspended liquid particles which have a diameter of less than approximately 10 μm, whereas spray droplets have a diameter of greater than about 10 μm)[19,23] is highly undesirable. The unhealthy vapors of the sprayed product can be inhaled by the user and the mist can drift onto unprotected surfaces, such as furniture, clothing, skin, eyes, and so on.[23] The antimisting property, which thus has to be conferred to such products, stems from the elastic characteristic of the viscoelastic surfactant in these formulations.[22] Elasticity also causes the stream of the product to break apart and snap back into the container at the end of the pouring instead of forming streamers of syrupy and stringy liquid. Moreover, elastic fluids appear more viscous than is indicated by their viscosity.[20,24]

Sink cleaners, for example, should have a high surfactant concentration in order to solubilize grease and resuspend particulate material. Breaking up the films typical of stained sinks without producing deep scratches or a "gritty" feel necessitates the presence of 1–2 μm abrasive particles in the formulation. Worm-like micelles confer a creamy feel, but they can suspend the abrasive particles only when the system is gelled by the addition of soluble polymers. The polymers also enhance the feel and texture of the product.[25]

The need for viscosified products has led to the thickening of surfactant-based compositions. For instance, betaines function as surfactants and wetting agents in a variety of cleansing (hand dish-washing detergents, toilet bowl cleaners, etc.) and industrial (alkaline and acidic cleaners, heavy duty cleaners, metal finishing and electroplating formulations, and the like) compositions.[18] They are mild, high foaming, biodegradable, and compatible with high concentrations of essential adjuvants (such as detergency builders, electrolytes, alkalis, and acids). The compatibility of betaines with quaternary ammonium compounds and other germicides makes betaines useful in disinfectant and sanitizing cleaning agents and in antibacterial scrubs.[18] Thus, aqueous solutions containing alkylamidobetaines and certain water-soluble inorganic (for example, sulfates or carbonates) and organic (such as citrates or succinates) salts of metals (for instance, alkali metals) are the basis of a patent which is particularly useful for the preparation of a highly viscous light duty cleaner.[18] Some typical formulations based on alkylamidobetaines are shown in Table 18.1. Betaines can also function as thickening agents alone. They are used, for instance, to thicken bleach compositions (i.e., materials that lighten or whiten a substrate via chemical action).[26] These compositions are

TABLE 18.1
Typical Formulations Based on Betaines (Source: Ref. 18)

Ingredient	Formulation (in wt%)			
	A	B	C	D
Cocoamidobetaine	15.0	12.0	15.0	13.6
Tetradecylbenzylmethylammonium chloride, dihydrate	—	—	2.0	—
Alpha olefin sulfonate	—	—	—	2.2
Sodium bisulfate	20.0	—	—	—
Ethyl alcohol	—	—	0.5	—
Sodium citrate dihydrate	—	20.0	25.0	13.5
Sodium silicate, ($SiO_2/Na_2O=2.4$)	—	5.0	—	—
Calcite powder	—	—	—	10.0
Dye	0.0003	0.0003	0.0004	0.0003
Fragrance	—	0.2	—	0.15
Water	65.0	62.8	57.5	60.5
Brookfield viscosity (cP)	44,000	2,000	92,000	166,000
pH (as is)	0.13	11.5	7.28	8.6

Formulation A is a gel-form acidic toilet bowl cleaner. The viscous product adheres to the surface and stays in prolonged contact with it.

Formulation B is a viscous alkaline liquid which is well suited for the cleaning of soiled vertical surfaces (such as walls, house sidings, and so on). In diluted form it may be used to wash floors.

Formulation C is a viscous, clear detergent sanitizing gel which is suited for bathroom cleaning, because it clings to tile walls, toilet bowls, bathtub, and the like for a long time.

Formulation D is a hand cleaning gel which is nonirritating to the skin and is readily rinsed in water. The combination of an alkylamidobetaine, an anionic surfactant, and certain salts has a synergistic effect on the viscosity of the product.[18]

generally aqueous solutions of alkali metal and alkali earth metal hypochlorites. Bleach compositions function as cleaners, disinfectants, bactericides, and fungicides.[23] They are widely used in home and institutional laundering.[26] They are effective cleaners of dishes, glassware, and porcelain items (such as sinks and bathtubs).[23] Typical scouring cleaners designed to clean hard surfaces are composed of a finely ground abrasive, a synthetic detergent, auxiliaries, and chlorine bleaches (such as sodium or potassium salts of chlorinated isocyanuric acid and chlorinated trisodium phosphate).[26] In the context of wormlike micelles we are interested in the ways by which the bleach compositions are endowed with viscoelasticity. One way is to use a binary system of a betaine (most preferred cetyldimethylbetaines; CEDB) or sulfobetaine and an anionic organic counterion such as sodium xylenesulfonate.[21] Other thickening agents for bleaching compositions are:

1. Systems of quaternary ammonium salts (most preferred cetyltrimethyl-ammonium chloride, CTAC,[20,24] or cetyltrimethylammonium bromide, CTAB[23]) and organic counterions, such as alkyl or aryl carboxylates, alkyl or aryl sulfonates, and sulfated alkyl or aryl alcohols.[20,24] For example, a formulation for hard surface cleaning contains 4 wt% sodium hypochlorite (as a bleaching agent) and 0.9 wt% hexadecyl-trimethylammonium p-toluene sulfonate. Note that the typical salicylate counterion is replaced by p-toluene sulfonate and the surfactant tail is a saturated hydrocarbon chain, thereby imparting the required chemical stability in the presence of a strong oxidizer. At high shear rate the viscosity of this formulation is low so that the solution is readily dispensed out of a hand sprayer, whereas at low shear rate the solution becomes viscous and thus can stay on vertical surfaces and have enough time to clean them.[22] CTAB also participates in thickened and acidic liquid cleaning compositions for removing mineral deposits from surfaces. Conventional acidic cleaners suffer from several shortcomings:[27]

   i.   The strength of the acid is substantially reduced upon dilution.
   ii. The low pH destabilizes fragrances and dyes of the acidic cleaners and may endanger consumers.
   iii. Ingredients in the cleaning formulations and the packaging of the products may deteriorate due to the low pH.
   iv. Some thickeners lead to viscous products that are difficult to apply from conventional nozzle dispensers.

   A preferred formulation is based on a combination of citric acid and sodium citrate as a buffer, a system of CTAC + sodium xylenesulfonate (or alkyl diphenylether sulfonate) for viscosifying the acidic cleaner, fragrance agents, dye agents, solvents and the balance water. Thus, the composition has a pH of about 2 to 3.5 and it was demonstrated that the buffered composition removes hard water deposits more effectively than conventional unbuffered mineral acid formulations at the same pH values. Moreover, the buffered formulation can accommodate stable fragrances and dyeing agents.[27]

2. Bleaching compositions can also be thickened by amineoxides. In one patent, a trialkylamineoxide and an alkali metal soap of a saturated fatty acid form the viscosifying system for alkali metal hypochlorites in the presence of a small amount of a strong electrolyte (for instance, sodium chloride, trisodium orthophosphate, and similar compounds) and a phosphate or carbonate buffer system maintaining a pH of 10-12.5 at 25°C.[28]

Another patent, related to hypochlorite cleaning compositions, addresses the problem of the characteristic and often objectionable bleach (chlorine) odor. The thickening system is composed of tetradecyldialkylamineoxide and a hydrophobic organic counterion such as diphenyloxide sulfonate.

Viscosity can also be promoted by adding an electrolyte such as alkali metal salts (for example, phosphates, silicates, carbonates, etc.). Such formulations reduce the bleach odor when they are dispensed onto a contaminated surface.[27]

A system composed of a tertiary amineoxide, an alkylpolyglycoside, and a water soluble organic solvent (a monohydric or polyhydric alcohol, a glycol ether or an alkanolamine) forms a low-viscosity cleaning concentrate which would be viscosified upon dilution with water. The formulation also contains suitable alkalis (hydroxides, carbonates, and silicates of sodium or potassium), chelating agents (such as EDTA or nitriloacetic acid), and builders (inorganic such as phosphates, carbonates or borates and organic such as polymers and copolymers of acrylic acid). These formulations are particularly designed for the cleaning of large surfaces as encountered, for instance, in the food and beverage industry, in canteens, warehouses, swimming pools, slaughterhouses, and the like.[29]

18.3.2 DRAIN OPENING

Drain opening consists of the cleaning out or removal of congestion or obstructions that lead to clog or restrict drains. These are formed due to the gradual buildup of sludge, organic waste (such as food soils, grease and oil, hair, lint, soap scum, and the like) slime, and so on in plumbing.[27] Drain openers may act by any one of the following methods:[27]

1. Chemical reaction between the opener and the clog material in order to dissolve, disperse, or fragment this material.
2. Physical interaction (adsorption, absorption, heating (to melt grease), solvation, etc.) which tends to loosen the deposits of the contaminating materials, thereby releasing stoppage in the pipe work.
3. Enzymatically catalyzed reaction in order to render the blockage more water soluble or dispersible.

Drain unclogging is based on a variety of active ingredients, including acids[20,24] (for example, citric, acetic, or boric and dilute solutions of strong inorganic acids, such as sulfuric acid), bases (such as alkali metal hydroxides or sodium and potassium carbonates and silicates), oxidants (for instance, hydrogen peroxide and peracetic acids), reducing agents, solvents (such as saturated hydrocarbons, ketones, terpenes, carboxylic acid esters, and glycol ethers), enzymes (for example, proteases,[20,24] lipases,[19] amylases,[20,24] and cellulases[20,24]), thioorganic compounds (for instance, sodium thioglycolate), and surfactants (detergents such as taurates, sarcosinates, and phosphate esters and mixtures thereof).[20,24] This variety of active compounds is needed to remove the variety of typically encountered materials which can cause clogging or restrictions of drains. Thus, thioglycolates, for instance, help to break down hair and other proteins.[20] Alkali metal hydroxides and hypochlorites, separately or in various combinations, are particularly suitable for drain opening. As in the case of hard surface cleaning (*vide supra*), it is highly desirable to thicken the solution of the active material. However, some problems typical of drain opening should be addressed.

A pool of standing water is frequently formed in the clogged drain. The drain opener composition, no matter how effective it is, will be significantly diluted and thus will not be fully delivered to the clog. Some attempts have been made to increase the viscosity of the formulation and mitigate dilution.[20] An alleged suitable formulation includes:[19]

- An inventive viscoelastic thickening system composed of a tetradecyl-dialkylamineoxide and a hydrophobic organic counterion.
- An alkali metal hydroxide (preferably sodium or potassium hydroxide) that is present in an amount of between about 0.5% and 20%.
- Optionally an alkali metal silicate, having the formula $M_2O(SiO)_n$, where M is an alkali metal and n ranges between 0.5 and 4. When M is sodium, n is preferably 2.3. The alkali metal silicate may be present in an amount of about 0% to 5.0%.
- Optionally, an alkali metal carbonate, preferably sodium carbonate, present in amounts between 0% and 5.0%.
- A drain opening active compound, often alkali metal hypochlorites, separately or together with alkali metal hydroxides. The active compound is present with a preferred concentration range of approximately 4% to 8%.
- Sodium chloride or other similar salts may be added as densifying agents, that is, substances which impart a density greater than that of water to a particular formulation, thereby facilitating the penetration or flow of the formula via standing water.
- The drain opener may also contain conventional adjuncts, such as corrosion inhibitors, dyes, and fragrances.[19]

Some experiments were conducted in order to find the optimal composition and concentration of the organic counterion. A typical drain opener formulation used in these experiments contains (in wt%):[19] sodium hypochlorite, 5.8; sodium hydroxide, 1.75; sodium chloride, 4.5; sodium silicate, 0.11; water, balance. This

TABLE 18.2
Thickening as a Function of Phenylsulfonate Chain Length (Source: Ref. 19)

Sample No.	C (%)	m	C_{max} (%)[a]	Maximum Viscosity (cP)
1	85	16	0.51	585
2	95	16	0.84	2,400
3	90	16	0.45[b]	1,175
4	95	12	0.29	80
5	85	10	—	1

[a] C_{max} = concentration of the sulfonate at maximum viscosity.

[b] The surfactant in sample 3 may have contained 5% monoalkyldiphenyloxide monosulfonate as impurity.

composition is viscosified with a mixture of tetradecyldimethylamineoxide (AO) and alkyldiphenyloxidesulfonate (DS). In the following, results related to alkyldiphenyloxidedisulfonate are shown. Three parameters have been found to affect viscosity:

1. The carbon number m of the stated alkyl group of the sulfonate (see Table 18.2)
2. The concentration C in percentage of the stated alkyl group of the sulfonate, the balance being the percentage of other alkyl chains present in the surfactant (see Table 18.2)
3. The amineoxide/disulfonate weight ratio, C_{AO}/C_{DS} (see Table 18.3)

The viscosity values were measured at 20°C on a drain opener that has the above-mentioned formulation with 0.65% amineoxide. It is seen that the longer the chain length, n, the higher the viscosity (see, for instance, samples 2 and 4 in Table 18.2).

TABLE 18.3
Effect of the Amineoxide/Disulfonate Ratio on Viscosity (Source: Ref. 19)

Parameter	Formulation			
	A	B	C	D
C_{AO} at maximum viscosity (%)	0.52	0.49	0.57	0.65
C_{DS} at maximum viscosity (%)	0.18	0.32	0.40	0.80
C_{AO}/C_{DS} at maximum thickening	1.77	1.55	1.43	0.81
Maximum viscosity (cP)	272	278	400	2,400

Also, it is seen that the higher the concentration, the higher the viscosity (see, for example, samples 1 and 2 in Table 18.2).

The viscosity values were measured at 20°C on a drain opener that has the above-mentioned typical formulation. Again, it is seen that the viscosity increases with C_{AO} and C_{DS}. It also increased with decreasing C_{AO}/C_{DS}.

Another solution was suggested by Rader and Smith.[20,24] Their formulation is based on an aqueous solution of a drain opening active component, an alkyl quaternary ammonium compound or surfactant, a bleach-stable organic counterion, and a free amine. An appropriate amount of the drain opener is poured into a clogged drain. The viscoelastic thickener virtually holds the active components together and, thus, the solution travels via standing water almost without dilution. This thickener also increases percolation times through porous or partial clogs, thereby facilitating clog removal.

It was observed that viscosity alone will not lead to good performance. The relative elasticity, τ/G_0, where τ is the relaxation time and G_0 is the static shear modulus, determines the efficacy of the drain opener. Too elastic formulations are undesirable because some consumers do not like the appearance of elastic flow properties. However, purely viscous compositions cannot operate in standing water unless their viscosity is very high (above ca. 1000 cP). Such viscous formulations suffer from the following drawbacks:

- Difficult dispensing in low temperatures.
- Inefficient penetration into clogs.
- Too viscous compositions are not attractive to consumers.
- Reaching high viscosities is expensive.

Consequently, good performance is expected when the product is elastic and has some viscosity.[20]

The assertion that τ/G_0 plays a dominant role was verified using two parameters for measuring the performance of drain openers (Table 18.4):

1. Percentage delivery, that is, percentage of undiluted product passing through standing water.
2. Flow rate through a specific screen.

Table 18.5 lists the compositions of formulations A–E.
The following observations are noteworthy:[20]

- The nonviscoelastic (albeit moderately thickened) formulation A has a very low delivery value and a high flow rate.
- Formulation C has only about half the viscosity of formulation B and their relative elasticities are not much different. However, the performance of formulation C as displayed by the delivery parameter is much higher.

TABLE 18.4
Effect of τ/G_0 on Drain Opener Performance (Source: Ref. 20)

Formulation	Viscosity (cP)	τ/G_0 (sec/Pa)	Delivery (%)	Flow Rate (ml/min)
A	141	0.016	6	92
B	334	0.058	47	52
C	140	0.075	93	55
D	7	0.35	74	133
E	21	2.91	90	120

- Formulations D and E also have low viscosities, but as their relative elasticities are high enough, they demonstrate good drain opening performance.
- Rader and Smith[20] estimate that a preferred relative elasticity should be above about 0.03 sec/Pa. Values above about 0.05 sec/Pa are still better. The most preferred values are above about 0.07 sec/Pa.
- A preferred flow rate is less than ca. 150 ml/minute, whereas a preferred delivery percentage should be above about 50%, more preferred is above ca. 70%, and most preferred value is above about 90%.[20]

TABLE 18.5
Compositions (in wt%) of Formulations A–E in Table 18.4

Ingredient	Formulation (in wt%)				
	A	B	C	D	E
Sodium hypochlorite	4.5-6.0	4.5-6.0	4.5-6.0	5.8	5.8
Sodium chloride	-	-	-	4.55	4.55
Sodium hydroxide	1.2-1.8	1.2-1.8	1.2-1.8	1.5	1.5
Sodium silicate ($SiO_2/Na_2O = 3.22$)	0.1-1.1	0.1-1.1	0.1-1.1	0.113	0.113
Sodium carbonate	-	-	-	0.25	0.25
Myristyldimethylamine oxide	1.6	0.8	-	-	-
Cetyltrimethylammonium chloride	-	-	0.62	0.100	0.200
Sodium lauroyl sarcosinate	0.37	-	-	-	-
4-chlorobenzoic acid	-	-	0.09	-	-
Lauric acid	-	0.25	-	-	-
Primacor 5980*	0.03	-	-	-	-
Sodium xylene sulfonate	-	-	0.29	0.050	0.100
1-Naphtol acid	-	-	-	0.050	0.050

* A trade marketed product of the Dow Chemical Co., comprising a copolymer of acrylic acid and ethylene.

A recently addressed problem underscores the difference between full drain clogs (wherein flow is blocked) and partial (slow flowing) drain clogs.[30,31] Elastic compositions are effective on full drain clogs because—as has been formerly explained—they maintain the active concentration upon traversing the standing water at a higher delivery rate than a purely viscous composition of the same viscosity.[31] However, percolation times of such elastic compositions via partial (or porous) clogs are relatively short and, thus, the contact time between the partial clog and the actives does not suffice to effectively disintegrate the clog.[31] On the other hand, viscous formulations have longer residence times on partial clogs, thereby enhancing their breakdown or dissolution, but they are not as effective in traversing standing water.[30] Suitable formulations are supposed to treat both types of clogs. One patent uses a mixture of hexadecylamineoxide and tetradecylamineoxide for the thickening system,[30] whereas another patent uses several mixtures, including C_{14}-C_{16} alkyloxylated or alkyl sulfates.[31]

18.3.3 PAINTS

Another application for wormlike micelles may be found in paints. For storage stability, paint at rest should behave like a solid (gel) in order to prevent settling (i.e., formation of hard sediments at the bottom of the container) of its pigment and binder particles. This behavior is due to the open network structure of the paint that is based on weak interparticle interactions.[32,33] On the other hand, for uniform painting the paint should behave like a fluid (sol) so that when brushed it flows in the bristles. This goal is achieved as the paint flocs are sheared by shaking, stirring,[32] or applying. Indeed, during application the paint is subjected to a shear rate range covering more than 6 decades. Thus, the shear rate $\dot{\gamma}$ equals 10^5 s^{-1} at spraying and 10^3–10^4 s^{-1} at brushing, whereas at leveling (smoothing and flattening of a painted surface so as to impart gloss and aesthetic appearance to it) or when paint is scooped out of its container, $\dot{\gamma} = 10^{-1}$ s^{-1} or less. Obviously, too thin a paint cannot form even layers on vertical surfaces as the force of gravity causes the coating to sag off or run down the surface. A good paint should then maintain its original gel-like structure (or, rather, recover it after application) without compromising its easy applicability.[33]

18.4 PERSONAL-CARE PRODUCTS

18.4.1 COMPARISON WITH HOME-CARE PRODUCTS

Intensive research has been directed to personal-care products. Shampoos (see Chapter 17), shower gels, or bath additives are just a portion of the wide range of grooming aids and products designed to answer the consumer requirements and to get marketing advantage. The principles underlying personal care formulation based on wormlike micelles are similar to those applied in home care products. Certainly, in personal care compositions the safety and health related aspects are more emphasized. These products should comprise "safe and effective amounts"[34] of pharmaceutically acceptable active components. The phrase "safe and effective amount" can be defined as an amount of an active ingredient "high enough to modify the condition to be treated or to deliver the desired benefit, but low enough to avoid serious side effects, at a reasonable benefit to risk ratio within the scope of sound medical judgement."[34]

18.4.2 Hair Bleaching and Oxidative Dyeing

The use of an oxidizer as the active ingredient in hair bleaching and oxidative dyeing highlights the similar role of viscoelasticity in both home-care and personal-care products.

A bleaching agent is a material that lightens or whitens a substrate via chemical action.[26] In the case of hair, the bleaching process may be defined as the irreversible destruction of melanin pigments by oxidation.[35] Most effective oxidizers are toxic or potentially harmful and, therefore, are not suitable for treating human hair. Thus, the possibilities are narrowed to sodium hypochlorite and hydrogen peroxide,[37] but the utilization of sodium hypochlorite is rather limited.[38] Hydrogen peroxide is one of the most common bleaching agents, a nonpolluting oxidant and the least expensive source of active oxygen commercially available.[36] It is used in many applications, including hair bleaching and oxidative dyeing. Although hydrogen peroxide is more stable in acidic solutions, the melanin granules are attacked more readily in alkaline environment.[35]

Oxidative dyeing is similar to bleaching and is the best way to achieve a permanent hair color, that is, its shade should resist the action of sunlight, shampoos, perspiration, and mechanical abrasion, and it should last until the hair grows.[36] In such systems a water-insoluble colored material is produced inside the hair fiber by oxidation of colorless precursors (first intermediates) and deep penetration of the dye into the hair cortex rather than just coating the hair fiber.[35,36]

A bleach or an oxidative coloring agent is generally a two-part system: a lotion and a developer.[39,40]

1. Lotion (tint). It contains a variety of ingredients. The following is a preferred embodiment disclosed by recent patents.[39,40]
    a. A surfactant system [most preferably from 3.5 to 10 wt% of a nonionic surfactant]. This system provides wetting, spreading, and viscosity control (*vide infra*).
    b. A solvent system [preferably from 15 to 27 wt% of a compound such as a lower alkanol or a lower polyol]. The function of the solvent system is to solubilize a sufficient amount of surfactant in order to ensure that a thin enough liquid is obtained for the lotion[40,41] and also to solubilize the dye intermediates.[36]
    c. An alkalizing agent [preferably from 9 to 15 wt% of monoethanolamine] is added[39,40] in order to promote the oxidation reaction and lightening.[36] Other roles of the alkalizing agent are to raise and buffer the pH of the developer (*vide infra*) that produces active bleaching species and to aid fiber penetration.[39,40]

Such formulations can also function as a hair coloring lotion. In such a case, it will contain between 0.2 and 3 wt% of oxidative hair dye precursors. These precursors are activated by the oxidant which is present in the developer and react with further molecules to form a larger complex in the hair shaft.[39,40] This size enlargement is largely responsible for fixing the dye in the hair.[35] Many aromatic diamines, aminophenols, and their derivatives can be used as precursors and couplers.

2. Developer. Essentially, the developer is an aqueous solution of an oxidizer. In the above-mentioned disclosed patent embodiment,[39,40] the preferred oxidants are H_2O_2 and the peroxymonocarbonate ion, formed most preferably from ca. 1 to ca. 8 wt% of a bicarbonate ion and from ca. 2 to ca. 5 wt% of a source of hydrogen peroxide. This ion is particularly effective in combination with a source of ammonia or ammonium ion, because it improves hair coloring, efficiently softens the hair without leaving a foreign residue,[37] and reduces adverse side effects, such as odor, skin and scalp irritation, and damage to hair fibers.[39,40] The action of the developer can be promoted by a powder activator composed of persulfate salts.[41] The pH of the developer has preferably a value between 2.4 and 3.1 prior to its use.[39,40] The formulations also contain an alkalizing agent (ammonium ion), a radical scavenger (species that reacts with reactive radicals, such as carbonate radicals, and transforms them into a less reactive species), a chelant such as ethylenediaminedisuccinic acid or aminophosphoric acid that serves as stabilizer or preservative and decreases hair damage during bleaching or dyeing,[39,40] and an antioxidant (erythorbic acid,[39,40] sodium metabisulfite,[36,37,39,40] tocopherol and tocopherol acetate[39,40]) to prevent a premature oxidation either during the manufacturing process[36] or in use.

Other adjuvants, such as conditioning agents, fragrance, carriers and hair swelling agents may also be added.[40,41] However, here we will only focus on the principal requirements that a desirable bleacher and oxidative dyeing agent should meet:

a. The efficacy of the product depends, to some extent, upon the oxidant amount. The hair coloring or bleaching lotions are required to cause little or no sensory irritation when mixed with a developer containing oxidizer sufficient enough for effective hair bleaching or oxidative dyeing.[39,40] In a broader sense, the product must not be injurious to the general health and must have no ill effect on the hair.[37]

b. The lotions should be stable to ensure a reasonable shelf life.

c. The mixture of the lotion and developer should be readily distributed throughout the hair mass when applied by utilizing a brush or with fingers, and must not drip or run from the hair during the color development period, when no stress is applied.[39,40] This requirement is important because the bleaching compositions must be applied only to the new growth of hair. Repeated treatment of the same portion of hair brings about serious damage to the texture of the hair.[37] The way to keep the formulation in place is to use viscous products, but the consistency of such compositions may hinder the development of the lightening color.[37] Again, viscoelasticity of the formulation is useful here.

d. The formulations should be readily rinsed from the hair with water.[39,40]
e. When the lotion and developer mixture is applied to the hair, the dye precursors will diffuse rapidly into the hair fiber.
f. It is desirable that the viscosities of the lotion and developer will be comparable in order to facilitate mixing.[39,40]

The major thickening mechanism in commercial hair colorants and bleachers is due to the formation of wormlike micelles (leading to an isotropic and viscous gel) or lamellar phases (leading to an anisotropic and viscous cream or gel) for application to hair.[39,40] For both cases, the thickening is suppressed in the original containers by adding more than 20 wt% of solvents to the lotion formulation. The lotion viscosifies when it is mixed by the consumer with a predominantly aqueous developer container to trigger the thickening *in situ* immediately, prior to application to hair. However, it has been observed that hair bleaches or colorants based on wormlike micelles are more skin irritating than those based on lamellar phases.[39,40] We assume that this difference in behavior stems from the higher viscosity of lamellar phases. Therefore, this drawback could, in principle, be eliminated by tuning the viscosity of the wormlike micelles.

18.4.3 Skin Cosmetics

This topic is far too rich to be covered exhaustively in this review. Therefore, we have focused on just a couple of illustrative examples, namely shower gels and sunscreens. Obviously, the ingredients of these products can be defined as items of skin cosmetics. Such materials are utilized in the care, cleaning, and decoration of human skin. All these functions concern the outermost layers of the epidermis (outer skin).[42] However, before delving in these two examples, we will survey some considerations regarding personal care formulations.

18.4.3.1 Common Personal-Care Formulations

The major components of personal-care formulations are surfactants. Their principal task is to reduce the surface tension of water so that via wetting, the physical and chemical bonds between dirt and substrate (skin or hair) are broken and the dirt can be solubilized and removed from the substrate.[35] Consumers associate cleansing power with lathering ability (the generation of a foam or lather when the surfactant is mixed with water and mechanically agitated).[34] Therefore, one of the most required properties of such surfactants is the capability to lather well even when charged with dirt, grease, sweat, and sebum.[42] In the past, personal care products were based on rather concentrated solutions of mostly anionic surfactants, as they have excellent foaming and solubility parameters.[43] Among the anionic surfactants used, the most important are: (1) Alkyl sulfates, which are resistant to hard water (in contrast to plain soaps, whose carboxylic groups form insoluble precipitates in hard water).[44] They also have a good detersive and degreasing ability, but they strongly defat skin and hair.[42] (2) Alkyl- and arylalkyl sulfonates which, because of their very strong defatting action, are more suitable and worldwide used for household cleaning agents.[42] In Western Europe, fatty alcohol ether sulfates (FAES) with a C_{12}-C_{14} alkyl chain and an average degree of ethoxylation of 1 to 3 moles of ethylene oxide per mole[42,45] and sulfosuccinates[46] are more important.

The zwitterionic surfactants carbobetaines[46] and alkylamphoglycinates and amphopheopionates[42] are also used due to their compatibility with the skin and mucous membranes of the eyes.[45,46]

The viscosity of compositions based on these compounds can be tuned relatively easily by the addition of sodium chloride or ammonium chloride and optionally also fatty acid diethanolamides. Such formulations display satisfactory foaming and are inexpensive. Thus, they could be utilized, for example, in shampoo formulations where copious foaming and some softening and conditioning of the hair are considered among the expectations from the product.[43] However, as was formerly mentioned, these compounds are very irritating to the skin and, moreover, they may contain traces of the dangerous contaminant, N-nitrosodiethanolamine. Thus, milder surfactant preparations, without nitrogen, that can also be thickened easily, are needed. However, such systems can be sufficiently thickened neither with electrolytes nor with water soluble polymers (which impair foam quality).[45] Combinations of alkyl poly (or oligo)glycosides with carboxymethylated alkenol and alkanol oxyethylates were observed to be thickened using conventional electrolyte concentrations in contrast to the behavior of their individual components.

The desired rheological behavior of personal-care products is characterized by the following four processes:[47]

1. At rest, the products should have a stable structure in order to prevent settling of suspended ingredients.[47]
2. During outflow, the structure should be broken (shear-thinning) since relatively low viscosity enables the consumer to discharge the product easily from the container as well as ensure no package residue.[47] The synergistic viscosifying effect described above leads to the desired viscosity.[45,46] The shear rates should then be between 3 and 10 s^{-1}. These values correspond to approximately those encountered when a fluid flows out of a container (such as a plastic bottle) with an average orifice,[45,46] though the experienced shear rate of a product upon dispensing is largely influenced by its package nozzle dimension and

flow rates.[45] This shear thinning effect is significant. For instance, the viscosity of a non-Newtonian bath gel has decreased by nearly 700 Pa.s upon increasing the shear rate from 0.05 to 0.5 s^{-1} (Ref. 16). The slow flow from the container indicates a high active content.[43] However, too viscous fluids suffer from uneasy pourability as well as a tendency to form very long rejuvenating threads, thereby leading to formation of cobwebs.[43,46] This behavior can obviously be a nuisance in handling.[16] In viscoelastic products (such as those based on wormlike micelles) the elastic contribution that usually increases with rising forces should also be considered. When dispensing high elastic fluid, a relatively unappealing oscillation of the shapelessly thick liquid thread is formed at the neck of the container, and the liquid can be drawn back (even after it has spilt over the container's rim)[4] by abruptly returning the container to its initial position.[4,46] (See Chapter 17, Sections 17.1.1 and 17.4.1.1.)

3. Clearly, the product should be sufficiently viscous so as to remain in the hand (or on a washcloth or sponge) as it is being dispensed.[48] A good matching between the viscous and elastic properties of the formulation also allows the simple spreading of an optimal dosage over the skin or hair.[43] This occurs under the action of higher forces, usually at shear rates above 100 s^{-1} (Ref. 46). Too elastic compositions make the application of the product all over the body more difficult.[46]

4. After dispensing, the structure should be recovered rapidly so as to ensure the stability of the product over its consumer lifetime.[47]

18.4.3.2 Shower Gels

Shower (or bath) gels are an increasingly going part of the body-cleaning category which includes an ever-burgeoning number of cleansing articles, such as soap bars, bath foams and liquid hand soaps,[48] creams, lotions,[34] and wipes, that is, absorbent sheets impregnated with topical compositions for various purposes.[34,49] These skin care actives deliver therapeutic, prophylactic, and chronic benefits to the skin and hair. They include, for instance, anti-acne, anti-wrinkle, anti-skin atrophy, and skin barrier repair actives, nonsteroidal soothing actives, artificial tanning actives and accelerators, skin lightening agents, sebum inhibitors, anti-itch ingredients, desquamating enzyme enhancers, and so forth.[34] It should be stressed that most of these benefits are not included in the formulations of the traditional body cleaners discussed here, because of the inherent problem of balancing cleansing efficacy against delivering a skin care benefit.[34] To compete in today's market, that is, to be acceptable to consumers, a cleansing formulation should satisfy a number of criteria, including cleansing effectiveness, mildness to skin, hair and ocular mucosae,[34] skin feel,[34,48] rinsability,[48] foaming[34] (particularly in hard water),[48] acceptable fragrance impact,[48] prevention of overly dryness after frequent use,[34] and product aesthetic attributes that consumers care about, such as texture and color (which are perceived by the consumer, especially if the product is marketed in clear packaging),[16] fragrance, and packaging.[48] We will now outline briefly the more important properties of shower gels.[48]

18.4.3.2.1　Viscosity

In today's market there is a wealth of such products spanning the entire range of the viscoelastic spectrum, from simple viscous fluids to highly elastic structured fluid systems with a variety of physical appearances and deformation manifestations.[16] It should be noted that most products currently on the market are not "true" permanent gels, but rather free flowing liquids having standing viscosities from 3,000 to 7,000 cP.[48]

The aforementioned viscosity differences stem from factors such as the strength and number of effective molecular interactions between product components, the presence of gellants and rheology modifiers, component concentrations, and the volume fraction of suspended solids.[16] The ability to effectively customize the rheological behavior of the product is determined by the ratio G''/G' [also known as $\tan(\delta)$],[16] where G'' is the loss modulus and G' is the storage modulus, and by the magnitude of G'. When $G'' \gg G'$, the product is predominantly viscous and unstructured.[16] When $G' \gg G''$ (especially at low strains) the product is a highly structured, elastic fluid. It has the following attributes:

- Upon increasing the strain, G''/G' increases, namely the elasticity and internal structure decrease. In quiescent state (no deformation), however, this loss is frequently fully recovered.[16]
- Such products may be designed to resist flow under gravitational forces and to appear thick and full bodied.[16]
- One of the targeted consumer-perceived product features is the ability to sustain a suspension of an immiscible phase. Elasticity can also be correlated to this attribute.[16]
- Highly elastic fluids that lack sufficient strain sensitivity suffer from processing and filling problems in manufacturing and in handling operations, due to their being stringy.[16] Consumers do not appreciate such behavior.[48]

Viscosity adjustment can be achieved in several ways:

- Blending of secondary surfactants, for example, alkanolamides and betaines such as cocoamidopropylbetaine, which is also known for its ability to reduce eye sting and irritation, caused generally by the primary anionic surfactant.[48]
- Addition of a slurry of a polymer, such as hydroxyethylcellulose and other cellulose polymers, cross-linked acrylic polymers (carbomers) and polyethylene glycols thickeners, to the surfactant solution early in a batch-making process.[48] These polymers provide residual skin feel

benefits. They are more expensive than the salts used as viscosity modifiers, but they may be indispensable when a true gel consistency is needed.[48]

- Adjustment of the viscosity at the end of a batch-making process can be accomplished by adding inorganic (such as sodium chloride, ammonium chloride or sodium sulfate) or organic salts. Electrolytes bring about the enlargement of micelles[42] and, in certain systems, the creation of wormlike micelles is possible.

18.4.3.2.2   Lather

As was formerly mentioned, in a number of applications, including skin and hair cleansing, intensive foaming has become associated in consumers' minds with high detersive power[50,51] and low sudsing detergents are conceived as inefficient. Thus, the lather formed should be *rich*, creamy, and long lasting rather than loose and watery. Whereas thin lather communicates "detergent-like" cleansing and harshness, dense foam imparts a perception of softness, cleansing, gentleness, and mildness to the consumer.[48] It should be noted that a direct relationship between foaming and effective cleaning cannot be established at all in many cases.[50] Furthermore, in machine laundering operations, for example, too copious foam is detrimental to the mechanical action required for effective cleaning.[51] Excessive foaming is a disadvantage also in hair bleaching and oxidative dyeing.[41]

Considering this demand for high foaming, it is clear that a shower gel, which is intended to clean the entire body, must foam intensively, delivering a large amount of foam per dose.[48] Foam boosters (or stabilizers), usually varying in amount from 2 to 5 wt%, are added to improve lather quality. These boosters, usually fatty acid mono and dialkanolamides, strengthen the surfactant film enveloping each bubble so that the quality and longevity of the foam is improved.[48] Yet, as was formerly alluded, alkanolamides contain free diethanolamine which is thought to be linked to the formation of nitrosamines in the presence of nitrosating agents. Amineoxides suffer from the same problem. Besides monitoring nitrosamines content and inhibiting their formation when using N-containing foam boosters, this problem can also be solved by using substituents such as fatty acid polyglycol ester sulfates,[50] organo sulfobetaine zwitterion, siloxane based surfactant compounds,[51] polymers, protein-fatty acid condensates, and proteins.[35]

18.4.3.2.3   Skin feel

A shower gel should not leave a perceived sticky residue or dry film which contributes to tight or dry skin feel.[48] The formulation should be designed so that skin feel will be smooth and slippery, but neither greasy nor slimy in use, whereas after use, once the skin has dried, it should feel moist and soft.[48] Emollients, such as lanolin and its derivatives,[48,52] sterols,[52] phospholipids,[52] hydrocarbons[52] and fatty acids, alcohols, and esters[52] may be used to achieve these desired properties.[52]

18.4.3.2.4 Fragrance

Fragrance is a very important component in shower gels. As with many other personal-care products, shower gel product choice appears to be influenced considerably by fragrance choice and preference.[48] Also, fragrance is added to mask the fatty base odor of the raw materials and provide a consumer-acceptable olfactory signal.[48] Addition of fragrance oils to a liquid formulation often results in a change of product viscosity. Usually, a decrease is observed, but certain fragrance components will cause a significant increase in a formula's viscosity. Therefore, it is essential to include fragrance as early in the product development process as possible.[48]

18.4.3.2.5 Aesthetic additives

The production of commercially attractive shower gels requires the addition of some auxiliaries such as colors and clarifying, opacifying, and pearlizing agents in order to enhance or improve their visual aesthetics.[42,47]

Aesthetic adjuvants may also be exploited to communicate higher quality and additional benefits (in some cases, they are included in a product solely as promotional components).[48] For instance, moisturizers are incorporated in small polymer beads which break and dissolve when rubbed on the skin.[48] In order to suspend such additives in a shower gel formulation, the viscosity of a product should approach that of a true gel, but then the product becomes difficult to dispense and the foaming process is slowed down considerably.[48] It seems to us that viscoelastic wormlike micelles can help in both tuning the desired viscosity and delivering their contents upon squeezing.

18.4.3.2.6 Stabilizers

In this category we include minor ingredients intended to ensure the integrity of shower gel formulations throughout their lifetime.[48]

Preservatives are added to these formulations to prevent microbial or fungal spoilage.[42,48] Naturally, such additives are expected to be active at low concentration, to be effective over a wide range of pH and microorganisms, to be stable to heat and storage, and to be nontoxic, odorless, and colorless.[53] In general, the more hydrophobic a preservative, the greater is its chance to be trapped in surfactant micelles, where it becomes ineffective.[42] Chelating agents (for instance, EDTA) function as sequestrants for metal ions (such as calcium, magnesium, or iron and other heavy metals) which participate in decomposition reactions[42] and often are responsible for the gradual discoloration of a product.[48] Antioxidants are added to slow the rate of oxidation of autooxidable materials,[53] such as soaps, vegetable oils, fatty acids, lanolin and its derivatives—that are present in shower gels.[48] Antioxidants are expected to be soluble in both water and oil, nontoxic, nonallergenic, odorless, effective in low concentrations, and stable under conditions of use.[53] A typical formulation of a shower gel is shown in Table 18.6.

TABLE 18.6
Basic Shower Gel Formulation (Source: Ref. 48)

Ingredient	Weight (%)
Primary surfactant	10 to 20
Secondary surfactant	0 to 8
Foam booster	2 to 5
pH adjusting agent	As needed
Viscosity adjusting agent	0 to 3
Aesthetic additives	
color	As needed
pearlizing agents	0 to 2
clarifying agents	As needed
Fragrance	0.5 to 2
Stabilizers	0.05 to 1
Specialty additives	0 to 4
Water	Added to 100

18.4.3.3 Sunscreens

Sunscreens are products that provide protection against the harmful effects of light on the human body, mainly the skin—including hair and nails—and eyes.[54] These effects of light can be grouped into four classes:[54]

1. Acute skin damage. Ultraviolet (UV) radiation in the UV-B range, that is, from 280–290 nm to about 315–320 nm,[57] is the primary cause of sunburn.[55–57] However, the development of erythema (reddening of the skin) is not coincident with insolation (continued exposure to UV radiation), and a latent period of several hours elapses before the erythema starts to develop.[56]
2. Chronic skin damage. Extensive insolation brings about several processes: (i) Tanning, that is, skin darkening through oxidation of melanin;[56] (ii) premature photoaging of skin;[58–60] and (iii) in more severe cases of overexposure, formation of precancerous tumors or lesions[53,54] and, ultimately, malignant tumors.[61,62]
3. Light-induced skin disorders and ocular damage. UV-A radiation is the major cause for exogeneous or endogeneous dermatoses, whereas visible and UV radiation may injure the conjunctiva, retina, and lens (cataracts).[54]

In fact, the only practical way to prevent UV radiation damage is to decrease the exposure of the skin to sunlight. This can be achieved by using the physical protection provided by blocking substances such as zinc oxide, titanium oxide, or mica[54] as well as diatomaceous silica, alumina, or magnesium carbonate.[56]

These are opaque materials that act as filters due to their ability to reflect or scatter the UV radiation. However the most important protection device is "chemical" UV absorbents included in sunscreens.[63] These absorbents are incorporated in a suitable base and penetrate the outer layers of the stratum corneum, producing a layer, 5–30 μm thick, that specifically absorbs UV-B and possibly also UV-A light.[54] The basic requirements for successful sunscreens are resistance to chemical and photochemical structural changes,[62,63] sufficient solubility in permissible cosmetic vehicles,[56] and absence of toxic, irritating, or sensitizing ingredients.[56]

When purchasing a sunscreen, consumers consider many factors, including the sun protection factor (SPF), the duration of the product after its application over the skin, the shelf life of the product and more importantly the product's form,[55] that is, whether it is marketed in the form of hydroalcoholic lotions,[56] cream lotions,[56] aerosol foams,[56] sprays,[55,56] and so forth. As was formerly demonstrated, the viscoelastic properties of wormlike micelles make them excellent candidates for such products. Customers also want a sunscreen that endows a feeling of softness and "silkiness" and can be applied in a smooth, continuous film over the skin.[55] This is achieved by incorporating suitable skin feel additives, such as synthetic polymers (preferably microscopic nylon beads) or silicones.[55] The larger cores of wormlike micelles have more available space for incorporating the required additives, such as emollients that provide a softening or soothing effect on the skin surface and also help control the rate of evaporation and tackiness of the product compositions (for example, several types of oil, cyclomethicone or dicapryl maleate),[55] pH adjusters, moisturizing agents (for instance, glycerol or polyethylene glycol),[55] and the like.

18.5 DRUG DELIVERY

Enhancing the therapeutic performance of (hydrophobic) drugs via innovative delivery strategies is one of the main issues in pharmaceutical research.[64,65] Drug delivery systems have many tasks including:[66]

1. Solubilization of poorly soluble drugs.
2. Protection of drugs (such as peptides and proteins) from enzymatic degradation or hydrolysis.
3. For enhanced efficacy, drug delivery should be targeted, that is, with a specific affinity for the affected organ or tissue, and have prolonged lifetime in the circulation so that the drug will be able to accumulate sufficiently in targets with low antigen concentration and/or diminished blood supply.[67] In the case of colloidal carriers, targeted delivery of drugs can be reached by conjugating a specific vector to the carrier.[66]

For example, poly(ethylene oxide)-poly(ethylethylene) (PEO-PEE) copolymers can use their hydroxylic distal end-group to attach ligands or antibodies, generally via free amino groups.[68]

4. Inhibition of the toxic effect of the drug until its release at the targeted tissue.[66]

5. Controlled release of the drugs.[66]

Many drug delivery systems, such as liposomes (i.e., phospholipid vesicles),[69] emulsions,[70] microemulsions,[64,71] nano and microparticles,[65] organogels[64] and hydrogels,[72] and a variety of lyotropic liquid crystalline phases (lamellar, hexagonal, cubic, etc.)[73] can provide matrices for sustained and controlled release of drugs. Micellar systems have many advantages, such as low viscosity, small aggregate size, simple preparation technique, and long shelf life.[64] In addition, they can permeate tissues with small pores, reduce toxic side-effects, enhance the therapeutic efficiency of the drug, and also provide platforms for targeting.[65] However, they suffer from a serious drawback, that is, they have a very low drug loading.[64] Using wormlike micelles seems to mitigate the severity of this problem as they provide larger core volume (per carrier) for drug loading and are able to flow readily via capillaries and pores due to their small cross section and flexibility.[74] In addition, wormlike micelles have the following benefits:[69] biocompatibility and selective permeability to solutes; the capability to retain internal aqueous compounds and control their release; and the ability to deform while remaining relatively tough and resilient.[69]

However, ordinary wormlike micelles formed from small amphiphiles are relatively unstable in dilute aqueous solutions and under shear stresses.[75] They cannot then survive injection as intact aggregation into the circulation of an animal.[69] Therefore, the resort to copolymers is deemed logical (see Chapter 14, Section 14.6.1). Copolymer assemblies can be much more stable relative to small surfactants, as they have much lower critical micellar concentration and slower dissociation rate.[74] Depending on the weight fraction of the hydrophilic block relative to total copolymer molecular weight, various structures may form. For instance, PEO-based diblock copolymers (widely used due to their biocompatibility)[74] in aqueous solution will assemble into wormlike micelles when the weight fraction of the ethylene oxide group W_{EO}, is ca. 45–55%. Lower W_{EO} gives vesicles, and higher W_{EO} yields spherical micelles.[75,76] Experimental methods that have led to changing W_{EO} will be shown below. Copolymer-based spherical micelles, which function as nanoscale carriers for drug delivery, have the following advantages:[77]

- They increase the solubility of hydrophobic drugs.
- They reduce toxicity to healthy tissues brought about from excess dosage.
- They achieve more sustained release profiles of drugs.

Yet, copolymer-based wormlike micelles have two additional features that make these assemblies more suitable for drug delivery:

- Their loading capacity is higher than that of copolymer-based spherical micelles. It is noteworthy that worms and spheres composed of the same copolymers will apparently have a similar area per copolymer due to the shared interfacial chemistry. Therefore, they will also have common interfacial energy density (that is, the interfacial tension, γ).[74] However, the wormlike micelles have a volume to surface area ratio larger than that of the spherical micelles by a factor of 1.5, as can be derived from simple geometric considerations. Consequently, there is 50% more drug carrying capacity in the core of wormlike micelles.[74]
- Their flexibility allows them to permeate via nanoporous gels (and presumably also via tissues).[77]

In the following list we will describe some promising experiments regarding these polymeric wormlike micelles.

1. By blending and polymerizing inert PEO-PEE and cross-linkable PEO-PBD (PBD = polybutadiene) block copolymers, micelles up to tens of micron long were formed with persistence lengths that continuously span from 50 nm to 100 μm.[75] The kinetic barrier to further aggregation and the strong association of the large hydrophobic blocks tend to stabilize the wormlike micelles. For instance, they do not fragment under flow-imposed tensions (~ 1 mN/m).[75]

   More important is the observation that these wormlike micelles do not stick to blood cells because the blood plasma is composed of many surface active agents, such as proteins and fatty acids. These proteins generically adsorb to artificial surfaces and mediate clearance by the attachment of immune system cells.[76] The inertness of wormlike micelles (and similarly, of polymer liposomes)[76] in blood plasma is undoubtedly due to stealthiness imparted by the brushy PEO corona that tends to sterically stabilize these wormlike micelles and minimize interactions with other wormlike micelles and surfaces.[75] A similar effect is observed in "stealth" liposomes which are coated with, for example, PEO. Such protected liposomes have increased longevity in the blood circulation relative to conventional liposomes.[67] It is expected that vesicles or wormlike micelles loaded with an appropriate drug will take advantage of the extended circulation times and find their way into targeted cells.[76]

2. Stable wormlike micelles were formed in water from PEO-PEE and PEO-PBD diblock copolymers via the selection of different-sized structural units and chemical fixation of unsaturated butadiene bonds. These wormlike micelles mimicked the flexibility (bending rigidity)

of various cytoskeletal ubiquitously expressed biopolymers, running the gamut from intermediate filaments to F-actin (microtubules).[78] It was shown that wormlike micelles can be grouped into nematic domains (structurally similar to active cytoskeletons), but they can also be encapsulated in bilayer vesicles when the W_{EO} of the polymer is close to that for the bilayer (vesicle)-wormlike transition. Such encapsulation is aimed at emulating both the elasticity of a cell's scaffolding and the isolation of the structure from the external environment.[78] Similar encapsulations of filaments in lipid vesicles are typical of cells. These observations may lead to mimicry of functions that might range from muscle contraction for device actuation to artificial viruses for drug delivery.[78]

3. Phase transitions may be induced by altering control parameters. For example, in solutions of poly(acrylic acid)-poly(butadiene) diblock copolymers low pH and high salt (NaCl, CaCl$_2$) favor the formation of vesicles. Decreased salt and neutral pH lead to the formation of multibranched cylinders and highly stable, but fluid and flexible, wormlike micelles. Further increase of pH (and intracoronal repulsion) generate spherical micelles by fragmentation and pinch-off at the wormlike termini.[79]

Another strategy is to increase W_{EO} by hydrolytic cleavage of the copolymer. For instance, hydrolytic degradation of PEO-PCL (PCL = polycaprolactone) wormlike micelles leads to a cylinder-to-sphere transition.[80] The hydrolytic degradation of PCL, which triggers this morphological transition, is governed by an end-cleavage mechanism and can be tuned by varying parameters such as temperature or pH.[80] A similar phenomenon has been observed in the system PEO-PLA (PLA = polylactic acid). The hydrolytic cleavage of the PLA leads to the increase of W_{EO} and to the transition of worm to sphere.[77] This observation may be utilized in drug delivery since, as was formerly described, wormlike micelles have larger core volume for drug loading and an ability to flow readily via capillaries and pores due to their cylindrical shape and flexibility. It would be advisable to start with wormlike micelles and then let them progressively degrade into spherical micelles as desired. The rate of degradation can be tuned by varying parameters such as temperature or pH.[80] Moreover, whereas in this system the hydrophobic core can load drugs barely soluble in water, the outer hydrophilic corona maximizes biocompatibility and helps micelles rapid clearance by the liver and spleen, after intravenous administration, thereby prolonging circulation in the blood.[74]

It is worth noting that a solubilized drug can impede the formation of wormlike micelles. This is the case regarding timolol maleate (TM) which is used in the therapy of glaucoma.[81] An ophthalmic TM gel has two advantages: (i) it has to be applied once a day instead of twice daily as when using an ophthalmic aqueous solution of TM, and (ii) enhanced bioavailability is achieved due to longer contact time between the viscous gel and eye and the drug is controllably delivered on contact with lachrymal fluid.[81] Furthermore, since the gel is sensitive to shear, it momentarily liquefies every time the eye blinks, and thus the gel is able to spread evenly over the entire eye.[82] In the system lecithin/isopropylmyristate/water/TM, unloaded reverse micelles form worms, but when loaded with TM, the drug interacts (through hydrogen bonding) with the phosphate group of lecithin, leading presumably to the formation of a sponge-like phase.[81]

Lecithin-based reverse wormlike micelles in nonpolar liquids (such as cyclohexane or n-decane) can be formed without any added water by using a bile salt (for instance, sodium deoxycholate) in trace amounts.[83] Drugs encapsulated in reverse micelles can show good biological activity, but the stability of these biomolecules is often inversely related to the water content of the formulation.[83] The advantage of these new waterless reverse wormlike micelles is clear.

Another reaction path is taken in the case of the cationic surfactant propranolol. It is an efficient multidrug resistance modulator and a nonselective β-blocker. Propranolol is used in drug delivery studies in order to treat systemic hypertension since this drug is thought to reduce hypertension by the decrease of the cardiac frequency and thus blood pressure.[84] Propranolol interacts with egg phosphatidylcholine vesicles. The gradual transition from liposome to micelle proceeds as follows: propranolol disrupts the lipid bilayer by perturbing the local organization of the phospholipids. This stage is followed by the formation of wormlike micelles and ends with the formation of globular micelles.[84]

4. PEO-PEE copolymers were end-biotinylated, namely their OH end group was replaced by the small vitamin biotin, which functions as a targeted molecule.[68] Biotinylated wormlike micelles are capable of loading, retaining, and delivering cell-viable hydrophobic dyes (that are important in diagnostic applications) and cytotoxic drugs to intracellular organelles. In a feasibility experiment, a well-known antiprofilerate drug, Taxol, that binds to and stabilizes microtubules inhibiting cell division and proliferation. Taxol is also coated onto vascular stents to block proliferation and arterial restenosis.[68] The micelles can be attached to specific receptors on aorta-derived smooth muscle cells. The capability of biotinylated wormlike micelles to deliver cytotoxic drugs has a significant therapeutic value since proliferating tumor cells overexpress various receptor targets. The proposed mechanism for the operation of Taxol in this case is as follows: The wormlike micelles are internalized by receptor-mediated endocytosis either via packing or fragmentation. The hydrophobic Taxol is finally repartitioned to its destination. Biotinylated wormlike micelles containing Taxol were shown to kill 75% of smooth muscle cells in culture, whereas pristine; nonlabeled Taxol-loaded worm micelles did not kill any cells above control levels.[68]

18.6 CONCLUSIONS

As this book demonstrates, wormlike micelles have garnered a great deal of attention from both theoretical and practical points of view.

Conspicuous features of wormlike micelles that are exploited in many applications are their viscoelasticity and the relatively large loading capacity of their cores, but we should not ignore other properties of these unique micelles. Thus, we may add the enhancement in fracture resistance and toughening of curable flame retardant epoxy resin compositions that self-assemble into wormlike micelle morphology.[85-87] This invention is particularly useful for preparing electrical laminates, but other uses of these compositions may be expected.

This review chapter has dealt with three types of applications: the use of wormlike micelles in the rather mundane home-care products, the exploitation of these micelles in the less dull personal-care formulations, and the utilization of wormlike micelles in the exciting area of drug delivery. This last application may be linked with possible use of wormlike micelles in nanobiotechnology, that is, the embedding of bioinspired chemical functions into amphiphiles (see, for example, Ref. 88 and leading references cited therein). Biological function units can then be inserted in wormlike micelles.[65,89,90]

Unfortunately, commercial utilization of wormlike micelles in drug delivery and nanobiotechnology is hardly within reach at present. It is hoped that the realization of these applications will occur in the near future.

REFERENCES

1. Chevalier, Y. *Current Opinion Colloid Interface Sci.* 2002, *7*, 3.
2. Schubert, B.A., Kaler, E.W, Wagner, N.J. *Langmuir* 2003, *19*, 4079.
3. Walker, L.M. *Current Opinion Colloid Interface Sci.* 2001, *6*, 451.
4. Hoffmann, H., Ebert, G. *Angew. Chem. Int. Ed. Engl.* 1988, *27*, 902.
5. Won, Y.-Y., Davis, H.T., Bates, F.S. *Science* 1999, *283*, 960.
6. Won, Y.-Y., Paso, K., Davis, H.T., Bates, F.S. *J. Phys. Chem. B* 2001, *105*, 8302.
7. Hamley, I.W., Pedersen, J.S, Booth, C. Nace, V.M. *Langmuir* 2003, *17*, 6386.
8. Kroy, K. *Current Opinion Colloid Interface Sci.* 2006, *11*, 56.
9. Cates, M.E. *Europhys. Lett.* 1987, *4*, 497.
10. Rehage, H., Hoffmann, H. *Mol. Phys.* 1991, *74*, 933.
11. Rodriguez-Abreu, C., Aramaki, K., Tanaka, Y., Lopez-Quintela, M.A., Ishitobi, M., Kunieda, H. *J. Colloid Interface Sci.* 2005, *291*, 560.
12. Maitland, G.C. *Current Opinion Colloid Interface Sci.* 2000, *5*, 301.
13. Yang, J. *Current Opinion Colloid Interface Sci.* 2002, *7*, 276.
14. Boek, E.S., Jusufi, A., Löwen, H., Maitland, G.C. *J. Phys. Condensed Matter* 2002, *14*, 9413.
15. Ridley, B. *Structured Fluids and their Use in Home & Personal Care Applications, Cosmetics and Colloids Conference*, Ciba Specialty Chemicals, 1/2005; www.soci.org/SC1/groups/col/2005/reports/pdf/gs3257_ridley2.pdf
16. Rounds, R.S. *Cosmetics & Toiletries* 1995, *110*, 52.
17. Bavouzet, B., Chapon, P. US Patent Application 2005011940 (6.2.2005).

18. Rubin, F.K., Van Blarcom, D. US Patent 4,375,421 (1.3.1983).
19. Choy, C.K.-M., Argo, B.P. US Patent 5,728,665 (17.3.1998).
20. Rader, J.E., Smith, W.L. US Patent 5,336,426 (9.8.1994).
21. Smith, W.L. US Patent 5,389,157 (14.2.1995).
22. Rose, G.D., Teot, A.S. In *Structure and Flow in Surfactant Solutions*, Herb, C.A., Prud'homme, R.K., Eds., ACS Symposium Series No. 578, American Chemical. Society, Washington D.C., 1994, p. 352.
23. Rose, G.D., Teot, A.S., Foster, K.L. US Patent 4,800,036 (24.1.1989).
24. Rader, J.E., Smith, W.L. US Patent 5,833,764 (10.11.1998).
25. Goodwin, J.W. *Colloids and Interfaces with Surfactants and Polymers — An Introduction*, John Wiley & Sons, New York, 2004, pp. 1–26.
26. Farr, J.P., Smith, W.L., Steichen, D.S. In *Kirk Othmer Encyclopedia of Chemical Technology*, Kroschwitz, J.I., Howe-Grant, M., Eds., 4th edition, Vol. 4, John Wiley & Sons, New York, 1992, p. 271.
27. Kong, S.B. EP 0 606 712 (1.11.1993).
28. Schilp, U. US Patent 4,337,163 (29.1.1982).
29. Ouzounis, D., Nierhaus, W. US Patent 5,906,973 (25.5.1999).
30. Ajmani, I., Tin, L., Choy, C.K. US Patent Application 20050272630 (8.12.2005).
31. Chan, S., Kong, S.B. US Patent Application 20050079990 (14.4.2005).
32. Van der Gucht, J. *Equilibrium Polymers in Solution and at Interfaces*, Ph.D. Dissertation, Wageningen University, The Netherlands, 2004.
33. Schramm, G. *Introduction to Practical Viscometry*, Haake, Germany, 1981.
34. Albacarys, L.D., McAtee, D.M, Deckner, G.E. US Patent 6,338,855 (15.1.2002).
35. Schwan-Jonczyk, A., Lang, G., Clausen, T., Köhler, J., Schuh, W., Liebscher, K.D. In *Ullmann's Encyclopedia of Industrial Chemistry*, Bohnet, M. Brinker, C.J., Cornils, B., Evans, T.J., Greim, H., Hegedus, L.L., Heitbaum, J., Herrmann, W.A., Keim, W., Kleemann, A., Kreysa, G., Laird, T., Löliger, J., McClellan, R.O., McGuire, J.L., Mitchell, J.W., Mitsutani, A., Onoda, T., Plass, L., Schubert-Zsilavecz, M., Stephanopoulos, G., Werner, D., Yoda, N., Zass, E., Eds., Vol. A12, Wiley-VCH, Online Edition, 2002.
36. Pohl, S., Varco, J., Wallace P., Wolfram, L.J. in *Kirk Othmer Encyclopedia of Chemical Technology*, Kroschwitz, J.I., Howe-Grant, M., Eds., 4th edition, Vol. 12, John Wiley & Sons, New York, 1994, p. 881.
37. Wall, F.E. In *Cosmetic Science and Technology*, Balsam, M.S., Sagarin, E., Eds, 2nd edition, Vol. 2, Wiley Interscience, New York, 1972, p. 279.
38. Shevlin, E.J. In *Cosmetic Science and Technology*, Balsam, M.S., Sagarin, E., Eds, 2nd edition, Vol. 1, Wiley Interscience, New York, 1972, p. 223.
39. Glenn, R.W., James, D.A. US Patent Application 20050283925 (29.12.2005).
40. Glenn, R.W., James, D.A. WO 2006/002362 (5.1.2006).
41. Casperson, S., Larkin, M., Lenzi-Brangi, A.M. US Patent 7,056,497 (6.6.2006).
42. Schneider, G., Gohla, S., Kaden, W., Schönrock, U., Schmidt-Lewerkühne, H., Kuschel, A., Pape, W. In *Ullmann's Encyclopedia of Industrial Chemistry*, Elvers, B., Hawkins, S., Russey, W., Schulz, G., Eds., Vol. A24, VCH, Weinheim, 1993, p. 219.
43. Balzer, D., Varwig, S., Weirauch, M. *Colloid Surf. A* 1995, *99*, 233.
44. Powers, D.H. In *Cosmetic Science and Technology*, Balsam, M.S., Sagarin, E., Eds, 2nd edition, Vol. 2, Wiley Interscience, New York, 1972, p. 73.
45. Balzer, D. US Patent 5,100,573 (31.3.1992).
46. Balzer, D. US Patent 5,965,502 (12.10.1999).

47. McMahon, D., Motyka, A., Broze, G. pp. 331-348 In *Dynamic Properties of Interfaces and Association Structures*, Pillai, W., Shah, D.O., Eds., AOCS Press, Champaign, Illinois, 1996.
48. Brassard, M. *Cosmetics & Toiletries* 1989, *104*, 53.
49. Pung, D.J., Hedges, S.K., Lin, E., Evans, M.W. US Patent 6,753,063 (22.6.2004).
50. Engels, T., Hensen, H., Raths, H.-C., Fabry, B., Seipel, W., Tesmann, H., Kahre, J., Bernecker, U. US Patent 6,235,696 (22.5.2001).
51. Lo, S.J., Snow, S.A. US Patent 4,879,051 (7.11.1989).
52. Strianse, S.J. In *Cosmetic Science and Technology*, Balsam, M.S., Sagarin, E., Eds, 2nd edition, Vol. 1, Wiley Interscience, New York, 1972, p. 179.
53. Rieger, M.M. In *Kirk Othmer Encyclopedia of Chemical Technology*, Kroschwitz, J.I., Howe-Grant, M., Eds., Vol. 7, 4th edition, John Wiley & Sons, New York, 1993, p. 572.
54. Ippen, H. *In Ullmann's Encyclopedia of Industrial Chemistry*, Bohnet, M. Brinker, C.J., Cornils, B., Evans, T.J., Greim, H., Hegedus, L.L., Heitbaum, J., Herrmann, W.A., Keim, W., Kleemann, A., Kreysa, G., Laird, T., Löliger, J., McClellan, R.O., McGuire, J.L., Mitchell, J.W., Mitsutani, A., Onoda, T., Plass, L., Schubert-Zsilavecz, M., Stephanopoulos, G., Werner, D., Yoda, N., Zass, E., Eds., Vol. A 24, Wiley-VCH, Online Edition, 2002.
55. Sanogueira, J., Fuller, J. US Patent 6,830,746 (14.12.2004).
56. Kreps, S.I., Goldenberg, R.L. In *Cosmetic Science and Technology*, Balsam, M.S., Sagarin, E., Eds, 2nd edition, Vol. 1, Wiley Interscience, New York, 1972, p. 241.
57. Bonda C.A., Pavlovic, A.B. US Patent Application 20050191249 (1.9.2005).
58. Matsumoto, K., Tsuruoka, H., Fujiwara, N., Neshimori, Y., Kenjo, Y. US Patent 6,338,854 (15.1.2002).
59. Bonda, C.A., Pavlovic, A.B. US Patent Application 20060002869 (5.1.2006).
60. Bonda, C.A., Steinberg, D.C. *Cosmetics & Toiletries* 2000, *115*, 37.
61. Osborne, D.W. Protectants US Patent Application 20050186156 (25.8.2005).
62. Serpone, N., Salinaro, A., Emeline, A.V., Horikoshi, S., Hidaka, H., Zhao, *J. Photochem. Photobiol. Sci.* 2002, *1*, 970.
63. Maier, H., Schauberger, G., Brunhofer, K., Hönigsmann, H. *J. Invest. Dermatol.* 2001, *117*, 256.
64. Bagwe, R.P., Kanicky, J.R., Palla, B.J., Patanjali P.K., Shah, D.O. *Critical Reviews in Therapeutic Drug Carrier Systems* 2001, *18*, 77.
65. Date, A.A., Patravale, V.B. *Current Opinion Colloid Interface Sci.* 2004, *9*, 222.
66. Yang, L., Alexandridis, P.A. *Current Opinion Colloid Interface Sci.* 2000, *5*, 132.
67. Torchilin, V.P., Levchenko, T.S., Lukyanov, A.N., Khaw, B.A., Klibanov, A.L., Rammohan, R., Samokhin, G.P., Whiteman, K.R. *Biochim. Biophys. Acta* 2001, *1511*, 397.
68. Dalhaimer, P., Engler, A.J., Parthasarathy, R., Discher, D.E. *Biomacromolecules* 2004, *5*, 1714.
69. Discher, D.E., Dalhaimer, P. U.S. Patent Application 20050180922 (18.8.2005).
70. Nakano, M. *Adv. Drug Deliv. Rev.* 2000, *45*, 1.
71. Lawrence, M. Rees, G. *Adv. Drug Deliv. Rev.* 2000, *45*, 89.
72. Peppas, N.A. *Current Opinion Colloid Interface Sci.* 1997, *2*, 531.
73. Drummond, C.J., Fong, C. *Current Opinion Colloid Interface Sci.* 2000, *4*, 449.
74. Geng, Y., Discher, D.E. *Polymer*, 2006, *47*, 2519.
75. Dalhaimer, P., Bates, F.S., Discher, D.E. *Macromolecules* 2003, *36*, 6873.
76. Discher, D.E., Eisenberg, A. *Science* 2002, *297*, 967.

77. Kim. Y., Dalhaimer, P., Christian, D.A., Discher, D.E. *Nanotechnology* 2005, *16*, S484.

78. Dalhaimer, P., Bermudez, H., and Discher, D.E. *J. Polymer Sci. B* 2004, *42*, 168.

79. Geng, Y., Ahmed, F., Bhasin, N., Discher, D.E. *J. Phys. Chem. B* 2005, *109*, 3772.

80. Geng, Y., Discher, D.E. *J. Amer. Chem. Soc.* 2005, *127*, 12780.

81. Mackeben, S., Müller, M., Müller-Goymann, C.C. *Colloids Surf. A.* 2001 *183-185*, 699.

82. Dagani, R. *Chem. Eng. News* 2007, *75*, 26.

83. Tung, S.-H., Huang, Y.-E., Raghavan, S.R. *J. Amer. Chem. Soc.* 2006, *128*, 5751.

84. De Carlo, S., Fiaux, H., Marca-Martinet, C.A. *J. Liposome Research* 2004, *14*, 61.

85. Dean, J.M., Verghese, N.E., Pham, H.Q., Bates, F.S. *Macromolecules* 2003, *36*, 9267.

86. Bates, F.S., Dean, J.M., Pham, H.Q., Verghese, N.E. WO 2004/108826 (16.12.2004).

87. Dean, J.M., Bates, F.S., Pham, H.Q., Verghese, N.E. US Patent 6,887,574 (3.5.2005).

88. Berti, D. *Current Opinion Colloid Interface Sci.* 2006, *11*, 74.

89. Bombelli, F.B., Berti, D., Keiderling, U., Baglioni, P. *J. Phys. Chem. B* 2002, *106*, 11613.

90. Bombelli, F.B., Berti, D., Pini, F., Keiderling, U., Baglioni, P. *J. Phys. Chem. B* 2004, *108*, 16427.

Index